THE HUMAN STRATEGY

THE HUMAN STRATEGY

An Evolutionary Perspective
on Human Anatomy

John H. Langdon

University of Indianapolis

New York   Oxford

OXFORD UNIVERSITY PRESS

2005

Oxford University Press

Oxford   New York
Auckland   Bangkok   Buenos Aires   Cape Town   Chennai
Dar es Salaam   Delhi   Hong Kong   Istanbul   Karachi   Kolkata
Kuala Lumpur   Madrid   Melbourne   Mexico City   Mumbai
Nairobi   São Paulo   Shanghai   Taipei   Tokyo   Toronto

Published by Oxford University Press, Inc.
198 Madison Avenue, New York, New York, 10016
www.oup.com

Oxford is a registered trademark of Oxford University Press

Library of Congress Cataloging-in-Publication Data

Langdon, John H. (John Howard)
    The human strategy: an evolutionary perspective on human anatomy / John H. Langdon.
        p. cm.
    Includes bibliographical references and index.
    ISBN-13: 978-0-19-516735-1
    ISBN 0-19-516735-X
    1. Human evolution.   2. Human anatomy.   3. Human skeleton—Evolution.   4. Physical anthropology.
5. Evolution (Biology)   I. Title.
    GN281.L355 2005
    599.93′8—dc22

                                                                    2004050101

Printing number: 9 8 7 6

Printed in the United States of America
on acid-free paper

CONTENTS

PART II: MUSCULOSKELETAL SYSTEM

3   BONES, JOINTS, AND MUSCLES   36

4   THE SKULL   50

11   THE HUMAN STRATEGY: BRAIN SIZE   150

12   THE SPECIAL SENSES   160

TABLES

PREFACE

This book tells a story: Once upon a time there was an animal that was different from other animals because it could reflect upon itself and ask where it came from. If this book had been written by a bird, it would have explored the origins of feathers and of flight. If it had been written by a kangaroo, it might have focused on adaptations to grazing, peculiar stages of infant development, and a unique pattern of bounding across the plains. Instead, it is written by a human and concentrates on human attributes—not because they are superior to those of other species, but because they are of special interest to us. Anatomy and "human nature" may be studied to learn what we are and where we are going; here they are examined to help us understand where we have been. *The Human Strategy* is an exploration of human anatomy that takes an evolutionary approach to help readers better understand the structure of the human body.

This textbook bridges disciplinary boundaries and brings new perspectives into classes on both human anatomy and human evolution. Anatomy students may approach the book with little prior exposure to evolution or anthropology. As they study descriptive anatomy, they can turn here for a different perspective on the same structures described in an anatomy text or atlas. They will be stretched to ask why rather than simply what. Students of human evolution will find that the anatomy is presented at a level that is accessible without a prior background in that field. Basic concepts are reinforced by the more than 250 illustrations and an extensive glossary, which should help all students venture more easily into unfamiliar territory. Paired with a text that looks at the span of the fossil record, *The Human Strategy* offers a more complete discussion of the functional and possible adaptive significance of anatomical changes, including those of soft tissues that are rarely addressed in anthropology courses.

The chapters are laid out in nearly the same sequence of material as in most anatomy texts, organized by body system and progressing through bones, muscles, the nervous systems, the visceral systems, and reproduction. A few topics, such as brain size and life history, go beyond the coverage of typical anatomy texts but nonetheless help us to understand the biological context in which our bodies evolved.

The "human strategy" is not a conscious set of tactics for playing the game of life. I refer instead to those aspects of our anatomy and relations with our environment that most profoundly distinguish our species. Our "strategy" is a set of anatomical tools, abilities, and behaviors that define a particular range of ways by which we can solve the fundamental challenges of life—obtaining food and water and shelter, tolerating the environment, avoiding predators, and replacing ourselves before we die. The five strategies that have merited separate chapters in this book are bipedalism, an expanded brain, an eclectic diet, a high activity level, and an extended life history. These adaptations—closely tied to one another—have ramifications throughout our anatomy for both body form and behavior.

I have attempted to be concise and direct. For the most part, chapters may be read and understood out of sequence. With each chapter, I identify the adaptive significance of those traits that make humans distinct from other vertebrates. Some set us apart at the species level; others, at the family, order, or class level. Our lower limbs, for example, have a unique anatomy befitting our upright stance and mode of walking, but our circulatory system, at the gross anatomical level, differs in only trivial ways from those of other mammals. Thus different chapters explore adaptations and contrasts at different taxonomic levels within the vertebrates, while some topics (e.g., cell and tissue levels, neuron functioning) are omitted altogether.

This book has grown out of my own desire to share a fascination with our species' history. It is my hope that students of both human anatomy and human evolution will find in it an exciting and different way of approaching and understanding their subjects. The numerous references are included to assist them in exploring topics of interest in greater depth.

I am very grateful for the support I have received from my colleagues at the University of Indianapolis, particularly from Stephen Nawrocki and Christopher Schmidt. I wish to thank the editorial staff at Oxford University Press—especially Jan Beatty, Talia Krohn, and Leslie Anglin—for their faith in the book and their help in making it a reality. Many reviewers contributed comments that helped steer this project in the right direction, including Steven Churchill, Duke University; Eric Delson, Lehman College; Ralph Holloway, Columbia University; William Jungers, SUNY at Stony Brook; Gail Kennedy, UCLA; Lyle Konigsberg, University of Tennessee; Andrew Kramer, University of Tennessee; Daniel Lieberman, Harvard University; Jeff McKee, Ohio State University; Jeff Meldrum, Idaho State University; Brian Richmond, George Washington University; and Carol Ward, University of Missouri. I want to acknowledge the artistic talents of Bobby Starnes of Electragraphics, Inc., and my daughter Katherine Langdon.

Most of all, I am grateful for the eternal patience and support of Terry Langdon, who alone knows how long this project has been evolving.

1

EVOLUTION AND ADAPTATION

The study of anatomy is ancient in its origins. There is a traditional approach of observing and describing. Certainly it appears to a student entering the field that anything in the body that can be named has been given a name, but descriptive anatomy by itself does not produce an understanding of how the body acquired the form that it has or why it has that form. Anatomy requires no theoretical perspective if no interpretation is attempted. The "how" and "why" questions give the field meaning. We can try to address them through functional and evolutionary answers.

A functional explanation describes the modern function of a structure. It considers the consequences of the design and, if desired, the clinical consequences

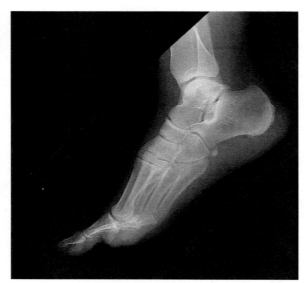

Figure 1.1 The arch of the human foot is a unique feature that reflects both adaption to its current function and evidence of different functions in its evolutionary history.

of an error or injury relating to that structure. The question "why does the foot have a longitudinal arch?" may be answered in terms of the shock-absorbing and stress-reducing role of the arch (Figure 1.1). Such a functional explanation helps us to understand why feet that have difficulty forming or maintaining the arch are prone to fatigue and pain.

An evolutionary explanation tries to comprehend the origins of the form and broadens the scope of the question. Why does the foot need reinforcement against such stresses? Why do we have so many little bones and joints within the foot? When did the foot acquire this form, and what was it like before that? Why don't we walk on our toes like the mythological satyrs? Is this form of our foot adaptive—does it meet our needs better than other possible forms, especially the anatomy of our ancestors? Although revealing accurate stories of adaptation is a notoriously difficult exercise—it is far easier to propose explanations than to test them—this concept that the anatomy and behavior of species adapt to their niches is one of the cornerstones of evolutionary biology. These questions can be best answered by considering the history of our body design, its past roles, and the changes that have occurred over time. The deepest explanation for the form of our body lies in its evolution.

WHAT IS EVOLUTION?

ev·o·lu·tion [L. *evolution*, an unrolling or opening, from *evolutus*, pp. of *evolvere*, to unroll;-, out, and *volvere*, to roll] 1. the act of unfolding or unrolling; a process of development, formation, or growth. 7. in biology, (a) the development of a species, organism, or organ from its original or rudimentary state to its present or completed state; phylogeny or ontogeny; (b) the obsolete theory that the germ cell contains the fully developed individual in miniature form; the theory of preformation; (c) the theory, now generally accepted, that all species of plants and animals developed from earlier forms by hereditary transmission of slight variations in successive generations. (From *Webster's New Twentieth Century Dictionary of the English Language Unabridged*, 2nd ed. William Collins and World Publishing Co., Inc. 1978)

Biological evolution is one of the fundamental tenets of the life sciences, but it is sadly misunderstood outside of those fields. The simplest meaning of the word *evolution* is an "unfolding of events"—change through time. Biologists understand that life changes through great expanses of time. Old species become extinct, and new species come into existence. A more specific definition of the evolutionary process can be made with reference to the subfield of genetics. *Evolution* is a change in gene frequencies within a population through time. The concept of evolution offers biologists a powerful tool with which to comprehend and explain the diversity of life.

Change occurs all around us. Each one of us changes as we grow, becoming taller, maturing, aging, and perhaps declining in senility. The seasons change during the year as we observe the vegetation of spring mature in the summer, produce seed in the autumn, and wither in the winter. These are cyclical changes, obvious to all observers but not very meaningful in the perspective of great expanses of time. To most people in the past, living in closed societies in which the world of the child was the same as the world of the parent, this was the only type of change apparent in an eternal universe.

In the past few centuries, Western society has become aware of a different type of change. Culture has evolved in a noncyclical, directional manner. Languages, political structures, technologies, and the very fabric of society change in such a way that they will never be the same again. As the dictionary entry cited earlier reveals, even the form and meaning of the word *evolution* have evolved. We are aware that the pace of cultural change has been quickening through modern history.

Once humans perceived that our world was on a road with no return, we began to study change itself, to try to understand it, perhaps even to harness it. The notion of historical change permeated the sciences, so that the eighteenth and nineteenth century naturalists became comfortable with the idea that the world itself has not always been the way we see it. Rocks, fossils, and living organisms bear witness to a different past. Natural history became, indeed, a field for historians of nature, charting the course from whence the modern world came and attempting to explain its transformation. For the notions of geological and biological evolution to take hold in scientific theory and popular imagination, mechanisms of change had to be proposed. These were the contributions of Charles Lyell and Charles Darwin and their colleagues. Darwin, in particular, should be remembered not for introducing the idea of evolution but rather for presenting a model for it to the scientific and lay public alike with such clarity of explanation and such a mass of supporting evidence that it was quickly accepted.

THE THEORY OF EVOLUTION

The Theory of Evolution asserts that life does evolve, that all living species are descended from a common ancestral species, and that life arose from nonliving materials. These are momentous ideas. As we explore them, it is important to keep in mind that the Theory of Evolution has received broad confirmation from all areas of biology. Although science cannot accept any theory as final irrefutable truth, the accumulated evidence that evolution has occurred is overwhelming. On the other hand, evolutionary theory—the body of numerous specific hypotheses concerning the course and mechanisms of evolution—remains a field of lively debate among professionals.

Life evolves. Populations of living organisms change through time. Evolutionary studies have redirected the focus of biology from individual organisms to populations. Populations have different properties than do individuals; most important is their intrinsic variety, which can change with each generation. To state that life evolves is simply to observe that the traits expressed among individuals within a population differ or exist in different proportions when followed across generations. To state that life evolves tells us nothing about either the mechanisms or the long-term consequences of such evolution.

The diversity of living species is descended from a common ancestral species. This, to Darwin, was the greatest mystery of evolution. Accordingly, he titled his great essay *On the Origin of Species.* Inspired by the geological principles of Lyell, Darwin extrapolated small observed changes across a few generations into major transformations of species through geological time. Thus not only did species change but also they could increase in number as disparate populations evolved in different directions. Working backwards through time, taxonomists and paleontologists today attempt to identify common ancestors of modern species and thus to clarify relationships among them. Biologists believe that ultimately all known living organisms have descended through diverging but unbroken lineages from a single ancestral life form.

Life arose from nonliving materials. Any modern scientific view of the universe holds that life could not have existed when the universe was formed and therefore had to arise from nonliving materials. Because the similarities in chemistry among organisms are greater than chance would permit, it is probable (but we can never prove it) that life arose only once on earth. This claim applies only to earthly life for the simple reason that we are in complete ignorance of any other kind.

THE EVIDENCE FOR EVOLUTION

The evidence for evolution comes from many different sources. Most important, evolution is a recurring theme that makes sense out of the cumulative knowledge of natural history.

All life is related. All living organisms share some characteristics. They are composed of cells. Their genetic material is passed from one generation to another in the form of DNA. They build proteins from the same subset of amino acids. Chemical details of membrane formation, metabolic pathways, and even individual genes are shared. Any theory about life that does not assume divergence from a common ancestor can neither predict nor explain such universally shared properties.

Organisms form natural relationships. Centuries of studies of anatomy and embryology permit us to state that some species more closely resemble one another than they do other species. This is the basis for the modern hierarchical classification of species. Humans comprise a single species because all human populations are similar enough to one another to permit interbreeding. Humans are grouped with apes, monkeys, and similar animals as primates because they all have opposable thumbs, nails on the digits, color vision, and other anatomical details. Primates are classified with other mammals because they share a tooth and jaw structure, three ear ossicles, homeothermy, hair, and lactation. Mammals belong with other vertebrates because they share a skeleton, a tail, and a hollow spinal cord. More important, all organisms have their places within this hierarchy. Their relationships are confirmed independently by studies of anatomy, development, and molecular biology. The hierarchy reflects common ancestry and patterns of evolutionary descent. No other theory of life could predict that organisms would fit consistently into a hierarchical classification.

Species do not have clear boundaries. A *biological species* is defined as an interbreeding population that is reproductively isolated from other populations. This is the best way to group like organisms within boundaries while recognizing individual variation. However, there are many specific cases where the species concept is hard to apply. Asexual species do not interbreed; thus defining their boundaries involves some arbitrary judgment. Some species do interbreed with other species or can be encouraged to under artificial conditions. "Sibling species" cannot interbreed but are so similar anatomically that they are indistinguishable. Lineages traced in the fossil record reveal sequences of species that grade into one another over time. Problems such as these make no sense in a theory that each species was created independently, but they make perfect sense if we view species as dynamic, changing their characteristics over many generations.

Organisms exhibit vestigial structures. In the details of anatomy and development, organisms are similar to their close relatives in ways that are neither functional nor adaptive. For example, our ears have several muscles potentially capable of wiggling them, although only some individuals have effective control of them (Figure 1.2). We can easily observe how dogs

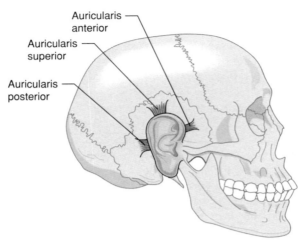

Figure 1.2 Vestigial structures. The three auriculares muscles to the ear no longer serve an adaptive function in humans but remain as evolutionary vestiges. Some individuals have voluntary control and can wiggle their ears, but others cannot.

and other mammals orient their ears to better distinguish sounds, and we infer that our ancestors once could do the same. Occasionally, small tails make appearances on the human body, reminding us that our body contains a blueprint for structures that are no longer important. Similarly, birds have genes for producing teeth and whales show rudimentary formation of hind limbs, even though these features are never functional. Vestigial structures are documents of the history of species and reveal that their ancestors were quite different in the past.

The peculiar distribution of species on the earth reveals a historical process. Species are not randomly distributed across the globe, even within ecosystems. For example, living marsupials are restricted to Australia and South America, except for one recent immigrant to North America, the opossum. The only placental mammals in Australia are those that could fly or travel with humans. Most Pacific Islands had no mammals at all until humans reached them in the last 2000 years. Biogeography, the study of the geographic distribution of species, reveals that the present distribution is the result of the species' places of origin, later migrations, and local extinctions. This discontinuous pattern is what evolutionary theory predicts.

Fossils also form natural relationships with living species. Newly discovered fossils can be placed within the hierarchical classification of living species. We do not expect a fossil to appear that cannot be classified with known groups. In fact, such a discovery would be important evidence against evolution. There are numerous examples of past "oddballs" of the fossil record eventually finding a position as we better understand them—after much wonder at its strangeness, the Cambrian *Hallucigenia* was placed among the velvet worms in Phylum Onychophora when more fossils were discovered; odd fossils called conodonts were only recently recognized as only parts of larger animals, in this case early chordates. In a very few cases, anomalies led to the revelation of hoaxes, as with the notorious Piltdown skull.[1] Likewise, the distribution of fossil species is consistent with the biogeographical patterns of their living or extinct relatives. Significantly, many fossils bridge gaps between living taxa so that the continuity of life is even smoother than appears from modern species alone.

Fossils reveal predictable sequences through time. When species and higher taxonomic groups become extinct in the geological record, they do not appear in later rocks. Communities of contemporary organisms reveal change through time such that communities closer together chronologically are more similar to one another in composition than are communities farther separated in time. These clusters of species are found so consistently that they may be used to date layers of rock. Thus we see that evolution proceeds in a linear manner.

Evolution explains and predicts. Details of natural history may be baffling out of historical context but are understandable within an evolutionary framework. The theory predicts that new discoveries of living species, fossils, genes, and anatomical quirks will obey

---

[1]The story of the fraudulent Piltdown specimens is discussed in detail in Spencer (1990). The hoax was revealed because the supposed fossils did not match the predicted range of anatomical possibilities (Langdon 1992), but such expectations are only possible if we have a theoretical framework, such as evolution, within which biologists can operate.

the principles of classification, biogeography, chronological sequence, and anatomical form that are defined by evolution. Theodozius Dobzhansky wrote, "Nothing in biology makes sense except in light of evolution."

Today biologists are no longer concerned with proving that life has evolved but rather with understanding the mechanisms and limits of change and with reconstructing the pathway that life has taken. Part of that reconstruction is the subject of this book.

DARWINISM: EVOLUTION BY NATURAL SELECTION

The most fascinating aspects of living organisms, and the subject of wonder from the earliest writers to the present, are the ingenious ways in which species are adapted to their particular modes of life. Under the name of evolution or function, adaptation is the theme of this book; adaptation is one of the few perspectives by which we can understand why our body has the

form it does. Natural selection is the key to the process of adaptation.

As stated earlier, Darwin was not the first to propose evolution. He was the first to articulate a mechanism, natural selection, by which it may occur that has become widely accepted by the scientific community. Natural selection is a process by which the greater success that some individuals enjoy in surviving and reproducing leads to a change in gene frequencies in the population. Darwinism is the theory that organisms evolve gradually through natural selection.

NATURAL SELECTION

Natural selection occurs whenever three conditions are met: (1) variation exists among individuals in a population, (2) that variation has a heritable basis, and (3) that variation correlates with the varying success of individuals in surviving and reproducing (Figure 1.3). Given these conditions, the occurrence of natural

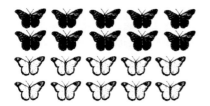

ancestral population

selection

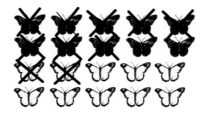

descendent population

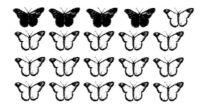

Figure 1.3 Natural selection.
A hypothetical example shows how natural selection can change the frequency of genes in a population. The parent population of butterflies shows a heritable variation in color. Predators preferentially catch and eliminate dark butterflies before they have a chance to reproduce. The new ratio of black to white butterflies among the parents is reflected in the offspring generation, even though the population size remains constant.

selection is beyond dispute, but its significance for evolution must be demonstrated.

Variation. Darwin was one of the first naturalists to appreciate the significance of variation within species. Just as previous generations failed to recognize that the world was changing, previous naturalists had often dismissed variation as a distraction obscuring the "true" characteristics of a species. Darwin saw in variation the raw material of evolution. All species are variable. Some individual members are larger than others, more brightly colored, more agile, or more resistant to disease. This variation may be observed at the anatomical, behavioral, or molecular level. Evolution may occur when frequencies of different variants change relative to one another in a population.

Heritability. Some, but not all, variation has a genetic basis. Examples of nonheritable variation might be an amputated limb, a suntan, or an education. Such acquired characteristics cannot be passed on biologically to the next generation and therefore are not relevant to organic evolution. Heritable traits may be transmitted to offspring through genes.

Reproductive success. Reproductive success is simply the number of surviving offspring of an individual. An individual's fitness, or evolutionary success, is a measure of its reproductive success relative to that of other members of the species. If a heritable trait helps an organism to obtain food, evade a predator, resist a disease, locate and attract a potential mate, compete for that mate, produce viable offspring, or have larger or more frequent broods, then that trait will increase the reproductive success of the organism. When an individual has more surviving offspring, he or she has passed on proportionately more copies of his or her genes than did a competitor.

When these three conditions are met—variation, heritability, and differential reproductive success— natural selection occurs, and traits that increase fitness increase in frequency within the population. This is a reflection of the fact that all individuals of a population are in competition with one another for limited resources. All species can produce more offspring than the environment can support with food, space, shelter, or other necessities. Some individuals succeed in gaining access to the resources that they need; others must

die. Differential survival and reproductive success are the outcomes of that competition.

Let us consider, as an example, the selection acting on infant birth size. A successful birth process is a critical event for both parent and offspring in the continuance of the species. In modern society we are aware of a number of factors that can affect birth weight, including rate of fetal growth and length of gestation. There are a number of potential health insults and other environmental factors at play, but there is genetic variation as well. Physicians recognize the importance of birth weight as an important predictor of the health and survivorship of the infant. Particularly in a traditional society in the absence of sophisticated medical care, an underdeveloped infant with low birth weight has a poor chance of surviving. At the other extreme, a very large infant has a high risk of complications at delivery that threaten both infant and mother. Selection should act against genotypes that tend to produce very small or very large infants and favor birth weight around an optimal intermediate value that is linked to the size of the birth canal (Figure 1.4). Thus observed human birth weights represent the outcome of a stabilizing selection that keeps infants close to an optimal weight.

A second example occurs at the molecular level. Individuals in tropical regions are under intense selection from the deadly falciparum malaria. The parasites that cause this disease spend part of their life cycle in human red blood cells. Some of the variant forms of the molecule hemoglobin in the blood cells confer some resistance to malaria, making infected cells and their parasites more likely to be destroyed by the body. Relatively recently (perhaps 1000 years ago), a variant form called hemoglobin C appeared in West Africa. This trait improves the health of people who possess it while it appears to have no negative effects. This trait satisfies all three of our conditions for natural selection: it represents a variation within the population; it has a genetic basis and therefore is heritable; it improves the chances for survival and reproductive success. We may consider malaria to be an agent of selection against individuals who possess the normal hemoglobin A in favor of the gene for hemoglobin C, and the latter is increasing in frequency in malarial areas.

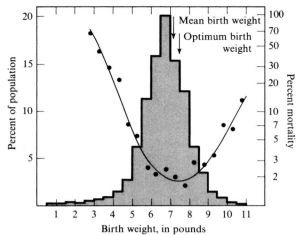

Figure 1.4 Stabilizing selection. Much natural selection acts against change in a population if the extreme variations are less well adapted than the average population. In this example, the birth weight of human children is subject to stabilizing selection because of increased mortality for both high and low birth weight infants. Low birth weight is associated with underdevelopment of the fetus and a variety of risks to survival. Excessive birth weight relates to complications of birth. The birth weight distribution and mortality curve depicted here are based on a study of 13,730 children. (From *The Genetics of Human Populations* by L.L. Cavalli-Sforza and W.F. Bodmer. © 1971 by W.H. Freeman and Company. Used with permission.)

EVIDENCE FOR NATURAL SELECTION

Natural selection has been demonstrated to operate in nature. The question for evolutionary biologists is how important natural selection has been for organic evolution. The evidence in its favor is both observational and experimental.

Natural selection is the best model to explain adaptiveness. Adaptiveness is the ability of an individual or a population to overcome the challenges of surviving and reproducing in a given environment. It is clear that any living species possesses the necessary structures and metabolic processes to survive and reproduce, and especially to exploit the niche in which it is found. Natural selection predicts that fit. More important, natural selection explains how that process

of adaptation may occur and may respond dynamically to changes in the environment.

Natural selection has been observed under natural conditions. There are numerous studies demonstrating natural selection in insects, birds, bacteria, grass, and other species leading to genetic change and adaptation in an observable time span. Some of the most detailed studies have involved finches in the Galapagos Islands (e.g., Larson 2001; Weiner 1995). These islands were colonized by a single species of finch that have since evolved into an array of different species. Many of these can be distinguished by their beaks that have become specialized for different dietary niches. For example, a drought that limited food supply in the 1980s selected for birds with deeper and stronger beaks. The form of the beak relates closely to the ease with which hard seeds may be cracked open. When a drought caused a food shortage and the population diminished by more than 85%, birds that could more easily eat a wider range of seed size and hardness were more likely to survive (Figure 1.5).

In 1986, one author catalogued 171 observations of natural selection involving more than 140 species (Endler 1986), and research continues to identify new cases. These observations include many examples for *Homo sapiens*, ourselves. Correlations with fitness were reported for human tooth size, birth weight, gestation length, height, body shape, hemoglobin type, and blood enzymes. Since that time the number of observed cases have increased greatly, especially in response to human activity (Palumbi 2001). Given the obvious constraints under which biologists work, observed changes are generally subtle and not likely to represent the creation of a new species. Nonetheless, they demonstrate that selection is acting to shape natural populations.

Natural selection can be simulated under artificial conditions. Humans have been experimenting with evolution for thousands of years. By permitting some individuals to breed at the expense of others, humans have produced grains that store more calories, sheep that are more docile, dogs that run faster, cotton that produces more fibers, cows that produce more milk, horses that carry heavier loads, bacteria that digest petroleum, and flowers with more

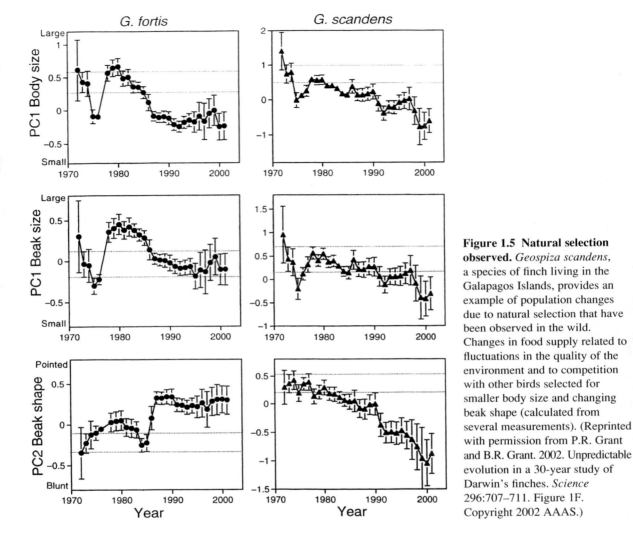

Figure 1.5 Natural selection observed. *Geospiza scandens*, a species of finch living in the Galapagos Islands, provides an example of population changes due to natural selection that have been observed in the wild. Changes in food supply related to fluctuations in the quality of the environment and to competition with other birds selected for smaller body size and changing beak shape (calculated from several measurements). (Reprinted with permission from P.R. Grant and B.R. Grant. 2002. Unpredictable evolution in a 30-year study of Darwin's finches. *Science* 296:707–711. Figure 1F. Copyright 2002 AAAS.)

beautiful colors. Plant and animal domestication has been practiced on numerous species for a wide range of traits (Figure 1.6). Until the past century, such breeding was done in ignorance of the concept of species evolution and required only the common sense of farmers and herdsmen who understood their jobs. This process, in which humans directed the course of genetic change, is termed *artificial selection*. Artificial selection requires the same principles and conditions as natural selection but can be made to occur at a faster rate.

THE ORIGIN OF HERITABLE VARIATION

After the publication of *On the Origin of Species*, Darwin was challenged by critics on a crucial question that he could not answer. What was the origin of heritable variation? Nineteenth century theories of inheritance could not explain the appearance of new traits or forms of traits, even though domestic animal and plant breeding clearly generated new varieties. Natural selection accounts only for the elimination of less-fit traits.

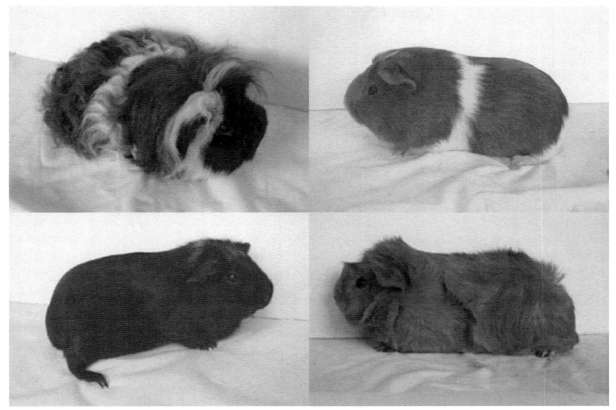

Figure 1.6 Products of artificial selection. Artificial selection has been used by animal and plant breeders to speed and direct the course of evolutionary change to produce exotic varieties in species such as these guinea pigs.

Darwin never was able to solve that problem; it was not resolved until the twentieth century with the development of modern genetics. We now understand that the molecule of inheritance, DNA, can spontaneously change or mutate to produce new forms. Although most detected mutations simply disrupt normal functioning of a gene, most DNA does not code directly for genes; mutations outside of a gene may go undetected. Noncoding regions may have a variety of functions, including regulation of gene activity and guidance of the folding of a chromosome. However, it is believed that long stretches of the DNA do not code for critical information.

An important source of new genetic material is the result of duplication of genes (Zhang 2003). Extra copies may be altered without threatening continued function of the original form. Because these copies

begin with a working function, there is a higher probability that random changes in them may improve the function of the protein for which they code or permit a new function. A recently identified example involves the ribonuclease enzyme in a leaf-eating monkey, the douc langur (Zhang et al. 2002). Ribonuclease is a pancreatic enzyme that helps to digest RNA in food. In this species, the gene was duplicated, and one copy evolved to produce a variant enzyme that is more effective in digesting bacterial RNA. Like many leaf-eating species, the langurs use a bacterial colony in the gut to help digest leaves; thus this adaptation is particularly valuable.

Sexual reproduction contributes further to variation by permitting genes or parts of genes from two parents to be brought together in the offspring in new combinations. Sometimes this recombination simply adds to the number of genes that can be expressed, as in the immune system where a great diversity of antibodies helps the body fight off disease. Because genes may interact with one another, novel pairings can also produce new variations on which natural selection may act. Errors in recombination can help generate duplicate genes, as in the ribonuclease example just given. Nonetheless, at the gene level, recombination can only help existing variation to express itself. Genetic mutation must be regarded as the ultimate source of all variation and a necessary condition for biological evolution.

HOW VARIATION TAKES FORM

The easiest way to study the effects of genes on the body is to look at the isolated products of them— proteins—in the bloodstream, where they are easily extracted. Thus genetic studies of humans and other animals were dominated over the past century by studies of blood types and similar individual molecules. It has been possible to use such studies to map human variation, study population movements, and explore a few individual cases of selection and evolution in our species. The roles of hemoglobin variants such as sickle cell, hemoglobin C, and the thalassemias are standard subjects of genetics and evolution textbooks. These are important studies, but they leave a significant gap of understanding between evolution at the molecular level and change in gross body form that we associate with the origin of species.

Another line of approach to understanding evolutionary changes has been the study of comparative embryology (Arthur 2002). Embryology helps us to understand how a fertilized egg develops into an adult. Observing when and how embryos of different species diverge in their developmental pathways provides hints at the mechanisms that determine body form. In recent decades such studies have advanced to the molecular level of controls. Two patterns of evolutionary differences between species may be usefully addressed in developmental terms: differences in degree and differences in direction of development.

Differences in Degree of Development

Some traits appear to differ among related species primarily by developmental timing or stage. For

example, various types of kidneys in living vertebrates may be observed as different stages in a developmental sequence for mammals. At a gross level, human brains are like those of apes that have continued to develop for a longer span of time, whereas the adult human face has a closer resemblance to that of an infant ape. Interspecies differences in rates and degrees of development are called *heterochrony* (Figure 1.7). There are several different mechanisms that can produce heterochrony, and not all such superficial resemblances are true examples of it. A developmental process may

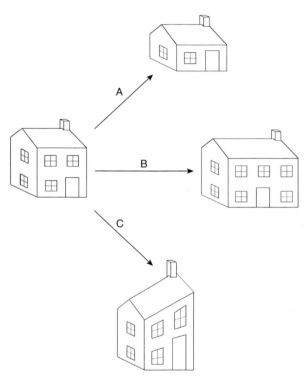

Figure 1.7 Heterochrony in evolutionary change. Heterochrony results from a change in developmental timing. We can use the analogy of a house plan to illustrate some types of these changes. *A*, Paedomorphosis results from an early termination of a developmental sequence so that the adult for or features of the ancestor do not appear. *B*, Peramorphosis is the extension of development beyond the ancestral pattern so that new or exaggerated features may appear. *C*, Different changes in developmental timing in different parts of the anatomy may change the size or shape of the organism.

speed up or slow down; it may begin at an earlier or later age; it may terminate early or carry on to a greater degree. If such changes make a feature appear less developed or resemble a juvenile form, the process is called *paedomorphosis*. If overdevelopment occurs so that the development is exaggerated, the process is *peramorphosis*.

Simple heterochrony within the skeleton, for example, may produce changes in the size or shapes of bones and other body parts (Lovejoy et al. 1999, 2000). The short, flat human face may be understood as paedomorphic compared with the long face that is typical of other primates. Such patterns may be the result of slight changes in regulatory genes. In the adult forms, they may make one species appear underdeveloped (paedomorphic) and another over-developed (peramorphic) for one or more traits.[2]

Differences in Direction of Development

Other sequences of development produce quite different structures in related organisms. Embryos may be very similar up to a certain point in development, after which the species proceed in different directions. Consider the example of the lung. It exists in two forms in fish with completely different functions. In some fish and in all land vertebrates, the lung is an air-filled sac connected to the pharynx that provides a surface where gases may be exchanged between the bloodstream and the atmosphere. In other fish, it is a swim bladder, filled with gas but no longer connected with the pharynx. The volume of gas in the swim bladder may be decreased or increased to adjust buoyancy by shifting gas molecules to or from the bloodstream. The lung and swim bladder are clearly homologous structures, but the instructions for their development have been rewritten. While it has long been known that the bases of such developmental pathways lie ultimately in the genes, it has been poss-ible to unite the fields of genetics and embryology only in the past couple of decades.

---

[2]For an extended discussion of heterochrony, see Gould (1977) and Minugh-Purvis and McNamara (2002). Parker and McKinney (1999) provide a succinct summary.

THE LIMITS OF NATURAL SELECTION

A full understanding of the mechanism of evolution also requires an understanding of what cannot happen. Natural selection cannot produce every conceivable adaptation.

Selection can act only on existing variation. If a novel combination of genes does not lead to the appearance of a new trait, however useful that trait is to the organism, it will never be favored by selec-tion. This constraint introduces an element of chance into evolution that makes specific predictions about future adaptations nearly impossible. Similarly, we may never be certain why certain adaptations do not exist.

Directional selection will act only to improve fitness. This statement should be self-evident: selec-tion does not increase the frequency of less adaptive traits over more adaptive ones. It is useful to keep this principle in mind when we reconstruct complex evolutionary pathways. Recognizing that each incre-ment of change by natural selection must increase fitness helps us to frame questions about evolutionary history.

Evolution must respect physical and develop-mental constraints. The existing form of the organism and the physical properties of its components limit the directions in which natural selection can cause change. Evolutionary change can operate only by making small alterations in the complex program of embryological development. Thomas Huxley explained this with the metaphor of a billiard ball: A ball on the table can be pushed in any direction by the cue, but if that ball is a polyhedron instead of a sphere, its shape will limit its direction and the extent of its movement. Similarly, evolutionary change is constrained by the "shape" of the developmental pathway of a species. There are only certain kinds of changes within a developmental program that are possible. Furthermore, changes at one point in the developmental program are likely to have ramifications in many parts of the body. Nipples on a male and inguinal canals in a female, for example, are useless structures that are present because both men and women share a common developmental blueprint.

What has adaptive value in one must also appear in the other, even though it has no purpose.

Some evolutionary changes are not the outcome of selection. *Genetic drift* refers to the random changes in gene frequencies that appear in every generation in populations of finite size. Drift may result in unpredictable evolutionary changes in a population. Many variations may arise that do not alter the fitness of an individual. These are called *neutral mutations* and are invisible to and unaffected by natural selection. The frequencies of neutral variations are particularly subject to genetic drift.

SEXUAL SELECTION

Darwin puzzled over the presence of striking traits in many species that appeared to represent liabilities to individuals. Many of these traits seemed to be involved in mating displays to attract the opposite sex; the tail of a peacock is a classic example. To explain such traits, Darwin invented the concept of sexual selection.

Sexual selection may be understood as a special case of natural selection favoring traits that increase the chances of obtaining a desirable mate. Sexual selection may occur in several ways. If males compete by threat or combat, sexual selection may favor larger or more aggressive males. It may lead to the elaboration of such weapons as antlers or large canine teeth. On the other hand, sexual selection may favor whatever trait the other sex finds attractive, including bright colors (e.g., bird plumage), singing (the roar of a bull frog), solicitous behavior (a presentation of food), or something as unique and arbitrary as a rooster's comb. Sexually selected traits are often sexually dimorphic (appearing differently in one sex than in the other) and are indicative of unequal opportunities for mating among members of a sex.

There is an evolutionary logic to the appearance of sexually selected traits—such as advertisement of maturity, health, or ability to provide to offspring—but the origin of a given trait may be obscure. Sexual selection has sometimes been treated as irrational behavior by one sex, so that a wide variety of unexplained traits can be dismissed as "explained." This approach at best explains only a current function of a trait but not its orgins.

ADAPTATIONISM: AN EVOLUTIONARY PERSPECTIVE

Adaptationism is a theoretical approach that assumes the traits we observe in living organisms evolved under natural selection because they are or were adaptive. The task of a strict adaptationist is to discover the adaptive values that led to the evolution of a trait. The value and problems of such a perspective are worthy of some consideration.

This approach has several major shortcomings (e.g., Andrews et al. 2002; Konner 2002). First, in assuming that all traits are the result of natural selection, all other possible mechanisms are discounted. Characters that appeared by chance or as a secondary consequence of other traits and have no adaptive value themselves can never be correctly understood. Evolutionary explanations for the appearance of specific traits therefore have to be formulated as stories of adaptation. The adaptationist's justification for this is the premise that natural selection is by far the most important force in evolution. While this is likely to be true for many characters apparent at the gross anatomical level (the subject of this book), it may be especially misleading for details at the molecular or behavioral level.

Second, the hypotheses of an adaptationist rarely can be adequately tested, and so are difficult to disprove. It is always possible that the best or most reasonable explanation will in the future be supplanted by a better idea or by new information. On the other hand, nonselection explanations can rarely be supported by anything but negative evidence, the lack of a recognized adaptive value. Often this is more a statement of our ignorance than of our understanding.

A third problem is the difficulty in correctly identifying evolutionarily relevant "traits." Broadly speaking, a trait is any observable characteristic of an individual, whether anatomical, physiological, molecular, or behavioral. However, not all traits are likely to have evolutionary significance, and some adaptive alterations to the developmental pathway may have multiple effects in the body. Take as an example the small parietal foramina, which are a pair of small holes usually present in the parietal bones on either side of the midline of the human skull. These foramina conduct emissary veins that carry blood from the scalp to passages surrounding the brain, and they may contribute to the regulation of temperature inside the skull. Thus the presence or absence of the foramina becomes an evolutionarily significant trait. But is the location of these foramina important enough to demand an evolutionary explanation? Would they be more or less effective if they were located in the frontal or occipital bones? Is the foramen one trait or many? Actually there are several similar emissary foramina in the occipital region. These are far more variable in number and symmetry. Can we assume that selection has acted or is acting independently on each possible passage, and that each should be studied for its adaptive significance? Or is their expression the result of functionally and adaptively insignificant variations in branching patterns of blood vessels and local differences in blood flow? This problem is even more acute for behavior. Are day-to-day decisions and actions independent traits? If I refrain from stepping off a curb in front of a truck, am I exhibiting an adaptive trait of avoiding moving vehicles or the adaptiveness of common sense? Which trait, if either, has an evolutionary history behind it?

The chapters that follow assume a cautious adaptationist perspective. They describe how features of anatomy are functional and increase the adaptiveness of the human body. Competing hypotheses are often examined to find the more reasonable explanations. Obedience to the following principles will give us greater confidence of being "reasonable."

Explanations should be framed in terms of the selective constraints of the past. Conditions of the past were not necessarily those of today. As we try to understand the interaction of an organism with its environment, we must envision the past environment rather than the present one and the limits of natural selection as they applied at that time.

The principle of parsimony states that fewer assumptions are better. Simple hypotheses that make few unsupported assumptions are more likely to be correct than are complex hypotheses with many assumptions. This is a good rule to follow only when we have considered enough of the data. We will encounter many simple hypotheses that sound parsimonious and reasonable when applied to a single trait but require many assumptions as additional data are considered.

Similarly, some models appear to be parsimonious in that they explain a great many traits; nonetheless, each of the "explained" traits requires additional assumptions (e.g., Langdon 1997). While **parsimony** is a good rule of thumb in all branches of science, there is no guarantee that evolution is always parsimonious.

Simpler hypotheses may be easier to understand, but this does not mean that they are true. Ideas that are easily understood may not be parsimonious in application. For this reason weakly supported adaptationist models, including sexual selection explanations, are often more easily accepted than are non-Darwinian explanations not involving selection. In broader society, this problem has enabled many problematic and unsupported ideas to flourish, from creation science to conspiracy theories, by the simple fact that people are more likely to accept ideas they can understand quickly than they are to accept ideas that involve difficult concepts.

Understanding the evolution of a uniquely human trait should implicitly address the question "Why did other species not evolve in this way?" If a trait seems self-evidently adaptive for humans—a larger brain, for example—it should be equally adaptive for other species; we must ask why humans are different. A fuller explanation for the expansion of the human brain must identify the unique circumstances favoring or enabling that trait in the human lineage that are not present for other taxa.

It is always possible that some traits are simply not adaptations. This is because the traits are neutral; or they are consequences of other constraints or adaptations; or they represent developmental, not genetic plasticity; or they have been interpreted from the irrelevant perspective of cultural or racial bias.

OTHER PATHS TO ADAPTATION

Some traits, especially behaviors, appear to us to be adaptive but are not the results of natural selection; thus even a clear identification of the value of a trait does not necessarily indicate how the trait arose. Three examples of alternative origins of adaptations are described and should serve as a further caution in applying adaptationism.

Exaptation. As Gould and Vrba (1982) have pointed out, we use the term "adaptive" loosely to refer both to traits that have evolved by natural selection for a specific function and to traits that acquired a functional significance after they evolved. They propose the term *exaptation* for the latter. For example, we can observe that human bipedal posture frees the hands for extensive use of tools. Because it appears that bipedalism evolved before a dependency on material culture, we can say that hominin posture is an exaptation for tool use (hominins are humans and their extinct close relatives). This does not imply that upright posture evolved for the purpose of tool use or because of it. That would have been impossible, a violation of evolutionary theory. Instead, freeing the hands was simply fortuitous. It might be that bipedalism made material culture possible, or at least more likely; however, we must look elsewhere to explain the appearance of bipedalism. Unfortunately, as we are so rarely certain of the causal relationship between traits and their functions, such a distinction of terminology has more theoretical significance than practical use.

Rational behavior. Rational behavior represents a chosen path of action by an individual that has a greater benefit than cost—or at least so it appears to the individual. Included under rational behavior are countless decisions we make every day, from choosing which brand of gasoline to buy to how we greet an acquaintance. Calculating cost-benefit analyses may appear very mercenary; but when we are permitted to include such variables as personal motivation, cravings, conscience, and aesthetics into the equation, "calculation" can reasonably well describe our lives. Certainly some of our assessments are wrong or short-sighted. On the whole, however, most individuals' behaviors are adaptive within their environments, helping them to achieve their goals.

If an acute businessman makes a fortune, elevates his social prestige and rank, and thus is potentially more successful in attracting one or more wives or mistresses, depending on his culture, his rational business decisions may be seen as adaptive. His behavior may even fulfill two of the three criteria for natural selection. However, the specific decisions he made do not represent heritable traits and therefore cannot be considered the products of natural selection. Such an example may appear trivial, but issues can become far more complex when behaviors that are not so culture laden or behaviors of other species are considered. Other animals do show rational behavior, such as when a chimpanzee uses a stone to crack open a nut or terns drive people away from their nesting ground. However, it may be difficult to determine what genetic bases might or might not underlie such actions.

Indirect selection. Some traits may be the product of natural selection but only in an indirect way relative to the context in which they are observed. The decisions described earlier by the businessman and the chimpanzee were rational decisions made within certain environments. The actions were not built into the brain by a genetic program. However, the brains themselves are anatomical organs that have evolved through natural selection and are capable of making such decisions. The actions may be adaptive but not evolved; the capacity for decision-making is both adaptive and evolved.

This distinction is crucial but often ignored or confused by writers. Natural selection has indirectly given us a taste for candy because sugar signals digestible calories in ripe fruit. Natural selection indirectly favored brains that sought out sexual gratification because individuals with such brains were more likely to have children. Does a gourmand seek food to satisfy her hunger or her appetite? Does the businessman seek a second sexual partner for reproduction or for pleasure? The cost-benefit analyses these people perform value gratification, not reproductive success, as their benefit. The outcomes may or may not enhance Darwinian fitness—they may just as easily lead to diabetes or a broken marriage. While some adaptive behaviors may indeed have been shaped by natural selection, we must be very cautious in selecting appropriate traits for examination.

IMPORTANT EVOLUTIONARY CONCEPTS

Homology. When a structure is transformed by evolutionary change, its ancestral and various descendent forms are said to be homologous. Homologies are most securely identified when we can trace the formation of structures in individual development. For example, all vertebrate embryos possess a series of paired pharyngeal arches at some stage. The second arch forms a series of small bones supporting the jaws in fish, but it develops into the styloid process, stapes, and hyoid bones in mammals (Figure 1.8). The

differences result from changes in the developmental programs. Nonetheless, we can identify these corresponding elements in fish and mammals as homologous to one another.

To cite a more common example, the pectoral fin of a fish, the forelimb of a reptile, the wing of a bird, and the upper limb of a human are all derived from the

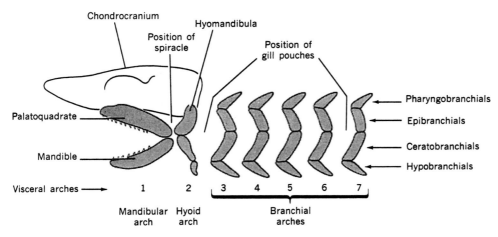

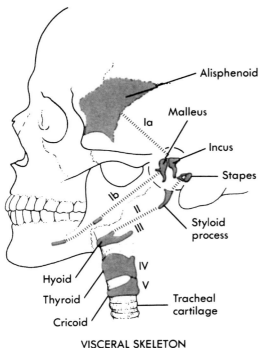

Figure 1.8 Homologous structures. The skeletal elements of the pharyngeal arches have assumed different positions and functions in different vertebrates. However, we can recognize the homology among them because they appear in similar form early in development and we can trace theirtransformation to the adult patterns. Above, the elements of the seven branchial arches (shaded) in a shark are compared with their homologies in an adult human skull. (Top: From *Analysis of Vertebrate Structure*, 3rd ed., by M. Hildebrand, Copyright 1988 John Wiley and Sons. This material is used by permission of John Wiley & Sons, Inc. Bottom: From G.C. Kent, *Comparative Anatomy of the Vertebrates*, 7th ed. Copyright 1987 Wm.C. Brown Communications, Inc. This material is used by permission of The McGraw-Hill Companies.)

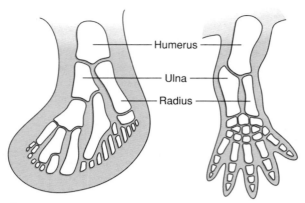

- Humerus
- Ulna
- Radius

Figure 1.9 Homologies in vertebrate fins and limbs. The homologies in the bones in the fins of certain fossil fish and in limbs of early amphibians help us to trace evolutionary change as vertebrates began to occupy a terrestrial habitat. On the left are bones of a sarcopterygian fish; on the right are those of an early labyrinthodont amphibian.

same ancestral structure. They are therefore homologous with one another at a gross structural level. Moreover, individual structures within the limbs, such as bones, may be identified as homologous with corresponding elements in the other species (Figure 1.9). The wing of an insect is not homologous with that of any vertebrate, because it was derived completely independently by an ancestor with no appendages. Such nonhomologous structures that perform similar functions are said to be analogous. While some homologies, such as those of tetrapod limbs, are trivially obvious, others, such as the cusps on different mammals' teeth, can be extremely difficult to sort out. Identification of true homologies is an important tool in determining evolutionary relationships among organisms.

Primitive and derived traits. Primitive traits are those that are present in an ancestral species or an earlier stage in evolutionary history. Derived traits are those that were later acquired by the descendants of that species. For example, quadrupedalism (four-legged locomotion) is an ancestral state for all land-dwelling vertebrates. Later vertebrates have derived a variety of alternative behaviors, including human bipedalism, the flight of birds, and the legless crawling of snakes.

Defining a trait as primitive or derived may depend on one's perspective. Hair is a derived trait of mammals, because the reptilian ancestor of mammals did not possess hair. However, hair is a primitive trait for primates, because the common ancestor of primates did possess hair. When we attempt to understand the evolution of mammalian adaptations, we must address the question of why hair evolved on mammals. The presence of hair is not important in understanding why primates evolved differently from other mammals, because hairiness was a preexisting state. Instead, the relative loss of hair in humans is a derived state that poses an interesting problem.

Identifying primitive and derived traits can be important and sometimes difficult. For example, certain small South American monkeys are unique among higher primates in having claws instead of nails. Are their claws primitive or derived traits? Certainly the ancestor of primates had claws (primitive). Was there an ancestor of these monkeys that had nails and later lost them, or do these species represent a lineage of primates that never evolved nails? After some debate, anthropologists now agree that claws were secondarily reacquired and therefore are a derived trait for these primates.

Common ancestry. All species had a common ancestor among the first cells. As we attempt to reconstruct the pathway of evolution, it is of greater interest for us to identify or reconstruct the last common ancestor between any two species. This would enable us to determine, for example, whether chimpanzees are more closely related to orangutans or to humans. Ancestry is determined by analyzing primitive and derived characters. A common ancestor should possess the primitive characters present in its descendent species but none of the derived adaptations that differentiate them.

Parallelism and convergence. Structures that share similar forms or functions may be explained by parallel or convergent evolution. Parallel evolution occurs when selection acts in similar ways on two closely related organisms. For example, thick enamel on the teeth appears to have evolved independently in early hominins (including our ancestors) but also among more distant relatives, such as the fossil hominoid *Sivapithecus*. It is likely that these two groups were responding to similar selective pressures in the environment—perhaps a drier habitat that resulted in tougher food.

Convergence occurs when similar traits evolve in two more distantly related species. Color vision has evolved independently in many groups of vertebrates, including birds, fish, primates and some other orders of mammals. Because these taxa and their ecological roles are diverse, the selective pressures acting on them and the evolutionary pathways involved were not the same. Hence we consider color vision to be an example of convergence rather than of parallelism.

2

A PLACE IN NATURE

Humans in an Evolving World

A PLACE IN TIME

The discovery in the eighteenth century that the earth was very old did not provide science with any means of determining just how old it was, only with an appreciation that there was much time in which the rocks and life could have evolved. Charles Darwin daringly speculated that the deposition of rocks on the earth perhaps required millions of years. Only in the twentieth century have we been able to estimate its absolute age based on the decay of radioactive isotopes. We now understand the earth to be approximately 4.6 billion years old. Life itself may be as old as 4.0 billion years.

Nonetheless, the lack of tools for measuring age did not prevent geologists from studying the order and sequence of the rocks and fossils. They erected a framework of eras, periods, and epochs so that we may speak of relative ages. Each division of geological time represented an apparent discontinuity from the next, observed in the nature of the rocks of the fossils contained within them. Table 2.1 provides a very brief overview of geological time and of the history of life. An appreciation of the immensity of time preceding the appearance of humans on earth helps us to gain some perspective on the relatively brief role we have had in its history and to see that an approach through comparative biology can shed much light on ourselves.

A PLACE IN LIFE: BIOLOGICAL CLASSIFICATION

One of the important themes of this text is that we should view ourselves not as a unique species to be studied and understood in isolation but as a species that fits into an evolutionary context. There is no single feature of human anatomy that sets us apart from other animals—a fact that has been repeatedly rediscovered in the past two centuries. Rather, we are distinguished by such features as a relatively larger brain, relatively longer lower limbs, habitual rather than occasional bipedalism, and partial reduction of body hair. The approach of comparative anatomy helps us to identify the functionally important and evolutionarily interesting features of our body and to formulate better questions concerning how and why they exist as they do. A taxonomic perspective, which classifies us within the animal kingdom, gives us a context for such comparisons.

We classify living organisms through an elaborate hierarchy of categories, or taxa (Table 2.2). Each type of organism is given a unique name and a relationship with every other organism. The most important level of classification is the species. The species is the only level that approaches some degree of objective definition—a population of interbreeding organisms. When two populations are unable to interbreed with normal fertility, we must regard them as separate species. Higher taxonomic levels—genus, family, order, etc.—are semi-arbitrary groupings of species based on common ancestry, similarity of ecological adaptation, and similarities in body design. Taxonomic groupings should also mirror evolutionary genealogy, but they are arbitrary in the sense that there are no objective guidelines to tell a taxonomist when a group of related species comprises, for example, a family as opposed to an order. A family or an order or a phylum has no meaning in the biological world—only in our study of it.

Even the seemingly objective definition of a species presents many problems. Interbreeding between populations is not an all-or-nothing event. Two populations as distinct as lions and tigers are separated by

Table 2.1 Geological Time

Age	Geological Name	Major Events
4.6 billion to 570 million years ago	Proterozoic Era	Origins of life, photosynthesis, aerobic respiration, eukaryotic cells, sexual reproduction; first multicellular organisms; first animals
570–225 Mya	Paleozoic Era	
570–500 Mya	Cambrian Period	Origin of vertebrates; all major living phyla of animals present
500–435 Mya	Ordovician Period	Agnathans (jawless fishes) and large invertebrates dominant in the seas; land invaded by plants, fungi, and invertebrates; first vascular land plants
435–395 Mya	Silurian Period	Jawed fishes evolve, become major aquatic predators; arthropods invade land
395–345 Mya	Devonian Period	Sharks, insects appear; fish dominate the sea; gymnosperms create first forests
345–280 Mya	Carboniferous Period	Vertebrates invade land; amphibians diversify, reptiles appear; swampy forests of the Carboniferous give rise to modern coal deposits; possibly first flowering plants
280–225 Mya	Permian Period	First conifers; reptiles flourish and diversify; therapsids dominant land animals; Permian ends with largest mass extinction in the history of the earth
225–65 Mya	Mesozoic Era	
225–195 Mya	Triassic Period	Reptiles continue to diversify and dominate; first dinosaurs and mammals
195–136 Mya	Jurassic Period	Dinosaurs dominant on land, pterosaurs in the air, and marine reptiles in the sea; first birds
136–65 Mya	Cretaceous Period	Angiosperms became dominant plants, introducing fruit and flowers as major dietary niches; dinosaurs dominant land animals; Cretaceous ends with major extinction on both land and sea
65 Mya to present	Cenozoic Era Tertiary Period	
65–54 Mya	Paleocene Epoch	Mammals diversify and become dominant; most modern orders of mammals, including primates, present
54–35 Mya	Eocene Epoch	Mammals, including primates, are notably more modern looking; first anthropoids appear in Old World
35–26 Mya	Oligocene Epoch	New World monkeys present
26–5 Mya	Miocene Epoch	Modern families of mammals present; hominoids diversify in Africa, Europe, and Asia; Old World monkeys appear; hominin line diverges late in epoch
	Quaternary Period	
5–2 Mya	Pliocene Epoch	First bipedal hominins appear in Africa; *Homo* evolves at the end of the Pliocene
2 Mya to present	Pleistocene Epoch	*Homo* migrates out of Africa into the other continents; anatomically modern humans appear

Mya = millions of years ago.

Table 2.2 Biological Classification (Example: Humans)

Taxon Rank	Example	Common Name	Selected Characteristics
Kingdom	Animalia	Animals	Eukaryotic multicellular heterotrophs without cell walls
Phylum	Chordata	Chordates	Animals with notochords, slitted pharynx, dorsal nerve cord
Subphylum	Vertebrata	Vertebrates	Chordates with cranium and vertebrae of bone
Class	Mammalia	Mammals	Vertebrates that suckle young; dentary squamosal joint
Subclass	Eutheria	Placental mammals	Mammals that form placentas
Order	Primates	Primates	Mammals with nails and grasping hands and feet, postorbital bar, petrosal bulla
Suborder	Anthropoidea	Anthropoids	Primates with postorbital closure, fused mandibular symphysis
Infraorder	Catarrhini	Catarrhines	Old World anthropoids; two premolars in each jaw
Superfamily	Hominoidea	Apes	Catarrhines without tails; flattened thorax and climbing adaptations
Family	Hominidae	African apes and humans	Semiterrestrial hominoids of Africa
Subfamily	Homininae	Hominins	Bipedal hominoids
Genus	*Homo*	Humans	Hominins with expanded brains
Species	*Homo sapiens*	Modern humans	Hominins with modern size brain

vast distances and never have the opportunity to mate with one another in the modern wild. However, they can and do interbreed in a zoo. We regard this as an artificial condition and continue to distinguish the cats as separate species. When two populations share the same habitat under natural conditions, are morphologically distinct, and avoid interbreeding, they are considered distinct species; yet even then, occasional healthy hybrids may appear.

These problems are, of course, directly related to the process of evolution. When two populations of one species are isolated from one another over long periods of time, they grow increasingly distinct. Someday, reproductive barriers may become absolute; in the meantime, selecting the point at which we choose to recognize two species instead of one is arbitrary. For example, dogs have been artificially selected to differ from their wolf ancestors in anatomy and behavior. Even though they will still readily mate with one another and produce fertile offspring, we consider them distinct species (*Canis lupus* and *Canis familiaris*). When populations are sampled across time, the problem of defining species becomes even more complicated. When the fossil record reveals a changing population, at what point are descendants considered new species?

Taxonomists must use all the data available to them —including morphology, genetics, behavior, geographical distribution, and a great deal of considered judgment—to define modern species. Paleontologists must focus on skeletal anatomy plus distribution of fossils in time and space to sort specimens. It is no wonder that there is so much argument over names and categories. Despite these problems, classification is a necessary and powerful tool for organizing the never-ending observations that naturalists make and for storing and extracting information. Each grouping, or taxon, shares anatomical and genetic characteristics that indicate common ancestry. To label a human as a mammal, for example, is to imply that a human (or a human's ancestor) has hair, bilateral symmetry, complex teeth, four limbs, and other features that all mammals share. Even as taxonomists pursue lively disagreements about details of classifying organisms, they largely agree on the important concepts of how species evolve.

COMPARATIVE BACKGROUND OF THE HUMAN SPECIES

KINGDOM ANIMALIA

Biologists currently recognize three major divisions of organisms. The domains Archaea and Bacteria contain prokaryotic organisms—single-celled life forms lacking nuclei and other organelles. Domain Eucarya, containing eukaryotes possessing organelles and chromosomes, is believed to have evolved primarily from the archaeans, although some genes may have been "captured" from bacteria. There are four kingdoms of eukaryotes. The Protista are single-celled or colonial organisms. Nearly all organisms in the kingdoms Plantae, Fungi, and Animalia (plants, fungi, and animals) are multicellular.

Animals may be further distinguished as heterotrophs that lack cell walls. Heterotrophs are unable to trap free energy from sunlight to use in manufacturing essential carbohydrates, proteins, and fats. Therefore heterotrophs, such as fungi and animals, must be predators, consuming chemicals in the bodies of other organisms. The lack of cell walls distinguishes animals from both plants and fungi and removes some of the constraints on locomotion and physical support for individual cells. The domains and kingdoms are formally and more precisely defined by subtle characteristics at the subcellular and molecular levels. For example, only animals produce the protein collagen or use *Hox* genes to direct the development of body form.

We cannot clearly identify the origins of the different kingdoms in the fossil record, because the earliest representatives of each were small and lacked the hard tissues such as shell and bone that are most likely to survive as fossils. Single-celled organisms sometimes appear as fossils, but they are rare. Molecular and geological evidence indicate that animals arose and diversified before 570 Mya (million years ago), during the Proterozoic Era (Knoll and Carroll 1999). Remarkably, the earliest fossil evidence of animals consists of microscopic forms about 590 million years old (Chen et al. 2004). Although some extraordinary collections of strange soft-bodied animals exist from the late Proterozoic Era, the first appearance of hard tissues occurs in the Cambrian Period when the major living phyla of animals make their first appearance.

PHYLUM CHORDATA

There are many phyla in the animal kingdom. Humans are members of phylum Chordata. Chordates arose as small animals that filtered plankton from the water. For at least part of their life cycle, chordates are free-swimming. The phylum is most clearly defined by the presence of three anatomical characteristics: a slitted pharynx, a notochord, and a dorsal hollow nerve cord (Table 2.3; Figure 2.1).

The slitted pharynx is an adaptation to filter-feeding. Water enters the oral cavity through the mouth and is expelled from the pharynx through a series of narrow slits in its wall. The narrow slits filter out small particles of food to be conducted to the stomach. Among fish, the increase in body size and changes in feeding strategies led to a new adaptation of the pharynx for respiration. Fish reduce the number of slits and separate them with skeletal elements, the branchial arches, commonly referred to as the gills.

The notochord is a stiff rod consisting of a thick fluid surrounded by a tough sheath of connective tissue. It provides an anchorage for muscles and permits the animal to swim by sweeping its tail from side to side. The notochord granted a potential increase in the speed of early chordates and gave them power for such activities as burrowing into the sea floor. The notochord is present only in larval or embryonic forms of most living chordates, disappearing or being replaced by bone in the adult form.

The nerve cord develops as a hollow tube dorsal to the notochord by an infolding of the ectoderm, the outermost layer of cells. This tube will differentiate into a brain and spinal cord. The cavity within it, the neurocoel, persists in the adult.

There are three groups, or subphyla, of chordates (Figure 2.2). The vertebrates are the most diverse and successful. The other two groups consist of tunicates, or sea squirts, which are small filter feeders that anchor

Table 2.3 Chordate Characteristics

Slitted pharynx
Notochord
Dorsal hollow nerve cord
Motility during some part of the life cycle
Bilateral symmetry
Metamerism (segmentation) of the body wall but not the viscera
True coelom separating body wall from viscera
Cephalization with an anterior concentration of sense organs
Tail extending posterior to anus
Ventral heart that pumps blood anteriorly and dorsally through aortic arches
Complete gut with mouth and anus

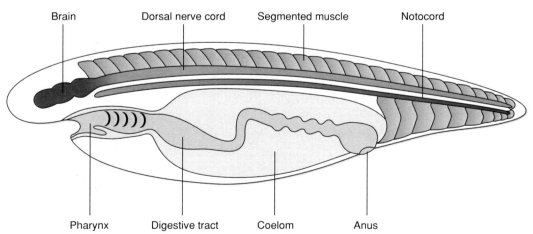

Brain   Dorsal nerve cord   Segmented muscle   Notocord

Pharynx   Digestive tract   Coelom   Anus

Figure 2.1 Archetypal chordate structure. The characters that define the chordate phylum are the starting point for the vertebrate body plan. The body wall is covered with segmented muscles, or myomeres. Deep to this, the notochord, nerve cord, and visceral plan are apparent. Water entering the mouth on the left filters through the slits of the pharynx and leaves the body. Food extracted from the water passes into the intestine.

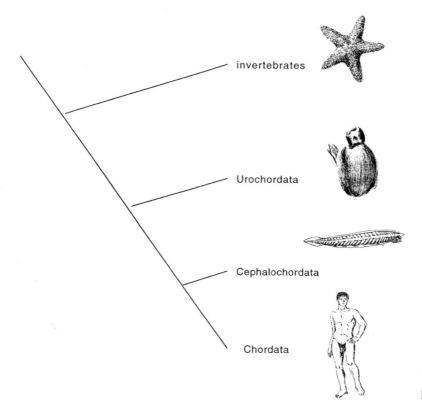

invertebrates

Urochordata

Cephalochordata

Chordata

Figure 2.2 Chordate diversity.

themselves to rocks, and the small fish-like lancelet, *Branchiostoma*. Both of these are found on the floor of shallow marine habitats. The chordates, like the other major animal phyla, apparently arose and diversified early in the Cambrian Period. Fossil relatives of all three divisions can be found in that time span.

The three defining characteristics may not persist in the adult stage, but they will be present at least in the larval stage of all chordates. Adult humans do not have recognizable pharyngeal slits or arches, but these are clearly recognizable in the embryo. We can trace the arches and their supporting muscles, arteries, and other tissues in fetal development and relate them to such structures as facial musculature and hyoid bone. Chordates also share many traits that are present in some other phyla of animals. The free-swimming stages of the chordate life cycle are bilaterally symmetrical (typical of most motile animals). They possess a head and a tail. Cephalization, the development of a head, represents a concentration of special sense

organs at the leading end of the animal so that an individual can scan and analyze the environment. This probably contributed to the development of a complex brain. The tail may be defined as an extension of the body wall posterior to the anus and the body cavity. The body wall itself is built from segment of repeating units. Each segment follows the same general plan, consisting of bone, muscle, blood vessels, and nerves. Developmental differentiation in individual segments then leads to structural and functional variation and specialization.

SUBPHYLUM VERTEBRATA

Ancestral vertebrates shifted from filter-feeding to become active predators. To be successful, they also became larger and maintained their mobility throughout their life cycles. The elaboration of the internal skeleton supported larger muscles and active locomotion. True cartilage and bone, which supplement or replace the notochord, are unique vertebrate developments. Specifically, vertebrates possess a cranium and a vertebral column. Enhanced sensory organs connect to a distinct brain within the cranium. Two pairs of appendages assist in locomotion.

Larger size also brought about changes in most of the physiological systems of the body to increase the efficiency of energy processing (Table 2.4). The circulatory system is closed, in that the vessels form a continuous loop and blood cells never need to leave the system. The heart possesses multiple chambers for more efficient pumping of blood. The pharynx becomes further elaborated for the functions of respiration. The digestive tract acquires accessory glands—the liver, pancreas, and gallbladder. All of these gross anatomical changes support adaptation for energy utilization at the cellular and molecular levels.

The habitat of the earliest vertebrates and whether they were marine or fresh water animals have been the subjects of a long-running debate. The important difference between the two habitats is the relative concentration of salt and other electrolytes in the water. The most primitive living vertebrates, hagfish, are marine and make little effort to regulate their own water balance and osmotic pressure. On the other hand, most other aquatic vertebrate groups have some means of shedding excess salt and regulating their water balance. Depending on which type of living fish is believed to model the ancestral condition, vertebrates may have appeared either in the sea or in fresh water habitats. Alternatively, they may have been anadromous—able to function in a wide range of salinity. An early habitat in river mouths and coastal waters, for example, might have offered a reasonable nutrient supply, reduced danger from predators, and sufficient variability in temperature, oxygenation, current, and salinity to stimulate natural selection and the pace of evolutionary change.

Of the seven classes of living vertebrates (Figure 2.3), three are fish. The most diverse of all the classes is Osteichthyes, the bony fish. This group populates both marine and fresh water habitats, encountering a variety of ecological and physiological challenges and evolving adaptations to overcome them. Among the land-dwelling vertebrate classes, or tetrapods, the paired fins were transformed into two pairs of limbs. It is believed that the limbs became stronger originally to facilitate maneuvering on the bottom in shallow water. As the earliest amphibians began to spend more time partially or wholly out of the water, the skeleton had to be strengthened to support body weight. Limb muscles increased in number for more complex movement.

Table 2.4 Vertebrate Characteristics

Motility throughout the life cycle
Large body size typical, permitted by elaboration of the
   musculoskeletal system
Notochord reinforced by the formation of vertebrae (segmented
   skeletal arches)
Cranium develops in the head to protect the brain
Internal skeleton of true bone (calcium phosphate)
Two pairs of appendages, containing skeletal elements
Pharynx assumes an important respiratory function
Branchial arches supported by cartilage, muscle
Neural crest cells (embryonic) give rise to nervous system
   melanocytes, and other tissues
Brain developed by elaboration of anterior segments of nerve cord
Approximately 12 paired cranial nerves
Special senses concentrated at the cephalic end (image-forming
   eyes, olfactory organ, lateral line system, and inner ear)
Complex endocrine system
Accessory digestive organs present—liver, pancreas, and
   gallbladder
Chambered heart
Closed circulatory system
Hemoglobin contained in red blood cells
Hepatic portal system present draining blood from the gut into the
   liver
Nonsegmented kidney derived from mesoderm
Kidney constructed from multiple glomeruli and tubules
Nonsegmented gonads
Two layers to integument

TETRAPODS

The tetrapods comprise an informal category for "four-footed" vertebrates, distinguishing them from the fish. Tetrapods include amphibians, reptiles, dinosaurs, birds, and mammals. The transition to life on land required many adaptations in body structure and physiological function (Table 2.5). It is between the fish and the tetrapods that the most dramatic changes in vertebrate body shape occur. The skin became thicker to resist the loss of water, and the kidneys also

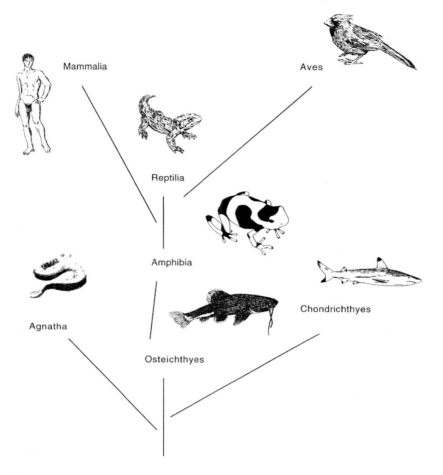

Figure 2.3 **Figure 2.3 Vertebrate diversity.**

became more efficient in water conservation. Lungs replaced gills for breathing. The lateral line system of fish, which detects vibrations in the water, became irrelevant and was lost. On the other hand, the ability

to hear air-borne sounds became useful, and sensitive ears proved adaptive. Lids and tear glands protect the eyes. The neck became longer and more flexible so that turning the head for feeding did not require reorienting the entire body.

The development of limbs occurred in the Devonian Period, and the fossil record contains diverse species with many variations on fins and limbs. There was not a single lineage that made an evolutionary leap, but a number of related species experimented with limb-like fins (Ahlberg and Milner 1994; Clack 2002a; Zimmer 1998). Terrestriality probably did not occur until the Carboniferous Period (Clack 2002b). Apparently limbs evolved for bottom-walking in shallow water.

The amphibians (frogs, salamanders, and their relatives) are intermediate in the adaptation to

Table 2.5 Tetrapod Characteristics

Redesigned limbs for locomotion on land
Differentiation of limb muscles
Stronger skeleton to support body weight
More flexible and elongated neck
Dependence on lungs with loss of gill slits
Skin resists water loss
Eyes protected against drying by eyelids and tear glands
Sense of hearing enhanced
Cerebral hemispheres expand on the brain
Internal fertilization
Additional protective membranes around egg

terrestrialism. Their structure resembles the other tetrapods. Their physiology and especially reproduction are more similar to those of fish in many ways. Eggs are laid in water and fertilized there by free-swimming sperm. Early stages of development are adapted to this temporary aquatic habitat. The other tetrapod classes have achieved a more complete independence from water with the development of internal fertilization (inside the female) and an amniotic membrane and shell to protect the egg and permit it to survive and develop out of the water. The distinction between amniotes and anamniotes may be considered a more significant adaptive shift in physiology than that between tetrapods and fish.

The amniotes split very early into two major lineages. The earliest reptiles gave rise to modern reptiles, the extinct dinosaurs, and birds. The other line of early amniotes gave rise to the mammals. The evolutionary transitions between amniote classes have been closely studied and are well documented in the fossil record.

CLASS MAMMALIA

The mammals demonstrate numerous adaptations to distinguish them from other vertebrates (Table 2.6). These certainly accumulated gradually over tens of millions of years. The term "mammal" refers to the mammary glands that nurse the young. Mammalian parents give much more attention and care to their infants than did earlier vertebrates. Of greater significance to paleontologists is the redesign of the jaws and teeth. We formally define mammals as those vertebrates possessing a mandibular joint between the dentary and squamosal bones. The dentary is the bone we now refer to as the "mandible," whereas the squamosal is now a region of the temporal bone. This seemingly trivial point of anatomy is the consequence of adaptations for more efficient feeding and strengthening of the jaw. It also relates to more complex dentition and to the unique arrangement of three ossicles in the middle ear.

The fundamental mammalian strategy is homeothermy (constant body temperature) achieved through endothermy (internal generation of heat by specialized tissues). Homeothermy permits the metabolic processes of the body to adapt to a single temperature range and thereby increase efficiency. Unlike those of many

Table 2.6 Mammalian Characteristics

Homeothermy and endothermy
Mandibular joint formed by dentary and squamosal bones
Secondary palate for simultaneous chewing and breathing
Complex occlusion of teeth
Heterodonty—teeth of different forms and functions
Diphyodonty—two sets of teeth (deciduous and permanent)
Mandible composed of a single bone, the dentary
Limbs placed vertically under the body
Adult bones form by the fusion of multiple parts
Regional differentiation of the spine into cervical, thoracic, lumbar, sacral, and caudal vertebrae
Head movement determined by paired occipital condyles and atlas rotating on axis
Pectoral limb girdle freed from axial skeleton
Sacral vertebrae may fuse to strengthen pelvic girdle
Coxal bone formed from fusion of three pelvic bones
Ribs reduced or eliminated in all regions except thorax
Hair
Vibrissae
Sweat glands
Subcutaneous fat
Diaphragm
Lungs divided into alveoli
Turbinals in nasal passages
Four-chambered heart
Enlarged cerebrum
Three ear ossicles in middle ear
Mammary glands and lactation

reptiles, mammalian muscles do not need extensive warm up time every morning to make them effective. Energy can be more readily released to the brain and other body tissues on demand. Nearly every body system has adapted to produce or to take advantage of homeothermy.

The increased need for food and the availability of energy evolved with a much greater level of activity and a more efficient skeletal design for movement at higher speeds. The limbs are positioned underneath the trunk for more direct support rather than oriented outward away from the trunk. Similarly, mammalian joints are reoriented to operate in a sagittal (fore-aft) plane. The spinal column increases its mobility by flexing in a sagittal plane during locomotion instead of laterally, as in lower vertebrates.

The generation of heat and energy by mitochondria within body cells implies a much greater flow of energy through the body as a whole and requires a more efficient level of processing food. Homeothermy was a gradual acquisition. Many gradations of thermoregulatory ability may be observed today among living vertebrates. Mammals, more than any other vertebrates, chew food to begin the digestive process in the mouth. They have developed a secondary palate between the nasal and oral cavities that permits them to eat and breath simultaneously. The teeth are complex in form in order to slice or grind food more effectively. Such chewing requires precise occlusion between teeth. To maintain this occlusion, mammals replace their teeth only once in a lifetime.

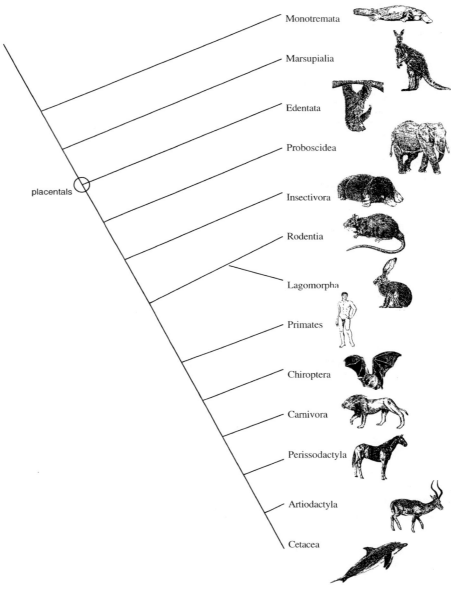

Figure 2.4 Mammal diversity.

The body systems that promote homeostasis (maintaining a balanced physiology) have had to change as well. Mammals are insulated against the loss of body heat through hair and subcutaneous fat. Sweat glands permit the elimination of excess heat. The appearance of a true diaphragm and more complex lungs enhances the ability to take in oxygen. The heart has four chambers with a complete separation of blood flow to the lungs from that to the rest of the body. The brain is enlarged and supports more complex behavior. The larger brain more effectively regulates metabolism and satisfies the increased demand for resources.

We divide living mammals into three subclasses —Subclass Prototheria, the egg-laying monotremes; Subclass Metatheria, the marsupials or pouched mammals; and Subclass Eutheria, placental mammals. Metatherians and eutherians bear live young (vivipary). Placeal mammals, including ourselves, are the most diverse and widely distributed group (Figure 2.4). Their defining trait is the development of a synergy between the uterine lining and a placenta to support the fetus with greater quantities of nutrients as it develops. Of the many living orders of eutherians, we belong to one of the more conservative, the primates.

The lineage from early tetrapods to mammals is reasonably well documented in the fossil record. Mammals are descended from the synapsids that date back to the late Paleozoic Era. The best known of the synapsids is *Dimetrodon*, the sail-backed "lizard." The sail is constructed from tall, elongated dorsal spines of the vertebra covered by a highly vascularized skin. This arrangement may have represented a primitive thermoregulatory strategy—turning the large surface area of the sail into or out of the sun and wind could raise or lower body temperature. If this interpretation is correct, *Dimetrodon* reveals that a concern for thermoregulation lay at the base of the mammalian lineage.

The Permian descendants of the synapsids were the therapsids, or "mammal-like reptiles." Therapsid carnivores and herbivores included the largest and most conspicuous animals on the landscape in their time. Some of them possessed features that we associate with the mammals, including elongated canine teeth, sexual dimorphism, and the beginnings of upright limb posture. Therapsid dominance was lost at the end-Permian extinction, when most animal and plant species died. As life recovered in the Triassic Period, dinosaurs flourished and the descendants of the therapsids, mammals, appeared.

Early mammals of the Mesozoic were small and probably initially were active at night, consuming invertebrates. Although mammals were scattered across the continents, they are scarce in fossil record until the dinosaurs became extinct at the end of the Cretaceous. In the periods that followed, mammals diversified into the modern orders.

ORDER PRIMATES

The primates include lemurs, monkeys, apes, and humans (Figure 2.5). According to the fossil record, primates, along with many other mammalian orders, arose shortly after the end of the Cretaceous Period, 65 Mya. Probably from the beginning, primates occupied a tropical arboreal habitat, and many of their defining characters relate to life as arboreal predators (Table 2.7). Several of the traits that we think of as distinguishing humans are also present to varying degrees throughout the order. These include enlarged brain size, complex behavior, erect sitting posture, and the ability to stand and walk bipedally.

Climbing probably was a common activity of the earliest mammals, but primates developed new specializations for it. Primate hands and feet can grasp

Table 2.7 Primate Characteristics

Housing of the middle ear formed from the petrosal bone
Emphasis on vision, including enlarged the eye and related brain centers
Three-color vision
Frontally directed orbits reinforced by a postorbital bar
Stereoscopic vision
Decreased emphasis on smell
Grasping hands and feet with opposable first digits
Nails rather than claws on at least some digits
Erect sitting posture
Facultative bipedalism
Increased brain size
Prolonged life history stages—gestation, infancy, immaturity, life span
Complex social behavior and communication
Arboreal habitat
Tropical distribution

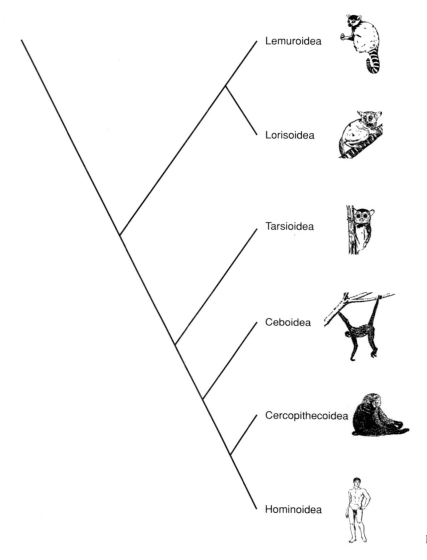

Lemuroidea

Lorisoidea

Tarsioidea

Ceboidea

Cercopithecoidea

Hominoidea

Figure 2.5 Primate diversity.

objects with opposable digits, allowing holding and fine manipulation of objects during feeding and other activities. At least some of the digits bear nails instead of claws to support the fingertips. Primates have emphasized the sense of vision over that of smell. The eyes are larger and directed more anteriorly, whereas the snout is reduced. Primates have independently evolved both three-color vision and depth perception.

Primates first appear in the Paleocene epoch, looking very much like other primitive placental mammals. In the Eocene, two modern-looking families emerged. One family (Adapidae) contained monkey-like arboreal fruit-eaters that probably are related to modern lemurs and other prosimians. The second family (Omomyidae) included insectivores whose descendants include the modern tarsier. From the omomyids or possibly from a third, unknown group, the anthropoids arose by the late Eocene. Anthropoids, the familiar monkeys and apes, are distinct because of the complete closure of the back of the orbit and a specific dental pattern. The earliest certain anthropoids lived in Egypt about 35 Mya. However, a growing collection of anthropoid-like fossils from Asia at about the same time and monkeys from South America a few million years later suggest an earlier time for the first members of the group. The fossil record in Africa and Asia for the early Cenozoic is unfortunately very poor and obscures the origin of the anthropoid lineage.[1]

In the Miocene epoch, Old World monkeys and hominoids diversified and scattered across Africa, Europe, and Asia. Initially the hominoids were more varied and occupied a wide variety of niches (Begun 2003). Old World monkeys probably specialized initially for a partially ground-dwelling existence. They tend to be fast runners and leapers. Hominoids, including modern apes, remained arboreal specialists, climbing more slowly to make use of resources in the smaller branches of trees and developing hanging postures to better distribute their weight (Temerin and Cant 1983). Hominoids have eliminated tails and reoriented their upper limbs for this purpose. Suspensory feeding permitted some of the hominoids to increase body size.

The surviving species of hominoids are mostly larger animals, although fossil forms included smaller species as well. Later in the Miocene and after, the relative diversity of hominoids and monkeys reversed. Hominoids appear to be more sensitive to habitat loss through climate changes and (most recently) human interference. Relatively slow rates of reproduction make it difficult for their populations to recover (Jablonski et al. 2000). Monkeys have proved to be more adaptable and are numerous and widespread across the Old World, whereas apes are restricted to a handful of endangered tropical forest species. Humans are the significant exception.

FAMILY HOMINIDAE

Family Hominidae has been redefined since the 1970s to include our closest relatives, chimpanzees and gorillas, along with ourselves. This grouping is based firmly on similarities at the molecular and genetic levels. While Asian hominoids, gibbons and orangutans, diverged from our lineage in the early or middle Miocene, the three African genera—chimpanzees, gorillas, and humans—have separated only in the past 6 to 8 million years (Figure 2.6). The fossil record in Africa for that time period is still fragmentary and does not document the divergence, nor does it tell us much about the evolution of the chimpanzee and gorilla.

It is usual to assume that the last common ancestor of Hominidae is most similar, although not identical, to chimpanzees. Chimpanzees are so like humans that they have become the baseline by which we identify and measure changes that have occurred in human anatomy, behavior, and ecological niche. Measures of genetic similarity produce figures of up to 99.4% identical DNA between our species, leading some to argue that this relationship is so close that chimpanzees should be included in genus *Homo* (Wildman et al. 2003). What, then, distinguishes human beings from all other living organisms? The answer is no single feature but many different trends (Table 2.8). This book emphasizes four different complexes of traits that have contributed to unique adaptive strategies:

Table 2.8 Human Characteristics

Relatively and absolutely large brain size
Highly developed intellect, awareness, consciousness, abstract
    thought, personality
Diverse facial expressions
Habitual use of language
Extended life history
Habitual bipedalism and its anatomical correlates
Longitudinal arch of the foot
Reduction of hair, increased sweating, more subcutaneous fat
Shortening of the face, reduction of tooth size
Thickened enamel
Reduced canine size
Food sharing
Food production
Extensive material culture
Use of clothing
Use of fire
Complex social behavior and organization
Gender division of labor
Extensive division of labor within society
Long-term pair-bonding
Pursuit of sexual behavior out of the context of reproduction
Reshaping of the habitat

---

[1]For discussion and debate on the earliest anthropoids, see Ducrocq 1998; Gebo et al. 2000; Gunnell and Miller 2001; Jaeger et al. 1998; Kay et al. 1997; Ross 2000; and Simons 1994.

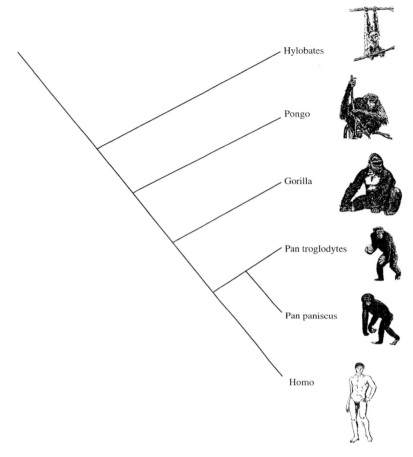

Figure 2.6 Hominoid diversity.

bipedalism, an enlarged brain, the ability to sustain high metabolic activities, and a unique life history strategy that depends on complex social relationships. The most conspicuous anatomical features relate to our upright posture and bipedal gait. These have implications for structural changes throughout the body. Nonetheless, the most remarkable feature of our species is our relatively large brain and the complex behavior that it makes possible.

THE EMERGENCE OF HUMANS

The fossil record of human evolution has been intensely scrutinized and endlessly debated (Table 2.9). The public is treated to the uncertainties and disagreements in the field more so than for any other academic discipline except perhaps medicine. This fact reflects

the high level of interest in our past and present nature. The disputes also reflect the fact that the questions we are able to ask address a level of detail and demand for certainty about taxonomy, diet, behavior, and migrations that are never applied to other lineages of organisms.[2]

Individual biologists and especially paleoanthropologists concerned with classifying species often find themselves characterized as "lumpers" or "splitters." A lumper recognizes fewer species and other taxonomic categories. This approach emphasizes similarity and

---

[2]Selected overviews of hominin fossils and classification: Cela-Conde and Ayala 2003; Curnoe and Thorne 2003; Johanson and Edgar 1996; Wood and Collard 1999; Wood and Richmond 2000.

Table 2.9 Overview of Hominin Species

Species	Appearance	Phylogenetic Position
Sahelanthropus tchadensis	6 Mya; Chad	Close to the origin of the hominin line. Known from one site; limited material makes placement uncertain.
Orrorin tugenensis	6 Mya; Kenya	Postcrania show bipedalism; possibly ancestral to later genus *Homo*, but distinct from *Australopithecus*. Some authors would recognize a separate genus for this lineage, *Praeanthropus*. Limited craniodental material makes direct comparisons difficult.
Ardipithecus ramidus	5.5–4.4 Mya; Ethiopia	Primitive dentition, some evidence for bipedalism; close to ancestor of *Australopithecus*.
Australopithecus anamensis	4.1 Mya; Kenya	Oldest member of the genus; probably ancestral to *A. afarensis*.
Australopithecus afarensis	3.6–3.0 Mya; Tanzania, Ethiopia	One of the two well known species before *Homo*. Ancestral to later hominins of East Africa.
"Kenyanthropus platyops"	3.5 Mya; Kenya	Flat face seems distinct, but distortions of fossil makes it difficult to distinguish from *A. afarensis* with certainty.
Australopithecus bahrelghazali	3.2 Mya; Chad	Only specimen of western lineage of australopithecines. Many anthropologists find it indistinguishable from *A. afarensis*.
Australopithecus africanus	3.0–2.0 Mya; South Africa	South African contemporary of *A. afarensis* known from hundreds of specimens and a few partial skeletons.
Australopithecus (Paranthropus) aethiopicus	2.5 Mya; Kenya, Ethiopia	Early robust australopithecine, showing similarities to both *A. afarensis* and *A. boisei*. Many place the robust species in a separate genus *Paranthropus*.
Australopithecus (Paranthropus) boisei	2.5–1.0 Mya; Tanzania, Kenya, Ethiopia	Robust lineage of East Africa, contemporary with early *Homo*. Placed by some in separate genus *Paranthropus*.
Australopithecus (Paranthropus) robustus	2.0–1.4 Mya; South Africa	Robust lineage of South Africa, contemporary with early *Homo*. Placed by some in separate genus *Paranthropus* but may have descended independently from *A. africanus*.
Australopithecus garhi	2.5 Mya; Ethiopia	As a late surviving nonrobust australopithecine, it is suggested to be close to origin of *Homo*. Limited material makes placement uncertain.
Homo habilis	2.0–1.5 Mya; Tanzania, Kenya, Ethiopia; possibly South Africa	Earliest member of *Homo*. May fit better as *Australopithecus*.
Homo rudolfensis	2.5–1.8 Mya; Kenya, Tanzania	Side branch of early *Homo*. These specimens overlap in form with those of *H. habilis*.
Homo ergaster	1.8–1.4 Mya; Kenya, Ethiopia; possibly South Africa and Georgia	Probably the ancestor of *H. heidelbergensis* and modern humans. This species is considered by some to be an African population of *H. erectus*.
Homo antecessor	0.6 Mya; Spain	Last common ancestor of *H. heidelbergensis*/Neanderthals and modern humans. Juvenile status of type specimen and limited material makes distinctiveness uncertain.
Homo heidelbergensis	0.4–0.2 Mya; Europe and Africa	Ancestral to Neanderthals and modern humans. This species has been considered a European population of *H. erectus*.
Homo erectus	1.6–<0.1 Mya; China, Java, and probably Africa	Asian hominins of Middle Pleistocene; no descendants. A broader interpretation of the species considers it to be the only Middle Pleistocene hominin and intermediate between early *Homo* and *H. sapiens*.
Homo neanderthalensis	250,000–30,000 BP; Europe, Middle East	Usually described as a distinct species contemporary with modern humans. An alternate interpretation views Neanderthals to be a regional subspecies or race of *H. sapiens*.
Homo sapiens	200,000 BP-present; Africa to worldwide	Modern humans, including a "archaic" transitional population in Africa.

Mya = million years ago; BP = years before the present.

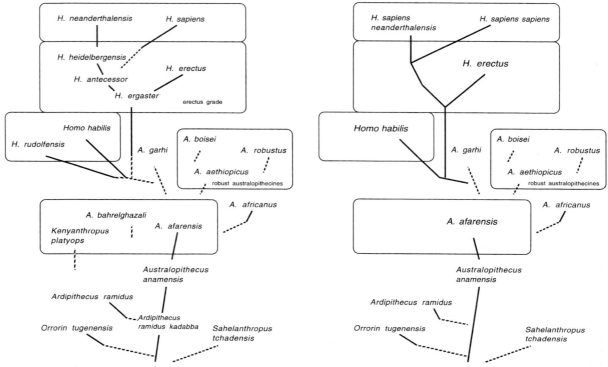

Figure 2.7 Different approaches to hominin classification. Differences in classification may result from different understandings regarding interspecific and intraspecific variation. (*Left*) A "splitter" tends to recognize more species with finer distinctions. (*Right*) A "lumper" tolerates more variation within a species and recognizes fewer species. These two phylogenies represent these different interpretations of the same hominin fossils.

continuity across variable populations. A lumper tends to be concerned with broader themes of evolution, such as ecological niche and adaptation. A splitter is interested in detailed points of anatomy that can distinguish populations and identify ancestor–descendent relationships. This approach tends to recognize more species and often more taxonomic ranks in the hierarchy. These extremes and all the positions in-between are reflections of individual philosophy and interest. If taxonomy is at least partly an arbitrary tool, then the different schemes of classification reflect differences in how the biologists want to use this tool.

The human lineage, as distinct from that of the apes, comprises its own subfamily, Homininae, and its members are referred to by the common name hominins. Current models recognize up to seven genera within it. Two extreme interpretations of species are represented in the phylogenies of hominins shown in

Figure 2.7. The version on the left recognizes many different extinct species of hominins. On the right, fossils are more simply interpreted as samples of variable populations evolving through time. The differences in interpretation are more significant in the later hominins because larger collections of specimens allow us to observe variations that overlap between populations and because the samples are not as clearly separated in time.[3] Some of the better known hominins are shown in Figure 2.8.

[3]Although this classification is now widely accepted, much previous and current literature uses the term "hominids" to describe humans and our fossil relatives. This accords with an older tradition of classification that excluded living apes from Hominidae. By the current classification, apes would also be considered hominids, but not hominins. To reduce confusion, this book avoids using the common term "hominid."

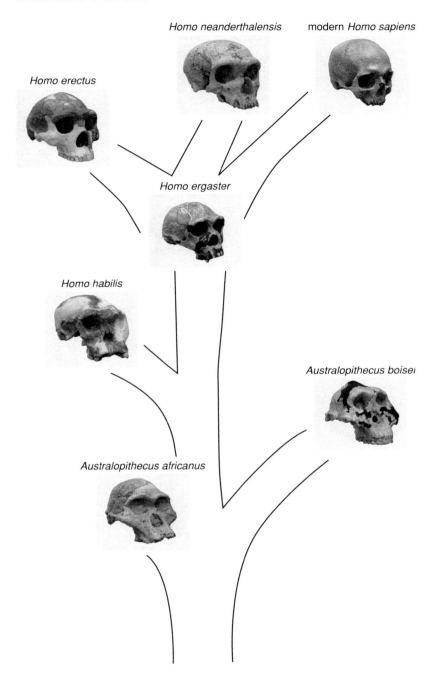

Homo neanderthalensis   modern Homo sapiens

Homo erectus

Homo ergaster

Homo habilis

Australopithecus boisei

Australopithecus africanus

Figure 2.8 Portraits of selected fossil hominins (casts).

Taxonomic squabbles aside, anthropologists are developing a better understanding of the chronology of hominin evolution (Table 2.10). Twelve species of australopithecines and other early hominins over-lapping with one another in time have been named among fossils in eastern and southern Africa. They can be divided roughly into two groups. The early species—*Sahelanthropus tchadensis* from Chad about

Table 2.10 Time Frame for Human Evolution

8–6 Mya (Late Miocene)	Human, chimpanzee, and gorilla lines become distinct according to molecular evidence.
	First hominin fossils about 6.0 Mya.
4.0 Mya (Pliocene)–1.0 Mya	Australopithecine fossils well known from eastern Africa, from Ethiopia to South Africa.
	Brain is still ape sized (about 500 cm^2).
	Clear evidence of bipedalism.
2.4 Mya	First appearance of genus *Homo*, in East Africa.
	Start of brain expansion (about 800 cm^2).
	Stone tools and evidence of meat consumption appear in the archaeological record (Lower Paleolithic culture).
1.8 Mya (Pleistocene Epoch)	*Homo* displays modern skeletal proportions, in Kenya.
	Continuing brain expansion (about 900–1000 cm^2).
1.8–0.3 Mya	Migrations of *Homo* to occupy Africa and Eurasia
	Increasing cultural complexity: fire, shelter, possibly clothing.
250,000–100,00 BP	Appearance of archaic *Homo sapiens* in Africa and Neanderthals in Europe.
	Cranial capacity approaching our own (about 1400 cm^2).
	Tools are more diverse and sophisticated (Middle Paleolithic culture).
160,000–100,000 BP	Appearance of anatomically modern humans in Africa.
30,000 BP	Anatomically modern humans present throughout Africa, Eurasia, and Australia.
	Cultural complexity and diversification accelerate.
25,000–12,000 BP	Humans enter North and South America.
10,000 BP	Ice Ages end; extinction of large animals on most continents.
	Lifestyles shift from big game hunting to specialized and intensive hunting and gathering.
8000 BP	Appearance of food production; domestication of plants and animals.
	Society becomes more sedentary and more complex (Neolithic).
5000 BP	Civilization appears in the Near East, Egypt, and China.

Mya = million years ago; BP = years before the present.

6 or 7 Mya, *Orrorin tugenensis* from Kenya about 6.0 Mya, and *Ardipithecus ramidus* from Ethiopia about 4.4 Mya—were all discovered and named quite recently (1994–2002). They have extended our record of hominins or of the human–ape split farther back in time. Each presents an interesting and inconsistent combination of primitive (apelike) and advanced (resembling later hominins) arrays of traits and may be understood as a side branches or possibly transitional forms on our own lineage. Unfortunately none of these species is well known.

The australopithecines (including *Australopithecus*, *Paranthropus*, and "Kenyanthropus") are found about 4.0 to 1.0 Mya. Although they have distinctive anatomical features, the australopithecines are understood to include our direct ancestors. Their postcranial anatomy shows some variation from East to South Africa, with the southern species displaying more ape-like proportions of the limbs and bodies. All these species were clearly bipedal but possibly spent a significant amount of time in the trees. The skull

and teeth of *Australopithecus* were neither ape-like nor human but displayed some unique traits, including expanded molars. Absolute brain size was comparable to that of the great apes (about 500 cm^2), but slightly expanded relative to body size.

Numerous changes occurred between 2 and 3 Mya in Africa. In both the east and south, the climate shifted. After a long preceding period of climatic and habitat fluctuation, the environment became more consistently drier and open. Although there were many patches of woodland, there was also a more consistent presence of grasses and of animals that prefer open country. During this time, the australopithecine species show a hyperdevelopment of the teeth, especially the molars, and of the muscles and facial structures supporting them. These later species are commonly referred to as the "robust" australopithecines to differentiate them from their earlier "gracile" relatives. It is likely that the robust australopithecines were adapting to a diet of tougher plant foods in the more arid habitat.

Also about 2.5 Mya, a population of *Australopithecus* gave rise to the genus *Homo*. Interestingly this same time period witnessed the first stone tools and the first clear evidence that hominins used those tools to butcher animal carcasses. Our own genus is characterized by increasing brain size over the past 2 million years, while tooth size has decreased through time. The identification of different species of human in the fossil record is a major point of contention among paleoanthropologists. Several species of early *Homo* appear to have coexisted briefly in Africa about 2 Mya, but it is difficult to assign all of the known fossils with confidence to these very similar populations, and some authors would not include all of them within *Homo*.

Homo habilis is generally recognized as the most primitive and australopithecine-like of these species. It has variously been defined broadly to encompass nearly all early fossil *Homo* specimens before *H. erectus*, or more narrowly broken into multiple species, including *H. rudolfensis*. Some authors would place *H. habilis* among the australopithecines (Wood and Collard 1999). The place of *H. habilis* as the ancestor of later *Homo* is thrown into doubt if multiple contemporary species are recognized in East Africa and Asia. These include a better candidate for our own ancestor, *Homo ergaster*.

Formerly it was argued that a single species, *Homo erectus*, occupied the Old World between the early African hominins and modern *Homo sapiens*. The more commonly used model today divides this population into *H. ergaster* in Africa, *H. erectus* in eastern Asia, and *H. heidelbergensis* in Europe and later Africa, with additional species possible. Such divisions are confused by the discovery of specimens at Dmanisi, Republic of Georgia, with traits intermediate between those of *H. ergaster* and *H. erectus* and by East African specimens from Olduvai Gorge, Awash Valley, Ileret, and Olorgesailie in Africa that more closely resemble *H. erectus*.

By whatever name, individuals of this "*Homo erectus* grade" had made some important changes in the postcranial skeleton, including lengthening of the lower limbs to modern proportions and modification of the hip and foot in particular to increase the efficiency of walking and running. Shortly after 2.0 Mya, they left Africa and began to appear in Asia. By 1 Mya, *Homo* was widely scattered across the two continents. Humans developed a highly adaptable culture, enabling them to occupy in a wide range of climates and habitats. Populations in different continents diverged somewhat and acquired their own morphological patterns.

Later in Europe, a distinctive population called the Neanderthals (*Homo neanderthalensis*) evolved from *H. heidelbergensis* by about 200,000 years ago. These are the best known of the fossil hominins and have left behind a rich record of their material culture. It is difficult for this author to view the evidence for their complex tools, subsistence, and possibly ritual life and not to consider them a part of our own species. However, in a relatively short time, about 35,000 to 30,000 years ago, Neanderthals and their culture were replaced by modern humans. Although there are some skeletons of this period that suggest interbreeding between the two populations, no genetic trace of the Neanderthals has been identified in modern peoples.

Our own human species, *Homo sapiens*, first appears among African fossils about 160,000 years ago. Genetic studies tell us that these people migrated out of Africa, perhaps in the past 100,000 years, and replaced all other forms of humanity by 25,000 years ago. The questions of how, whether, and how much those modern migrants interacted with and interbred with the resident populations in Europe and Asia are some of the most contentious issues in current anthropology.

It is hard for us to study the anatomy of the fossil species of *Australopithecus* and early *Homo* without viewing them merely as unfinished humans. Because this text is an attempt to understand modern human anatomy and not a thorough study of the fossils, it risks perpetuating that erroneous perspective. Evolutionary theory teaches us, however, that each of these species must have been adapted to its own environment in a unique and, at least temporarily, successful way. Similarities between *Australopithecus* and *Homo* must represent solutions that worked well for both genera and were not mere anticipations of modern human form. The anatomical differences between them represent alternative and sometimes unique strategies that must be appreciated in their own right.

3

BONES, JOINTS, AND MUSCLES

The skeletal system contains specialized connective tissues—cartilage, bone, and ligament—whose function is to give mechanical strength to the body. The skeletal tissues also support the mass of the body against gravity, anchor the muscles, protect the soft tissues, facilitate movement, and receive forces from the outside world. Muscle tissue is unique in its ability to shorten its length with force. The different tissues have their own properties and their own functional roles. As these tissues are assembled to build the bones, joints, and muscular complexes, the recurring theme is compromise. There is no perfect structure. Increasing mobility weakens structural integrity. Reinforcing structure adds mass and generates greater forces but encumbers movement. The body is remarkable in its ability to meet so many conflicting demands, but we must understand that it has material and evolutionary limitations.

STRESSES AND STRAINS

Skeletal design seeks to provide stability in the face of externally or internally generated forces. A *stress* is a measure of the intensity of a force applied to a structure, comparable to pressure. If we wish to quantify the term, stress may be expressed in force per unit area or Newtons per square meter (N/m^2).[1] A *strain* is the deformation experienced by a material in response to a stress, expressed as a unit of distance or an angle of twist. When stress on a bone causes it to bend, for example, the extent of that bending is the strain. The ability to resist strain is called *stiffness*. Excessive strain may break or tear a tissue; thus, the ideal design

of the body should minimize the amount of strain or deformation it is likely to experience.

There are several different patterns of stress (Figure 3.1). A *compressive stress* is one applied at two points on opposite sides of the material. *Tension,* or a *tensile stress*, tends to stretch a material. Other stress patterns create a complex array of compression and tensile stresses in different places. *Shear stress* occurs when opposing forces are applied to two points that are not aligned. *Bending stress* produces compression on one side of the material and tension on the opposite side. *Torsional stress* occurs when one part of a material is twisted with respect to another.

Bone and cartilage are good at tolerating compressive stress. Bone is stiffer and deforms less under compression. Cartilage absorbs energy better by deforming. Neither accept tensile stress well.

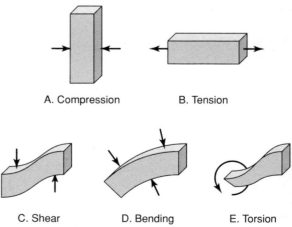

A. Compression   B. Tension

C. Shear   D. Bending   E. Torsion

Figure 3.1 Types of stresses. The different patterns of forces that may create stress on an object are best visualized by the distortions, or strains, they create. Various body tissues will respond to different stresses in different ways.

---

[1] A Newton is the amount of force needed to accelerate 1 kilogram of mass 1 meter per second per second.

Consequently, the body uses these tissues where compressive stresses are common. Collagen fibers, present especially in ligament and tendon but also in many other tissues, is much better under tension. Because collagen is flexible, it has little stiffness against other types of stress. Shear, bending, and torsional stresses are destructive for all body tissues. Much of the design of individual bones and joint systems reflects an attempt to minimize these forces.

Elasticity is the ability of a material to tolerate strain without permanent change of shape. A completely elastic material returns to its original shape when the stresses are removed. Most tissues have a certain degree of elasticity, following the principle that it is safer to yield to common forces than to try to resist them. *Plasticity* is the ability to accept strain as a permanent change of shape. Commercial plastic is so named because it can be molded into any shape desired. Plasticity is rarely a desirable physical characteristic of body tissues.

Developmental plasticity refers to the responsiveness of a developing or regenerating tissue to various environmental influences. Developmental plasticity is critical for bone, for example, so that it can realign itself to new compressive forces as behavior changes.

CARTILAGE

Cartilage is a connective tissue that contains chondrocytes suspended in a matrix of their own making (Figure 3.2). The definitive component of this matrix is a complex molecule of protein and polysaccharide called chondromucoprotein, which contributes both tensile strength and resilience. The matrix also incorporates a varying number of protein fibers of collagen and elastin. The cartilage is surrounded by a membrane called a perichondrium, from which new chondrocytes are produced.

The essential properties of collagen are stiffness and elasticity. Chondromucoprotein attracts large numbers of water molecules. The retained water is incompressible and therefore confers considerable stiffness. Cartilage can be stiff enough to support active muscle. One class of vertebrates (Chondrichthyes, or "cartilaginous fish," including sharks) builds the entire skeleton of cartilage. Elasticity permits cartilage to

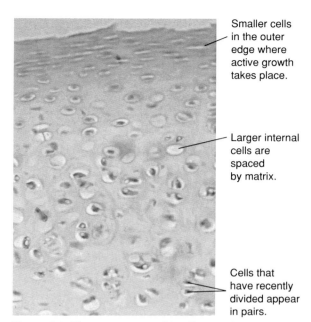

Smaller cells in the outer edge where active growth takes place.

Larger internal cells are spaced by matrix.

Cells that have recently divided appear in pairs.

Figure 3.2 Tissue structure of hyaline cartilage.

absorb energy and return it without permanently changing shape. Thus, cartilage makes a vital contribution to shock absorption in the human body.

One of the more serious limitations of cartilage is its poor nutrient supply. Cartilage does not contain blood vessels, so nutrients must diffuse through the matrix from the periphery to supply the chondrocytes. Consequently the activity level of internal chondrocytes is low and their ability for cell division, growth, and repair is extremely limited.

Hyaline cartilage is the most common form in the human skeleton, having relatively few fibers. It is a fetal tissue, laying down the framework on which bones will develop later. Hyaline cartilage is present on the joint surfaces of mature bones as articular cartilage. Its ability to deform enables articular cartilage to absorb shock and to enhance the fit, or congruence, between bones. Hyaline cartilage also persists in the costal cartilages joining ribs to the sternum, where its natural elasticity permits respiratory movements of the ribcage, and in other special locations, including the larynx, trachea, and bronchi.

Variations in fiber content give cartilage different properties. Large quantities of collagen embedded in the matrix produce fibrocartilage with a great tensile strength (resistance to stretch). This fibrocartilage functions to attach bones, between vertebral bodies and at the pubic symphysis, and as intra-articular discs, or menisci. Elastic cartilage has larger quantities of elastin fiber and can tolerate much greater bending and distortion than other forms. Elastic cartilage is present most notably to give flexible form to the outer ear and the nose.

BONE

Bone is a connective tissue whose matrix contains a crystalline mineral, hydroxyapatite (Figure 3.3). Hydroxyapatite is composed of calcium phosphate in repeating units with the chemical formula $Ca_{10}(PO_4)_6(OH)_2$. The calcium and phosphate required in the human diet are major components of bone. Other materials are present as well. Because phosphates and calcium are vital participants in many physiological processes, it has been argued that bone evolved in vertebrates as a mineral reservoir in the skin before it was used structurally within the body. Although more recent findings suggest that the first skeletal tissues of chordates may have been functional teeth (Sansom et al. 1992), bone continues to serve this reservoir function. Thus the skeleton should be viewed not as inert tissue but as a dynamic participant in maintaining homeostasis in the body. This flow of minerals into and out of bone both satisfies the chemical needs of the body and permits the skeleton to adjust its form in response to changing stresses.

Unlike chondrocytes, osteocytes are well supplied by the bloodstream. Lying within the mineral, these bone cells perform a regulatory function, releasing calcium and phosphates into the bloodstream as needed. Other bone cells are able to perform more drastic deposition or resorption to remodel the bone in response to changing needs. The responsiveness of bone shape to both evolutionary adaptations and ontogenetic factors (having to do with an individual's developmental history) makes the skeleton a uniquely long-lasting record of the individual's sex, age, geographic derivation, and health history.

A bone typically has a dense outer layer of cortical bone and an inner region of spongy or trabecular bone (Figure 3.4). Trabecular bone is a network of struts and braces capable of transmitting and distributing stresses with less mass than cortical bone. Trabecular bone contains marrow, a soft material consisting of fat cells and vascular tissues. A vascular membrane called a periosteum surrounds the bone to provide nourishment and growth.

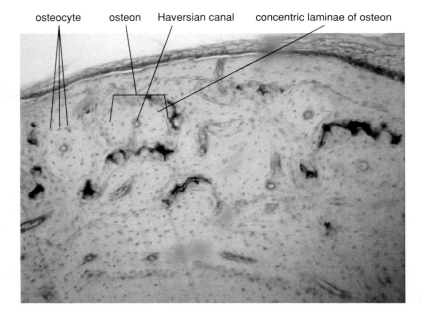

Figure 3.3 Tissue structure of bone.

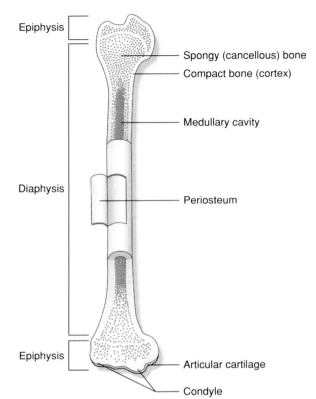

Figure 3.4 **Anatomical features of a long bone.**

DEVELOPMENT AND GROWTH OF BONE

Embryologically, bone develops from either of two different tissues: membrane or hyaline cartilage. The process of ossification that occurs in these areas is similar, and mature bone from these two sources is indistinguishable. Membranous bone forms in dermal layers of the scalp and contributes the flat bones of the skull vault. Endochondral bone forms via the transformation of hyaline cartilage. It accounts for the rest of the skeleton.

Once the skeletal elements are laid down in the fetus as cartilage, growth continues through the active division of chondrocytes on the periphery of these structures and their secretion of new matrix. The process of ossification begins as blood vessels and connective tissues invade from outside the cartilage and begin to displace the degenerating cartilage. An active

type of bone cell called an osteoblast invades and lays down the mineral matrix of true bone. Ossification thus proceeds from both the center of the cartilage and from the periosteum. Eventually these centers of ossification will meet as the cartilage is entirely consumed (Figure 3.5). As individual osteoblasts become encased in the mineral, they assume the less active regulatory role of osteocytes. Later on, bone may continue to grow or decline in thickness by continued activity under the periosteum.

Larger bones may have multiple centers of ossification to accelerate development. These may also be determined by factors beyond size alone. If the bone has evolved by the fusion of several different bones in an ancestral form, those distinct bones may appear briefly as separate centers of ossification in the fetus. If there are several places on the bone that are mechanically critical at an early stage, these may ossify first to provide a firmer support for muscular action.

A long bone of a limb that has a joint at either end commonly begins ossification in the center of the shaft, or diaphysis, and in the epiphysis at either end,

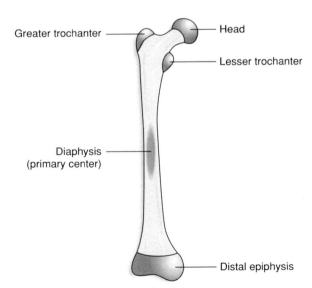

Figure 3.5 **Centers of ossifications.** Many bones develop initially by independent ossification from multiple centers of ossification within the cartilage. As these centers expand, they will eventually fuse.

deep to the articular cartilage. As ossification spreads along the shaft, the cartilage continues to grow between the shaft and epiphyses. This growth of this cartilage is the sole contribution to growth of the length of the bone. Once the centers of ossification have met and fused, displacing the cartilage entirely, growth must stop. After about age 20, human bone growth is possible only under the periosteum, where the diameter, but not the length, is increased. From this time, the skeleton is mature but not inactive.

After growth is completed, bone continues to be an active tissue. Bone is continuously remodeled in response to new patterns of stresses or to increased loads or to repair damage. These processes are under

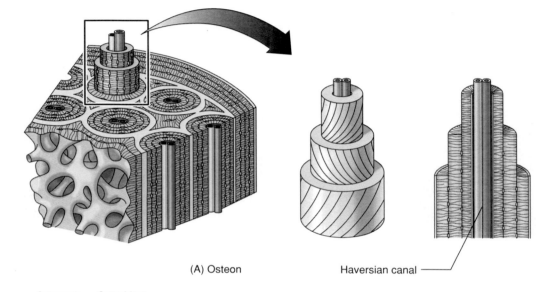

(A) Osteon

Haversian canal

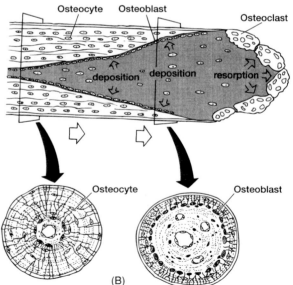

(B)

Figure 3.6 Remodeling bone. Bone resists compressive forces most effectively when the osteons are aligned parallel to those forces. To achieve this, bone must be remodeled by rebuilding osteons as stresses change due to growth or different behaviors. *A*, Remodeled bone contains osteons aligned with habitual stresses. (Also see Figure 3.2.) *B*, To create a new osteon or replace an old one, osteoclasts hollow out a cylinder in the direction of stress by resorbing the mineral within the old bone. Beyond this cutting edge, osteoblasts lay down new bone around the sides of the cylinder, creating concentric rings of hydroxyapatite. The active bone cells are nourished by a blood vessel that remains behind in the central Haversian canal. (From *Vertebrates: Comparative Anatomy, Function, Evolution* by Kenneth V. Kardong. ©1998 by the McGraw-Hill Companies. Reproduced with Permission of McGraw-Hill Education.)

the influence of a number of hormonal, mechanical, metabolic, and nervous factors and may also be influenced by disease and diet.

REMODELING OF BONE

One of the structural units of mammalian bone is an osteon, which appears as a series of concentric rings of mineral surrounding an Haversian canal (Figure 3.6). The osteon is specifically a product of the bone-remodeling process. Under appropriate conditions, bone cells called osteoclasts dissolve the mineral and carve a cylindrical cavity within the bone. Then osteoblasts replace the osteoclasts and proceed to lay down new bone in the cavity. As new matrix is created, the osteoblasts become encased in their own secretions and turn into less active osteocytes. The new layers of bone take the form of concentric cylinders, one inside the other, until only a narrow tube, the Haversian canal, is left for the passage of blood vessels.

Newly created bone is amorphous in that its mineral and fiber components are not organized in any particular way. Osteoclasts attack amorphous bone and transform it into more organized patterns, which

may include osteons. As the collagen fibers and apatite crystals are realigned in the layers of the osteons, lamellar bone is created. This endless activity repairs defects and microfractures of the bone. Moreover, the adjustment of orientation of individual osteons and apatite crystals helps to fine-tune the mechanical properties of bones.

The ability of bone to respond to changing mechanical stimuli by remodeling is a crucial property. Because bone is used to stand up to compressive stress, it is important that the crystals of hydroxyapatite and the collagen fibers in bone align themselves with those stresses. Thus, bone functionally remodels in an infant learning to walk or an adult learning to golf.

It has been noted since the nineteenth century that the trabecular lattice found within certain bones of larger animals shows great resemblance to the architectural designs of engineers (Figure 3.7). Both align structural elements (struts of bone or steel) parallel to the anticipated stresses. This observation fostered the development of simple theories for the design and restructuring of bones according to loads placed upon them. Further examination of a wider range of bones and species shows the picture to be more complicated, if only because the properties of

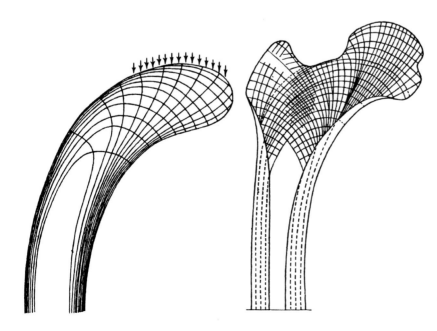

Figure 3.7 Trabecular bone as an engineered structure. An engineer first noticed the close relationship between orientation of force lines generated in a crane (*left*) and that of trabeculae in the proximal end of a femur (*right*). The two structures have similar stresses. The structure of the bone resembles the engineering design of the crane. (From Thompson, D. 1961. *On Growth and Form.* Reprinted with permission from Cambridge University Press.)

bone and the nature of the stresses are more complex than once thought. Nonetheless, it appears that the basic principle of bone design is the alignment of osteons in an attempt to convert all stresses into compressive stresses.

For much of the skeleton, normal strains (deformations of the bone) reach magnitudes up to about 0.3%. This suggests a mechanism for regulation of the remodeling response (Skerry 2000). When strains exceed that level, the bone is too weak for the forces being generated. The excessive strain on the bone is the signal for further deposition and strengthening. When bone is not being strained, its strength exceeds requirements. The lack of a signal for strengthening may permit resorption.

How is this accomplished? At some level, the strains have to be "perceived" by the bone so that an appropriate response may be generated. According to one model, an important mechanism is the creation of a stress-generated electrical potential. The exact mechanism by which this potential, or flow of ions, is created is not clear, but it appears to signal bone cells to deposit new bone under compressive strain (Curry 1985). Alternatively, bone cells under strain may release a chemical signal that stimulates osteoblasts (Skerry 2000). Tensile strain on bone is often associated with resorption, possibly through the absence of a trophic signal. Pressure and tension on the periosteal membrane have the opposite effect. Tension here, such as that caused by the pull of muscle or joint ligaments, stimulates deposition of mineral on the surface of the bone, whereas pressure may cause resorption.

ADAPTATIONS OF MAMMALIAN BONE

The bone of most reptiles and other poikilothermic ("cold-blooded") vertebrates grows slowly, in keeping with their generally lower metabolism and energy budget. The bones grow in dimension through steady additions of laminae to the external surfaces. These layers of bone have an organization that is adequate to support the activities of the animals. Bone growth is continuous throughout life. A crocodile or tortoise that lives for many decades can continue to increase in overall body size.

Endothermic animals with higher metabolic rates grow in briefer, but faster, spurts. Bones reach an adult size and then cease growth. Rapidly produced bone is disorganized in orientation and is not adaptive to any particular pattern of use. Because homeothermic animals may sustain long-term high-level activity, the stresses experienced by the bone are greater. Therefore it is necessary to reshape the primary bone and to provide for a continuing possibility of remodeling. The Haversian system is one strategy for this remodeling and is typical mostly of fast-growing, endothermic vertebrates, including mammals and birds. (Haversian bone is also found in dinosaur fossils. Investigators have debated whether it indicates dinosaur endothermy or a different growth strategy.)

The higher stresses generated at mammalian joints are not compatible with the growing cartilaginous regions found in reptiles. A more precise and less compliant articular end is needed. The mammalian solution is separate ossification centers at the epiphyses. This permits the ends of the bone to achieve correct shape and ossify at a young age without interfering with overall growth of the bone.

LIGAMENT AND TENDON

Anatomical terminology has an unfortunate tendency to use the term "ligament" very loosely. In a literal sense, a ligament is any binding tissue. Named ligaments may consist of fascia, folds of serous membrane, degenerated blood vessels, or nearly any other form of connective tissue. Skeletal ligaments and tendons are classified as dense regular connective tissue. They consist primarily of bundles of collagen fiber arranged in a parallel fashion with collagen-secreting fibrocytes scattered among them (Figure 3.8). The fibers penetrate bone a short distance to anchor the ligaments. Long-term application of tension above a certain threshold stimulates the synthesis of new fibers. Thus, ligaments are also able to respond adaptively to use. Because skeletal ligaments are poorly supplied with blood, injuries heal slowly. Tendons have a similar structure but are used to anchor skeletal muscle.

Collagen is an important structure found throughout the animal kingdom. It consists of a class of proteins

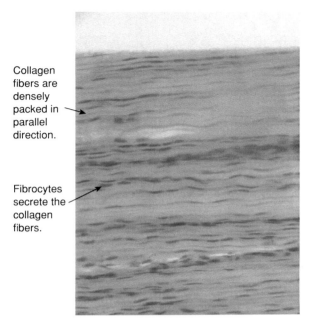

Collagen fibers are densely packed in parallel direction.

Fibrocytes secrete the collagen fibers.

Figure 3.8 Tissue structure of dense connective tissue (tendon).

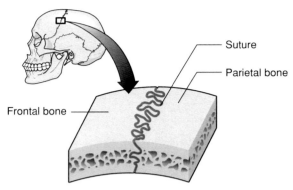

Suture

Parietal bone

Frontal bone

A. Fibrous joint

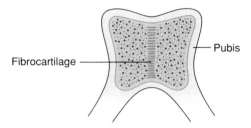

Fibrocartilage

Pubis

B. Cartilaginous joint

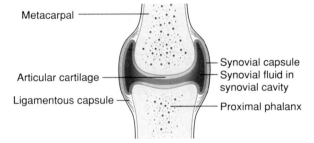

Metacarpal

Articular cartilage

Ligamentous capsule

Synovial capsule

Synovial fluid in synovial cavity

Proximal phalanx

C. Synovial joint

Figure 3.9 The three classes of joints. *A,* Fibrous joints bind bones directly together and generally allow little motion. The sutures of the skull often disappear as the bones fuse together. *B,* Cartilaginous joints place cartilage between articulating bones, such as at the pubic symphysis. *C,* Synovial joints are the most mobile because they connect bones only at the sides of the articulation.

that link together in long helical chains, or fibers. Bundles of collagen have a tremendous strength in tension. Their resistance to stretch is related to cross-linkages (molecular bonds formed between individual fibers). Disruption of cross-linkage permits greater stretching without necessarily destroying the collagen fibers themselves. Such changes occur adaptively in certain circumstances, most notably during the birth process, when maternal joints and soft tissues around the birth canal must stretch without giving up their normal strength.

JOINTS

A joint, or articulation, occurs whenever two bones are in contact or interact. Joints are classified according to their structure (Figure 3.9). Bones may be attached at joints by fibers, cartilage, or a synovial cavity and membrane.

At fibrous joints, the bones are connected tightly together with fibers of collagen. Some of these permit slight movement, as in the joints between the distal radius and ulna, whereas others, such as the sutures of the skull, are immobile and destined to fuse completely.

Cartilaginous joints place fibrocartilage or hyaline cartilage between bones. Fibrocartilage may give strength to mobile joints such as the intervertebral disc; however, the parallel construction of the fibers does not permit rotation or sliding of the joint. Epiphyses and diaphyses of long bones are connected by temporary cartilaginous joints until the bone completely ossifies.

Diarthroses are synovial joints and are characterized by a significant potential for motion. Bones at a synovial joint are separated by a cavity that is enclosed by synovial membrane (synovium) that contains a lubricating synovial fluid. As for all joints, specific constructions represent a compromise between the demands for stability and the need for mobility.

Important considerations as we interpret the design of a joint are congruence, transmission and absorption of forces, friction, stability against dislocation and unwanted movement, and limits to the range and direction of motion (Figure 3.10). *Congruence* refers to the similarity of shapes and the extent of physical contact by opposing surfaces of articulating bones. Increased congruence enhances stability and helps to define the pattern of motion. It also reduces pressure at the joint, since pressure is calculated as the transmitted force per unit area of contact. Congruence is increased by pressure, because the cartilage deforms under load.

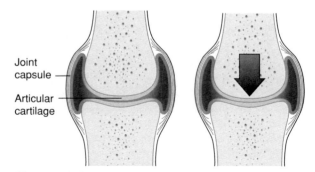

Joint capsule

Articular cartilage

Figure 3.10 Role of articular cartilage. Articular cartilage deforms when the joint is loaded to absorb shock and increase stability of the joint. An unloaded joint has only a small area of contact between opposing articular surfaces. Increasing pressure causes the cartilage to flatten and the area of contact to increase. A greater area of contact, or greater congruence, better distributes forces across the joint.

Forces acting at a joint may damage tissues. Over the course of a lifetime, articular cartilage wears down. For example, it is very common for wear in the knees of elderly people to expose the bone of the femur or tibia within the joint capsule. The body's defense to slow down this wear is to minimize pressure by distributing forces over a larger articular area and to minimize friction. The need to distribute forces explains why lower limb joints tend to be larger than upper limb joints. Forces can also be relieved through strain. The temporary deformation of the cartilage absorbs some of the energy. The cortical bone underneath articular cartilage is typically thin and underlain by fat cells in the marrow; these tissues also absorb some force.

Synovial joints have an extremely effective lubrication. Synovial fluid is more viscous than water and has unusual properties of elasticity so that it maintains a thin film separating the two articular cartilages. This is also enhanced by the spongy nature of the cartilage. Articular cartilage "soaks up" fluid when the joint is not loaded. As it deforms, the cartilage squeezes out fluid into the space between the bones. As a consequence, the fluid is always present between the moving surfaces. A synovial joint has a coefficient of friction only 15% of that of a lubricated bearing used in industry.

Stabilization of the joint, in the sense of preventing unwanted motion, and free movement are inherently conflicting demands. Every joint makes its own compromises based on the needs of movement and on the magnitude and nature of destabilizing forces. Joints of the lower limb must be more secure because they bear body weight, while upper limb joints can afford to be more mobile because they do not. This argument helps us to understand some of the differences in joint design and function.

Conventional classification of synovial joints refer to the shapes of the articular surfaces and assumes that the bones are in contact during joint motion. Every joint movement is a rotation in space about an axis of rotation. The axis is an imaginary line attached to a moving bone such that the line does not move when the bone moves. Hypothetically, the shapes of the articular surfaces determine the motions of the joint. Joints may then be described according to the

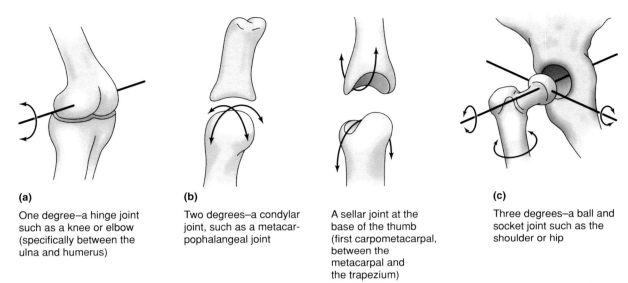

(a)
One degree–a hinge joint
such as a knee or elbow
(specifically between the
ulna and humerus)

(b)
Two degrees–a condylar
joint, such as a metacar-
pophalangeal joint

A sellar joint at the
base of the thumb
(first carpometacarpal,
between the
metacarpal and
the trapezium)

(c)
Three degrees–a ball and
socket joint such as the
shoulder or hip

Figure 3.11 Degrees of freedom. Synovial joints may be described and classified according to the number of degrees of freedom at the joint.

number of axes of potential movement, or the number of degrees of freedom (DOFs) (Figure 3.11). The maximum number of DOFs is three, observed in ball-and-socket joints such as the shoulder and hip. An oval-shaped condylar joint has two DOFs, as seen in the wrist and at the base of the fingers. A joint such as the elbow or knee, that is restricted to a single type of rotation, or movement in a single plane, has a single DOF and is considered a hinge. Some synovial joints without regular curvature may be capable of gliding movements on flat or irregular surfaces. This characterizes the joints among the carpals and metacarpals in the hand and the tarsals and metatarsals in the foot.

ACCESSORY AND CONJUNCT MOTIONS

The typical classifications of synovial joints as ball-and-socket, condylar, etc., are ideal more than real. Because real joints are never geometrically regular, our descriptions of motions and our classifications of joints are always approximations. We can measure an instantaneous axis that tells us about motion of the

joint at any given moment or for any part of the range of excursion, but the instantaneous axis change during excursion. The overall joint axis is an average of an infinite number of varying instantaneous axes. Furthermore, bone shape is relevant to motion only if the bones are in contact with a fair degree of congruence. Soft tissues (ligaments, muscles, etc.) also influence motion by their ability to hold the bones in contact. Consequent deviations in the actions of joints from the ideal are described as conjunct and accessory motions.

Conjunct motions occur because of the curvatures of the articular surfaces. In addition to the predicted primary rotation, the bone will also rotate about an additional axis. For example, the knee appears to be a hinge (double condylar) joint with one DOF. However, in the last phase of extension, the tibia pivots about its long axis under the femur (Figure 3.12). (To observe this, sit on a table with your feet dangling freely. Hold your knees together and extend them to bring your feet up to the level of the table. Your feet should have been together initially, but the toes move apart as the knees extend.) All joints have some normal conjunct rotation. In many cases, including

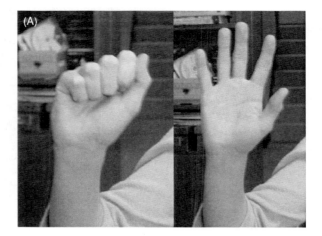

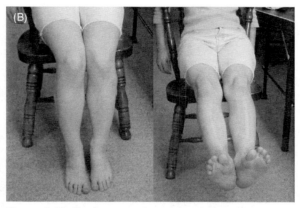

Figure 3.12 Conjunct motions. Most joints display some degree of conjunct motion accompanying the major axis of motion. *A*, The fingers tend to adduct when flexed into a fist and to abduct when extended in a relaxed manner. *B*, The lower leg has an outward rotation as the knee is fully extended.

that in the knee, the rotation contributes significantly to the stability of the joint because the rotation must be reversed before the joint can move.

Accessory motion is outside the defined axes of rotation and occurs when the joint surfaces are sufficiently separated that they no longer determine or interfere with motion. Unlike conjunct motion, accessory motion must result from external forces acting when the joint is not loaded. (If the muscles are active, they will load

the joint and bring the bones together.) You may easily visualize it by wiggling your jaw (temporomandibular joint) from side to side with your hands or the head of your fifth metacarpal back and forth.

MUSCLE

Muscle tissue provides the power to move or stabilize the skeleton. Neither skeletal form nor the muscle system can be fully understood without the other. "Muscles" are groupings of muscle fibers bound by fascia and thus anatomically distinct. Anatomical divisions do not necessarily correspond to functional divisions. Several muscles may share a common function, or different parts of one muscle may have different functions (Figure 3.13). We describe potential function from the anatomical design and attachments of a muscle. However, actual function depends on its use by the nervous system. Thus, our anatomical list of muscles is arbitrary and represents only a crude understanding of neuromuscular function.

Skeletal muscle cells, or fibers, are elongated polynuclear cells containing contractile proteins. These proteins, actin and myosin, form filaments that are able to interact with one another so as to slide in parallel, but opposite, directions. This sliding shortens the cell and causes powered contraction. The strength of muscle results from large numbers of these complexes within each fiber, aligned and working in synchrony, and from large numbers of parallel muscle fibers in each organ. The power of contraction is additive. The number of actin–myosin complexes working at a given time determines the strength of a muscle; and thus strength is proportional to the cross-sectional area of a muscle perpendicular to the direction of the fibers. The extent to which a muscle can shorten, or its contractile distance, is a proportion of the length of the fibers, usually about 50% of resting length or two thirds of its maximum length.

Muscles display different arrangements of fibers depending on their functions. Muscles that must contract over a great distance need long parallel fibers. If the muscle fibers insert obliquely into the tendon, it is possible to include more fibers but shorter ones. Such pennate ("feather") muscles have greater potential power but contract over shorter distances (Figure 3.14).

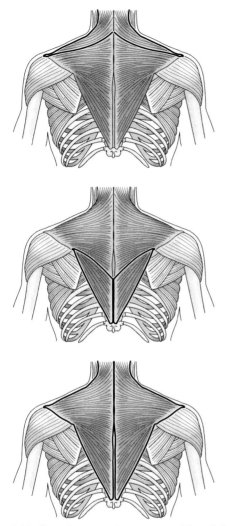

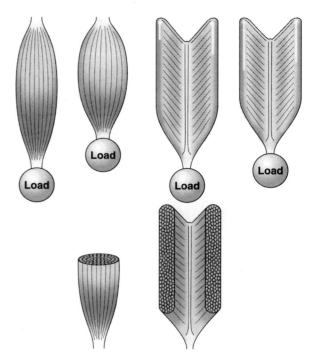

Figure 3.14 Pennate muscles increase power. Converging fibers in pennate muscles shorten the potential distance of contraction but increase power because they contain more individual fibers and a greater effective cross-sectional area.

Figure 3.13 Complex muscles are partitioned by the nervous system into independent units with different functions. The parts of trapezius, shown here, can work together or separately. The superior fibers elevate the scapula; inferior fibers depress it; and the middle fibers or all fibers acting together depress it.

In a waking person, muscles maintain a resting tone, or tonic contraction. They tense without actually moving joints. Because there is no "slack," tone permits the muscles to respond more quickly and effectively to reflexes or to voluntary commands. Muscle contraction can be isometric, concentric, or eccentric.

Isometric contraction maintains the same length of the muscle but increases the tone or resistance to stretch. Isometric actions are important for stabilizing the joints of the body against external forces and for maintaining posture and equilibrium. *Concentric* contraction shortens a muscle and effects movement. Although we usually describe muscles in terms of their functions under concentric contraction, this may be the least essential pattern for many muscles.

Eccentric contraction may occur as a muscle is lengthened due to an outside force. That force may result from the actions of the antagonist muscle or from gravity. Eccentric contraction enables a muscle to control the rate of joint movement, to stabilize a joint, or to resist collapse of a limb. They are at least as important as concentric motions in walking and other supportive actions of the lower limb.

CONCEPTS OF FUNCTIONAL DESIGN

The design of the musculoskeletal system follows the constraints and mechanical principles outlined above. In addition to reducing stresses in the body, there are other general objectives: minimization of mass and maximization of energetic efficiency.

REDUCTION OF MASS

Mass is expensive. Bone has to be built with calcium, phosphorus, and other minerals; muscle, with protein. Construction of these tissues requires investment in both nutrients and energy. Extra tissue has a metabolic cost of maintenance, keeping the cells healthy and the matrix in good repair. The body must expend additional energy to move it, especially if that mass is on the limbs. Design should reduce mass to the extent compatible with structural strength. There are several mechanisms by which this objective is achieved.

Bones are designed to minimize material. The placement of mineral on a bone resembles the solutions that an engineer might devise. Bone is reinforced where stresses are concentrated and can be remarkably thin elsewhere. For example, the blades of the scapula and the ilium and certain facial bones are so thin as to be translucent. Bone is actually absent at the obturator foramen of the pelvis where a membrane can substitute and in the facial sinuses where it is not needed.

Tissues under stress can be reinforced. The body avoids constructing a skeleton and muscular system for every contingency. Instead, use guides development even in fetal life, and changes in activity patterns can be met with the synthesis of new bone and muscle as needed. Because tissue production responds only when stresses exceed a certain threshold, it will cease when the new demands are satisfied. Patterns of stress that occur over long periods of time, such as those relating to subsistence activities or to certain occupations, are sometimes identifiable in the skeleton.

Unused tissues are reduced in mass. Usually, bone under tension will be resorbed. Muscle that is not exercised will atrophy. The body is continually trimming its budget.

Muscles act on multiple joints in the body. Particularly in the distal limbs, but also in the axial skeleton, long muscles span several joints. This makes it possible, for example, for a single muscle to flex the two joints of the wrist and three of each of the fingers. Although there may be other reasons for this design, it is more efficient in the use of muscle mass than many smaller muscles would be.

ENERGETIC EFFICIENCY

Energy is the currency of the body. The musculoskeletal system has many economizing measures to increase efficiency in the expenditure of energy. Some examples are listed next.

Stability may be maintained with noncontractile tissues. Although muscle tone and postural adjustment represent a constant expenditure (which also help to generate body heat), it is also possible to use noncontractile tissues such as ligaments to stabilize joints. Noncontracting tissues require no additional energy beyond keeping them healthy.

The hip joint is stable in extension because its fibrous capsule is twisted taut, holding the head of the femur in the acetabulum and preventing further movement. A person standing on one hip uses body weight to maintain that tension and thus expends no energy to remain erect at that joint. Only muscular effort can mobilize the hip again. This position of the joint in which the bones are pressed tightly together with a high degree of congruence and the ligaments are taut is called the *close-packed position*. The knee close-packs in extension as well. The conjunct rotation of the knee described earlier locks the knee into a stable configuration to bear body weight without muscular effort. Although there is disagreement whether every joint has only one close-packed position, it is important that weight-bearing joints have at least one.

Energy may be stored by tendons. Locomotion is one of the most expensive and predictable motor activities. Proper design may reduce energetic costs dramatically. One strategy is to employ the force of gravity, using muscle only to harness and redirect that force. The elasticity of tendons of weight-supporting limbs may temporarily store the energy of downward movements produced by gravity. The limbs may recover that energy and apply it to aid in forward progression.

Limb muscles are positioned proximally. The energy needed to move a limb is a function of both mass and distribution of the mass along the limb. The farther from the joint mass is located, the longer is the load arm and the greater is the force needed to move it. For that reason, muscle bellies tend to be located proximally on the limbs. Tendons of origin may be very short, whereas tendons of insertion are long and slender. In running animals that need long limbs and fast movements, the feet tend to be small and minimize the use of tissue.

4
THE SKULL

Few fossil discoveries generate as large a headline as the recovery of a new hominin skull. A small-brained heavy-browed skull has become a symbol for human evolution, as the skull represents the most recognizable and diagnostic part of skeletal anatomy. We can trace changes in the hominin line through the transformation of the skull and especially the expansion of the brain. Furthermore, the teeth and the bones in which they are embedded are the most dense, durable, and recognizable of body parts. For these reasons, a bias toward cranial remains skews both our collections and the direction of research and debate. This is also true for earlier periods of vertebrate evolution. The postcranial skeleton may allow us to sort fossils by class and order, but it is usually only the jaws and teeth that are diagnostic of genera and species, especially for mammals.

Why do vertebrates have a head at all? Other animals, such as starfish and clams, manage well without one. The answer lies in the highly mobile lifestyle of our ancestor. The head is a region that houses the brain, organs of special senses, and the proximal ends of the digestive and respiratory tracts. That these lie together is not a coincidence. In an actively swimming animal, the head is the first part of the body to encounter and explore a new environment and the first to overtake food. Body structure is, in part, a system designed to move the mouth forward. Sense organs appeared to scout out the environment ahead, and the brain evolved to support them. Animals that do not have heads simply are not chasing prey and are not using their long distance senses to tell them where to go. If the features of the head are the most specialized and distinctive features of many vertebrate taxa, it is because species are distinguished largely in the way they encounter and interact with their environments.

THE HUMAN SKULL

The human skull consists of the cranium and the mandible (Figure 4.1). We can best picture its design in terms of several overlapping functional areas, each of which responds to a different set of developmental and mechanical constraints (Table 4.1).

- The facial skeleton, including the orbits, cheek bone, nasal cavity, and paranasal sinuses, simultaneously supports the muscles of mastication and the special senses of vision and smell. It also helps define the upper respiratory tract.
- The braincase houses and protects the brain. The basicranium, the floor of the braincase, houses the ear and supports the brain. The details of its structure reflect the shape of the brain and the

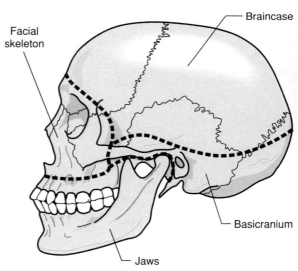

Figure 4.1 Functional regions of the human skull. The skull includes the cranium and the mandible.

Table 4.1 Characteristics of the Human Skull

Vertebrate characteristics
    Encephalization, including concentration of sense organs at the
        anterior end of the body
    Skeletal protection of the brain, usually involving bone
    Formation of the skull from dermal, cartilaginous, and branchial
        elements
Mammalian characteristics
    Expansion of a lateral fenestra
    Expansion of the braincase
    Transfer of muscle origins to side of braincase
    Secondary palate
    Mandible reduced to a single paired bone, the dentary
    Three ossicles in the middle ear
Primate characteristics
    Orbital convergence
    Postorbital bar
    Shortening of the face
    Flexion of the face relative to the basicranium
Anthropoid characteristics
    Postorbital closure
    Fusion of the mandibular symphysis
Hominin characteristics
    Chewing involving significant lateral motions of the mandible
    Expansion of the brain and braincase
Modern human characteristics
    Anterior protrusion of the chin
    Development of the mastoid process
    Flexion of the basicranium and retraction of the face
    Forehead replacing supraorbital tori

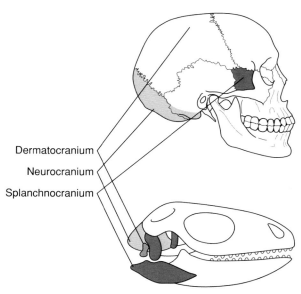

Figure 4.2 Composition of the vertebrate skull. The vertebrate skull contains bone from three different evolutionary and developmental origins. The dermatocranium contributes primarily to the roof of the skull; the neurocranium forms the base of the braincase, and the splanchnocranium contributes to the jaws and a few other bones.

connections of the brain and supporting tissues with the rest of the body.

- The jaws—mandible and maxilla—support the teeth and reflect the mechanical demands of chewing.

EVOLUTION OF THE MAMMALIAN SKULL

ORIGINS OF THE VERTEBRATE CRANIUM

The cranium was probably the first part of the skeleton to evolve in vertebrates, aside from the notochord itself. Its evolution, inferred largely from the embryological development of living vertebrates, represents a convergence of three structural components: dermal elements, the dermatocranium; cartilaginous elements

protecting the brain, the chondrocranium or neurocranium; and derivatives of the visceral elements, the splanchnocranium.

The dermatocranium evolved as plates of bone in the integument, possibly for protection, or possibly as a reservoir for storing essential minerals such as calcium. In some early fish, these plates extended over considerable portions of the body (Figure 4.2). This dermal bone is embryologically distinct from other bone in the body because it forms from membranes, rather than cartilage. Dermal bone contributes the greater parts of the cranial vault, face, and jaws of the mammalian skull. The teeth also have their derivation from this dermal armor by a more complex route, as discussed in a Chapter 6. In early vertebrates, the dermatocranium forms massive armor; however, underwater its weight is not as serious a burden. As described later, the evolutionary history of land vertebrates reveals a considerable reduction of dermal bone and, in the skull, a simplification of many small elements into a relatively few larger ones.

The chondrocranium may have been the first part of the skull to evolve. It arose as a cartilaginous support for the brain and special sensory organs. Its elements contributed to the inferior and lateral portions of the braincase and the skeletal housing of the eyes, ear apparatus, and olfactory membranes. The chondrocranium originally lay deep to the dermal plates and was separated from them by the muscles of mastication and other tissues. As the dermal plates lightened and reduced in size, the bones of the chondrocranium were exposed in a more superficial position. In the human skull, the chondrocranium contributes the central and inferior bones, including parts of the occipital; petrous and mastoid portions of the temporal; ethmoid; and the body and pterygoid plates of the sphenoid.

The splanchnocranium derives from the cartilaginous supports of the branchial arches. These arches formed in the slitted walls of the pharynx in ancestral chordates and the gills of modern fish. Most vertebrates have seven arches. They appear in the early human embryo, but their skeletal support is either lost or transformed into other elements. In the adult human skull, derivatives of the splanchnocranium are limited to the hyoid bone, the styloid process of the temporal bone, the three ossicles of the middle ear, and some cartilages of the neck. In nonmammalian vertebrates, the splanchnocranium also contributes to the bones of the lower jaw.

Different neighboring tissues influence the development of these three elements within the embryo (Schilling and Thorogood 2000). The chondrocranium is guided by the developing brain, and the dermatocranium is closely linked to the epidermis. The branchial arches comprising the splanchnocranium are segmental in pattern and incorporate muscle, blood vessel, and nerves, as well. Signals from mesodermal tissues link together and guide the development of these structures. Because developmental pathways involve the integration of different elements and different organ systems, interpreting the evolutionary pathways is not a straightforward exercise. Selection or mutation involving one system will affect the others. Conversely, a change in any one structure may reflect an adaptation within another system.

MAMMALIAN MODIFICATIONS OF THE TETRAPOD SKULL

The mammalian skull has experienced significantly transformation with respect to that of the early tetrapods. As we trace this evolutionary pathway, we can better understand both the development and anatomy of the modern human skull. Major changes in the mammalian skull may be summarized as a series of six functional shifts.

1. Fenestration. The roof of the skull of primitive tetrapods was a solid mass of bone, broken only by the openings for the eyes and nostrils. These bones shrank in mass in the descendants of these amphibians and stem reptiles. Often areas of bone under low stress thinned into simple membranes, creating fenestrae, or "windows" in the skull. Such a thinning or elimination of bone in the center of a broad muscular attachment may be observed elsewhere in the body (e.g., the obturator foramen and iliac fossa in the pelvis and the infraspinous fossa in the scapula) and indicates a site where bone can be spared to save on materials and mass without sacrificing mechanical function.

We tend to identify the diverging lineages of reptiles according to the number and locations of holes that appear in the bony shield. The synapsid reptiles, the lineage leading to mammals, developed a fenestra between the parietal and postorbital bones (Figure 4.3). This fenestra continued to expand in the mammal-like reptiles and the mammals, eventually merging with the orbit and leaving only the zygomatic arch as its lower border. As it reduced skull weight, fenestration exposed the muscles of mastication that underlay and arose from the dermal shield. The membrane over the synapsid fenestra continued to support the origins of the muscles. The membrane is present today as the temporalis fascia, a tough layer of connective tissue covering the temporalis muscle.

2. Rearrangement of the jaw musculature. The jaw muscles in the primitive tetrapod skull arose from the underside of the bony roof and occupied a considerable portion of the volume within the skull. In the evolving mammalian lineage, effective musculature for chewing became increasingly important because endothermy requires more efficient processing of larger quantities of food. As the cranium reduced its mass,

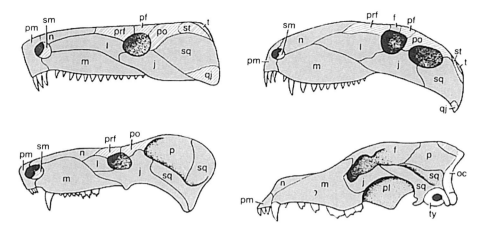

Figure 4.3 Expansion of the fenestra. The crania of four representative species in the lineage leading to mammals are shown in lateral view. A postorbital fenestra appears behind the eye and expands as the dermal bone is reduced in size. *Upper left*, a labyrinthodont amphibian showing the primitive tetrapod condition. *Upper right*, an early synapsid reptile with a fenestra opening posterior to the orbit. *Lower left*, a therapsid (advanced mammal-like reptile) with the fenestra enlarged to expose part of the braincase. *Lower right*, a modern mammal (dog). (From A.S. Romer and T.S. Parsons. 1986. *The Vertebrate Body*, 6th ed. Copyright by Saunders College Publishing. Reproduced with permission from Thomson Learning.)

the musculature increased its relative area of attachment, spreading to the wall of the chondrocranium, and differentiated into several distinct muscles (Figure 4.4). These muscles created forces in slightly different directions and were able to direct their action more onto the teeth and less onto the joint.

3. Expansion of the brain and braincase. The greater size of the mammalian brain was necessarily accompanied by an increase in the size of the chondrocranium. This expansion provided increased surface area to support a greater proportion of the muscular attachments. This permitted further reduction in the

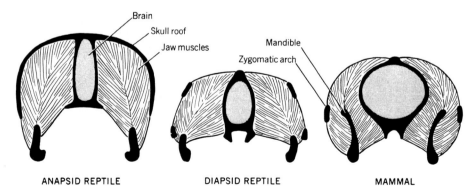

Figure 4.4 Redesign of the jaw musculature. A coronal section through the cranium shows the enlargement of the fenestra and the braincase and the accompanying change in muscular attachments. *Left*, early tetrapod form without a fenestra; *middle*, reptile with two fenestrae; *right*, modern mammal. (From M. Hildebrand. 1988. *Analysis of Vertebrate Structure*, 3rd ed. Copyright © 1988 John Wiley and Sons. This material is used by permission of John Wiley & Sons, Inc.)

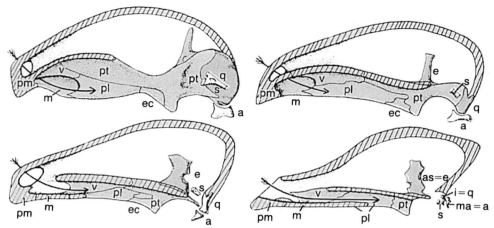

Figure 4.5 Development of the secondary palate. A sagittal section through the cranium reveals the progressive expression of the secondary palate dividing the oral and nasal cavities. Specimens are as given in Figure 4.3. (From A.S. Romer and T.S. Parsons. 1986. *The Vertebrate Body*, 6th ed. Copyright by Saunders College Publishing. Reprinted with permission from Thomson Learning.)

dermal bone and expansion of the fenestrae. Ultimately, the chondrocranium formed the smooth continuity with the dermatocranium that we observe in modern mammals.

4. Development of the secondary palate. The expansion of the secondary palate was an adaptation for prolonged chewing of food. The primary palate, primitive for tetrapods, separated the mouth cavity from the brain itself. The secondary palate interposed an additional wall of bone. The two palates isolated the nasal air passages from the oral food passage. As a consequence, mammals are able to continue to breath as they chew. The expanded palate is extended posteriorly by soft tissues so that the food and air passages remain separated as far as the pharynx (Figure 4.5).

5. Strengthening of the mandible. The increased food processing demands of endothermy places a greater stress on the jaws and teeth. Numerous changes were made in the design of the dentition, but the mandible itself had to be improved.

The mandible of fish and lower tetrapods developed from the first branchial arch of the splanchnocranium. The branchial arches are each composed of several individual units of cartilage that appear in reptiles as various bones on each side of the mandible. An additional bone, the dentary, developed from the dermatocranium, bearing denticles as the precursor

of teeth. The articulations among these bones of the mandible represent potential weak spots where chewing stresses might damage the bone. In the ancestors of the mammals, the smaller posterior bones gradually shrank in size and escaped from direct lines of stress. Eventually, the mandible itself was reduced to a single paired bone, the dentary. Some of the other units migrated across the articulation of the skull to become integrated into the hearing apparatus (Figure 4.6).

6. Development of the middle ear ossicles. The synapsid jaw joint occurred at the articulation of the quadrate bone on the cranium and the articular bone on the mandible. The articular was one of the small bones fated to be lost from the jaw. As the dentary expanded posteriorly, it established a parallel joint with the squamosal area of the cranium. The quadrate and articular bones, now redundant in function, continued to shrink in size. The articular eventually migrated across the joint and, with another mandibular bone, the angular, helped to create the modern ear.

The names of these bones are different in their new functions. The dentary–squamosal joint, which is diagnostic of fossil mammals, occurs in humans between the temporal and mandible. The squamosal merged with other elements to form the temporal, and we drop the term "dentary" because it is now synonymous with "mandible." The quadrate and articular

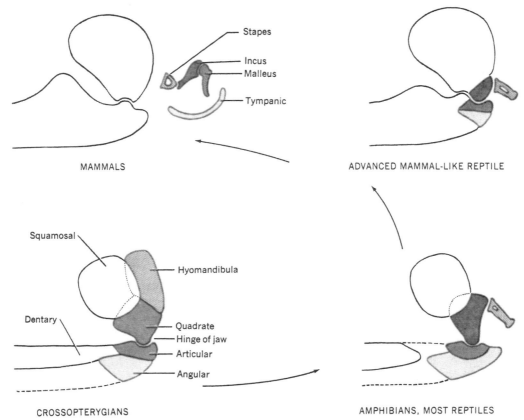

Figure 4.6 Development of the mammalian middle ear ossicles. As the mammalian lineage simplified the structure of the mandible and the jaw articulation, smaller bones of the region became incorporated into the hearing apparatus. *Lower left,* crossopterygian fish, ancestral to tetrapods; *lower right,* amphibian; *upper right,* advanced mammal-like reptile; *upper left,* mammal. (From M. Hildebrand. 1988. *Analysis of Vertebrate Structure,* 3rd ed. Copyright 1988 John Wiley and Sons. This material is used by permission of John Wiley & Sons, Inc.)

prove to be homologous with the incus and malleus of the middle ear, respectively. The angular is homologous with the tympanic bone (fused in humans with the remainder of the temporal).

CHANGES IN THE PRIMATE SKULL

THE PRIMATE FACE

From the basic mammalian design, various orders of mammals have further modified the cranium according to their particular needs. The bones of the face among primates tend to be lightly built except when reinforcement is needed to withstand specific forces, primarily those relating to chewing. Many are hollow and contain sinuses, which are air pockets lined with mucous membrane. The most obvious specializations of the primate face relate to a repositioning of the orbits and a shortening of the face.

Orbital Convergence

Primates rely heavily on a highly developed visual system. In addition to having improved resolution and good color discrimination, primates have emphasized stereoscopic vision (also called binocular vision or depth perception). Stereoscopic vision requires an

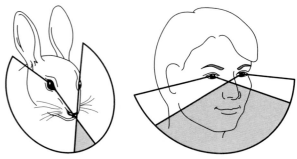

Figure 4.7 Orbital convergence. Placement of the eyes facing anteriorly increases the overlap between the visual fields.

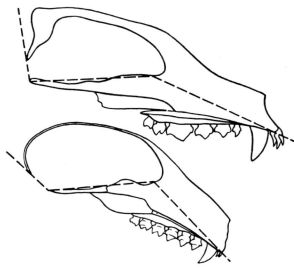

Figure 4.8 Flexion of the facial skeleton. Primates are characterized by a downward flexion of the facial skeleton relative to the brain case. *Above*, the outline of a dog skull. *Below*, a lemur, a prosimian primate. (From W.E. Le Gros Clark 1971. *The Antecedents of Man.* Quadrangle Books. Copyright 1971 by Edinburgh University Press. Reproduced with permission by Edinburgh University Press.)

overlap of the visual fields of the two eyes, particularly of the centers of the fields. Among the skeletal correlates of this change are the enlargement of the orbits and a shift of the orbits from the side to the front of the face (Figure 4.7).

Repositioning the orbits brings them closer together and narrows the bone between them in the face. Primates further tend to reduce the length of the snout and move it to a lower position on the face that does not obstruct the visual field. The downward movement of the face is reflected in the relative angle between the basicranium (floor of the braincase) and the snout (Figure 4.8). Compared even with other primates, humans have shorter and flatter faces so that the eyes are nearly in a plane with the front of the mouth. Making such changes in the face has other consequences, such as reducing the apparatus of smell and altering the mechanics of mastication.

The changing position and enlargement of the eyes made them more vulnerable to strain and injury from the chewing muscles and from external forces. The evolutionary response was to select for a protective skeletal housing. All living primates therefore have a postorbital bar formed by an extension of the frontal bone connecting the zygomatic arch. Certain other mammals that produce heavy chewing stresses, such as cows, also have a postorbital bar. The bar appears to resist torsion between the face and the braincase (Greaves 1985); however, the postorbital bar appears to be stronger than masticatory forces alone would generate (Ravosa et al. 2000). Therefore protection for

the eye may have been the primary selective value in primates.

Anthropoids carry orbital convergence to a greater degree than do the prosimians and provide additional protection for the eye by enclosing the back of the orbit with bone. Expansions of the zygomatic, sphenoid, and frontal bones contribute to this wall. Evidence suggests that this enclosure is also in part a response to stresses of mastication (Ross 1995).

Shortening the Face

Orbital convergence explains only part of human facial reduction, because facial shortening continued throughout the hominin lineage. Another important determinant in the length of the face is the length of the dental row, which in turn is influenced by tooth size and mechanics of chewing. The shortening of the face seen in hominins has the effect of bringing the biting areas of the teeth closer to the attachments of the masticatory muscles. This shortens the load arm and increases the power of the bite.

Shortening the face appears to come at a cost of functional area in the nasal cavity. An important function on the nasal surface is the sense of smell. The importance of olfaction appears to decline in importance in primates across the order. Reducing the snout sacrifices receptors in an apparent trade-off of olfactory sensitivity for visual acuity. This trade-off also appears in the brain, where the proportions of cortical area devoted to the two functions in primates differ from those in other mammals.

The nasal passages are places where air is warmed and moistened before it descends to the lungs. Regardless of the sense of smell, there would appear to be a point beyond which the nasal cavity can no longer be reduced. Humans exceeded that threshold, perhaps as they moved into the colder regions of Eurasia. To compensate, we developed an external nose that extends the passages beyond the skeletal structure.

THE CRANIAL VAULT

The shape of the modern braincase, particularly the contours of the cranial fossa under the brain, is a close mirror of the shape of the brain itself (Figure 4.9). As the cranium develops, the flat bones of the vault mold themselves about the expanding brain and its meninges. These bones meet in an infant along sutural lines. The sutures represent regions of growth and are maintained until the brain reaches its adult size. The

eventual shape of the vault is determined by the extent of growth at specific sutures and the relative timing of suture closure. These in turn are directly influenced by the growth of the underlying brain and its surrounding dura mater (Lieberman 2000). In primates, for example, the relative decrease of the olfactory regions of the brain anteriorly and expansion of visual cortex posteriorly cause corresponding changes in the shape of the braincase relative to those of other mammals.

The importance of maintaining open sutures to allow brain growth becomes even more critical during human birth. The average head diameter of a full-term fetus is slightly larger than the typical diameter of the birth canal. When uterine labor contractions push the fetus through the birth canal, the fetal head is forced to adjust its shape, becoming longer and narrower. This reshaping occurs by sliding and even overlapping of the ossified plates along the suture lines. Where the sutures intersect, fontanelles—soft spots unprotected by bone—persist into the neonatal period.

The expansion of the cranial vault in earlier hominins did not initially necessitate the high rounded braincase of modern humans. In other anthropoids and in other hominin species, the braincase lies well behind the facial skeleton. The increase in hominin brain size led to an increase primarily in length of the vault. Within *Homo sapiens*, a more drastic restructuring occurred within the basicranium that allowed for a greater breadth-to-length ratio of the braincase. The shape of the vault shifted to a more globular form and to a spatial overlap of the face and braincase (Ross and Ravosa 1993).

The rounded cranium is more typical of newborn mammals than of adults of almost any species. It represents early growth of the brain relative to the facial skeleton. A rounded head is more likely to pass smoothly through the birth canal without snagging. The idea of infants possessing round heads is so ingrained in us that we tend automatically to regard any being with a large rounded head as "childlike" and "cute." The retention of infantile characteristics into adulthood is called *neoteny* and is one way to achieve paedomorphosis (see Chapter 1). Humans appear to have several paedomorphic traits, including the short face, relatively large brain, extended period of behavioral development, and hairless skin. Possibly the rounded braincase is

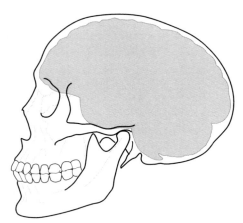

Figure 4.9 The brain within the cranial fossa. The shape of the brain follows the contours of the cranial fossa.

merely part of this pattern. However, paedomorphosis is not an evolutionary goal nor an adaptive trait in itself; it is merely a description of a developmental process by which a species may achieve a more specific adaptive character. In this case, the cranial shape may be a consequence of prolonged brain development and expansion (Lieberman et al. 2002).

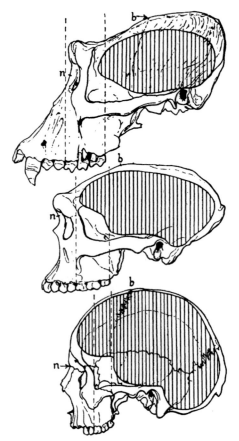

Figure 4.10 Reduction of the hominin face. The hominin lineage shows a progressive shortening of the face, expansion of the braincase, and flexion of the basicranium. From top to bottom: ape (gorilla), fossil hominin (*Homo erectus*), and modern human skulls in lateral view with the brain superimposed. (From F. Weidenreich. 1941. The brain and its role in the phylogenetic transformation of the human skull. *Transactions of the American Philosophical Society* N.s. 31[5]:321–342. Reproduced with permission of the American Philosophical Society.)

The skull of anatomically modern humans experienced several related changes. The shape of the brain became more spherical, so that the braincase is now shorter and higher (see Figure 4.10). The masticatory apparatus greatly shrank in size and mass, particularly in its anterior dimension. The smaller face shifted under the front of the braincase by flexing the entire skull between the face and braincase (DuBrul 1977; Lavelle et al. 1977; Ross and Ravosa 1993). This resulted in a high forehead characteristic of modern humans and a measurable flexion, or bending, of the floor of the cranial cavity. Another important consideration of this transformation was the maintenance of balance in the head. Weight was reduced simultaneously from the face and the occipital region, making it possible to maintain posture in a bipedal stance with relatively little muscular effort.

Browridges

The skull of an ape or of an early hominin often has conspicuous browridges (supraorbital torus) that define the upper margins of the orbit. The expression and prominence of the brow are a reflection of the degree to which the braincase is separated from the facial skeleton. In the great apes and early hominins, a marked postorbital constriction distinctly separates the two parts of the cranium, so that the facial skeleton resembles a broad facade with little behind it (Figure 4.11).

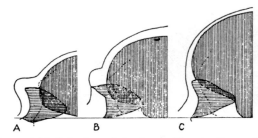

A   B   C

Figure 4.11 Relation of the browridge to the position of the braincase. The disappearance of the browridges coincides with the expansion of the hominin brain above the orbits. From left to right: chimpanzee, fossil hominin (*Homo erectus*), and modern human. (From F. Weidenreich. 1941. The brain and its role in the phylogenetic transformation of the human skull. *Transactions of the American Philosophical Society* N.s. 31[5]:321–342. Reproduced with permission of the American Philosophical Society.)

The reinforcement of bone over the orbit has been interpreted as a response to masticatory stress, but such stresses have not been observed in this region (Hylander et al. 1991; Picq and Hylander 1989). Facial length appears to be a more important correlate (Picq 1994; Ravosa 1988, 1991a; Weidenreich 1941). The modern human skull eliminates the separation between face and braincase by the expansion of the brain up over the top of the face. The forehead thus serves the function of defining and protecting the upper face, and there is no place for heavy browridges. The relative development of the brows may have little to do with behavior. The basic patterns of craniofacial development appear more important in determining browridge size than functional considerations (Lieberman 2000).

Basicranial Flexion

Primates, relative to other mammals, possess a flexion within the sagittal plane of the basicranium. The floor of the posterior cranial fossa anterior to the foramen magnum defines one plane of the angle against a second plane from the dorsum sellae or hypophyseal fossa. (An approximation of this flexion appears in Figure 4.10 as the creation of a second "step" in the countour of the underside of the human brain.) Different authors have used a variety of landmarks to measure this angle of flexion. Because it reaches extreme development in hominins, some researchers have attempted to link basicranial flexion with upright posture, increased brain size, reduced projection of the face, and other factors (reviewed in Strait 2001). Studies across species show that the angle tends to increase with endocranial volume or brain size (Straight 1999; Lieberman et al. 2000), but this is not a sufficient explanation for all hominoid species (Jeffery and Spoor 2002; Ross and Henneberg 1995).

One consequence of this flexion is to limit the length of the basicranium. The cranial cavity and brain therefore assume a more spherical shape. This may relate to a more efficient pattern of neuronal connections between the cerebral cortex and the diencephalon (Lieberman et al. 2000). Another consequence is to cause the face to be tucked under the front of the braincase; although whether this reflects selection for facial shape or for basicranial length or neither remains unknown.

The degree of flexion within the cranial base gained interest when it was a proposed indicator of the capacity of speech in fossil hominins. Lieberman and Crelin observed a greater degree of flexion between modern adult human crania on the one hand and those of human infants and a reconstructed Neanderthal skull on the other. They proposed that the shape of the cranial base correlates in modern species with the size of the pharynx and the position of the larynx. They then reconstructed the Neanderthal skull from La Chapelle-aux-Saints and the soft tissues of its pharynx, found that it had a relatively unflexed basicranium, and concluded that the Neanderthal, like modern chimpanzees and human infants, had a high larynx and a limited repertoire of vocal sounds. Neanderthals would have been unable to pronounce the sounds represented by the letters a, i, u, g, and k. Neanderthals, they stated, were incapable of "articulate human speech" (Crelin 1987; Lieberman 1998; Lieberman and Crelin 1971; Lieberman et al. 1972).

This study, had it been substantiated, would have had significant implications. It implied that speech is a very recent phenomenon, belonging only to modern humans. It emphasized the distinctiveness of the Neanderthals from modern humans and presented a possible explanation for the sudden replacement of Neanderthals by modern humans. Critics attacked this hypothesis on several grounds, including the reconstruction of the fossil and the nature of the correlation between basicranial shape and position of the larynx. While there is certainly some anatomical relationship between the basicranium and the shape of the pharynx, subsequent studies have shown that the functional performance of speech is probably independent of cranial form and that modern levels of cranial flexion evolved much earlier than the Neanderthals (Laitman 1983; Lieberman and McCarthy 1999; Lieberman et al. 2000). A further problem is that although the form of the pharynx is certainly important for the range of vocalizations that people use today, a narrower range would not have precluded spoken language. The question of Neanderthal speech capability remains unresolved (see Chapter 15).

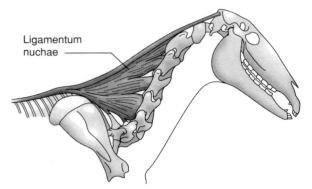

Ligamentum
nuchae

Figure 4.12 Ligamentum nuchae. The ligamentum nuchae in a quadrupedal mammal contains elastic fibers that help to support the weight of the head at the end of the neck.

MECHANICS OF HEAD BALANCE

The upright posture of the human body has also had an influence on the shape of the skull, and it offers a more efficient solution to the mammalian problem of supporting the weight of the head. Although lower vertebrates may rely on the buoyancy of water or on the firmness of the ground itself to relieve neck muscles of the task of supporting the head, mammals generally position the head at the end of a mobile neck. This cantilevered arrangement requires careful engineering to minimize the inevitable expenditure of energy. In a quadruped, long elastic ligamentous fibers anchored at the base of the neck and the occipital bone assist muscles in this support and form the ligamentum nuchae (Figure 4.12). The fibers of this ligament are under constant tension and their elasticity makes it easier for an animal to raise or sustain its head.

In an upright animal, vertical loading of the spine can directly support more of the weight of the head. Weight must balance at the occipital condyles, where the skull articulates with the spinal column. To protect the spinal cord, the condyles lie on either side of the foramen magnum. Thus the position of the foramen magnum and the condyles relative to the center of mass of the head may reflect the success of the animal in passive balancing of the head on the neck. This postural change is another phylogenetic trend of primates and is reflected in several adjustments to the cranium. The position of the foramen magnum is more anterior

in primates in general and in hominins in particular. The nuchal region of the occipital bone, to which the posterior neck muscles attach, changes its orientation from vertical in most mammals to more oblique in primates, and the area of attachment is reduced as balance is improved.

The face is still heavier than the back of the head in primates because of its length, but the postural muscles of the neck do not have as much load to bear. Shortening the face improves the balance. In the case of modern humans, the extreme reduction of the face resulted in a better balance than in other primates and required a reshaping of the braincase to shorten it posteriorly, as well (Figure 4.13). The downward pull of the posterior muscles needed to balance the skull is only 16% of the head weight in humans but 37% of head weight in other primates (Schultz 1955).

The human arrangement also created a new problem. When the neck extends to throw the head back, the center of gravity passes behind the occipital condyles. Muscular effort is required to return it to a neutral position. To accommodate this necessity, the mastoid process provides a unique leverage for the sternocleidomastoid muscle (Krantz 1963). From a flexed position, the muscle helps to extend the head, but from an extended position the muscle flexes it (Figure 4.14).

JAWS

During the course of hominin evolution, one of the most striking series of changes involved the development and loss of a powerful chewing complex in our ancestors. This pattern contradicted the long cherished but falsely held notion that evolution was linear, from an apelike ancestor to ourselves, and that any missing link would be something of an average between chimpanzees and humans. Instead we recognize that an evolutionary pathway can wander, following adaptive strategies unique to a particular era or species.

THE MANDIBULAR SYMPHYSIS AND THE CHIN

Primitively for mammals, the mandible consists of right and left dentary bones joined at the midline symphysis

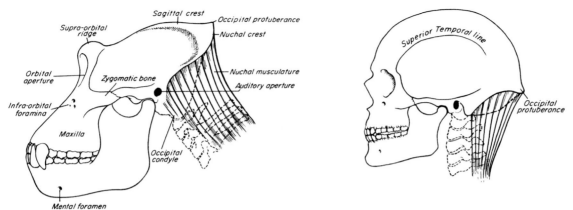

Figure 4.13 Mechanics of head balance in apes and humans. The occipital condyle, where the spinal column articulates with the skull, is the fulcrum for lever action of neck musculature. The longer skull and less upright position of the neck in the gorilla (*left*) require stronger action from muscles in the neck. Human muscles (*right*) are smaller and exert a more vertical pull. (From W.E. Le Gros Clark. 1967. *Man-Apes or Ape-Men? The Story of Discoveries in Africa.* Copyright 1967 by Holt, Rinehart and Winston.)

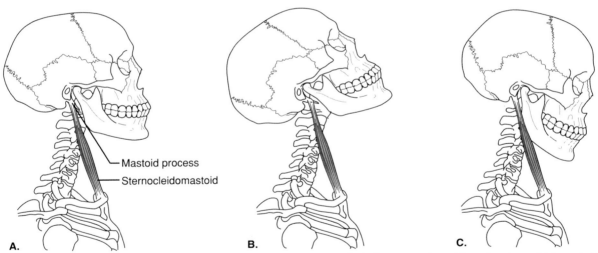

Figure 4.14 The role of sternocleidomastoid in balancing the head. *A*, In a neutral position, the sternocleidomastoid muscle inserts on the mastoid process slightly posterior to the articulation between the skull and the cervical spine; therefore, it can help to extend the head. From either a fully extended (*B*) or flexed (*C*) position, the muscle can use the leverage of the mastoid process to help to guide the head back to a more neutral position.

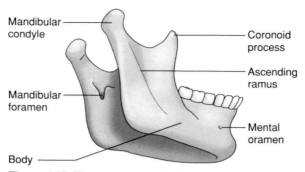

Figure 4.15 The human mandible.

Labels: Mandibular condyle, Coronoid process, Ascending ramus, Mandibular foramen, Mental oramen, Body

by ligaments (Figure 4.15). Such a joint can provide some movement, if desired, but it can also weaken the jaw under heavy chewing stresses. Thus, in anthropoids (and many other mammalian groups), the symphysis is fully fused and immobile. This appears to relate to the strain forces generated there by chewing tougher food items (Hylander et al. 1998; Ravosa 1999) and by the application of more lateral chewing forces (Lieberman and Crompton 2000).

Modern apes and monkeys typically reinforce the symphysis with ridges of bone on the inner curvature of the chin called a simian shelf (Figure 4.16). This reinforcement behind the symphysis resists the actions of the pterygoid muscles in pulling the condyles to toward one another as well as forces twisting the two halves of the mandible at the midline (Daegling 1993). Various primates have also strengthened that area by making the jaw deeper near the midline (Hylander 1979; Ravosa 1991b).

Modern humans have replaced the shelf with an external buttress, the chin. The prominent human chin distinguishes our species from other living primates and from all earlier hominins (Schwartz and Tattersall

2000). It has been argued that the development of a chin is the only character that consistently distinguishes anatomically modern humans from all archaic populations. In popular image, a prominent chin is mythically a sign of a strong, honest character, whereas a weak or receding jaw is stereotyped as indicating weakness and deviance. The adaptive explanation for the chin has been a matter of more academic consideration. Several explanations have been advanced, not all of them conflicting (reviewed by DuBrul and Sicher 1954). As the jaw shortened, it potentially "crowded" the tongue and floor of the mouth. The lower part of the mandible was therefore reduced less than the tooth-bearing region, leaving a jutting chin. Developmentally, these two parts of the mandible appear to be under independent controls (Biggerstaff 1977); thus, we can understand the superior portion as undergoing more reduction than the inferior. Chewing produces the same bending, twisting, and shearing stresses near the symphysis in humans observed in other anthropoids. Of these, the wishboning forces that tug at the condyles apart from opposite directions may be the most critical and may have favored some reinforcement and probably account for the reinforcement that constitutes the chin (Hylander et al. 2000). However, mechanical models of later hominin jaws do not show sufficient differences in properties to account for the addition of the chin (Dobson and Trinkaus 2002).

THE AUSTRALOPITHECINE CHEWING COMPLEX

One of the most striking features of the early hominins is the relatively large size of the molar teeth. Dental and mandibular size increases between 3.5 and 3.0 Mya in *Australopithecus afarensis* and reaches an extreme

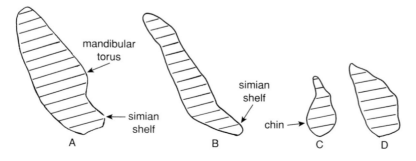

Figure 4.16 The profiles of the mandibular symphysis differ among primates. The bone is reinforced on the posterior surface of the ape jaw by the "simian shelf" and on the anterior surface on the human jaw by the chin. *A*, Gorilla. *B*, Orangutan. *C*, Modern human. *D*, *Australopithecus africanus*.

expression in the later robust australopithecine species (Lockwood et al. 2000). Particularly among the later species, this is accompanied by a tremendous hypertrophy of the muscles of mastication and the skeletal support for those muscles. The temporalis muscle is especially massive. Its area of attachment in males has expanded to the midline and forms a sagittal crest (Figure 4.17). The zygomatic arch flares widely to accommodate its bulk. The arch is thicker to support

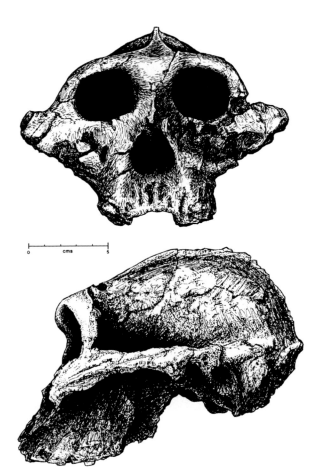

Figure 4.17 Cranium of *Australopithecus boisei*. The skull of *Australopithecus boisei* shows a massive construction of the face and and zygomatic arches and the presence of a sagittal crest. (Reprinted by permission of Waveland Press, Inc. from C.S. Larson, R.M. Matter, and D.L. Gebo. 1998. *Human Origins: The Fossil Record*, 2nd ed. Prospect Heights, IL: Waveland Press, Inc. 1998. All rights reserved.)

the origin of masseter. A strengthening of the face appears above the molars, and the mandible has been reinforced along the entire horizontal ramus to handle stresses generated by chewing.

Rak (1983) analyzed the structural features of the face that participate in this functional complex. Among the major adaptations he noted are prominent reinforced anterior pillars descending on either side of the nasal cavity. A shorter face brings the anterior temporalis, masseter, and medial pterygoid muscles more directly in line with the molars, resulting in the typical "dished" lateral profile. The strongly developed supraorbital ridges reinforce the attachments and actions of the temporalis muscles.

What was the purpose of such a development? What did the robust australopithecines eat? The evidence from the dentition (see Chapter 14) is that they were primarily vegetarians, subsisting on fruit, tubers, and similar food; while the form of the jaws, intermediate between those of "typical" carnivores and herbivores, suggests omnivory (DuBrul 1977). The large size of the teeth permitted the grinding of tougher food, which is typical of a drier savanna habitat, and a longer life of the dentition under more rapid wear (Wolpoff 1975). The cost of the large molars is a dispersion of the bite force over a larger area so that pressure (measured in force per unit area) is reduced (Demes and Creel 1988). To maintain the pressure necessary for chewing, the muscle mass had to increase. For example, the area of the cheek teeth of *Australopithecus boisei* to be 756 mm^2 compared with 334 mm^2 for modern humans (McHenry 1984). To maintain the same bite pressure, the australopithecine would need muscles with a cross-sectional area 2.26 times that of our muscles.

THE PROBLEM OF SKELETAL AND DENTAL REDUCTION

The short stout limb bones and thick cranial vault of the Neanderthals have contributed to an indelible impression of the archetypal "caveman" as strong but brutish. Although a century of discoveries has produced a more favorable image of our ancestor's cultural state, the robustness of the bones remains. However, rather than marvel at their strength, anthropologists have instead wondered why our skeleton, in contrast, is so light.

It was not only the Neanderthal skeleton that showed this robusticity but extinct species of *Homo* in other continents as well. The last 100,000 years of human evolution has witnessed a marked reduction in the strength of the hominin skeleton. This trend affected the entire skeleton—the size of the teeth, the length of the jaws, the robusticity of the limbs, and the height of individuals. It determined the final proportions of modern humans, including the relatively flat face and the chin. The cause of this transformation, whether adaptive or not, remains a topic of speculation and debate, although cultural changes are most commonly cited.

Between the earlier and later phases of the Upper Paleolithic culture in Europe, where the skeletal and archaeological records are most complete, skeletal reduction coincides roughly with trends of increasing complexity of tools, with tool kits becoming more specialized and sophisticated. If culture and changes in behavior made life less physical, anatomical changes might follow. At least part of the anatomical change would be expected to follow from normal developmental plasticity responding to less demanding activity patterns (Ruff et al. 1993; Trinkaus et al. 1994, 2000). Leach (2003) considers this alteration of the skeleton in an "artificial" cultural context to be domestication, similar to the changes humans have induced in dogs, horses, and other animals and plants. Interestingly, less skeletal change is observed at the Middle/Upper Paleolithic boundary (about 35,000 years ago in Europe) when far greater cultural changes occur (Trinkaus 1997; Holliday 2002; Holliday and Falsetti 1995). Tooth size in European fossils shows a sharp decline in size beginning about 35,000 years ago and continuing into the agricultural period. To further confuse the issue, the contrasts in body form may be exaggerated by cold adaptations of a stockier and more robust body for Neanderthals evolving in glacial Europe versus modern *Homo sapiens*, which immigrated from Africa (Pearson 2000a, 2000b).

An intuitive model of evolution argues that unused or unnecessary characters in a species would in time be eliminated. Darwinian evolution has been able to give this idea a firmer theoretical ground with the argument that developing and maintaining non-adaptive characters would use energy and material resources so that unused anatomical parts were necessarily maladaptive. There would be a positive selective advantage in their elimination. Perhaps as lifeways became less demanding, robust bones and large teeth became an unnecessary burden on the body (Brace 1967, 1971, 1992; Brace et al. 1991; Frayer 1981; Greene 1970). However, applying this very logical argument to real biological cases has been difficult because advantages and disadvantages of stronger or lighter bones are very hard to assess.

Brace offered a different mechanism for a similar model that he called the *probable mutation effect* (Brace 1964; also see Wolpoff 1969). According to his model, disuse of a character results not so much in directional selection against the character as in a loss of stabilizing selection to maintain it. Under normal conditions, most mutations are deleterious and selected against. A new mutation pertaining to an unused character, however, would not have such a negative impact. Brace suggests that because random mutations are more often disruptive to normal function, their effects will probably be the elimination or reduction of a given character. For example, if the need for larger dental occlusion area were suddenly decreased, a random mutation would have a higher probability of causing a reduction of tooth size than an increase. When this reduction proceeds to the point where teeth have the minimal necessary area, there will again be stabilizing selection against further reduction. The probable mutation hypothesis makes very simplistic assumptions about the nature of genetic controls of body form and ignores the considerable developmental interaction among body parts. Thus, it is not at all clear that a functional disruption of genes for trait size (if such exist) will result in smaller size.

The disuse and probable mutation models have much in common, including the fact that they are difficult to test. The transitions between cultural traditions are not sudden, as often implied, and the co-occurrence with skeletal changes is not perfect, especially outside of Europe. Alternative hypotheses posit adaptive reasons for directional change. For example, Brues (1966) argues that facial reduction lightened the head and increased its mobility and reaction time. Smith (1982) interprets dental reduction as an adaptation for a different diet. Others cite factors such as changing climate and/or different nutritional levels, particularly in the context of increasing population density (Jacobs 1985; Formicola and Gianneccini 1999; Macciarelli and Bondioli 1986). Calcagno (1989) argues for a "selective compromise effect" that recognizes the importance of several variables, including diet, nutritional demands, and disease, determining tooth size through a pattern of selection that may be unique in any given population.

Although there is no solution that is beyond dispute, only very fine studies of a reasonably complete fossil record are likely to tell us whether or not skeletal and dental reduction coincided as a single transformation event. One interpretation is that there is a systematic component that indicates bodywide effect (Churchill 1996). However, that does not account for all of the observed change in robusticity, and the upper limb, in particular, appears to be responding differently. This implies multiple factors are determining skeletal robusticity. Tooth size is generally regarded to obey strict genetic determination, unlike the rest of the skeleton. Probably a combination of both systematic and specific explanations and both genetic and nongenetic factors will be needed to understand these trends.

5

TEETH

Teeth, the hardest of all body structures, are made from the same mineral as bone. Because they must withstand the large forces that chewing generates day after day for the lifetime of the individual, they must be harder and more durable than regular bone. Nonetheless, the attrition of the teeth is one of the most conspicuous signs of the irreversible ageing of the body.

Teeth also have a special place in mammalian paleontology. They preserve well in the fossil record and the molars have complex shapes that are unique to each species. Some of these minute variations are functional; some are not. Like fingerprints, molar teeth permit ready identification of skulls and fossils. The dentition also reveals aspects of an animal's lifestyle. Shapes of the teeth and patterns of wear help us to reconstruct ancient diets, whereas other aspects enable us to discern rates of growth and maturation, episodes of disease or nutritional stress, and even mating patterns.

TOOTH DEVELOPMENT

The human tooth contains two types of mineralized tissue: dentin and enamel (Figure 5.1). Both of these, like bone, are connective tissues with a mineral matrix of hydroxyapatite. Unlike bone cells, the cells that lay down the mineral in teeth are excluded from the tissue. Thus, mature teeth are mostly nonliving material.

Embryologically, a tooth develops from two layers of tissue in the gums of the mouth. An outer layer of ameloblasts secretes columns of a nearly pure crystalline matrix of hydroxyapatite. The cells retreat from the mineral as it is laid down and they are lost when the tooth erupts (Figure 5.2). This is the enamel, the coating on the crown of the tooth. The inner layer of cells, the odontoblasts, secrete their mineral imme-

diately deep to the enamel. This becomes the dentin layer. Again the cells remain outside the dentin but underneath the tooth. Because enamel is nearly pure mineral, it is extremely hard and sustains compressive forces well. However, the crystalline structure is also brittle. Dentin has a significant component of collagen fibers interlacing the matrix. These fibers soften the dentin slightly and permit it to erode much more quickly under the wear of chewing, but they also give it a greater resistance to fracturing. Because enamel and dentin lack internal cells and the vascular

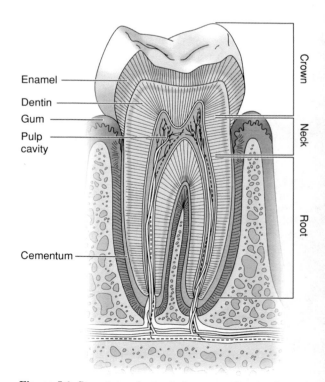

Figure 5.1 Structure of a typical mammalian tooth.

66

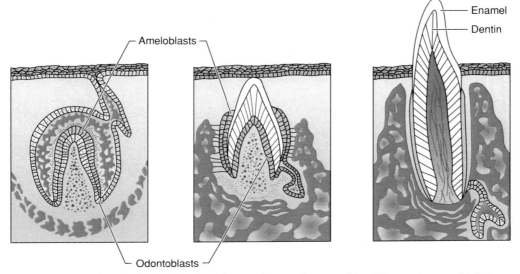

Figure 5.2 Development of tooth tissues. Dentin and enamel are each secreted by different tissues, which then retreat and are not part of the tooth itself. Odontoblasts produce dentin and persist within the pulp cavity of the tooth. They may continue to produce dentin after the tooth has erupted. Ameloblasts, which lay down enamel, are lost at eruption and can no longer contribute to tooth formation.

passages necessary to nourish those cells, they are both more dense and harder than bone.

When the tooth matures and erupts through the gums, the ameloblasts are lost. After eruption, no additional enamel can be added to the tooth. Wear is therefore irreversible and structural damage cannot be repaired. Odontoblasts survive and are able to add to the dentin on the internal surface of the roots of the tooth, reducing the size of the pulp cavity. However, the odontoblasts are largely unable to repair dentin already laid down. The longevity of the tooth is therefore dependent on the initial thickness of the enamel and the rate of wear.

The pulp cavity contains blood vessels that nourish living tissue there. It also contains nerve fibers bearing pressure and pain receptors. The pressure receptors give critical feedback during mastication to inform the brain when to relax pressure on the food.

In humans, teeth are held in their sockets, or alveoli, with cement. Cement has a mineral content very similar to that of bone. In conjunction with a periodontal ligament, cement anchors the tooth in place, although a small amount of movement is tolerated.

EVOLUTION OF VERTEBRATE TEETH

ORIGINS OF TEETH

Teeth closely resemble the dermal bone of the earliest vertebrates and the dermal scales of modern fish and some tetrapods. Both teeth and early dermal bone, as best exemplified by the denticles covering shark skin, are built of successive layers of enamel, dentin, and bone. One of the earliest vertebrate groups, the conodonts, lacked bony jaws and dermal bone but possessed denticles. Some agnathan (jawless) fish also possessed teeth. Although we have assumed in the past that teeth evolved from remnants of dermal armor on the jaws, the sequence of development of these various structures is now in doubt (Smith and Coates 2000).

Two important subsequent adaptations fixed the denticles in place on the jaw and provided for their replacement. Attaching the denticles in place permitted them to function as teeth, in gripping and tearing prey. Replacing the teeth lost during feeding ensured lifelong function for the animal. In most lower vertebrates, teeth are attached to the edge of the jaw rather than in a socket. Although this leaves them more vulnerable to being broken off, new teeth are continually appearing to replace them. The jaws of many lower vertebrates thus resemble serrated blades with the serrations irregular and variable in size. Teeth in opposing jaws will not necessarily mesh well or fit closely together. Such dentition is best adapted for grasping prey rather than for chewing it. Predators such as sharks, crocodiles, and snakes swallow their food whole or in large chunks, relying on the lower digestive tract to break it down the food over a long period of time.

MAMMALIAN DENTAL ADAPTATIONS

The mammalian adaptive complex requires rapid processing of larger quantities of food to support a higher metabolism. Mammals have innovative dental adaptations for extensive oral preparation of their food. In general, their teeth are thecodont, diphyodont, and heterodont and occlude, or interlock, in a complex pattern. *Thecodont* teeth are those that are anchored to the jaws in sockets. Such an arrangement strengthens the teeth against loss.

Diphyodont teeth erupt in two waves—deciduous and permanent. Deciduous teeth ("milk teeth") represent a smaller set of teeth suitable for the small size of an immature mouth, while permanent teeth are larger for adult mouths. The mammalian pattern allows for no replacement of permanent teeth that have been lost. The advantage of diphyodonty is that teeth are stabilized in their size and position in the mouth. Precise occlusion is possible and chewing is more efficient. Disadvantages are equally apparent— the teeth must be stronger and more resistant to wear if they are to last a lifetime.

Heterodonty describes teeth that have different shapes and functions in different parts of the jaw. There are four classes of mammalian teeth, but variable numbers of each of them in the mouths of different species. These types are, from front to back, incisors, canines, premolars, and molars (Figure 5.3). The anterior dentition consists of incisors and canine teeth. Incisors generally have simple blade-like crowns. Positioned as they are at the front of the mouth and farthest away from the masticatory muscles, incisors

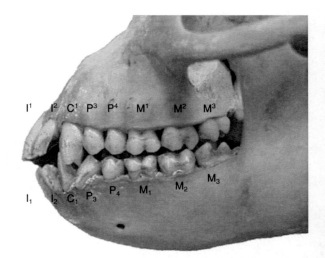

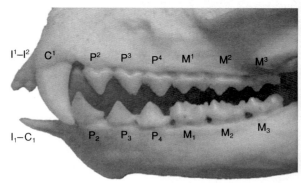

Figure 5.3 Types of mammalian teeth. Mammal teeth are differentiated into four types: incisors, canines, premolars, and molars. These different types are illustrated on two primates, *Lemur*, above, and *Macaca*, below. The lower anterior teeth on *Lemur* are specialized to form a tooth comb.

are less likely to be used in powerful chewing. They are more useful in procuring food for more extensive processing in the back of the mouth. Canine teeth are usually the longest, projecting beyond the rest of the tooth row. Often the back side is honed into a sharp blade. These teeth may be used by predators to penetrate and grasp prey or by fighting animals to slash an opponent. The posterior or cheek teeth are the premolars and molars. Together, they perform the finer processing of food. Premolars lie between the canine and molar teeth and are often intermediate in size, form, and function. Molars are the largest teeth in area and the most complex in shape, performing slicing and grinding action on food. It is the precise occlusion of opposing molars that makes mammalian chewing so effective.

THE TRIBOSPHENIC MOLAR

Modern mammalian molar patterns evolved from a basic form called the tribosphenic molar (Crompton and Hiiemäe 1969). This primitive molar had three cusps, or vertical projections that meshed with and sheared against a complimentary pattern in the opposite jaw (Figure 5.4). In the upper molar, the three primary cusps form a triangle with its base aligned with the cheek, or buccal margin. In the lower molar, the cusps form a triangle with the base along the tongue, or lingual margin. These triangles interlock tightly in a zig-zag pattern so that upper and lower teeth alternate with one another. The high, steep sides of the triangles interact like the blades of scissors to shear food. The addition of basins to oppose the cusps of the occluding teeth permits crushing of food as well as shearing. Complexity of the tribosphenic molar increases considerably with the addition of extra cusps, ridges, and basins. For example, the upper molar acquired a fourth cusp, the hypocone, in many lineages of

mammals that makes the tooth more quadrangular and increases its efficiency in crushing food against the lower cusps (Hunter and Jernvall 1995). It is these finer details, whose individual functional significance is slight or unknown, that make the molar teeth of each species unique.

The tribosphenic pattern is the basic design of primitive mammalian and primitive primate teeth. It is adaptive for eating insects and other foods that need much shearing action. With the evolution of larger body size, anthropoid primates and many other mammalian groups changed to more vegetarian or omnivorous diets, and their dentition adapted accordingly. Bulkier fruit, leaves, and poorer quality plant materials require much more crushing and grinding. For occluding teeth to glide across one another, it is better to have a molar tooth row that is much more uniform in height. In practice, this simplified molar structure. Ape and human molars, which have a fairly generalized pattern, have a ring of four or five low cusps (Figure 5.5).

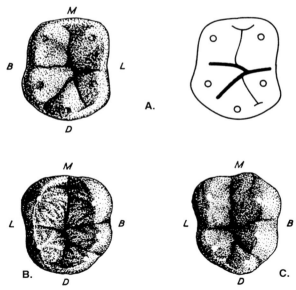

Figure 5.5 Cusps of the hominoid molar. The arrangement of the five principle cusps in the hominoid lower molar is derived from the tribosphenic pattern. *A*, Pattern as seen in an early Miocene hominoid, *Dryopithecus. B*, Chimpanzee. *C*, Modern human. (From M.H. Day. 1977. *Guide to Fossil Man*, 3rd ed. Copyright 1977 by University of Chicago Press. Reproduced with permission of University of Chicago Press.)

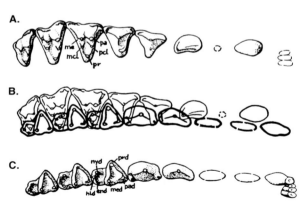

Figure 5.4 The tribosphenic molar pattern. The primitive mammalian tooth series shows a derived tribosphenic molar pattern. The tall shearing cusps form interlocking triangles, while the protocones of the upper molars crush food in the hypoconid basins of the lower molars. Molars of modern mammals are derived from this pattern. (From A.S. Romer and T.S. Parsons. 1986. *The Vertebrate Body*, 6th ed. Copyright by Saunders College Publishing. Reprinted with permission from Thomson Learning.)

HOMININ TEETH

Hominin teeth are distinctive for their adaptations to omnivory and a long life span (Table 5.1). Although early hominins showed specializations for processing large amounts of tough food, modern people have relatively small and simple teeth. Nonetheless, our dental formula—a shorthand for the number and classes of teeth in each quadrant of the mouth—is the same as that of other higher Old World primates.

DENTAL FORMULAE

The dental formula is a shorthand notation describing the types and numbers of teeth that a species possesses. All Old World monkeys and apes, including humans, have the following dental formula for permanent dentition:

$$\frac{2.1.2.3}{2.1.2.3}$$

Table 5.1 Characteristics of Human Dentition

Vertebrate characteristics
  Teeth composed of enamel and dentin layers
Mammalian characteristics
  Teeth embedded in sockets in the jaws (thecodont)
  Two generations of teeth (diphyodont)
  Teeth specialized as incisors, canines, premolars, and molars
    (heterodont)
  Molar teeth have a complex pattern of occlusion
  Molars show a derived tribosphenic pattern with multiple cusps
Hominoid characteristics
  Dental formula 2/2, 1/1, 2/2, 3/3
  Sexual dimorphism expressed in size of teeth, especially in
    canines
  Keyhole pattern of enamel prisms
  P_3 sectorial, hones the upper canine
Hominin characteristics
  Enamel layer thick
  Modified keyhole pattern of enamel prisms
  Anterior dentition reduced in size
  Canines no longer project beyond the other teeth
  Canines wear at the tip
  All premolars bicuspid
  Molars greatly enlarged in early hominids, reduced and
    simplified in later ones
  Dental arcade parabolic in shape

This means that the upper and lower jaws (above and below the line) each have two incisors, one canine, two premolars, and three molars on each side of the mouth. The total number of permanent teeth is 32. Other primates and other mammals often have different formulae.

Individual teeth are identified by their homologies with the primitive mammalian condition. Some mammals have three pairs of incisors, but no primate has more than two. Primate incisors are believed to be homologous with the first two ancestral teeth. Hence, the upper incisors are designated I^1 and I^2 and the lower ones, I_1 and I_2. There is never more than one canine in modern mammals, C^1. The primitive number of premolars is four. The anterior two have been lost by the catarrhine ancestor; thus human premolars are designated P^3 and P^4. Three is the primitive number of molars, and they are designated M^1, M^2, and M^3.

The complete human set of 32 permanent teeth follow a set of 20 deciduous teeth, with a formula of

$$\frac{2.1.0.2}{2.1.0.2}$$

Deciduous teeth are smaller in size to accommodate the smaller mouth of a child. As the child grows, the deciduous teeth are gradually replaced by the set of permanent teeth erupting in a predictable sequence. The teeth present in a human skull may be used to estimate age at death. The sequence and timing of eruption in other species, including the great apes, differs from the human pattern. These differences have been the focus of an extended debate concerning the rate of growth in early hominins (see Chapter 21).

ANTERIOR DENTITION

The incisors of living apes tend to be broad and spatulate, particularly the upper medial incisors (I^1). The four incisors in each jaw form a continuous cutting blade. The upper incisors sit slightly apart from the upper canines, leaving a gap, or diastema, for the lower canine (Figure 5.6). The canines themselves are long and sharply pointed in primates. Among anthropoids, the back edge of the upper canine hones against the lower premolar (P_3). This action constantly renews a sharp posterior border on the tooth.

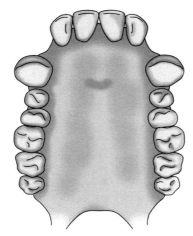

A. *Gorilla gorilla*

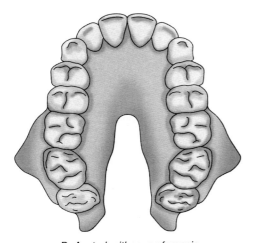

B. *Australopithecus afarensis*

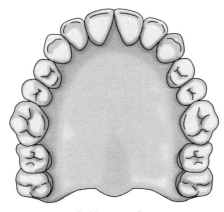

C. *Homo sapiens*

Figure 5.6 Upper dentition of gorilla, *Australopithecus*, and humans compared. The shape of an ape dental arcade differs from that of a human in length and compactness of the teeth. The gorilla molars lie in long parallel lines. There is a gap (diastema) in front of the large ape canines, while human anterior teeth are reduced and squeezed together. The teeth of *Australopithecus* show the same pattern of anterior reduction and crowding as in modern humans, but the cheek teeth are greatly enlarged.

The canines show pronounced sexual dimorphism, being conspicuously larger in males than in females. These may be formidable weapons against predators or rivals and are probably maintained through sexual selection. Even a casual observer of troops of baboons or macaques in a zoo will see the males expose their canines in threatening gestures. Precisely because these are such dangerous weapons, they are rarely used in combat. Threatening displays are usually enough to deter a leopard or a rival animal.

Hominins have markedly reduced the anterior jaws and the teeth they bear. The incisors have the same spatulate shape but are much smaller in size and are more closely crowded. They are also sit more vertically in the jaws. The canines are reduced nearly to the size of the incisors and do not project beyond the rest of the tooth row. The diastema, or open space, between upper canines and incisors is lost. The canines have further lost their honing occlusion and participate in the continuous cutting edge of the incisors. In hominins, canines wear from the tip instead of from the back. Because the small human canine has a disproportionately long root, it is clear that our ancestors had a larger tooth, more typical of apes. Furthermore, the human canine bears a pointed tip when it erupts, although it quickly wears to a more spatulate shape.

Darwin (1871) noted the importance of long canines in defense and concluded that their reduction in hominins followed the adoption of tools as weapons, relieving teeth of that function. A related interpretation is that canine reduction followed changes in social organization that selected against aggressiveness (Holloway 1967, 1968), although observation of human behavior might not see us as any less belligerent. A third possibility relates canine reduction to changes in chewing mechanics—a small canine is less likely to interfere with lateral chewing strokes.

POSTERIOR DENTITION

The premolars and molars of apes lie in long parallel rows. The lower third premolar, which hones the canine, has a special sectorial shape. It bears a single low cusp and is elongated anteroposteriorly. Other premolars are bicuspid; that is, they have two cusps of subequal size, one lingual and one buccal. The upper molars generally bear four cusps and the lower, five. In each jaw, these cusps are arranged in two parallel rows, continuous with the cusps of the premolars.

Hominin posterior dentition is again crowded. Teeth are in close contact with one another. The sectorial form of P_3 is lost and the tooth is bicuspid. The molar cusps are blunt and tend to wear quickly to a flat surface. In modern humans, the premolars and molars are packed tightly together, reduced in size and simplified. Frequently, the lingual cusp on the premolars is reduced in size. The molars are shortened and rounded in shape, increasing their breadth to compensate for the occlusal area lost by reducing length. The molars also tend to reduce the size and number of cusps.

Early hominins possessed the dental characteristics of modern humans but greatly expanded the size and role of the posterior dentition. Although incisors and canines were absolutely reduced in size relative to those of apes, the premolars and molars reached massive proportions in robust australopithecines (Lockwood et al. 2000; McHenry 1984; Robinson 1956; Suwa et al. 1994; Turner and Wood 1993). Individual molar teeth of *A. boisei* were as large as one centimeter square. Later australopithecines apparently experienced selection for a greater grinding surface that incorporated not only all five of the cheek teeth but also the anterior teeth. All teeth in the jaws may show significant wear from the occlusal surface toward the roots.

THE DENTAL ARCADE AND THE CHEWING CYCLE

The differences between ape and hominin dental arcades are immediately obvious in a comparison of intact jaws. An ape dental arcade is roughly "U" shaped or three-sided. The molar rows are parallel, and the canines define corners of the jaw. In contrast, the hominin dental arcade is parabolic. The molar rows diverge posteriorly and the anterior jaw is smoothly rounded, while the entire jaw is shortened from front to back relative to its width. Along with the elimination of projecting canines, this form facilitates the lateral movements of the mandible needed for grinding food. Chewing movements in animals with large canines are mostly constrained to vertical displacement of the jaw, but mastication in humans occurs in three dimensions. The lower jaw describes a circular motion during a single chewing cycle. As the mandible elevates on the chewing side of the mouth, it also moves medially. Contact between the teeth proceeds from the outside of the upper molars to the lingual side and from the back of the tooth row to the front.

Wear patterns on human teeth confirm this. Molars wear in a flat surface that inclines inferiorly from lingual to buccal. The changing direction of mandibular movement in later phases of occlusion causes the more anterior teeth to contact one another in a shifting plane. Thus the planes of wear become more horizontal anteriorly. This pattern is referred to as the *helicoidal plane of occlusion*.

HOMININ ENAMEL ADAPTATIONS

The structure of the enamel layer is crucial to determining the function and longevity of a tooth. Changes in the arrangement of crystals, thickness of the enamel, and size and shape of the cusps reflect adaptations to different dietary demands. We assume that the teeth in a given species are adapted to ensure that they continue to function for the anticipated lifetime of its members. There are three principal ways enamel can be destroyed: fracturing and chipping, abrasion from mastication, and chemical erosion. The chemical attack results commonly in human mouths from acids produced by bacteria and from acidic beverages.

When enamel is worn through, the softer dentin is exposed. Because the dentin wears more quickly, the tooth will maintain a rough edge along the enamel–dentin boundary (Figure 5.7). This edge gives the tooth a greater ability to process food beyond that of a smoothly worn but intact layer of enamel. The number and size of the cusps add to this effect. However, total loss of the enamel cap results in a more rapid and complete erosion of the entire crown of the tooth.

Enamel consists of apatite crystals arranged in prisms. These prisms lie perpendicular to the dentin

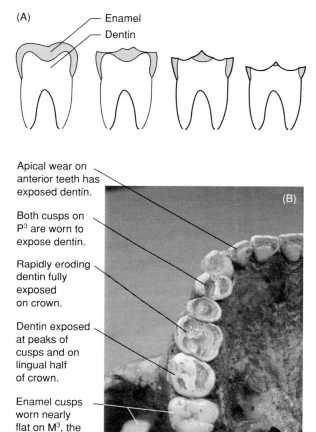

(A) Enamel
Dentin

Apical wear on anterior teeth has exposed dentin.

Both cusps on P^3 are worn to expose dentin.

Rapidly eroding dentin fully exposed on crown.

Dentin exposed at peaks of cusps and on lingual half of crown.

Enamel cusps worn nearly flat on M^3, the most recently erupted tooth.

(B)

Figure 5.7 Pattern of hominin dental wear. *A,* As a tooth wears, erosion of the enamel exposes the dentin. The softer dentin erodes more quickly, leaving a ridge at the edge of the enamel that is more effective for shearing food. *B,* Human tooth row showing various stages of wear.

surface and are packed closely together. Their orientation is ideal for receiving the compressive stresses of normal biting. The shapes of the prisms vary in different groups of species (Gantt 1982, 1983; Hillson 1986, 1996). Modern and extinct apes use a pattern in which prisms have a "keyhole" shape, with each prism possessing a head and a tail (Figure 5.8). Humans have a variant of this shape in which the interprismatic partitions are less complete, although

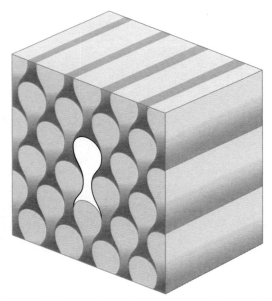

Figure 5.8 Prism design of human enamel. Enamel prisms are long structures running across the tooth, which lie side by side. In cross section, the enamel prisms of hominins and other higher primates show a "keyhole" pattern in which the interlocking prisms tend to prevent cracks from growing.

horses have hypsodont teeth, which grow continuously for the lifetime of the animal. Enamel may be also secreted continuously below the gumline (but it may not, of course, be added to the occlusal surface of the tooth). A third strategy is increasing occlusal surface area, as exemplified by the australopithecines. A fourth is to increase the number of teeth.

Hominins have thicker enamel than do any other living primates. Thickness increases the longevity of the tooth by withstanding greater wear. In the course of human evolution, we may understand this initially as an adaptation for a coarser diet that accelerated the rate of wear. In later stages, long-lasting enamel becomes significant in keeping up with the increased human life span. The appearance of thick enamel in several branches of the hominoid lineage has led to controversy over whether it represents a primitive trait or evolutionary convergence. Chemical erosion of the enamel is a consequence of the postagricultural diet high in sugars and starches and is not of great concern for other species. Caries, or dental cavities, are rare before the domestication of grains. They represent one of many health problems that appear with agriculture and that may be used to identify the transition from gathering to agriculture in the archeological record.

there is great variation in the way it has been described in the literature (Maas and Dumont 1999), reflecting differences between teeth and differences in the angle of view. When tooth enamel fractures, it tends to break along the partitions between the prisms. Fractures in human teeth must cross the prismatic material itself and thus encounter greater resistance. This form may add strength to the enamel.

Mammals have displayed several different strategies for increasing the longevity of enamel (Janis and Fortelius 1988). The primitive mammalian diet, consisting of insects and other high-quality foods, was relatively undemanding on the enamel. Later specializations to herbivory incorporated increasing amounts of abrasive fiber and grit in the diet and accelerated the wear. One approach, just described, is to increase durability by making the enamel harder and more fracture resistant. Some nonprimate species such as

DENTAL ADAPTATIONS OF THE NEANDERTHALS

Although reduction of jaw length and anterior dentition appears to be a trend throughout the history of the hominins, the Neanderthals appear to have temporarily reversed that pattern. The "classic" Neanderthals from Europe show an increased facial length that is caused by enlarged anterior teeth (Figure 5.9). The lower face is also enlarged to withstand greater stresses generated at the incisors. Examination of the teeth themselves shows heavy wear (at the tip of the crown) that is unusual for modern humans (Figure 5.10). What were the Neanderthals doing with their teeth?

Modern hunter-gatherers also show unusual incisor wear, but generally not in the same pattern or to the same extent. A likely cause of such wear is using the mouth as a vise in anchoring tools and other cultural materials. The best modern analogies

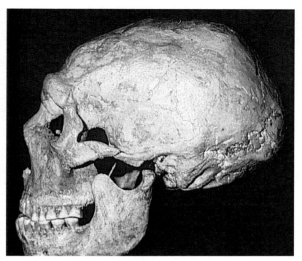

Figure 5.9 Neanderthals are characterized by an enlarged lower facial structure. Photograph of Shanidar I specimen. (Copyright Erik Trinkaus. Reproduced with permission.)

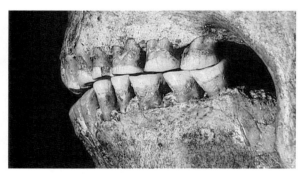

Figure 5.10 The incisors of some Neanderthal skulls, such as this one from Shanidar, show an unusual wear at the tips of the teeth. (Copyright Erik Trinkaus. Reproduced with permission.)

to the Neanderthals, both in climatic habitat and tooth wear, are traditional Inuit of North America. In a cold climate where appropriate clothing makes the difference between survival and death, Inuit spend hours chewing animal skins to break down its natural fibers and soften into wearable leather. Aspects of Inuit facial structure and pathology reflect

such behaviors (Hylander 1977; Merbs 1983; Spencer and Demes 1993), although they do not show the prognathism of the Neanderthals. The Neanderthals probably had habitual uses for their teeth in a range of cultural behavior.

MODERN TRENDS AND VARIATIONS IN HUMAN DENTITION

The modern Western diet is less physically demanding than that of prehistoric and earlier historical times. Consequences include reduced tooth wear and less stimulation for the growth of the jaws. Because tooth wear reshapes teeth according to chewing patterns, it has been argued that unworn teeth may be less effective and contribute to such problems as temporomandibular joint (TMJ) disorders (Kaifu et al. 2003).

The primary change in human dentation over the past 100,000 years has been a reduction in jaw and tooth size (see Chapter 4). Because tooth size is determined before eruption, diet will have little effect on that parameter. However, reduction in jaw length may reflect both genetic factors and the stresses exerted during mastication. A shortened jaw contributes to a high frequency of complications for the eruption of the third molar. In modern Western society, dentists are frequently called upon to extract crowded or improperly erupting "wisdom teeth." In a significant number of people, the tooth fails to develop altogether. As many as 30% of some populations are missing one or more of them.

Cultural patterns of tooth use can result in distinctive wear. Hunter-gatherers frequently show more occlusal wear on the incisors due to diet and to the use of those teeth as tools (Hinton 1981), as suggested by the extreme example of the Neanderthals. The prevalence of overjet in our society, in which the upper incisors lie anterior to the lowers is a recent development over the past two centuries. It may relate to changes in culinary practices and the use of a knife and fork at the dinner table (Brace 1986).

The dentition of modern populations provides extensive ground for research on human variation and adaptation (e.g., Cadien 1972; Turner 1989). Asian and Native American populations exhibit a relatively high frequency of "shovel-shaped" incisors in which the vertical margins are thickened. This may represent a reinforcement of the teeth, an adaptation met in other populations by lengthening the roots. Upper and lower molar teeth vary independently in the number of cusps. These variations have the effect of increasing or decreasing crown size and may be responsive to trends in jaw length. Simplification of the crown pattern may also resist caries formation by removing spaces in which bacteria may reside (Van Reenen 1966).

6

THE SPINE

The axial skeleton consists primarily of the spine, the ribs, and the skull (Figure 6.1). Although it maintains a similar pattern of segmentation throughout vertebrate evolution, the axial skeleton has undergone profound functional changes that have necessitated accompanying alterations in its structure and in the design of the entire body (Table 6.1). The human spine must meet the two conflicting requirements of supporting body weight against gravity and providing substantial mobility to the trunk. In addition, it anchors muscles and protects the spinal cord. Because of this, the structure of our spine is a compromise that

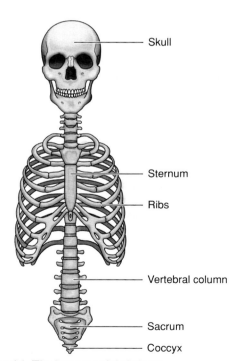

Figure 6.1 The human axial skeleton.

Labels: Skull, Sternum, Ribs, Vertebral column, Sacrum, Coccyx

Table 6.1 Characteristics of the Axial Skeleton

Vertebrate characteristics
  Bony elements alternating with intervertebral discs at each segment, supporting axial musculature
Tetrapod characteristics
  Single pair of ribs with each segment of the trunk
Mammalian characteristics
  Differentiation of regions of the spine
  Elimination of ribs from cervical and lumbar region
  Standardization of number of cervical vertebrae at 7, of postcervical vertebrae at 22
  Double articulation between occipital bone and atlas
  Odontoid process on axis
  Incorporation of sagittal flexion of lumbar spine in locomotion
  Fusion of sternum into a single bone
Hominoid characteristics
  Expansion of sacrum by incorporation of more segments
  Reduction of tail to coccyx
  Flat, broad ribcage
  Thoracic vertebral bodies placed well into thoracic cavity
Hominin characteristics
  Lumbar lordosis
  Sacral angle
  Relative increase in size of vertebral bodies
  Barrel-shaped, rather than conical ribcage

occasionally fails to meet the demands we place on it (e.g., Putz and Müller-Gerbl 1996).

STAGES IN THE EVOLUTION OF VERTEBRATE BODY DESIGN

The vertebrate body displays an array of strategies using different mechanical designs for body support and locomotion. Each is a response to the mechanical demands of body size and gravity in and out of water. The chordate ancestors of the vertebrates possessed notochords. The notochord is a flexible rod containing fluid that runs the length of the body. This rod provides stiffness and elasticity, permitting muscles to anchor to it. Side-to-side flexion of the body and tail permit swimming and burrowing. The notochord is present in the embryos of all chordates, including vertebrates, and may be observed in the adults of the chordate *Amphioxus* and in lampreys and hagfish.

FISH

The notochord quickly reaches mechanical limits as body size and activity increase. The compressive forces produced by the muscles of a swimming fish require a stronger material. Thus the notochord is replaced by a series of bony or cartilaginous vertebrae. Vertebrae develop initially as partial rings alongside the notochord and then expand inwardly to pinch it off. Bony arches also surround the spinal cord and fuse to the bodies of the vertebrae. The series of loosely linked elements provides mobility by summing the slight movement permitted at each successive joint.

Fins evolved early in vertebrate history to increase maneuverability in the water. Forward propulsion comes from a lateral sweeping of a strong tail. As the musculature grew in bulk and strength, the need for strong attachments increased. Adjacent segments of muscle, myomeres, joined together by attaching to the same tendinous partition, the intermuscular septum. Septa were strengthened by the addition of bony elements projecting laterally from the vertebrae. These became vertebral spines and ribs (Figure 6.2).

EARLY TETRAPODS

The skeleton of early land vertebrates faced new challenges of supporting the body against gravity. Among fish, the skeleton of the trunk serves to anchor muscles and facilitate lateral sweeping movements of the tail. In tetrapods, the ribs in particular became much more important for strengthening the body wall. The ribcage supports the body weight of an animal lying on the ground and prevents the collapse of the lungs. To stabilize the ribs, tetrapods have evolved a sternum, a series of bony elements to which ribs are anchored. Relative to the spine and sternum, the ribs are able to move, changing the volume of the body cavity during respiration.

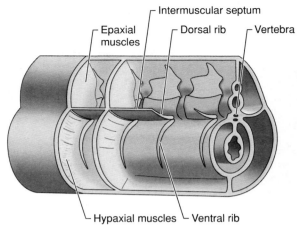

Figure 6.2 Axial skeleton and musculature. The axial skeleton, including the vertebrae and ribs, evolved as adaptations to support the trunk muscles used in swimming.

The paired appendages, having evolved into limbs in the water, strengthened and served to elevate the trunk from the ground and to provide some of the locomotor propulsion. However, locomotor design and the motor program in the brain remained essentially fishlike. As seen in most salamanders and lizards, walking involves fishlike lateral sweeping of the trunk and tail (Figure 6.3). The limbs are short, and flexion of the trunk serves to increase their stride length. The limbs also splay laterally, a design that is perhaps more efficient for swimming, when water supports much of body mass. This body design is adaptive for large tetrapods that spend much time in shallow water, as did the earliest amphibians and do modern crocodilians. However, this limb posture is energetically less efficient for supporting the trunk on land; and such animals prefer not to stand for very long at a time. Salamanders and most lizards are small and the energetic consequences of this design are less significant, although the same does not hold true for all reptiles.

MAMMALS

Mammals overcame the limitations of the early tetrapod design by shifting the limbs underneath the trunk to a mechanically more advantageous posture. This

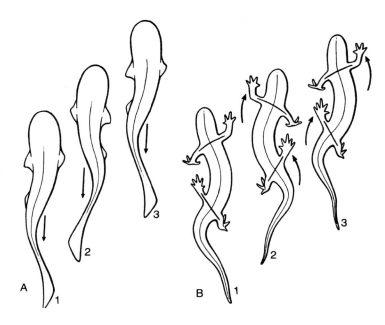

Figure 6.3 Spinal flexion in locomotion.
Lateral spinal flexion during locomotion
is similar in a fish and an amphibian. They
share a common musculature and neuromotor
pattern, even though propulsion is obtained by
different structures. (From A.S. Romer and
T.S. Parsons. 1986. *The Vertebrate Body*,
6th ed. Copyright by Saunders College
Publishing. Reprinted with permission
from Thomson Learning.)

lifted the trunk off the ground and suspended it from the spine, which was restructured accordingly. Many anatomists have modeled the spine of a quadrupedal mammal after a suspension bridge, receiving support from each of the limb girdles. However, Slijper (1946) demonstrated that the mechanics are better described by a strung bow, with the spine arched in tension maintained by the abdominal muscles (Figure 6.4). The spine is curved into a dorsal convexity. Any downward pull of the organs is translated into compression

Span 1,000 Feet, Width 81 Feet, Rise of Arch 220 Feet, Maximum Depth of Truss 140 Feet.
THE LARGEST ARCH BRIDGE IN THE WORLD. CROSSING EAST RIVER AT HELL GATE.

Figure 6.4 Architecture of the mammalian spine. *A,* Structurally, the spine of a mammalian quadruped has been compared with an architectural arch, such as the Hell Gate suspension bridge, New York. Weight is suspended in the middle but ultimately borne by the supports on either end. (From *Scientific American*, June, 1907.) *(continued)*

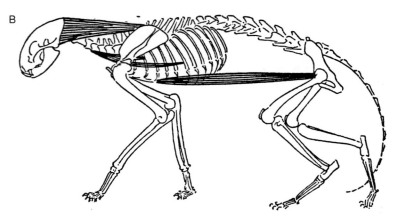

B

Figure 6.4 (*cont.*) *B*, The spinal skeleton arches upward, while tension is maintained by the tone of rectus abdominis and other muscles. The weight of the viscera is suspended from the spine and supported by the limbs. (From Everhard Johannes Slijper, 1946. Comparative biologic-anatomical investigations on the vertebral column and spinal musculature of mammals. *Verhandelingen der Koninklijke Nederlandsche Akademie van Wetenschappen, Afd. Natuurkunde*. Tweedie Sectie 42[5]:1–128. Figure 8. Reproduced by permission of the Koninklijke Nederlandsche Akademie van Wetenschappen.)

along the axis of the vertebral column, rather than into bending stresses. Elevating the head results in a concavity in the neck region.

Most significantly, this new arrangement permitted active flexion of the spine in a sagittal plane.

As the limbs also move in a sagittal plane, spinal flexion greatly increased stride length and speed in a full gallop (Figure 6.5). The inefficient lateral bending of the body used by lower vertebrates was discarded. To increase mobility of the spine, ribs were eliminated from all but the thoracic region and the other regions —cervical, lumbar, sacral, and caudal—became more specialized.

The assumption of upright posture and bipedal walking is a characteristic of primates to varying degrees. Humans are the only species to make this an habitual posture. It is achieved largely by introducing a second concavity into the lumbar spine (lumbar lordosis) and fully extending the hip joint— an anatomical transformation reenacted every time a toddler learns to walk. Otherwise the human spine maintains the characteristic design of other mammals. The erect posture places new and different stresses on the spine. Although the human vertebral column clearly is able to these demands for most individuals, our evolutionary history has not endowed us with the best of all possible designs.

STRUCTURE OF THE HUMAN VERTEBRAL COLUMN

Mammalian vertebrae are distinguished as cervical, thoracic, lumbar, sacral, and caudal, based on both form and function of those elements. Most simply, the thoracic vertebrae bear ribs and the sacral vertebrae

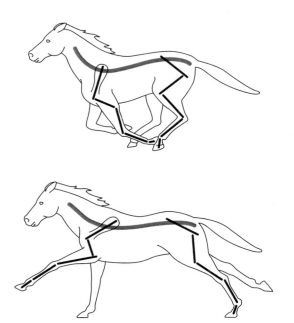

Figure 6.5 Spinal flexion in a mammal. The mammalian spine flexes and extends in a sagittal plane to increase stride length. This is particularly evident in a full gallop. Contrast with Figure 6.3.

articulate with the ilium. The cervical, lumbar, and caudal elements do neither and permit more spinal flexibility. There are 24 vertebrae in the human spinal column. The sacrum and coccyx (formed from the fusion of caudal vertebrae) represent about nine additional segments.

The numbers of segments within each of these regions varies among species, although the overall number is fairly standard among mammals. All mammals have seven cervical vertebrae and most species have about 22 segments caudal to the neck, not counting the tail. Comparative anatomists have been interested in the variations in distribution among these regions. What does such variation mean for functional adaptation? Increasing the number of segments is not the same as increasing the length of a region. Commonly, a region elongates or shortens simply by changing the lengths of individual vertebrae—the necks of a giraffe and of an owl each have seven cervical vertebrae. Adding segments, however, also adds joints, and more joints may mean more flexibility of the region. Variation in numbers may also occur within a species, including modern humans. Some individuals may have an extra set of ribs, producing one more thoracic and one less lumbar vertebrae (defined by the presence or absence of a rib). Others may fuse a lumbar vertebra into the sacrum. These variations tell us that the functional consequences of such shifts are subtle and that individual and species differences may not always reflect natural selection.

INTERVERTEBRAL JOINTS

Adjacent vertebrae connect with one another in two different types of joints. The vertebral arches contact one another at synovial joints with processes called *zygapophyses*. These facets do not align with the motion of the bones and limit rather than guide rotational movement. The bodies of the vertebrae make connections with one another via intervertebral discs. The discs consist of concentric rings of fibrocartilage, the annulus fibrosus, surrounding an inner gelatinous material, the nucleus pulposus (Figure 6.6). The rings of the annulus fibrosus limit both flexion and rotational movement. Fibers of successive rings have

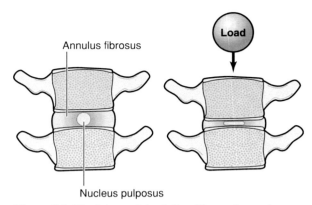

Figure 6.6 The intervertebral disc. The nucleus pulposus, supported by the anulus fibrosus, bears the weight of the upper body in the intervertebral disc.

alternating direction, slanting obliquely downward, first clockwise, then counterclockwise. Rotation of the spine in either direction will be quickly arrested as one set of fibers is pulled taught.

The nucleus pulposus is the remnant of the fetal notochord. Throughout postnatal life, it is gradually replaced by fibrocartilage. The fluid content of the nucleus is incompressible. As long as it is retained within the annulus, it will maintain the spacing between the vertebrae. With aging, its water content diminishes and the spacing, along with the height of the individual, is slightly reduced. The nucleus also serves to absorb shock, an important function every time we take a step.

These joints between vertebrae are supplemented by numerous ligaments, some connecting the spines and vertebral arches, and others running the length of the vertebral column connecting each segment to its neighbors. Thus, adjacent vertebrae are held firmly together and permitted very slight movement. The human spine is further reinforced by laterally positioned ligaments between the ribs and transverse processes (Jiang et al. 1995). These are absent in quadrupedal mammals. The summation of movement in all of the joints becomes significant and permits the flexibility with which we are familiar. The presence of elastic fibers in the ligaments ensures smooth movement and constant tension stabilizing the spine.

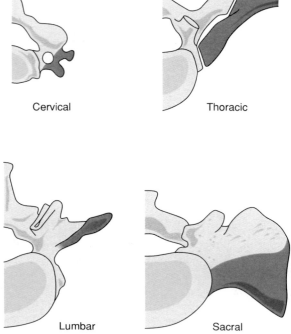

Cervical

Thoracic

Lumbar

Sacral

Figure 6.7 Fate of ribs in mammals. Ribs in the cervical and lumbar regions of the mammalian spine were reduced in size and fused into the vertebrae.

REGIONS OF THE HUMAN SPINE

There are seven cervical vertebrae in humans and all mammals. These have the special task of supporting and balancing the head. Mobility of the head independent of the body is very important for orienting the sensory organs, for feeding, and for using the teeth in fighting. Mammals have therefore eliminated the cervical ribs by greatly reducing their size and fusing them to the vertebrae, creating the transverse foramina where the two costovertebral articulations had been (Figure 6.7). Consequently, the neck is the most mobile—and most vulnerable—region of the spine. A large number of neck muscles gives it greater stability and maintains balance of the head.

The 12 thoracic vertebrae retain the primitive arrangement of articulated ribs. Each rib joins with the bodies of two adjacent vertebrae and with the transverse process of the more superior one. Because the ribs attach both posteriorly on the spine and anteriorly on the sternum, the thoracic spine is effectively immobilized by the ribcage. Only a small degree of extension and flexion from the natural convexity is permitted as respiratory movements. Thoracic vertebrae uniquely possess long, caudally directed dorsal spines, which support ligamentous attachments, and by the facets for articulations with the ribs.

The shape of the human ribcage differs markedly from that of living monkeys and apes (Figure 6.8). Quadrupedal mammals, including monkeys, have a deep and narrow thorax. This allows the limbs to be placed closer together and nearer the midline of the body. The scapula, or shoulder blade, lies on the lateral side of the ribcage and moves easily with the limb in a parasagittal plane. In contrast, the ape thorax is conical, much narrower at the top than at the bottom. It is also flattened from ventral to dorsal. This shape places the scapula on the back, rotating more in a coronal plane.

The human thorax more nearly resembles that of the great apes, except that it tapers less at the top. Thus it is commonly described as "barrel-shaped" rather than conical. In terms of respiratory function, the dynamic volume of the thorax is more important than its shape. Its shape is determined by the external constraints of limb girdle functions. The bottom of the ape ribcage is wide because the pelvis flares widely laterally. Because the abdominal wall is formed by muscles connecting the iliac crest to the ribcage, these two parts of the skeleton should be roughly congruent. The human pelvis is relatively deeper and more rounded; hence the lower ribs are also.

The five lumbar vertebrae have the largest centra since they bear a greater proportion of body weight than do the more superior segments. Their disproportionate size is a uniquely human adaptation to bipedalism. The lumbar spine also receives the greatest stress from our bipedal posture. The ribs of the lumbar region are very short and fused onto the vertebrae to contribute to the transverse processes (Figure 6.7). This arrangement permits considerable mobility in the region, particularly lateral flexion. At the same time, the transverse spines anchor muscles that move or help stabilize this region (Rose 1975; Shapiro 1993).

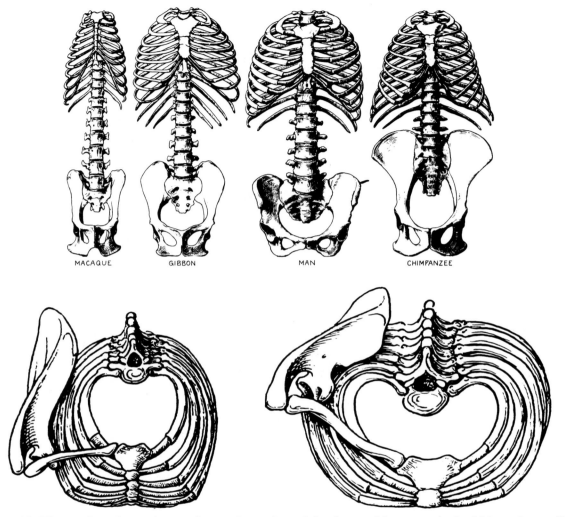

Figure 6.8 Ribcage shape in monkeys and apes. Comparison of the ribcage of humans and other higher primates. *Top*, frontal view indicating the more narrow monkey (macaque) condition and the conical shape of the chimpanzee ribcage. (From A.H. Schultz. 1950. The physical distinctions of man. *Proc Am Phil Soc* 94:438. Reprinted with permission from the American Philosophical Society.) *Bottom*, cranial view of the thorax of a monkey and human showing the difference in breadth. (From A.H. Schultz. 1957. *The Irish Journal of Medical Science*, 6th Series (380):341–356.)

Unlike the bodies of other vertebrae, those of the lumbar region are wedge-shaped, decreasing slightly in height from front to back. This wedging creates the lumbar lordosis and permits the sacrum to form an angle of about 60 degrees with the long axis of the spine (Figure 6.9). Unfortunately, the lordosis also means that a load carried down the spinal column will create bending stress in the lumbar region. If this load is artificially increased, as when a person carries a heavy load, there is a possibility of damage to the intervertebral disc.

The human sacrum is a single bone created by the fusion of five vertebrae. The segmentation is apparent in the presence of the five pairs of sacral spinal nerves

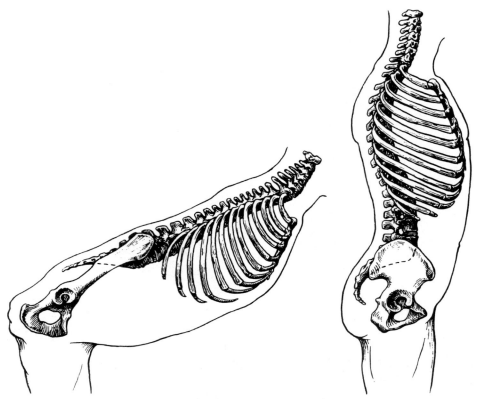

Figure 6.9 Spinal curvatures and sacral flexion. The curvatures within the spin contrast markedly in a gorilla (*left*) and a human (*right*). The lower spine in humans has a forward curvature, or lordosis, while the thorax has a normal backward curvature, a kyphosis. The human spine is sharply flexed just above the sacrum. (From A.H. Schultz. 1957. *The Irish Journal of Medical Science*, 6th Series (380):341–356.)

from the persisting intervertebral foramina. The number of segments that participate in the sacrum varies greatly in different species. Most mammals, including monkeys, have three sacral segments. This number increases among the apes as they assume a more erect posture. The large number (five) in humans reflects the exceptional demands of bipedalism, where the sacroiliac joint transmits the weight of the entire body above the pelvis. The hominin sacrum was also strengthened by enlarging it in all dimensions (Abitbol 1987a; Leutnegger 1977).

The coccyx consists of three to five rudimentary vertebrae fused together. These segments are homologous with the caudal (tail) vertebrae of other mammals, although they have lost identifying landmarks. While commonly described as an evolutionary vestige, the

coccyx has an important function. Ligaments and muscle fibers attaching to it play an important role in forming the pelvic diaphragm—the muscular floor of the pelvis. The upright posture of humans plus the large size of the human fetus relative to the pelvis place unique demands on the pelvic diaphragm in our species that require such reinforcing.

The absence of a tail is the most conspicuous feature distinguishing apes from monkeys. That characteristic extends to humans, as well. Of what use is the tail in primates? Why has the tail been lost in the hominoid lineage? Tails are generally very useful for arboreal animals. They help to balance the body on narrow supports and in leaping. The tail may brace a resting animal. Some South American monkeys and other mammals have prehensile tails capable of gripping a support so that they may hang by their tails while feeding among small terminal branches.

How, then, do we explain the loss of the tail in our lineage? We can only speculate, but evidently the adoption of a slow climbing pattern removed its adaptive value. The fossil record is not yet clear on the timing of this loss. Several remarkably complete skeletons of the early hominoid *Proconsul* (about 20 Mya) and *Nacholapithecus* (about 15 Mya) appear to lack caudal vertebrae, and the tapering of the sacrum and coccyx suggests that an external tail had already been lost (Nakatsukasa et al. 2004; Ward et al. 1991). Significantly, *Proconsul* represents a generalized skeletal anatomy and functional design similar to that which preceded the divergence of the modern monkey and ape patterns. Although it may have lost its tail, it had not committed to specialized climbing. Apparently the loss of the tail occurred early in the hominoid lineage, before the divergence of the gibbons.

CURVATURES OF THE SPINE

The most stable design of the bipedal spine might be a straight vertical column in the center of the trunk. However, given our evolutionary history and the need for movement, this was an impossibility. The spinal column lies dorsal to the visceral organs and thus off-center. As a partial compensation for this, the spine improves balance with regular curvatures in each region (Figure 6.9). These curves bring the spine closer to the center of the trunk.

The design of the human neck, in contrast to that of quadrupeds, reflects its function in supporting the weight of the head. The human cervical spine possesses a mild lordosis, or anterior curve, in a resting position. The middle intervertebral joints are slightly extended and the upper and lower joints are in an approximately neutral position. This places the head directly on top of a column of bone and loads the vertebrae in compression. All of these joints are able to contribute small amounts of both flexion and extension to the total range of neck motion. However, to prevent the heavy brain from placing too much strain on the neck, the atlantooccipital joint, between the skull and first vertebra, is restricted to only about 12 degrees of flexion/extension.

In contrast, quadrupeds have an "S"-shaped curvature to the neck to maintain continuity with the thoracic spine. In a resting position, the lower cervical joints are at the extreme range of extension and the upper spine is flexed. The strong but elastic fibers of ligamentum nuchae support the head in this position. Quadrupedal animals are unable to flex the lower cervical spine much beyond a neutral position and depend more on movement at the atlantooccipital joint to lower the head. The joint between the skull and C1 has a range of motion of approximately 90 to 105 degrees in quadrupeds, nearly eight times that in humans. Other cervical joints also have greater ranges of motion than in humans.

The thorax has a normal outward curvature (kyphosis) reflecting the need for a large thoracic volume. The rigidity of the ribcage significantly stabilizes this region of the spine. Nonetheless, the bodies of thoracic vertebrae sit well inside the ribs, closer to the center of the trunk. This placement reflects the hominoid broadening of the thorax (Larson 1998). In contrast, the deep narrow thorax of quadrupeds, such as monkeys, positions the vertebrae dorsally well away from the center of the trunk.

Most individuals also have a slight degree of lateral curvature in the thoracic region. If severe, this scoliosis may have clinical implications for posture, muscle and back pain, and even respiration. In the cervical or lumbar regions, such asymmetry would dangerously destabilize the spine. The reinforcement of the thorax provided by the ribs increases the structural tolerance for scoliosis, even though it may reach disfiguring proportions.

The lumbar spine is thrust forward, between the center of the thorax and the center of the pelvis, by a pronounced lordosis that is a unique characteristic of humans. The lower lumbar vertebrae and upper sacrum are aligned with the sacroiliac and hip joints so that the pelvis and these joints are placed in compression. Although the sharp lumbar lordosis is prone to injury, a straighter spine would be even more unstable where it joins the upper and lower parts of the spine. The curvature provides some shock absorption, as well. The lordosis is extended by the angle of the sacrum, which projects its lower part posteriorly. This brings the coccyx to the back of the perineum and minimizes interference with the birth canal.

FOSSIL EVIDENCE

Several partial or nearly complete spinal columns are now known for australopithecines and *Homo ergaster*. The lumbar lordosis is present in these species, as indicated by dorsal wedging of the vertebral bodies. A later increase in the size of human vertebral bodies better distributed weight on them. The bodies are also less dense, permitting them a degree of compression and shock absorption under loading. Bodies of the australopithecine lumbar vertebrae appear surprisingly small and not adapted for the habitual weight-bearing of normal human bipedality (Sanders 1998; Shapiro 1993). This is consistent with other unique features of australopithecine skeletal anatomy that suggest a unique, but poorly understood bipedal gait (see Chapter 9), possibly less well adapted for sustained travel. In contrast, the lumbar vertebrae of *H. ergaster* are closer to the modern condition in this regard. Like modern humans, australopithecines and *H. ergaster* apparently had five lumbar vertebrae (Haeusler et al. 2002), although earlier interpretations suggested six

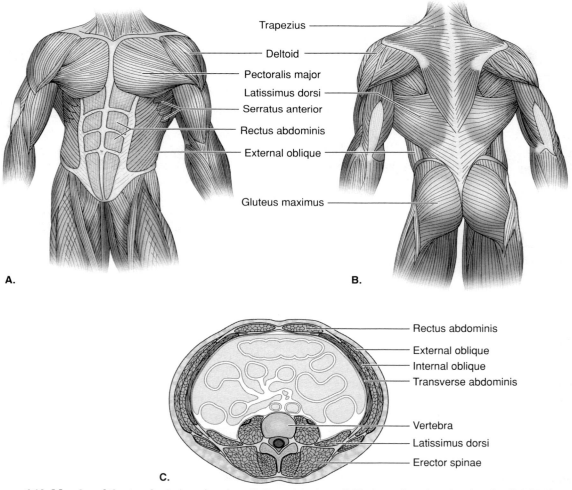

A.

Trapezius

Deltoid

Pectoralis major

Latissimus dorsi

Serratus anterior

Rectus abdominis

External oblique

Gluteus maximus

B.

Rectus abdominis

External oblique

Internal oblique

Transverse abdominis

Vertebra

Latissimus dorsi

Erector spinae

C.

Figure 6.10 Muscles of the trunk. *A*, Anterior view. *B*, Posterior view. *C*, Horizontal section showing the division between epaxial and hypaxial musculature (compare with Figure 6.2).

(Sanders 1998; Walker and Shipman 1996). The African great apes typically have 4 lumbar segments and 13 rib-bearing thoracic vertebrae.

MUSCULATURE OF THE TRUNK

The dorsal musculature of the trunk are the epaxial muscles (Figure 6.10), which act largely to extend the spine. The ventral muscles are hypaxial. They are include both trunk flexors and the musculature of the limbs.

The most important group of epaxial muscles is the erector spinae. Erector spinae consists of three paired columns of muscles fibers, ascending from the sacrum as high as the occipital bone. Because individual bundles within these columns arise and insert from each segment, the erector spinae should be considered a muscle group and not one or three discreet muscles. Acting together, the erector spinae muscles extend the spine. The muscles on one side laterally flex the spine in that direction. The more lateral columns act on the spine with a great lever arm and thus can flex the spine more easily. Deeper and smaller groups of muscles help to stabilize and rotate the intervertebral joints.

The abdominal wall is unprotected by bone and is maintained entirely by its muscular layers. The lateral abdominal wall consists three layers of muscle. Each layer orients its fibers in a different direction. The resulting muscular wall resembles a sheet of plywood, in which greater strength is obtained by alternating mutually perpendicular grains. Each of these muscles ends partly in a broad sheetlike tendon, or aponeurosis, which is anchored along a midline. The oblique direction of the fibers has also been interpreted as a "girdle," constricting the abdomen with normal tone (Figure 6.11).

The lateral muscles of the abdominal wall contribute to lateral flexion and especially rotation of the lumbar spine, while rectus abdominis is the most important flexor of this region. At least of equal importance is the ability of all of these muscles to create and regulate pressure in the abdomen (Aspden 1987, 1992; Morris et al. 1961). This pressure restores the diaphragm to its resting position during exhalation and provides the power of coughing and forced

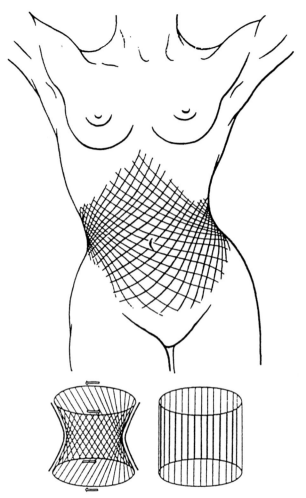

Figure 6.11 The "girdle" configuration of the fibers of the lateral abdominal muscles. Criss-crossing fibers narrow the waist of the abdominal wall where there is no skeletal support. This arrangement also maintains internal abdominal pressure. (From I.A. Kapandji. 1974. *The Physiology of the Joints. Vol. 3: The Trunk and the Vertebral Column.* 2nd ed. Copyright 1974 by Churchill Livingstone. Reproduced with permission from Elsevier Science.)

exhalation (e.g., blowing). Abdominal pressure also aids in defecation and birthing. The posterior pressure on the lumbar vertebrae supports the spine against collapse when it is heavily loaded.

7

THE UPPER LIMB

The forelimb is used prominently in locomotion by nearly all land-dwelling vertebrates. Almost 370 million years of evolutionary history have shaped our own upper limb since the first tetrapod walked on the bottom of a shallow pool. The pectoral fin or limb has enabled our ancestors to swim, to waddle, and to climb, and each adaptive phase has left its mark on its anatomical design, even though the assumption of an upright posture has changed its function dramatically. Freed from the need to support body weight, the human hand and the limb and brain behind it are retooled for manipulation and for all of the cultural potential that this implies (Table 7.1). Nonetheless, homologous of bones and muscles are readily identified among tetrapods (Figures 7.1 and 7.2).

EVOLUTION OF THE FORELIMB

THE ORIGIN OF TETRAPOD LIMBS

Paired appendages are a characteristic of vertebrates. They evolved as aids to steering and braking in the water, a function that can be observed among fish today. The fins of most modern bony fish consist primarily of rays—each a series of long stiff modified dermal scales—anchored into the musculoskeletal tissues along the side of the trunk. However, among the lobe-finned fishes (Subclass Sarcopterygii), there is a significant component of bone and soft tissue within the proximal part of the fin. It was from one of these lineages that the tetrapods evolved in the Upper Devonian Period about 370 Mya. A close similarity may be observed in the skeletal structure in fossil sarcopterygians and the earliest amphibians (Figure 7.3). From the combined evidence of fossils and developmental genetics, many, but not all, of the homologies between the bones of the sarcopterygian and tetrapod limb can be identified (e.g., Laurin et al. 2000).

Table 7.1 Characteristics of the Upper Limb

Vertebrate characteristics
> Two pairs of appendages

Tetrapod characteristics
> One proximal skeletal element (humerus)
> Two intermediate skeletal elements (radius and ulna)
> Two to three rows of carpal bones
> Division of musculature into extensor and flexor groups

Mammalian characteristics
> Reduction of the pectoral girdle to a scapula and a clavicle with increased mobility
> Loading of the limb bones in a vertical column
> Suspension of the thorax from the limbs in a muscular sling

Hominoid characteristics
> Placement of the scapula dorsally on a broad, flat thorax
> Orientation of the shoulder joint laterally
> Expansion of the scapula craniocaudally, especially in the infraspinous fossa
> Increased development of elevating muscles relative to propulsive muscles
> Humeral head larger, more rounded
> Humeral head rotated to face more medially
> Insertion of pectoralis minor on coracoid process
> Development of ridge separating distal facets on humerus
> Shortening of olecranon process, elevation of coronoid process of ulna
> Loss of ulnar articulation with carpal bones
> Reduction to eight carpal bones (African apes and humans)
> Sellar joint at the base of the first metacarpal

Human characteristics
> Reduced pisiform
> Well-developed opposition of the thumb to the pads of the other finger tips
> Lengthening of the thumb relative to other digits
> Power grip and precision grip
> Greater differentiation of thumb muscles, including flexor pollicus longus
> First metacarpal most robust

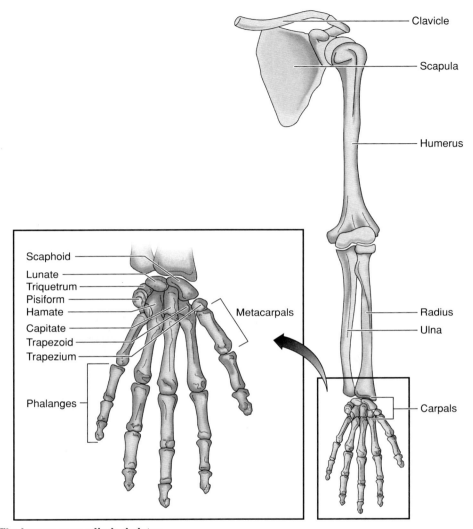

Figure 7.1 The human upper limb skeleton.

The evolutionary transformation of lobed fin to limb involved both a simplification and strengthening of the skeleton. The number of segments aligned longitudinally in the limb, and therefore the number of joints, was reduced (Coates et al. 2002). Multiple joints provide a flexibility desirable for swimming but are not very useful on land. Similarly, there was a reduction in the number of rays, or digits.

These early amphibians were moderately large animals with short limbs. It is likely that they inhabited shallow water environments and used their limbs to support the body and push off from the bottom rather than to walk on dry land (Eaton 1960; Edwards 1989). The forelimb evolved supportive capabilities earlier than the hind limb, perhaps to prop the head out of the water for breathing (Lebedev 1997; Thompson 1991). Such a scenario is consistent with a gradual evolutionary transition, in which incremental strengthening of the limbs would be adaptive. Reducing the number of skeletal segments helped to make the limb stiffer.

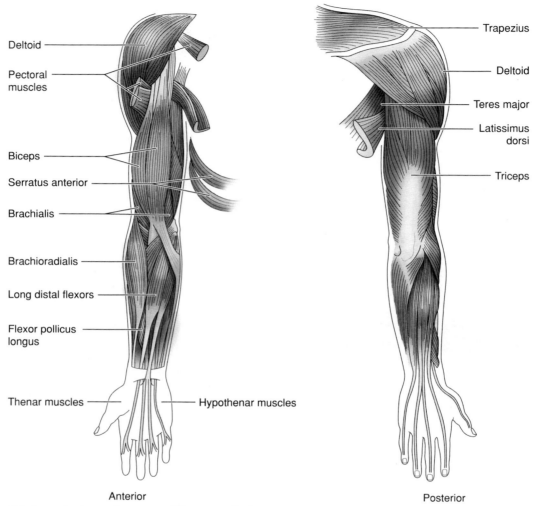

Figure 7.2 Important muscle groups of the upper limb.

The basic plan of the fore and hind limbs is the same in all tetrapods. The proximal limb segments (arm, or brachium, and thigh, or femur) each have one bone that articulates with the limb girdle. The intermediate segments (forearm, or antebrachium, and leg, or crus) each have two. These are surmounted by rows of many smaller bones called podials (carpals and tarsals), meta-podials (metacarpals and metatarsals), and phalanges. Tetrapods also share a fundamental division of the musculature and innervation into the dorsal (elevator or extensor) and ventral (depressor or flexor) groups.

The five digits that are characteristic of modern land vertebrates (excepting the many lineages that have secondarily reduced that number) was a chance inheritance. Fish extend the skeleton of a fin with numerous dermal rays extending the reach of the fin while maintaining is stiffness. The number of skeletal elements supporting these may be quite numerous. As the use of the limb shifted from stroking the water to support on the bottom, the number of elements decreased to varying degrees. Early fossil amphibians have been found with as many as 10 digits per foot.

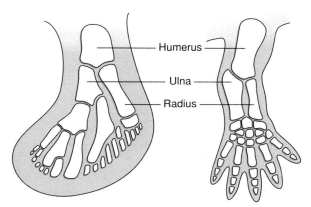

Figure 7.3 Skeletal elements in the fin of a fossil sarcopterygian fish (*left*) show clear homologies with those of the limbs in a primitive tetrapod (*right*).

The fact that only five-toed animals had surviving descendants, to our knowledge, is probably unrelated to the number of their digits (Gould 1991). Once that number was established, it was hard to change. In amniotes (reptiles, mammals, and birds) the developmental controls for the digits are intricately tied to other developmental processes (Galis et al. 2001); thus, a change in the number of digits would have extensive, and probably intolerable, consequences for different parts of the animal. Mammals that have fewer than five digits as adults start off with five in the embryo and eliminate one or more later. Amphibians, on the other hand, apparently have fewer constraints and display more variability in numbers.

Although increasing the strength of the bones and muscles of the limbs enabled early amphibians to spend more time walking on land, the body design for swimming limited the possibilities for locomotion. The limbs, like the pectoral and pelvic fins from which they evolved, were situated on the sides of the trunk and not underneath it. The locomotor pattern reflected in both the muscular design and the neural program was dominated by lateral sweeps of the tail. The shoulder girdle was rigidly anchored on the trunk, and the humerus was positioned mostly in a horizontal plane. As a result, movements of the shoulder contributed little to the stride length. Frogs and toads, of course, departed from this design with a radically different, but equally limiting, most of locomotion.

THE MAMMALIAN FORELIMB

The mammalian reorganization of the body greatly affected the design and function of the upper limb. Its bones were transformed from a lateral support to a column placed nearly underneath the trunk. The vertical alignment of bones and joints makes it possible to load them in compression (Figure 7.4). It also achieves far greater mobility, stride length, and power and speed of locomotion.

The redesign of the pectoral girdle was the key to increased limb mobility. In most of vertebrate history, the limb was firmly fixed on the lateral side of the trunk by an extensive skeletal bracing system. The placental mammalian lineage gradually reduced the girdle to

Figure 7.4 Limb posture in reptiles and mammals. The limbs of reptiles and amphibians (*left*) typically lie lateral to the trunk and the shoulder and hip joints move partly in a coronal plane. The mammalian limbs (*right*) were redesigned to be held in a supporting position under the trunk and moved in a parasagittal plane. (From *A Practical Guide to Vertebrate Mechanics* by Christopher McGowan. © Cambridge University Press 1999. Reproduced with permission of Cambridge University Press.)

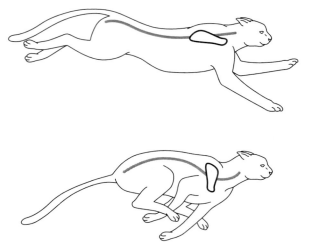

Figure 7.5 The movement of the scapula on the trunk may contribute to the stride length. Some animals adapted for great speed, such as the cheetah depicted here, lack a functioning clavicle. The scapula is free to glide along the trunk in a cranial or caudal direction. Such freedom increases the maximum length of stride.

the scapula and, usually, a clavicle. Other bones persist only as ossification centers that fuse to the body of the scapula. The scapula glides across the ribcage, separated from it by layers of muscle and fascia. Its cranial-caudal progression increases stride length (Figure 7.5). The clavicle is the only skeletal connection between the scapula and the thorax. When present, it spaces the shoulder joints. However, some cursorial mammals have functionally eliminated the clavicle in the interest of even greater mobility.

The reduction of bone led to a different strategy for supporting the trunk. A quadrupedal mammal suspends the ribcage between the two forelimbs. Although the clavicle may connect the scapula and trunk, actual support for the trunk is performed by muscles in tension. The pectoral group and serratus anterior in front suspend the trunk, while trapezius on the back holds the scapula in place (Figure 7.6). In this position, the trunk can glide fore and aft between the stationary limbs.

The rest of the limb may undergo various specializations to suit the size and activities of different groups of mammals (Figure 7.7). The bones may be elongated for speed, as in medium-sized ungulates,

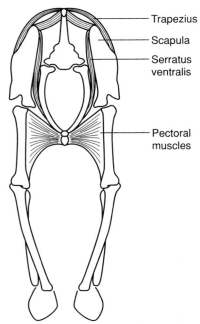

Trapezius

Scapula

Serratus ventralis

Pectoral muscles

Figure 7.6 The suspension of the trunk from the pectoral girdle. The trunk is suspended between the forelimbs in a muscular sling formed by the pectoral muscles and serratus ventralis (called serratus anterior in humans). Trapezius and other muscles hold the scapula close to the trunk.

or increased in diameter to support massive body weight, as for the elephant. Similarly, the hand, or manus, has been reshaped. A broad contact with the ground on numerous digits better distributes body weight. Mobility of the wrist accommodates uneven substrates and adds one more propulsive joint. However, speed is most compatible with reduced weight and narrow points of contact with one or two digits, as observed in ungulates.

HOMINOID ADAPTATIONS

THE PECTORAL GIRDLE

Among placental mammals, primates, including humans, are conservative in their retention and arrangement of bones. Most African monkeys have essentially adapted a terrestrial running pattern of locomotion to an arboreal setting, traveling along the tops of branches or leaping quadrupedally. In contrast,

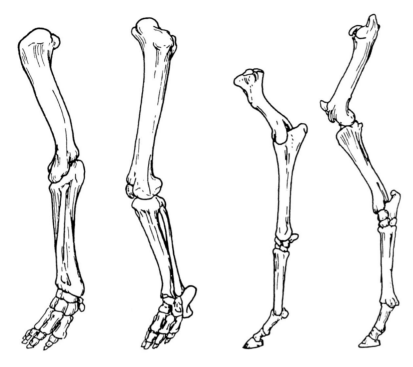

Figure 7.7 Limb specialization.
The limbs show specializations in different mammals for different demands of speed or weight. (*Left*) The elephant is built to carry a great load; thus, its limb bones are straight and thick and weight is distributed on five digits plus a pad of soft tissue. (*Right*) In contrast, the horse limbs reflect adaptations for speed: multiple joints with great ranges of flexion, lighter and thinner bones, and a single digit at the end. (From M.H. Wake, 1979. *Hyman's Comparative Anatomy*, 3rd ed. Copyright 1979 University of Chicago Press. Reproduced with permission from the University of Chicago Press.)

hominoids have redesigned the body to move vertically in the trees, reaching overhead to grasp branches and hoist themselves up by their hands. The adaptations that may be observed for climbers include a more upright (orthograde) trunk posture and greater mobility of the limb joints.

The scapula is a flat triangular plate of bone that lies against the ribcage. In running animals, the ribcage is compressed laterally and the scapula lies more or less vertically, in the plane of the weight-bearing bones of the limb. In the hominoids, the thorax has become broad and flat, and the scapula accordingly lies dorsal rather than lateral to the trunk (Figure 6.8). This orients the shoulder joint laterally, giving it a greater range of motion, especially superiorly and laterally. Neither the shapes of the bones nor the ligaments determine well-defined axes of motion. Instead, scapular motion is defined and limited by its relations and muscular attachments to the ribcage. The clavicle is retained for additional support and spacing of the shoulders.

The shape of the scapula reflects this change as well (Ashton and Oxnard 1964; Aiello and Dean 1990; Larson 1993). The scapula of a quadrupedal animal tends to be short from cranial to caudal and broad from vertebral border to the glenoid. This facilitates greater translation of the scapula as it slides craniocaudally. In contrast, the scapula of climbers is expanded caudally, reflecting the different development and attachments of its musculature. The glenoid fossa has a more cranial direction in climbers compared with the more lateral and ventral facing in quadrupeds. Such a scapula provides greater power to its rotators.

Quadrupeds derive a great deal of locomotor power from depression of the scapula toward the tail and extension of the shoulder. Power is provided especially by the pectoral muscles and latissimus dorsi. Hominoids place a greater emphasis on overhead positions of the limb, involving a powerful action of the muscles when the scapula is laterally rotated. The most important muscles for this are the trapezius, deltoid, and serratus anterior. This emphasis is

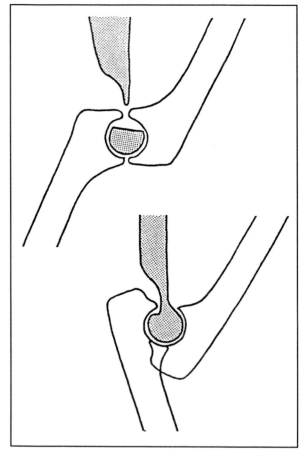

Figure 7.8 Elbow adaptation for climbing. Comparison of elbow structure and function in a climbing ape (*Gorilla*, above) and quadrupedal monkey (*Nasalis*, below). The distal humerus is shaded. The ulna is drawn in extreme flexion and in extreme extension. Note the different ranges of motion and the shortened olecranon process and higher coronoid process in the ape. (From M.D. Rose. Functional anatomy of the elbow and forearm in primates, in D.L. Gebo, ed., *Postcranial Adaptation in Non-human Primates.* © 1993 by Northern Illinois University Press. Used with permission of Northern Illinois University Press.)

reflected in the relative development of those muscles. The ratio of propulsive muscle mass (pectoral group, latissimus dorsi, and teres major) to the mass of limb elevating muscles (deltoid, trapezius, serratus anterior) shifts from about 2:1 in quadrupeds to nearly 1:1 in apes

(Ashton and Oxnard 1963; Aiello and Dean 1990). In hominoids, pectoralis minor has transferred its attachment from the humerus to the coracoid and assists in stabilizing the shoulder girdle. Human musculature resembles the hominoid pattern in distribution of mass, but the specific design of the trapezius and the lateral facing of the glenoid suggest an intermediate pattern, in which the limb is used more habitually with the shoulder lowered.

The humeral head is larger and more rounded in hominoids. This contributes to the increased range of motion at the shoulder. It is rotated relative to the humeral shaft to face more medial, rather than dorsal (or caudal) as in quadrupeds (Larson 1998). This represents the different delivery of forces across the shoulder joint, suspensory or laterally directed rather than weight-bearing.

THE ELBOW

The hominoid elbow also has several distinctive features that relate to climbing and suspensory locomotion (Aiello and Dean 1990). The elbow receives much greater torsional stresses in suspension than in the weight-bearing posture of a quadruped. Hominoids have developed a ridge on the distal humerus separating the trochlear and capitular facets that helps to stabilize the ulnar articulation. At the same time, the range of pronation and supination is increased for greater mobility in suspension (O'Connor 1979; Stern and Larson 2001). Accommodating this, the radial head is more rounded in apes and elliptical in monkeys.

The proximal ulna also shows specializations (Rose 1993). The olecranon process, which projects behind the joint, is long in quadrupeds to provide more leverage for triceps in propulsion. Suspensory behavior places more stress in extended positions and requires greater power in flexion. Consequently, in hominoids the olecranon is short, not extending proximally beyond the trochlear notch (Figure 7.8). The coronoid process rises higher above the ulnar shaft. Both of these processes are accommodated within fossae on the humerus so as not to interfere with a full range of motion. Brachialis is well developed and the insertion of biceps is placed farther away from the joint for greater power.

HANDS

Although human hands are superficially much like those of our primate relatives, they are the key to our material culture, which separates us with an immense gap from all other animals. The grasping hand, with adroit fingers and an opposable thumb, is part of our primate heritage, However, humans have perfected the grasp and motor control that permits us to manipulate tools.

Five digits is an ancestral trait for mammals. Many familiar species, such as horses and cows, have reduced the number of fingers and toes, but the more primitive ones have retained five. Moreover, some mammals, such as raccoons and squirrels, have the ability to hold and manipulate objects between their hands. Only in primates, however, do the thumb and big toe flex opposite the others and permit an effective single handed grasp. This ability is most likely an ancestral adaptation to arboreal foraging, where secure holds on both small supporting branches and dinner were essential. A key development was a flexible joint at the base of the metacarpal of the thumb. This is a sellar joint in humans, permitting two axes of motion. Folding the thumb across the palm while rotating it to face the tips of the other fingers is the act of opposition. Opposition is most effective in the human hand because of the greater length of the human thumb relative to those of the other fingers.

KNUCKLE-WALKING

Chimpanzees and gorillas have a common climbing heritage and continue to spend significant amounts of time in the trees but both species have had to find a compromise in the structure of the hand that permits them to walk quadrupedally on the ground. Both fold the hand into a unique posture that enables them to knuckle-walk.

Climbing requires long curved fingers that are capable of grasping overhead branches. However, in typical palmigrade (in which the palm is placed flat on the ground) or digitigrade walking (in which the weight is borne on the heads of the metacarpals and on the hyperextended fingers), the necessary hyperextension of long fingers places intolerable stresses on their joints. To prevent this, the fingers of the African apes are curled into the palm and weight is borne primarily on the heads of the proximal phalanges (Figure 7.9). African apes have several adaptations to support this unique weight-bearing pattern (Tuttle 1967; Richmond et al. 2001). The joint is stiffened by reducing its mobility. Wrist extension and ulnar and radial deviation are all limited by the shape of the distal ulna and radius and of the scaphoid, and weight-bearing is strengthened by expanding those articular surfaces. The heads of the metacarpals also have ridges preventing hyperextension of the digits. Even the skin over the knuckle is modified to bear protective knuckle pads.

Because the African apes appear to be as closely related to ourselves as they are to one another, it is

Figure 7.9 Knuckle-walking hand posture. The knuckle-walking gait of a chimpanzee places weight on the backs of the middle phalanges of the hand. (From J. Napier, 1980. *Hands,* rev. ed. Copyright 1993 by Princeton University Press. Reprinted by permission of Princeton University Press.)

necessary to ask whether knuckle-walking was also characteristic of our common ancestor. Humans do not display the features unquestionably identified as specializations for this behavior. Early australopithecines from East Africa do, however, show some similarities to the African apes in the radiocarpal joint that might be related to stabilization during knuckle-walking (Richmond and Strait 2000); however, South African australopithecines and *Homo* lack this design, and the functional interpretation of this feature has been challenged (Dainton 2001). Humans do share with the knuckle-walking apes a fusion of two carpal bones, the os centrale and the capitate (Corruccini and McHenry 2001), which Richmond et al. (2001) interpret as an adaptation to strengthen the weight-bearing hand. Other features of the carpal bones and joints are also consistent with the knuckle-walking pattern.

It is also not entirely clear whether the knuckle-walking pattern arose in a common ancestor of chimpanzees and gorillas that was not shared by humans or evolved in parallel in the two ape lineages. Differences in the developmental patterns of the wrist and hand between the chimpanzee and gorilla have been interpreted to indicate independent evolutionary pathways (Dainton and Macho 1999). Most recently, a dorsal ridge was identified on a distal metacarpal of *Kenyapithecus*, a 14-million-year-old African hominoid (McCrossin 2002). Although the relationship of *Kenyapithecus* to modern species is in doubt, the possible identification of yet a third knuckle-walking hominoid supports the idea that this mode of travel did evolve relatively early. Knuckle-walking behaviors, perhaps with only limited anatomical specialization for it, may well have characterized the common ancestor of humans and African apes.

HUMAN HANDS

ADAPTATIONS FOR MANIPULATION

The human hand is unique in its ability to manipulate tools. While tool use is widely documented in a number of animals, its extraordinary degree of development in human culture is made possible by adaptations of the hand as well as of the brain. Napier (1963) describes the two primary positions of the hand in tool manipulation, the power grip and the precision grip (Figure 7.10). The power grip is used to hold a hammer. The fingers and thumb wrap around the handle in opposite directions and grip it tightly against the palm. A tool held in this way becomes an extension of the forearm and can be used for hammering or other functions requiring power. The precision grip is used to handle a pencil. It involves the application of the thumb and one or more of the fingertips for finely controlled manipulation. The unique human ability to hyperextend the distal interphalangeal joint further permits close contact between the pads of the thumb and other fingers (Shrewsbury and Johnson 1983). These two fundamental hand positions have also been described as a clubbing grip and a throwing grip, respectively, with the suggestion of a more violent evolutionary context for the human lineage (Young 2003).

Because of the finger proportions in the hands of African apes, they are incapable of producing the human patterns of either precision or power grips (Napier 1993). Instead of a power grip, chimpanzees prefer to grasp objects in their curled fingers without engaging the palm. The human grip is enhanced by unique pads in the skin of the palm, by an increase in the robusticity of the fifth metacarpal, and by increased stabilization of the metacarpophalangeal joints (Marzke 1987, 1992). The relative shortness of the ape thumb interferes with opposition to the tips of the other fingers in a precision grip. Apes may instead grasp small objects between the tip of the thumb and the side of the index finger. This hold has inferior sensitivity and motor control. Young chimpanzees have been observed using a precision grip (Jones-Engel and Bard 1996). This does not equate human and chimpanzee manual dexterity, but it does suggest that finer behavioral distinctions between the categories of grip are necessary to differentiate species (Marzke and Wullstein 1996). Whether additional categories have distinctive anatomical correlates may be questioned (Susman 1998).

Human fingertips, especially that of the thumb, are marked by an expanded, spadelike development of the distal phalanx. Ape phalanges are narrow, but the human bones are broad and flattened. This shape better distributes pressure across the fingertip to reduce tissue damage and increase sensory discrimination.

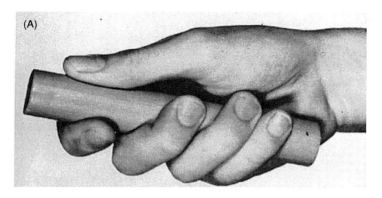

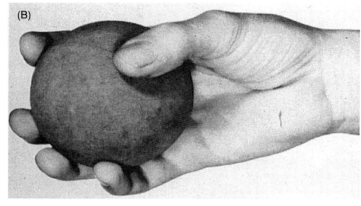

Figure 7.10 The human hand in power (A) and precision (B) grips. (From J. Napier, 1980. *Hands*, rev. ed. Copyright 1993 by Princeton University Press. Reprinted by permission of Princeton University Press.)

EVOLUTION OF THE THUMB

The opposable thumb, in the broad sense of the term, is a hallmark of the primates. While manual graspings of tree branches and food, for example, are basic survival functions for primates, there is still much variation in the form and function of the thumb. True opposition of the thumb, involving a sellar joint and conjoined rotation at its base, is present only in catarrhine primates (Napier 1961). In both monkeys and apes, the thumb is generally reduced in size and unable to contact the tips of the other fingers. This relative proportion is exaggerated in animals such as the colobus monkey and the orangutan that use elongated fingers in a hooklike grip to suspend the body in the trees. In these animals, a large thumb would interfere with rapid grasp and release and therefore has been greatly reduced. The human hand, freed of such concerns, has evolved a relatively longer

thumb in parallel with the relative enlargement of the first toe. These two evolutionary changes may be linked developmentally, as the digits of the hands and feet are influenced by the same genetic program (Webb and Fabiny 2001).

The musculature of the hand, and of the thumb, in particular, has some unique aspects in humans. One intrinsic group, the contrahentes, has been reduced to a small muscular slip, considered the deep head of flexor pollicus brevis (Day and Napier 1963). In many other primates, contrahentes inserts upon and adducts all of the digits, stabilizing the hand and facilitating a grasp in the center of the palm. Contrahentes may have been reduced in the African ape and human lineage in favor of a digital grasp. Other muscles have been enhanced. The thenar muscles (short muscles at the base of the thumb) and hypothenar muscles (at the base of the little finger) give more independent movement to the thumb and last digit. Both groups are important in manipulation (Marzke et al. 1998). The flexor pollicus longus appears as a tendon of the digital flexors in monkeys but becomes an important independent muscle only in humans (Hamrick et al. 1998). African apes are lacking both the long flexor and part of the short flexor to the thumb. These may be seen as developments for unique human uses of the digits.

The bones of the human thumb are increased in length and robusticity, more so than other metacarpals or phalanges. The head of the metacarpal is increased in size to handle the considerable forces than may be generated. These are certainly contributions to the human pattern of tool use. Several researchers have claimed that the uniquely human patterns of morphology may be used to infer tool-making in the fossil record.

FOSSIL TOOL-MAKERS?

In 1961, Louis Leakey discovered a partial hand skeleton at Olduvai Gorge and attributed it to *Homo habilis*. Because Olduvai at that time had yielded the oldest collection of stone tools, Leakey was eager to discover which hominin was responsible for making them. Can fossil anatomy identify tool-making ability?

Napier (1962) analyzed this particular fossil and concluded that the individual had had an unusually powerful grasp. Although hand proportions seemed human-like, the bones were more robust and the distal phalanges very broad. These observations suggest a more human-like than apelike use of the hand in grasping, but Napier thought that the trapezium did not indicate a precision grasp. Nonetheless, the identification of *H. habilis* as a tool-maker has been accepted with little question because tools and fossils attributable to early *Homo* appear at about the same time in the fossil record (about 2.4 Mya). It has been more parsimonious to consider our closer relatives to be toolmakers, but the question of whether *Australopithecus* also worked with stone remains open.

Since then, other sets of early hominin hand bones have been recovered. One collection from Hadar in Ethiopia belongs to *Australopithecus afarensis*. A second, from Sterkfontein in South Africa, belongs to *A. africanus*. A third from nearby Swartkrans has been attributed to *A. robustus*. The older specimens from Hadar, as well as the less complete Sterkfontein specimen, differ from those of modern humans in ways that are commonly interpreted as partly or wholly lacking the power and precision grip (Aiello and Dean 1992; Marzke 1986, 1992, 1997; Ricklan 1987; Shrewsbury and Sonek 1986; Susman 1994). However, these bones do show an insertion of flexor pollicus longus and enhanced wrist extension, both of which might indicate some selection for manual manipulation (Panger et al. 2002). The digits also appear to have had roughly modern human proportions in length, a requirement for power and precision grips (Alba et al. 2003).

The Swartkrans material better fits the human pattern (Susman 1988, 1994). In particular, these first metacarpals are closer to modern proportions and robusticity, including enlarged joint surfaces capable of transmitting greater forces. Since there are many more *A. robustus* specimens than *Homo* at the site, Susman also assigns both the thumb bones and toolmaking ability to the australopithecine. This assignment is not beyond dispute (e.g., Pickering 2001).

Is there a simple skeletal indicator for the type of manipulative ability necessary for tool-making? Marzke and Marzke (1987) proposed the styloid process of the third metacarpal. Susman (1984) proposed the robusticity of the thumb and its joint surfaces. However, as critics have pointed out, chimpanzees are not incapable of using stone tools in the wild (McGrew 1995; Hamrick and Inouye 1995; Ohman et al. 1995; Shrewsbury et al. 2003). Moreover, the morphology of the thumb in other primates, particularly gorillas, overlaps with that of humans in some of these characteristics. It appears that the invention of stone toolworking and the evolution of the tool-wielding hand must both be considered gradual processes.

8

THE LOWER LIMB

The most remarkable feature of the human lower limb is its commitment to bipedalism and upright posture. Although other groups of vertebrates also evolved bipedalism—most notably dinosaurs and birds—they did so independently of humans and in different ways. This chapter explores the anatomical characteristics and function of the human lower limb (Table 8.1; Figures 8.1 and 8.2).

Superficially, there are a great many similarities between the forelimbs and hindlimbs of tetrapods. The similarity of pattern extends to the skeleton, musculature, and other soft tissue systems and simplifies the learning of anatomy. The functional differences are

very pronounced, on the other hand. As the humerus became oriented to direct the elbow posteriorly, the femur directs the knee anteriorly. The forelimb is highly mobile and relatively unstable in mammals. Much of the stride length in running animals comes from forelimb movement, including the ability of the pectoral girdle to slide along the trunk. The extensors of the elbow and more distal joints provide much of the power. The hindlimb is more stable, with the pelvic girdle fixed to the spine. Thus it is better suited to bear weight. Power comes from action at the hip and the ankle. The mobility of the forelimb is consistent with its manipulative role in primates, just

Table 8.1 Characteristics of the Lower Limb

Vertebrate characteristics	Human characteristics
Two pairs of appendages	Obligate bipedalism
Tetrapod characteristics	Weight borne on gluteal fat and muscle in sitting position
Pelvic girdle anchored directly to spinal column	Sacrum wide and narrows inferiorly
Pelvis constructed from three paired bones (ilium, ischium, pubis) plus axial component	Sacroiliac joint closer to acetabulum
	Ilium shorter, broader, and deeper
One proximal skeletal element (femur)	Iliac pillar strengthens ilium for stresses of abduction
Two intermediate skeletal elements (tibia and fibula)	Ischium shortened
Two or three rows of tarsal bones	Birth canal enlarged
Five digits with metatarsals and phalanges	Gluteus maximus reoriented as an extensor of the hip
Division of musculature into extensor and flexor groups	Gluteus medius and minimus and tensor fasciae latae reoriented as abductors
Mammalian characteristics	Hip and knee capable of full extension
Expansion of sacral attachment of pelvis	Increased carrying angle of the femur so that knees and feet touch at midline
Femoral neck at a sharp angle to shaft	Femoral condyles elliptical
Limb joints operating in a single plane	Tarsals and other bones greater in relative volume with thinner cortex
Primate characteristics	Calcaneal tuberosity has medial and lateral inferior processes
Tendency for upright posture of trunk	Medial longitudinal arch present
Opposable hallux	Conversion mechanism of the midtarsal joint permitting both flexible and rigid states of the foot
Hominoid characteristics	Hallux robust and of similar length to other digits
Iliac blade flaring widely in coronal plane	Hallux permanently adducted parallel with other digits
Subtalar and transverse joints of foot mobile	Lateral toes shortened
Digits elongated; second toe longest	Metatarsophalangeal joints capable of significant hyperextension
Development of inferior process of heel tubercle	

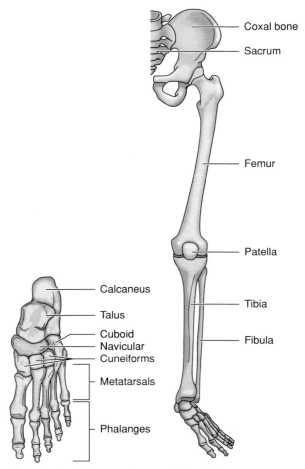

Figure 8.1 The human lower limb skeleton.

as the more stable hindlimb assumed a supportive function. In this way mammalian design invited the evolutionary possibility of bipedalism.

THE PELVIS

The pelvic girdle evolved by gradually increasing its strength to support muscle action. It first appeared as a simple pair of plates of bone supporting the caudal fins in fish. Contact between them at the ventral midline offered stabilization, but in order to support terrestrial locomotion, the bones needed to be made larger and stronger. Three elements, the ilium, ischium, and pubis, developed and expanded as long processes

to support various muscle groups (Figure 8.3). In mammals they are fused on each side of the body into a single coxal bone. The girdle is strengthened by increasing the length of the sacrum—the number of vertebrae in contact with the ilium. The limb joints are brought into the same sagittal plane by setting the femoral neck at a sharp angle to the shaft. They are thus able to work together to support body weight and push against the ground.

The human pelvis ("funnel") consists of two coxal bones plus the sacrum. The sacrum represents the bodies of five different vertebrae whose movement has been eliminated. The two coxal bones join on the midline in front at the pubic symphysis (Figure 8.4). The human pelvis is overall shorter, broader, and deeper than that of other primates and of mammals in general (Kummer 1975). Quadrupedal locomotion requires hip flexors attaching more cranially on the pelvis and powerful extensors more caudally. An elongated pelvis provides a greater lever arm and thus more power for those muscles. In a human, cranial-caudal length is less important, but locomotion depends on ventral-dorsal attachments of the muscles. For this reason, the ilium and ischium are both shortened. To gain some mechanical advantage, the coxal bone is not entirely aligned with the spine but is tilted significantly in a sagittal plane. The short, tilted ilium also brings the sacroiliac and acetabular joints closer together to reduce shear stresses between these joints.

ADAPTATIONS OF THE HIP JOINT FOR BIPEDALISM

The skeleton and musculature of the human hip and thigh show clear adaptations for bipedalism. A considerable proportion of the effort for balancing and shifting weight comes from the hip. Except for climbing behaviors, the most important hip movement occurs near full extension. In quadrupeds, full extension is impossible and gait movements center about a highly flexed position of the joint. Motion of the hip in abduction and adduction, although small in range, becomes far more crucial in a biped for balance than in a quadruped.

Hip stability. In the human hip, the fibers of the three ligaments of the joint spiral as they descend the

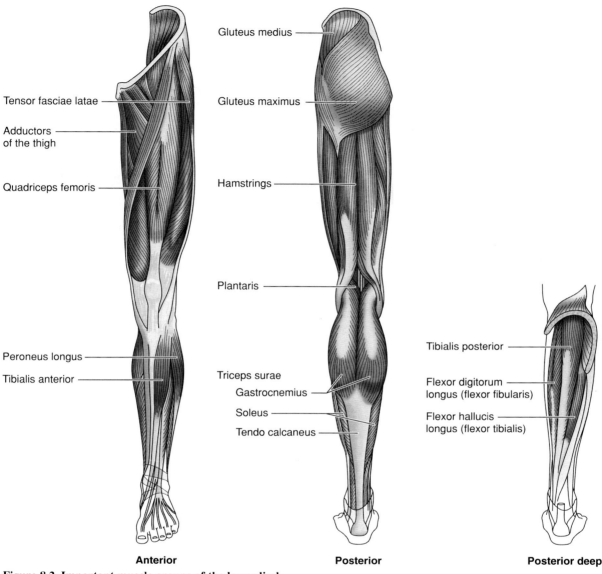

Gluteus medius

Tensor fasciae latae

Gluteus maximus

Adductors
of the thigh

Quadriceps femoris

Hamstrings

Plantaris

Peroneus longus

Tibialis anterior

Triceps surae

Tibialis posterior

Gastrocnemius

Flexor digitorum
longus (flexor fibularis)

Soleus

Flexor hallucis
longus (flexor tibialis)

Tendo calcaneus

Anterior   **Posterior**   **Posterior deep**

Figure 8.2 Important muscle groups of the lower limb.

neck of the femur in such a way that they are tightened as the hip extends (Figure 8.5). As they reach maximum tension, they check hyperextension of the hip. Simultaneously, the femoral head has been drawn most tightly into the acetabulum for optimal congruence. In this extremely stable position, the center of body weight lies slightly posterior to the acetabulum and the action of gravity inhibits any further movement of the joint. Such positions of joints, in which ligament tension, congruence, and stability are maximal and the joint is locked into position, are called close-packed positions. Close-packed positions play important roles for the lower limb, enabling a person to maintain a standing posture with minimal expenditure of muscular energy.

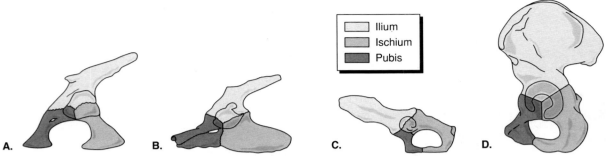

Figure 8.3 Mammalian coxal bone. The mammalian pelvic girdle was strengthened by the fusion of the bones and the expansion of the joint with the spinal column. *A*, Early amphibian. *B*, Synapsid reptile near the lineage of mammalian ancestry. *C*, Mammal (cat). *D*, Human.

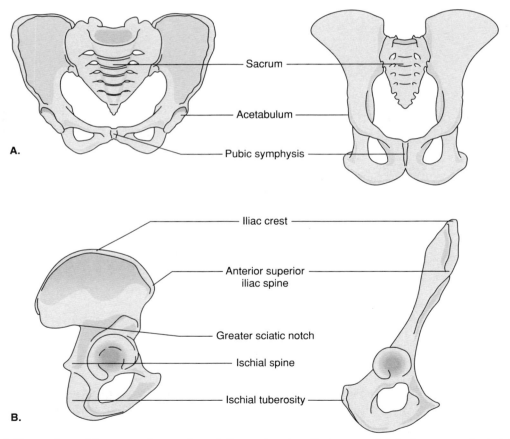

Figure 8.4 Comparison of human and ape pelves. *A*, In anterior view, the human pelvis is shorter superior to inferior and broader from ventral to dorsal. *B*, In lateral view, the differences in lengths of the ilium and ischium relate to the posture of the lower limb. A long ilium and ischium give the ape longer lever arms especially for powerful extension. Comparable levers on the reoriented human pelvis depend on its ventral-dorsal dimensions.

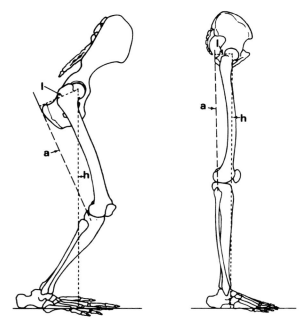

Figure 8.4 (*Cont'd*) In lateral view, the differences in lengths of the ilium and ischium relate to the posture of the lower limb. A long ilium and ischium give the ape longer lever arms especially for powerful extension. Comparable levers on the reoriented human pelvis depend on its ventral-dorsal dimensions.

Hip extension. The important extensors in humans are the hamstrings (semitendinosus, semimembranosus, and biceps femoris on the back of the thigh) and gluteus maximus. The three hamstring muscles are crucial extensors of the hip during normal gait. By propelling the trunk over the planted limb, they help maintain continuous and smooth movement. In addition, they cross the knee and flex it. The hamstrings arise from the ischial tuberosity at the base of the coxal bone so that the length of the ischium represents the lever arm for these muscles. The ischium is considerably shorter in humans than in the apes, and thus results in an increase in speed of hip extension at a cost of power.

Gluteus maximus may be the most powerful muscle of the human body, measured by its cross-sectional area (Figure 8.6). The muscle has an extensive origin from the back of the pelvis and underlies the fat on the buttock. Its insertion into the fascia lata is unique

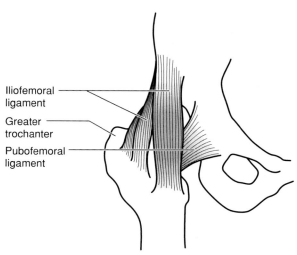

Figure 8.5 Ligaments at the hip. The ligaments of the hip joint, shown here in anterior view, are arranged to spiral as they pass from the pelvis to the femur. This causes them to tighten when the hip extends so that they pull the bones closer together and stabilize the joint.

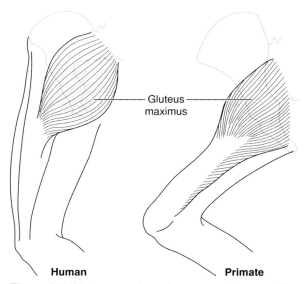

Figure 8.6 Gluteus maximus in ape and human. The gluteus maximus, shown in posterior view, has a lateral position relative to the hip joint in most primates and other quadrupeds, and functions as an abductor. In humans, it has shifted to a more posterior position to act as a hip extensor.

to humans. It is a powerful extensor but may not be consistently employed during normal walking. It is more important for extending the hip from a flexed position (as in bicycling or climbing stairs). In quadrupedal mammals, gluteus maximus lies on the lateral side of the pelvis and assists in the relatively less important function of abducting the hip. The human pelvis is broader and places gluteus maximus on the posterior side of the joint, where it can act as an extensor and also help to stabilize the trunk (Stern 1972). Consequently, maximus is proportionately a much larger and stronger muscle in humans than in other mammals.

Hip abduction. Abduction for quadrupedal mammals occurs occasionally to reposition the limb but rarely requires much power. The abductor muscles therefore are relatively small and arise close to the midline. Abduction movements in humans bear and balance body weight on a single limb at a time and therefore require power and control. Accordingly, the human iliac blade flares widely, shifting the abductors laterally away from the hip and closer to their insertion on the greater trochanter (Figure 8.7). The iliac blade as seen from the front serves as a lever arm for

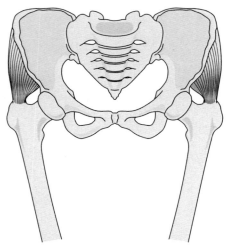

Figure 8.7 The lateral flaring of the ilium supports hip abduction. The external surface of the iliac blade provides a surface of attachment for gluteus medius and minimus in stabilizing. When the blade is extended laterally over their attachment on the greater trochanter of the femur, the muscles have more power for effective abduction of the joint.

these abductors. The length of the femoral neck and trochanter provide a mechanical advantage on the other side of the joint. The power produced by such an arrangement necessitates strengthening of the ilium. Thus, an iliac pillar is uniquely developed on humans in line with that stress.

The important abductors for humans are gluteus medius and minimus and tensor fasciae latae. These muscles arise from the blade of the ilium and thus takes advantage of both its lateral flare and anterior position. In quadrupeds, medius and minimus are more important as extensors, whereas tensor fasciae latae is a flexor of the thigh.[1] Only a bipedal animal needs to worry about maintaining balance on a single supporting limb. When one foot is raised from the ground, the body tends to sag toward the unsupported side. This collapse is equivalent to adduction of the supported hip. Action by the abductors is therefore critical in counteracting gravity and reestablishing equilibrium. They are activated not only as we change a static posture but also with each step of a walking gait.

THE THIGH

The femur is the sole bone of the thigh, articulating proximally at the hip joint and distally at the knee. From the head and neck at the hip joint, the shaft of the femur descends to the knee where two large condyles rest on the tibia. When the human femur is held upright so that the condyles both contact a flat surface, the shaft is seen to slant medially. The angle formed between the shaft and a vertical line is the carrying angle (Figure 8.8). The carrying angle compensates for the width of the pelvis and length of the femoral neck and aligns the acetabular joint vertically over the knee. The angle also permits the knees to approach one another in a comfortable and efficient stance.

The carrying angle is a distinctively human feature. The average angle in humans is about 11 degrees from the vertical. This compares with 4 to 6 degrees

---

[1]Although gluteus medius and minimus are generally described as the primary hip abductors (e.g., Greenlaw and Basmajian 1975), Gottschalk et al. (1989) argue on the basis of anatomical position that tensor fasciae latae has that duty while the gluteus muscles are more important as stabilizers of the joint.

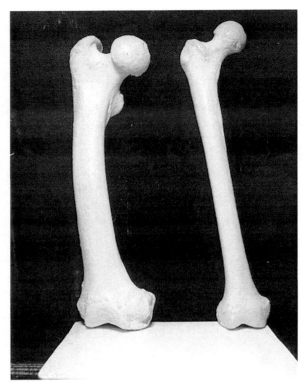

Figure 8.8 The carrying angle of the femur. The human femur (*right*) is carried at a distinct angle from the vertical compared with that of quadruped (gorilla shown on *left*). This brings the weight-bearing head of the femur over the lateral condyle of human knee but over the medial edge of the ape knee.

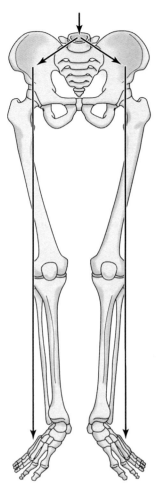

Figure 8.9 Transmission of body weight through the lower limb. The carrying angle of the human femur brings the knees and feet closer to the center of the body and contributes to a column of bone bearing weight from the hip to the ground.

in the apes and approximately 0 degrees in monkeys. Consequently, nonhuman primates standing or walking bipedally place their knees and feet conspicuously apart and are less stable when lifting one foot from the ground. The presence of a significant carrying angle among fossil hominins is a simple indicator of an adapted bipedal stance.

Despite the fact that the acetabulum lies significantly off the midline and that the neck displaces the femur further laterally, the transmission of body weight through the hip is well supported by a column of bone (Figure 8.9). The lumbar lordosis and inclination of the sacrum serve to shift the sacroiliac joint forward, close to the coronal plane of the acetabula. The viscera lie in front of the spine so that actual body mass is

more nearly centered between the hips. The sacrum is wedged between the coxal bones, to reduce the shear stress on the sacroiliac ligaments. The weight of body mass thus passes easily from the sacroiliac joint to the acetabulum. Forces placed on the head of the femur are transmitted to the medial side of the femoral neck and shaft, then to the lateral condyle of the knee joint. This seemingly oblique course through the femur approximates a vertical column of compressive stresses through the bone. Shear stress in the neck or any other part of the femur is greatly reduced.

THE KNEE

The knee is a compound joint involving three separate articulations—between the patella (knee cap) and the femur; between the medial condyles of the femur and tibia; and between the lateral condyles of the femur and tibia. Although they have independent developmental origins, these articulations function as part of a single complex. The knee joint permits a single degree of freedom, flexion-extension, that is accompanied by a small degree of conjoint rotation.

The femur and tibia contact one another in an extremely shallow articulation, giving them remarkably little stability for a critical weight-bearing joint. To compensate for this, the joint is secured by several strong ligaments and reinforced by surrounding tendons. The tibial and fibular collateral ligaments lie on either side of the knee. Both are lax in flexion and become taut as the joint extends. Between the condyles is another pair of ligaments, the anterior and posterior cruciate ligaments, so called because they cross one another as they ascend from the tibia to the femur. Each of these ligaments is a thick bundle of fibers with the notable property that some of the fibers in each ligament are taut in any position of the joint (Fuss 1989, 1991).

The quadriceps femoris is a group of four muscles that insert on a single tendon that crosses the anterior knee, envelops the patella, and inserts on the tibial tuberosity. The patella, or kneecap, is a large sesamoid bone that forms in the patellar tendon at the front of the knee. It plays an important role in protecting the underlying joint and in holding the tendon away from the joint for greater power. The patella fits into the intercondylar groove between the femoral condyles and thus helps to track the patellar tendon on the front of the knee.

ADAPTATIONS OF THE KNEE FOR BIPEDALISM

Bipedal stance and locomotion present several challenges to the knee joint. The greater static and dynamic forces crossing the joint threaten to destabilize it.

Consequently, the human knee has undergone modifications of both skeletal and muscle design to resist these forces and to enable the knee to bear weight efficiently.

The carrying angle of the knee directs the pull of the quadriceps group laterally, as well as superiorly. There is a danger of the patella being pulled out of the patellar groove and off to the side of the joint. To prevent this from occurring, the lateral lip of the patellar groove is much steeper in the human femur. In addition, the most inferior fibers of quadriceps have an oblique orientation and attach near the patella. They pull both laterally and medially and play an important role in holding the patella in place in the intercondylar groove (Figure 8.10).

Uniquely among primates, the human knee is habitually used in full extension. The extended and locked knee bears weight more efficiently than a partially

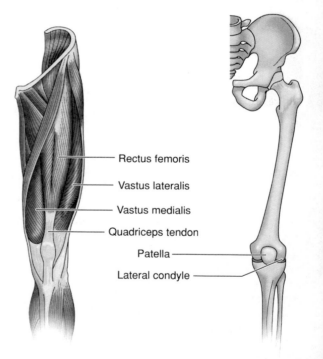

Rectus femoris

Vastus lateralis

Vastus medialis

Quadriceps tendon

Patella

Lateral condyle

Figure 8.10 Role of the vastus medialis in stabilizing the knee. The femoral condyles and the patella help to track the quadriceps tendon in the intercondylar sulcus. The lower fibers of vastus medialis, which approach the patella obliquely, prevent it from slipping laterally.

flexed one because little muscular effort is needed to support a straight column of bone. To further lock the joint into stable extension, the cruciate ligaments guide the tibia in lateral rotation under the femur during the last degrees of extension (Fuss 1992). Flexion of the joint requires an initial lateral rotation or unloading of the joint. A particular muscle, popliteus, performs this task. Because all ligaments achieve maximum tension in extension, this may be considered the close-packed position of the knee.

The femoral condyles are well rounded in lateral view for most higher primates. In humans, however, they are elongated from front to back and appear somewhat flattened from top to bottom. This elliptical shape increases the area of contact in extension, thereby reducing the pressure on the articular cartilage. It also shifts the axis of rotation more posteriorly on the condyles to increase ligamentous tension in the stable extended position. The patella consequently lies farther anteriorly from the joint and improves the lever arm for the quadriceps muscles.

THE LEG

The bones of the leg are the tibia and fibula. The head of the tibia forms a broad horizontal platform for the base of the femur, divided into medial and lateral condyles. Distally, tibia articulates with the talus at the ankle joint. This joint resembles a pulley, or trochlea, in shape. A process of the tibia, the medial malleolus, extends on the medial side of the joint, bearing the strong deltoid ligament of the ankle.

The fibula is a more slender bone. In humans, its head is tucked under the lateral condyle of the tibia and does not articulate with the femur. Distally, it extends inferior to the trochlear surface of the tibia to create a lateral malleolus that further helps to secure the ankle. The tibia and fibula meet one another both proximally and distally, where they are tightly joined by ligaments. Only a slight rotational movement is permitted between them. The length of the fibula is connected to the shaft of the tibia by an interosseous membrane. The fibula transmits a small proportion of body weight to the ankle (Lambert 1971), and the membrane stabilizes that bone against forces pushing it laterally away from the tibia.

Many groups of mammals have fused the tibia and fibula to create a single skeletal element of the leg. This arrangement produces greater stability but only at the cost of reduced mobility at the ankle joint. If the stresses occurring in the lower limb consistently occur in a single plane—such as those from running on the ground—it is possible to design the limb very economically to dissipate such stresses. In some mammals, most notably primates and carnivores, the limb is used in a greater range of positions and stress patterns are less predictable. Either the stress levels must be reduced (e.g., by limiting speed or body size) or the limb must be strengthened. Maintaining an unfused tibia and fibula permits primates and carnivores to use the foot in a wider variety of positions and activities. For primates, this permits the use of arboreal habitats. For carnivores, it facilitate agility and grappling with prey.

THE ANKLE AND FOOT

THE BONES OF THE FOOT

The bones of the foot include the tarsals, metatarsals, and phalanges. The numbers and arrangements of these vary considerable among different mammalian groups. In general, the primate foot maintains a primitive design and relatively great joint mobility. The human foot has seven tarsals, constituting the body of the foot and defining its important mobile joints (Figure 8.1). The two largest of the tarsals, the talus and calcaneus, lie at the back of the foot and participate in a complex set of joints. The talus articulates in the trochlear joint with the tibia and with the two malleoli. The malleolar facets are nearly vertical to restrain the talus within the joint. The calcaneus underlies the talus and supports body weight on the ground. The calcaneal tuberosity is the heel process and acts as a lever arm for plantar flexors of the ankle. This tuberosity developed an inferior process among hominoids and other climbing primates to enhance the action of short flexors of the toes. Humans have a unique second inferior process, more lateral, that increases the area of bone on which to bear body weight (Figure 8.11).

One of the striking characteristics of the human foot skeleton compared with that of other mammals is

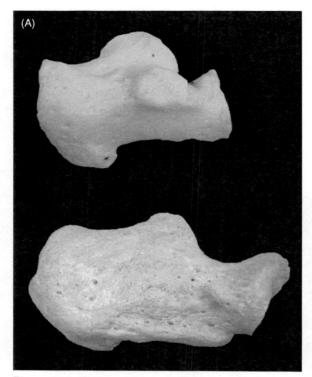

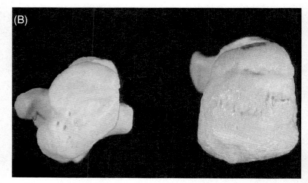

Figure 8.11 Heel processes of chimpanzee and human. The calcaneus of a chimpanzee and a human are compared in lateral (*A*) and posterior (*B*) views. The human heel rests on a much broader base of bone because of the development of a lateral tubercle on the undside. The human bone is also disproportionately larger because it contains more shock-absorbing fluids.

the relatively large size of individual bones. The tarsals of a chimpanzee, for example, appear compact and strengthened with a thick layer of cortical bone. Human tarsals appear puffed up and brittle. Cortical bone is so thin that it is easily eroded in a museum or teaching collection to expose the underlying cancellous bone. We can understand this contrast as two alternative strategies for dealing with stresses of weight-bearing. The thick cortical bone of the chimpanzee is dense and strong and resists the pressures that might fracture it. The increased volume of human trabecular bone, otherwise filled with fat, is better able to absorb and dissipate the greater compressive stresses delivered during normal bipedal gait. The thin cortex permits some deformation that is padded by the fat-filled interior. This same pattern is also found in many other parts of the human skeleton.

THE TOES

Human toes are relatively and absolutely small but play an important role in walking. The long grasping toes of climbing primates are at a disadvantage in walking. When an animal pushes off from the ground with the front of the foot, as humans do, the toes create powerful levers capable of overpowering the flexor muscles and destroying their own joints. Various species have found different solutions to this problem. The gibbon, which has very long digits, simply does not use the foot for push-off. The large orangutan curls the toes into a fist and tends to roll off the foot to the side rather than off the distal end. Chimpanzees and gorillas curl the toes under to shorten them and then collapse the very mobile midtarsal joint so that it, not the toes, accepts most

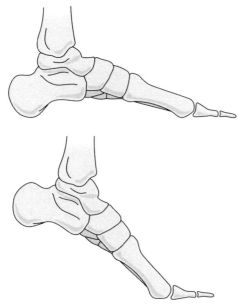

Figure 8.12 Hyperextension of the toes in gait. One unique aspect of gait involves the hyperextension of the toes when the heel comes off the ground. In contrast, the chimpanzee foot breaks at the midtarsal joint.

of the stress. Humans use the toes during push-off but have made them quite small.

The metatarsophalangeal joints at the base of the toes are single condyles, capable of both flexion and abduction. In this they resemble metacarpophalangeal joints of the fingers. Human toes have the uniquely developed ability to dorsiflex, or hyperextend, so that the metatarsophalangeal joint appears to collapse (Figure 8.12). The interphalangeal joints within the toes are double condyles with a single axis of motion. The distal, but not the proximal interphalangeal, joints have some ability to hyperextend.

These joints of the toes allow the pads of the toes to be planted firmly on the ground during walking, or even to grip a soft substrate, such as sand or mud. The toes help to extend the lever arm of the foot in the last phase of push-off. This increases the rate of acceleration and effectively shifts the foot to a higher gear (Carrier et al. 1994). Metatarsophalangeal joints are stabilized against these stresses by the contraction of the long flexor muscles and the numerous smaller intrinsic muscles of the foot.

The design of the human first toe is unique. It is a primate characteristic that the first toe is opposable like the thumb and has a suite of muscles similar to those in the hand. This is an important trait in the African apes, which use the toe to grasp supporting tree branches. Among bipedal hominins, the digit has been converted to another function—that of supporting the entire body in the final thrust of gait. Therefore, it has become larger and stronger and brought alongside the other toes. The curved joint surface between the cuneiform and first metatarsal at the base of the toe has been flattened to inhibit movement. Two large muscles that used to move the toe, tibialis anterior and peroneus longus, have been redesigned to insert on both the cuneiform and metatarsal so that and thus cannot adduct or abduct the digit (Figure 8.13). In addition, the shaft of the first metatarsal has rotated

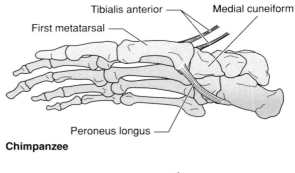

Chimpanzee

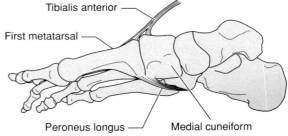

Human

Figure 8.13 The human first toe is immobilized. In contrast with the first toe of a chimpanzee (*top*), the human has immobilized the tarsometatarsal joint between the first cuneiform and the metatarsal to inhibit free movement. (*Bottom*) The insertions of two muscles responsible for abduction of the toe in most primates, tibialis anterior and peroneus longus, are modified in humans to stabilize the joint.

so that the phalanges now face the ground, as in the other toes.

EVOLUTIONARY TRENDS IN ANKLE AND FOOT MUSCULATURE

The musculature of the primate foot is subject to a great deal of minor variation within and between taxa (Langdon 1990). Some of these variations, described later, appear to have functional significance and to indicate adaptive changes in locomotor patterns of the hominoids. The general trend has been to make the joints more independent of one another for better control of the digital grasp in climbing.

The calf of the human leg is dominated by the triceps surae group, including the large soleus and gastrocnemius muscles. These two muscles blend to insert by a common tendon (tendo calcaneus or Achilles tendon) into the calcaneal tuberosity. They are powerful plantar flexors of the ankle. Normal gait is not possible without them. A third muscle, plantaris, is vestigial and functionally disposable. Soleus is a minor muscle for most other primates, but plantaris is usually well developed. The primitive insertion for plantaris is in the plantar aponeurosis, a tendinous sheet on the sole of the foot, where it can tense the sole of the foot as it plantar flexes the ankle. Hominoids have shifted this insertion to the calcaneus and made the aponeurosis independent of ankle motions.

The two long flexors of the toes, flexor hallucis longus and flexor digitorum longus, are derived from a primitive condition in which two muscles each supply flexor tendons to all five digits. In this arrangement, these muscles are named according to their origins flexor fibularis and flexor tibialis, respectively. The most common condition of flexor fibularis among primates is for the tendons to digits V and II to be reduced or lost. This enables the muscle to oppose the first toe against digits III and IV with a single contraction. In a substantial proportion of the human population, flexor hallucis longus continues to send a tendon to digit III.

Flexor tibialis tends to complement the distribution of flexor fibularis, sending its strongest tendons to digits II and V in most primates. The human pattern, in which flexor digitorum longus flexes digits II through V only,

is rarely observed in other primate species, except as a variation in chimpanzees. The human pattern of these two flexors corresponds to the loss of opposition by the first toe and an emphasis on collective toe flexion in gait rather than for grasping.

The numerous intrinsic muscles of the foot stabilize the joints of the foot, support the arch, and move the digits. Several are named according to their potential actions, but these names do not necessarily reflect their role in the human foot. For example, the abductor hallucis stabilizes the first metatarsophalangeal joint, but that joint is incapable of abduction. The distribution of the muscles, as well as their names, reveals a more complex pattern of movement among other higher primates that has been lost in our species.

Of the intrinsic muscles, flexor digitorum brevis shows an interesting trend among primates (Figure 8.14). Primitively, the bulk of the muscle arose from the long flexor tendons. Only the part of the muscle inserting on digit II arises independently, from the calcaneus. Occasionally among monkeys and in nearly all of the great apes, this independent part of the muscle produces a second tendon to digit III. In one group of New World monkeys (the atelids) and many of the great apes, a third tendon to IV is present. Nearly all human feet have tendons to three or four digits. The separation of the short flexor from the long flexors permits an individual to flex the toes without plantar flexing the ankle or to plantar flex the ankle without weakening the action in the toes. This independence of action is of greatest importance in animals that suspend themselves from small branches—atelids, great apes, and, apparently, our own ancestors.

THE LONGITUDINAL ARCH OF THE FOOT

The longitudinal arch of the foot is a uniquely human construction. It plays a critical part in normal gait. The skeleton of the arch is the row of bones along the medial side of the foot—calcaneus, talus, navicular, first cuneiform, and first metatarsal (Figure 8.15). Only the calcaneus and the head of the metatarsal normally bear weight, although the height of the middle of the arch is, and should be, variable. The longitudinal arch is a dynamic structure, increasing and decreasing in

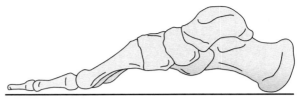

Figure 8.15 The medial longitudinal arch is a unique feature of the human foot. The medial view of the human foot skeleton shows a modest arch.

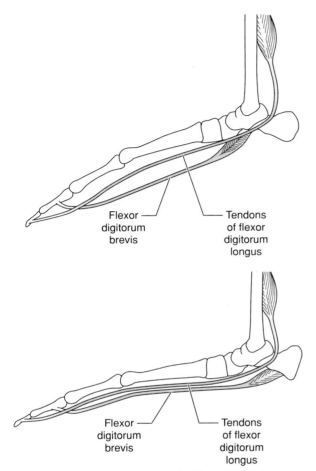

Figure 8.14 The digital flexors in higher primates give more control over the joints. In most primate feet (*top*), the short digital flexor (flexor digitorum brevis) is small, often supplying a single digit, and arises from the tendons of the long flexors. Thus, digit flexion is coupled closely with ankle plantar flexion. In apes and humans (*bottom*), the short flexor is better developed and, to varying degrees, has an independent origin from the calcaneus so that the toes can be flexed independently of the ankle.

its height and strength during gait. Although the foot should be pliable and supple during passive weight-bearing and weight-acceptance phases, the arch must be rigid and strong in push-off. The ability of the foot to alternate between these two functional states is due to the conversion mechanism built into the major joints.

CONVERSION MECHANISM OF THE FOOT

There are many articulations of the tarsal and metatarsal bones, but only a few have appreciable movements. Most are irregular gliding joints, permitted enough wiggle by their ligaments to relieve stresses and absorb shock, but not possessing identifiable axes of rotation. Two important exceptions are the subtalar joint and the midtarsal joint. On the other hand, the tarsometatarsal joint at the base of the hallux (big toe) is notable for its lack of movement.

The subtalar joint (also called the talocalcaneal or inferior ankle joint) exists between the talus and the calcaneus. The midtarsal joint (also called the transverse tarsal joint) divides the foot into anterior and posterior portions. On the proximal side of the joint are the head of the talus and the end of the calcaneus. Distally are the navicular and the cuboid. All of these joints are linked together so that when one moves, both are affected.

The talonavicular articulation has the form of a ball-and-socket joint with 3 DOFs. The calcaneocuboid joint is sellar, with 2 DOFs. However, these two parts of the midtarsal joint constrain one another, since movement on one side of the foot has to be accompanied by movement at the other. When the two articulations are aligned with one another, a small degree of flexion-extension is permitted in the midfoot—enough to flatten the arch. When the calcaneus is inverted under the talus, the axes of the two parts of the midtarsal joint are no longer in mutual alignment. No movement may occur and the foot is rigid. In addition, inversion of the calcaneus twists and tightens ligaments crossing the midtarsal joint, further stabilizing it. The foot is therefore able to shift back and forth between mobile and

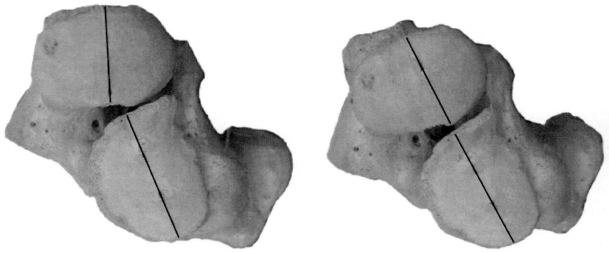

Figure 8.16 The conversion mechanism. The talus and calcaneus of the human foot are viewed from the front in everted (*left*) and inverted (*right*) positions to show the facets for the cuboid and navicular. The axis of dorsiflexion-plantar flexion at each joint are approximated by the black lines. Inversion of the foot disrupts the alignment of the joints and prohibits movement at the midfoot. Thus the motion of inversion and eversion permits the foot to convert from a rigid structure to a pliant one and back.

immobile positions (Figure 8.16). This design of the joints, plus the muscles and ligaments that prevent the bones from separating, may be referred to as the conversion mechanism of the foot (Mann and Inman 1964).

SUPPORT FOR THE ARCH

Four factors contribute to forming and supporting the arch: joint design, muscles, ligaments, and the windlass effect.

Joint design. The joints around the talus, subtalar and transverse tarsal, are critical in establishing the arch. The interactions of these creating the conversion mechanism are described above. The rigid supinated position of the joints corresponds to the highest position of the arch. Pronation loosens and lowers the arch. Normal variation in the design of these joints can affect the performance and health of the arch.

Muscular support. The arch is strengthened by long tendons of muscles active during push-off in gait —tibialis posterior, flexor hallucis longus, and flexor digitorum longus in the posterior compartment, and many of the smaller plantar intrinsic muscles. As these contract, they pull the bones more tightly together,

especially across the medial side of the transverse tarsal joint; and they resist passive dislocation of the joints.

Ligamentous support. In a loosely packed pronated position of the foot, the tarsals are permitted some motion about an axis that passes through both the talonavicular and calcaneocuboid joints. When the subtalar and midtarsal joints are supinated and locked, the head of the talus becomes close-packed against both the navicular and calcaneus. This joint is strongly reinforced by a tightening of the ligaments on the plantar surface of the foot and surrounding the head of the talus.

The windlass effect. A windlass is a crank used to hoist a heavy load, such as the anchor of a ship, over a pulley. The analogy in the foot refers to the passage of tissues under the metatarsophalangeal joints, which serves as a pulley, that draw the two ends of the arch closer together (Figure 8.17). The key structures in the windlass are the long and short digital flexor tendons and the plantar aponeurosis. Although these muscles are already contracting and in tension when the arch is created in gait, the toes can exert additional pull on the rear foot as they are passively hyperextended. This

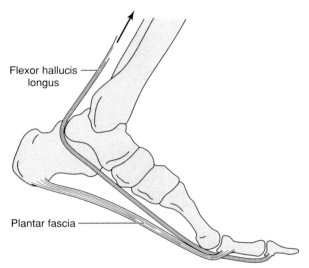

Flexor hallucis
longus

Plantar fascia

Figure 8.17 The windlass mechanism of the foot. When the toes hyperextend, the long tendons and ligaments attached to them are cranked around the heads of the metatarsals and pull on the posterior foot. This action heightens the arch and stabilizes the joints.

hyperextension is produced naturally by ground reaction forces as the heel rises. The pull around the windlass brings the back of talus and calcaneus closer to the metatarsals and heightens the arch. The windlass effect can be reproduced by wearing high heels —the arch heightens even if the muscles are resting— or simply by pressing the toes into hyperextension with your hand.

FUNCTION OF THE LONGITUDINAL ARCH

The significance of the conversion mechanism lies in the importance of both the mobile and rigid configurations. Most terrestrial mammals have a very rigid foot capable of generating greater speed (Figure 8.18). By reducing the number of bones and joints, ungulates, for example, have eliminated several potential weak points. Most cursorial animals contact the ground only with the heads of the metatarsals (digitigrady) or with the phalanges (unguligrady) instead of with the tarsals (plantigrady). Such foot positions convert the tarsals into a column of bone able to bear weight with less shear stress.

Climbing demands a mobile foot in hominoids that is capable of adapting to a wide variety of substrates. The feet of the great apes are therefore less able to sustain activities generating high stresses. The human foot, with its conversion mechanism, appears to represent a mosaic of features of climbing and cursorialism. The mobility inherited from the climbing adaptation is necessary to permit the foot to redistribute pressure on the ground. This enables us to balance ourselves when standing or walking on an uneven surface. In the rigid mode, most of the foot— the length of the longitudinal arch—serves as the load arm for rapid push off during walking.

Why is the rigid foot shaped like an arch? A flat beam is adequate for static postures, but it would sustain great shear stresses during push-off. The arch has the form of a strung bow, in which body weight is placed on the arch itself. Much of the force of body weight is converted to compressive stress in the bones, while the soft tissues supporting the arch are loaded in tension. The soft tissues under the arch also can absorb shock. Although it is generally expected that ligaments have little elasticity, it has been demonstrated that the plantar fascia absorbs and returns energy to the foot during running to enhance efficiency (Ker et al. 1987).

FUNCTIONAL VARIATION IN THE DESIGN OF THE ARCH

During World War I, the military draft provided physicians with the first opportunity for a large-scale survey of the state of health of American males. One alarming finding was a high frequency of flat feet with little or no longitudinal arch. Concern about flatfeet continued in our culture for decades and prompted widespread use of rigid high-arched shoes for children in an attempt to mold their feet into the desired shape. For example, Morton wrote in 1924,

> Disorder of the feet is the most common and most widely spread form of physical impairment among civilized peoples today (with the sole exception of dental defects), and has already reached a stage where over three-fourths of the nation's youth enter adult life with an acquired weakness that may at any time develop into a source of actual disability.

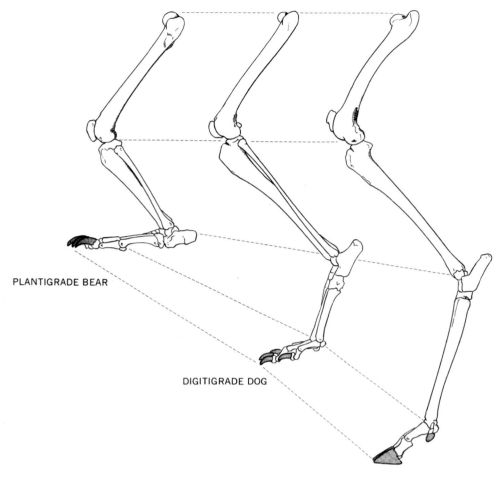

PLANTIGRADE BEAR

DIGITIGRADE DOG

UNGULIGRADE DEER

Figure 8.18 Plantigrade, digitigrade, unguligrade, graviportal feet. If the foot forms a long horizontal lever, the animal can run and jump faster. However, this length increases the shear stresses in the foot in proportion to the horizontal separation between the ankle joint and the point of contact with the ground. Therefore, foot design is often a compromise between the speed and the size of the animal. A long foot provides more speed in acceleration. A short foot or vertically oriented limb is more stable and is better suited for larger animals. (From M. Hildebrand. 1988. *Analysis of Vertebrate Structure*, 3rd ed. Copyright 1988 John Wiley and Sons. This material is used by permission of John Wiley & Sons, Inc.)

In retrospect, the medical fears were misguided. Physicians had failed to appreciate the dynamic nature of the arch. The variation observed did not reflect healthy versus unhealthy feet, nor did it observe differences in evolutionary development, as some interracial studies suggested. Persons with high arched feet observed standing still in a physician's office probably had worn shoes most of their lives, whereas the

flat-footed subjects had grown up barefoot. Rigid shoes act as a splint, immobilizing the foot and preventing normal stretching of the ligaments.

Pathological flat feet do exist. If the joints do not lock into a close-packed position or the ligaments are too lax to hold joints tightly together, the arch will be unable to support body weight for long. The muscles will fatigue and the patient will experience pain from

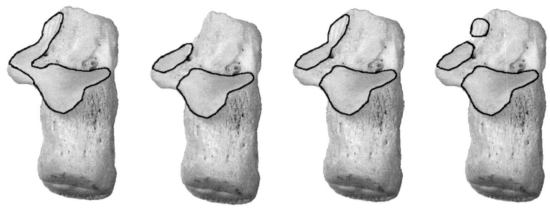

Figure 8.19 Variation in the human subtalar articulation. The articular facets are variable in shape and number, as is readily seen on these four human calcanei. These variations may correlate with mobility and clinical problems.

the stresses placed on the soft tissues. These disorders illustrate the importance of the conversion mechanism in strengthening the foot for normal walking.

Aside from shoe-wearing and pathologies, there is a normal congenital variation in the behavior of the longitudinal arch. This variation derives from individual differences in ligament tension and bone shape. One of the more important determinants is the form of the subtalar joint. This joint can involve one, two, or three distinct facets, although a single facet is extremely rare (Figure 8.19). Commonly there is one facet on the body of the calcaneus, articulating with the body of the talus. The sustentaculum bears two or one elongated facet to contact the head and neck of the talus.

A single sustentacular facet generally has a concave curvature. The talus rotates smoothly within it. This condition corresponds to a greater range of mobility. If there are two distinct sustentacular facets,

they tend to be relatively flat and set at an angle to one another. The talus does not readily rotate when set into such an angle; thus the three-facet subtalar joint is relatively immobile.

These two patterns mirror the difference found between the foot of apes, whose mobile feet are adapted for grasping, and that of Old World monkeys, whose rigid joints are better adapted to running. Although both of these represent normal and common forms of the human foot, they have different implications for function and clinical problems (Bruckner 1992; Hively 1999). The rigid three-facet joint provides a stronger lever for running and jumping. Such individuals are more likely to achieve success at sports involving such activities. On the other hand, the noncompliant ligaments are more likely to be strained. The mobile two-facet joint is more yielding to stresses. While not lending itself to athletic prowess, it provides greater protection against injury.

9

THE HUMAN STRATEGY

Bipedalism

Bipedalism may be the oldest gross anatomical adaptation distinctive of the hominins. In its impact on our body design, it is also the most far-reaching. This chapter explores the origin and significance of bipedalism for our species. This discussion also permits us to focus on those skeletal adaptations that may be specifically linked to the evolution of bipedalism in early hominins. Such indicators enable the fossil record to tell us when bipedalism evolved, but they have not answered the big question of why. Instead, the fossils have posed new questions. The early hominins were neither apelike quadrupeds nor human-like bipeds. They were clearly bipedal but displayed a unique anatomy. Were they bipedal because it was a better way to move about or because it freed the hands? Or, perhaps, because it was the most effective way for a climbing ape to get about on the ground?

THE TRANSITION TO BIPEDALISM

The preceding chapters have observed numerous unique adaptations of the modern human body to accommodate bipedal posture and locomotion (Table 9.1). Many of these involve changes in the skeleton itself and should be identifiable in the fossil record. It was long expected that the earliest members of our lineage would show an intermediate or transitional anatomy between those of humans and apes and that this would correspond to a transitional behavior. What we have discovered is that the australopithecines are neither apelike nor human-like nor transitional. They are unique. The analysis of their skeletal function must be undertaken with care because, as new discoveries are making clear, that skeleton is full of surprises. Different parts of the skeleton evolved at different rates, a phenomenon called *mosaic evolution*. The different parts did not evolve directly from ape to human form

but often took an indirect path, suggesting unique behavior patterns as well.

We have been slow to realize a further complication. Different australopithecine species show some important morphological and anatomical differences that we cannot fully observe at present. *A. afarensis* from Ethiopia and *A. africanus* from South Africa are the best known for postcrania, and discussions in the literature long considered their skeletons interchangeable. Only the rapidly growing collection from Sterkfontein Cave has thrown that assumption into question by revealing bones of *A. africanus* that are

Table 9.1 Musculoskeletal Adaptations of the Human Body for Bipedalism

Forward placement of the foramen magnum and occipital condyles
Mastoid process
Lumbar lordosis and wedging of the lumbar vertebrae
Sacrum wide and narrows inferiorly
Sacroiliac joint closer to acetabulum
Ilium is shorter, broader, and deeper
Iliac pillar strengthens ilium for stresses of abduction
Ischium shortened
Gluteus maximus reoriented as an extensor of the hip
Gluteus medius and minimus and tensor fasciae latae reoriented as abductors
Hip and knee capable of full extension
Increased carrying angle of the femur so that knees and feet touch at midline
Femoral condyles elliptical
Tarsals and other bones greater in relative volume with thinner cortex
Calcaneal tuberosity has medial and lateral inferior processes
Medial longitudinal arch present
Conversion mechanism of the midtarsal joint permits both flexible and rigid states of the foot
Hallux robust and of similar length to other digits
Hallux permanently adducted parallel with other digits
Lateral toes shortened
Metatarsophalangeal joints capable of significant hyperextension

Table 9.2 Australopithecine Features of the Pelvis and Proximal Femur

Apelike Features	Human-Like Features	Unique Features
Iliac blade projects laterally rather than anteriorly	Ilium shortened, broader	Ilium excessively broad; pubis long
	Sciatic notch well developed	Interacetabular distance increased
Ischium long	Sacrum wide	Small sacroiliac joint surface
Femoral head small	Sacrum posterior to acetabulum	Femoral neck long
Low neck angle	Cortical bone in femoral neck supports extended hip	

different in proportions and some details of anatomy from their more northern relatives. Any differences possessed by robust australopithecines are not well known and are less relevant, since they were contemporaries with early *Homo* rather than possible ancestors. The following discussion refers to the gracile species.

THE PELVIS AND HIP

The australopithecine pelvis has both human-like and apelike features, summarized in Table 9.2 (Aiello and Dean 1990). Overall, it has been shortened craniocaudally to resemble superficially the human condition (Figure 9.1). Most of this shortening has occurred in the ilium and reflects commitment to an upright posture with a lesser degree of habitual hip flexion. The ischium, too, is reduced in length (McHenry 1975). The ischium represents an arm of lever action for the hamstring muscles that is of advantage in climbing, during which the extensors more often operate from a highly flexed position. In *Homo*, the power for this lever is less important and can be mostly assigned to gluteus maximus. Shortening the ischium increases efficiency of hamstring design.

The iliac blade is broader in early hominins than in apes. Although it curves somewhat anteriorly in the modern human pattern, its breadth is remarkable so that the blade faces mostly posteriorly (apelike) rather than laterally (human-like). This orientation of the external surface area favors the gluteal muscles acting to extend the hip rather than those abducting it. The range of abduction may have been reduced relative to that of apes (MacLatchy 1996). This resembles humans as it compromises climbing ability.

The entire pelvis is broad mediolaterally. The pelvic inlet is correspondingly large and the pubic bones

elongated. The hip sockets are placed widely apart, and the iliac blade flares well beyond these. To match the ilium, the femoral neck is long, so that the greater trochanters lie far apart (Figure 9.1). This arrangement gives the abductors leverage for a trunk stabilization that is superior to that of humans (Lovejoy 1988; Wolpoff 1978). However, the widely spaced hips destabilize gait. When the trunk is supported on a single limb, both the lever arm (iliac blade breadth) and the load arm (distance to the center of mass) at the hip are elongated. Although the mechanical ratio may remain favorable, the force crossing the hip joint is unnecessarily increased. Hunt (1994) argues that the australopithecine pelvis is better suited for bipedal standing, supported by widely spaced feet, and he raises the possibility that bipedal stance played a more important role in early stages of hominin evolution than bipedal walking.

The human ischium moved closer to the sacrum, both medially and superiorly. The sacrotuberous ligament both shortens and expands so that it can play a role in stabilizing the sacrum and resist downward rotation of the lumbosacral joint. Australopithecines show an intermediate position in this feature between apes and humans.

Other features of the australopithecine hip region that appear to be less well adapted for human-like bipedalism than human bones are differences in the distribution of trabecular bone within the ilium (Macciarelli et al. 1996, 1999), smaller joint surfaces at the sacroiliac and acetabular joints, and a lower neck angle on the femur (Corruccini and McHenry 1980; Walker 1973). Small articular areas imply a narrower distribution of loads across those joints. A low neck angle, combined with a long femoral neck, creates greater shear stresses. In contrast, the human arrangement transmits forces in a nearly vertical path

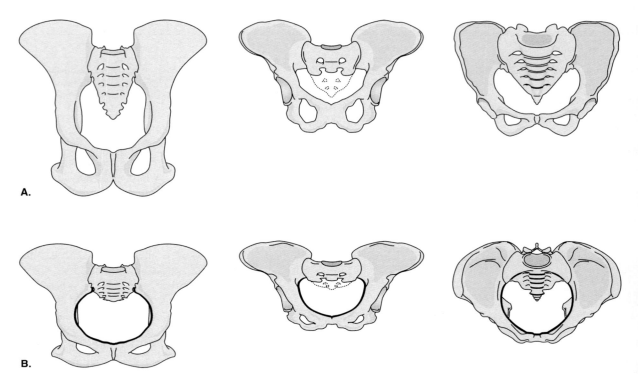

Figure 9.1 Comparison of human, ape, and australopithecine pelves. *A*, Superficially, the short australopithecine pelvis in the center resembles that of a human on the right much more than that of the chimpanzee on the left. (This figure also illustrates the significance of the carrying angle of the femur in determining posture of the lower limb.) *B*, A superior view of the pelves shows greater similarity in the alignment of the ilium of the fossil to that of the chimpanzee. Both fossil and living hominins share a laterally flaring ilium that anchors the abductor muscles (gluteus medius and minimus) to stabilize the hip. The breadth of the human iliac blade is actually intermediate between those of the chimp and of *Australopithecus*.

from the lumbar spine to the femoral shaft. The distribution of cortical and trabecular bone within the proximal femur, a good indicator of habitual stress patterns, shows little difference between the fossils and modern humans (Lovejoy et al. 2002; Ohman et al. 1997). Perhaps the seeming superiority of the modern design was necessary only in response to increasing body size and a greater absolute magnitude of forces in this area.

THE KNEE

The distal femora of modern humans and australopithecines agree in several key indicators of bipedalism but differ in some more subtle features that suggest to some authors functional differences (e.g., Berger and

Tobias 1996; Senut and Tardieu 1985; Stern and Susman 1983). Some specimens show less asymmetry of the femoral condyles and less development of the weight-supporting elliptical lateral condyle. In addition, the patellar groove, essential for stabilizing the patella in extension, is inconsistently developed (Tardieu 1981). These features are difficult to interpret, in part because of variability in the fossils, and have led to significant differences in view. Some authors infer that at least some australopithecines had not achieved the specializations of the modern skeleton; for example, Berge (1994) suggests that full hip and knee extension were not possible. To the contrary, Heiple and Lovejoy (1971) argue there is no feature "which is not fully commensurate with completely bipedal locomotion." Unfortunately, the

Table 9.3 Australopithecine Features of the Foot

Apelike Features	Human-Like Features	Unique Features
Hallux still capable of adduction in some specimens	Calcaneal tuberosity may have medial and lateral inferior processes	Phalanges of intermediate length
Peroneal groove on fibula wide and deep	Medial longitudinal arch present	Phalanges show slight curvature
Peroneal tubercle well developed	Hallux robust and of similar length to other digits	
	Adduction mobility of hallux reduced in some specimens	
	Metatarsophalangeal joints capable of hyperextension	

differences in form between fossils does not correspond to recognized species, time, or geographical groupings. Thus they are more likely to represent population variation than separate lineages.

THE FOOT

The ambiguity is continued in the foot (Table 9.3). Several features of the australopithecine foot clearly resemble the derived human condition and are adaptations for bipedal walking, yet others are quite different. Some key indicators of human foot function—the presence and dynamics of the medial longitudinal arch and the form of the heel process—are difficult to interpret in the available fossils and are subject to contradictory interpretations.

Compared with modern humans, the tibiotalar joint of *Australopithecus* has a greater range of motion, hinting at a climbing behavior or, at least, a climbing ancestry (Latimer et al. 1987). The distal fibula of several specimens also indicate increased range of ankle motion. A powerful peroneus longus muscle is indicated for the australopithecines by a wide tendon groove in the fibula and a large peroneal tubercle on the calcaneus (Latimer and Lovejoy 1989). Strong eversion provided by this muscle is important for supporting body weight on a support grasped by the medial side of the foot, as is observed in chimpanzees.

Phalangeal length and curvature are reduced for the australopithecines relative to apes. Although the metatarsal heads show the distinctly human capacity for dorsiflexion at the metatarsophalangeal joint, this joint continues to evolve in later *Homo* (Duncan et al., 1994; Latimer and Lovejoy 1990b). The hallux is robust and of similar length to other digits. It appears to be aligned with the other toes in some fossils from

Hadar (Latimer and Lovejoy 1990a), as it is in modern humans, and footprints from Laetoli that presumably were made by *A. afarensis* show an adducted great toe. However, specimens from South Africa shows variation. An older specimen of the medial foot of *A. africanus* indicates considerable mobility for the first toe (Clarke and Tobias 1995). The first tarsometatarsal joint in modern *Homo* is fully immobilized by a flattening of the articular surface. A similar flattening is observed in a first metatarsal attributed to *A. robustus* but not to the foot of *A. africanus* (Susman and Brain 1988).

These morphological changes indicate functional compromises in the australopithecines skeleton. Later reduction of ranges of motion in the ankle joints increases stability of those joints and tolerance of a greater magnitude of forces across them. Peroneus longus may be assisting in the stabilization of the first digit and of the midfoot in general in australopithecines or providing eversion during climbing. Although eversion is critical is modern human gait, the muscle is called upon in a smaller range of foot positions and it is no longer critical in positioning the first toe. Human toes provide an important propulsive force in the final stage of push-off in gait, but ground reaction forces threaten to destabilize the metatarsophalangeal joints. Reduction of toe length protects the toes as the forces increase with changes of gait or body size.

THE AUSTRALOPITHECINE BODY

The lower limbs of the best known australopithecine skeleton, "Lucy" (AL-288 from Ethiopia), are relatively short by modern human standards, whereas the upper limbs are of comparable proportions relative to the trunk to those of both humans and chimpanzees. It seems clear that the lower part of the body underwent far more significant transformation for bipedalism than did the upper body.

New material from Sterkfontein reveals an even greater limb disproportion in *Australopithecus africanus* (Berger 2002; McHenry and Berger 1998). The upper limb bones are large and robust relative to those of the lower limb and suggest more apelike proportions than in the better known Ethiopian material. Along with the difference in the toe structure, the contrasts suggest the South African australopithecines may have spent much more time climbing. These finds serve to highlight the fact that the shift to bipedalism in the earliest hominins was not a sudden adoption of modern behavior and my have retained a great deal of arboreality. It is likely that East and South African australopithecines had more profound diffences in anatomy, habitat, and locomotor behavior than has been previously supposed.

How did the australopithecines walk? While the difference between the fossils and the equivalent anatomy of apes is striking, the more subtle contrasts with humans has been more difficult to interpret. Some authors have argued for virtually no significant difference between australopithecine gait and that of modern humans, or even for greater efficiency for australopithecines (e.g., Kramer 1999; Lovejoy et al. 1973, 2002). However, the majority of recent opinion does not support such a view (e.g., McHenry 1986; Stern 2000; Stern and Susman 1983; Susman et al. 1984; Ward 2002). They would have been less efficient than modern humans in attempting a modern human gait. They probably could not sustain the speed, magnitude of stresses, or distances to which modern humans are accustomed. It is possible, as some have suggested, that they spent less time walking on the ground in favor of time in the trees. It is also possible that they used a unique pattern of gait.[1] Yet, there should be no reason for us to assume that

early hominins had to follow a simple continuum from ape to human. The broad pelvis and wide hips apparently pushed the structural compromise in favor of static stability rather than balance in motion. What we would consider inefficiencies would be less serious in smaller species. The fact that an australopithecine skeletal pattern in the pelvis was identifiable over a period of 2 to 3 million years assures us that their gait was not the desperate inefficient transition that has been supposed in the past.

AFTER THE AUSTRALOPITHECINES

Since the appearance of *Homo*, the postcranial skeleton has made many changes that can be interpreted as increasing locomotor efficiency. This must be understood as an evolutionary event independent of the beginnings of bipedalism in the earliest hominins. The lower limbs of later *Homo* elongated relative to trunk size. Increasing the length of the lower limb length increases stride length and potential speed; therefore, the elongation of the femur and tibia are important indicators of full commitment to a terrestrial habitat. Other modifications of the joints and muscle mechanics produced the modern human gait.

The fossil record for early *Homo* is ambiguous regarding exactly when the second locomotor transition occurred, in part because of the paucity of articulated material and especially because of the difficulty of assigning individual specimens among the various sympatric hominin taxa. Of the two earliest partial skeletons attributed to *Homo* from Olduvai Gorge, one shows limb proportions possibly less modern than those of the australopithecines (Hartwig-Scherer and Martin 1991), although accurate reconstruction of the fragmentary remains is impossible (Haeusler and McHenrg 2004; Richmond et al. 2002). A later skeleton and isolated bones of *Homo ergaster* and *Homo erectus* show more modern body proportions and body size (Ruff 1993).

It appears most likely that *H. ergaster*, or some contemporary species of *Homo*, was the first species to follow the modern locomotor strategy, that may be defined in part by increasing locomotor efficiency and speed. Since that time, the hominin postcranial skeleton has maintained a large body size, long lower

---

[1]For example, a "Groucho Marx" style of walking with lack of full extension at either the hip or knee has been proposed but contains inherent inefficiencies that make it unlikely for a habitual terrestrial biped: Crompton et al. 1998; Wang et al. 2003.

limbs, and a reasonably consistent mechanical design. The most obvious subsequent change has been a decrease in overall robusticity of the skeleton. This reduction probably reflects a continued increase in efficiency of the overall activity pattern, whether due to anatomical or cultural adaptations.

THE ADAPTIVE SIGNIFICANCE OF BIPEDALISM

Bipedalism has been recognized as the definitive trait of hominins. It can be related to many of the other major features of our species, including terrestrialism, freed hands, use of tools, carrying objects, and frontal sex. What, exactly, is the nature of this relationship? Did bipedalism predate the other features? Was bipedalism a consequence of the other features? Were they a consequence of bipedalism? Ultimately, why are we bipedal? Speculation on this question has filled the literature for more than a century. Currently two approaches are most frequently cited. One attempts to link bipedalism to cultural behavior; the other views it arising for its physiological advantages.

HISTORICAL SPECULATIONS CONCERNING THE ORIGIN OF BIPEDALISM

Because of the centrality of bipedalism to the human species, it has been at the center of nearly all models to explain the emergence of the hominin lineage. The proposed explanations not only provide a look at the imaginations of generations of anthropologists; they also serve as a mirror of how we have looked at ourselves.

Tools. Darwin (1871) proposed that the ability to use tools and weapons (for defense and hunting) would have given our ancestors a powerful selective advantage and could have led to bipedalism to free the hands. This idea has been repeated more recently. For example, Washburn (1960) wrote "tool use is both the cause and effect of bipedal locomotion."

The first stone tools appear about 2.5 Mya and become common by 2.0 Mya. Adaptations of the hand relating to tool use are difficult to identify in the fossil record, or even in living humans, with confidence, but such abilities may have been present in later australopithecines, as well as in genus *Homo*, after 2.0 Mya (see Chapter 7). Both of these lines of evidence long postdate the appearance of bipedalism in the fossil record 4.0 Mya. Perishable or unmodified tools may have been used earlier in the way that chimpanzees and other species use them. Thus we have no direct evidence linking tools and bipedalism. The hypothesis must suppose an early pattern or dependence unique to hominins that is not evident in the archeological record. If technology was so important, why have chimpanzees not become bipedal tool-users?

Darwin and many other scientists of his period were strongly influenced by the emphasis on technology within their society and projected it into their concepts of human evolution. Victorian society of nineteenth century England was exceedingly optimistic about the ability of science and technology to conquer the ills of society. Darwin and others saw human evolution as an upward spiral driven by material achievement. It is probably no coincidence that, for them, technology became the defining property of our species.

Climbing. In the 1920s, as evolution theory had blended with studies of comparative anatomy, several anatomists sought the origins of bipedalism in the behavior of modern apes (Gregory 1928; Keith 1923; Morton 1926). Treading close to the common error of seeking an ancestor among living species, they observed that the anatomical adaptations for climbing conflict with efficient quadrupedalism but may predispose a climbing ape to bipedalism on the ground. Perhaps bipedalism itself was the offshoot of an arboreal adaptation, as observed in gibbons. Gregory thought gibbons would be a good ancestor. Recently, Tuttle (1974, 1975, 1981) and Vancata (1978) revived the concept as the "hylobatian hypothesis," although in a less literal form.

There is no fossil evidence to support the idea that proto-hominins had such extreme specializations as those observed in gibbons. The link between arboreal vertical climbing, or "scrambling," and bipedalism, however, continues to stimulate interest (Fleagle 1976; Fleagle et al. 1981; Langdon 1985; Prost 1980; Rose 1984; Stern 1975). Scrambling involves the use of all four limbs, grasping supports with both hands and feet. The longer upper limbs are used in suspension rather than weight-bearing. The trunk is help in an upright, or orthograde, posture and the lower limbs support the weight of the trunk. Increasing specialization for climbing detracts from effective terrestrial quadrupedalism. In numerous ways, including musculoskeletal design and limb use, climbing prepares the body for bipedalism (e.g., Fleagle 1976; Fleagle et al. 1981; Prost 1980; Stern 1975). Electromyographic studies of muscles of the back and lower limbs also reinforce the similarity between recruitment during climbing and their use in human bipedalism (Hirasaki et al. 2000; Shapiro and Jungers 1988; Stern and Susman 1981).

Supporting this argument are field data showing that bipedalism among the climbing apes is much more common in the trees than on the ground. Although the link with climbing does not provide a complete explanation for hominin bipedalism, it sets the evolutionary stage on which hominins appeared. Doran (1993a) observed chimpanzees walking bipedally 1.2% of the time overall, but that figure rose to 5.8% for adult males in the trees. Hunt (1992) reported chimpanzees bipedal during 0.1% of bouts on the ground but 1.4% in the center of a tree and 2.9% in the terminal branches. Thus, an arboreal habit is a more likely setting in which bipedalism might have evolved.

Weapons. The "killer ape" hypothesis introduced by Dart (1957) and Ardrey (1961) viewed *Australopithecus* as a predator and the human lineage as defined by that behavior. To these writers, the ability to wield weapons would have been sufficient to cause the development of bipedalism. Others have seen bipedalism as an ideal posture from which to stalk and hunt (e.g., Merker 1984). As a modification of this thesis, Kortlandt (1979) suggests that the ability to stand and threaten predators with such weapons as thorny branches would have been important.

There really is no evidence for either frequent predation on large animals or for homicide before *H. ergaster* or *H. erectus*, at about 1.6 Mya. Humans are not unique among primates as predators, since chimpanzees and baboons frequently kill and consume smaller mammals.

The "killer ape" hypothesis is a particularly pessimistic view of human nature, since it explicitly argues that humans have an instinct for violence and for murderous competition among people, of which territoriality and ethnic identity are natural expressions. Such an about face from the optimism of the Victorians is set in the historical context of the horrors of the twentieth century. The carnage of trench warfare in World War I was followed by genocide and civilian bombing in World War II. The advent of nuclear weapons held civilization hostage to the fear of a nuclear holocaust during the Cold War. The 1950s and 1960s represent a low point in the morale of our species, as seen by anthropologists, and the "killer ape" expressed that.

Reviving the predation theme with a new twist, Eickhoff (1988) suggested that our ancestors, while still arboreal, stood upright on tree branches in ambush and used their hands to catch arboreal prey, such as monkeys. This resembles Cartmill's visual predation model for primate origins but has almost nothing to recommend it. It clings to the rejected ideas that protohominins were quintessentially carnivorous and assumes a degree of arboreality much greater than that of the African apes or any plausible ancestor. Finally, it requires a prey biomass density in the canopy that simply does not exist. Even if such a lifestyle were feasible, arboreal predation as observed among modern chimpanzees does not depend on bipedal postures.

Display. A visual aggressive display is effective as both threat and deterrence. Increasing one's apparent stature is a common strategy and could be enhanced by bipedalism (Jablonski and Chaplin 1993; Wescott 1969). However, displays observed in living primates are not sustained behaviors and generally refer to posture rather than locomotion. Many primates (e.g., gorilla, chimpanzees) are bipedal for brief times, especially in static posture, but do not incorporate this into their locomotion. Even in a society with chronic heightened tension, it is difficult to imagine a species responding by a permanent change of posture.

Ravey (1978) suggested that bipedalism evolved for defensive observation in tall grass. Temporary upright stance is a basic primate capability, but we must ask why no other species found it necessary to walk bipedally. This hypothesis, like several others, is rooted in the now-discarded belief that hominins and bipedalism first evolved in a savanna landscape.

Aquatic habitat. The aquatic ape hypothesis explains bipedalism as arising from a phase in which our ancestors lived along the edge of the sea and spent much time wading (Hardy 1960; Morgan 1972, 1982, 1990, 1997). The buoyancy of water might have eased the stresses of the postural transition, assuming our immediate ancestor was not already upright. However, upright carriage of the trunk is already a common hominoid pattern and does not require the assistance of water.

Shallow water foraging is common for proboscis monkeys and has been observed in an island population of Japanese macaques, neither of which seem bothered by the body design and locomotion shared with all Old World monkeys. Apes may wade bipedally to keep their upper bodies dry. They would have to change postures, however, to exploit bottom resources. The aquatic bipedalism model is built on the argument that the adaptation to bipedalism was too difficult to have occurred on land. Such a model must assume dependency on aquatic resources to favor a difficult transformation of the entire species population. However, there is simply no evidence that early hominins were dependent on aquatic habitats or foods and the model is unparsimonious (Langdon 1997).

Food carrying. Transport of food is nearly unique to humans and has profound implications for social and family structure and for a sharing economy (Hewes 1961, 1964). Perhaps the significance of carrying food led to the adoption of bipedalism. Again, however, since food carrying is observed in some primates (e.g., Susman 1984), why does bipedalism not appear in other species? Others have pointed out that human mothers may have needed to carry both young and their food (Cann and Wilson 1982; Lancaster 1978); certainly the use of tools suggests carrying might have been a useful ability (Helmuth 1985). There is no evidence for such behavior in the archeological record—although there are good reasons for why we should not expect to see such evidence—and the idea must be considered unsupported. More important, the implied economic model may not be a valid generalization even for our species.

The emphasis on food carrying coincided with a reexamination of sexism in scientific paradigms. Clearly previous models and hypotheses of "Man the Hunter" had built on the assumption that males were the active agents and that females were passive. This discussion occurred within the larger historical and social context of feminism and redefining the role of women in society. Anthropologists discovered in the 1960s that women in hunter-gatherer societies actually supplied the majority of calories for their families by gathering plant and other small food items.

Lancaster's (1978) model gave females a prominent role in the behavioral changes leading to human emergence and accompanied the feminist movement within paleoanthropology. In this social context, women are seen as a nurturing/civilizing counterweight to the violent tendencies in males. Secondarily, the entry of women into the work force raised fears of the deterioration of the family. By placing renewed emphasis on both family nurturing and economic roles, the hypotheses reaffirmed family values—a favorite political theme ever since.

Seed eating. In 1970, Jolly proposed the "seed-eating hypothesis," based on parallels with gelada baboons. Eating seeds in a grassy plain requires hand-to-mouth motions, dexterity, emphasis on heavy molar action, smaller canines, and a shorter snout. Bipedal shuffling would enhance efficiency, as observed in the geladas. In addition to gelada baboons, an extinct lemur in Madagascar named *Hadropithecus* shows a parallel suite of characters. While seed-eating might explain the hypertrophy of the molars and chewing musculature in *Australopithecus*, the microscopic wear patterns on the teeth reject such a diet. Nonetheless, Jolly's hypothesis was important in turning the focus on human evolution away from cultural models, which tend to project modern behaviors backwards in time, and onto living primates as models. Recently, Kingdon (2003) has revived the argument that squat-feeding for unspecified ground resources was important.

Feeding. Rose (1976) approached the issue with the question, "Why do nonhuman primates assume bipedal postures?" His fieldwork on baboons revealed that bipedalism occurs most commonly when baboons are foraging from trees or bushes at heights over their heads. He suggested that habitual foraging on such foods may have selected for more bipedalism as a more efficient posture and locomotion.

His model has been echoed more recently by Hunt (1992, 1994), who studied chimpanzees and observed similar behavior. Hunt's subjects were not observed travelling bipedally but spent 5.9% of their feeding time upright. A comparable study by Doran (1993b) observed that male chimpanzees traveled bipedally only 0.2% of the time and females not at all; yet males and females fed bipedally 5.4% and 4.6% of the time respectively. When in the trees, males spent 18.5% of their feeding time bipedal.

In response, we might ask why no other species, including baboons, who forage at such levels have become bipedal. Historically we can place Rose's hypothesis in the same context as Jolly's, coinciding with the studies of hunter-gatherers and primates in the 1960s and 1970s. During this time anthropologists became increasingly interested in ecology and ecological descriptions of primates. The more popular environmental movement reinforced the "noble savage" ideal that was explicitly a part of the hunter-gatherer models. The models remain attractive. However, unless we can develop a much more specific idea of early hominin diet, they are untestable.

Demographic model. Lovejoy (1981) built onto the food-carrying theme to develop a more extensive model for human evolution. This hypothesis has received widespread attention and inclusion in textbooks. Lovejoy's reasoning is as follows: The critical and limiting feature of the ape condition is a low rate of reproduction. This fact reflects the limited ability of a female to obtain enough nutrients to sustain repeated pregnancy and lactation. It also leads modern great apes into extinction. The hominin breakthrough occurred when males supplemented the food supply of the females by carrying food back to them. Such behavior increased the reproductive success of the individual male as well as the reproductive rate for the entire species. The model accounts for distinctive features of sexual division of labor and the pooled family economy, while building upon a model of monogamy in which female choice of mates strongly encouraged the fidelity and investment by the male. Lovejoy extended it to account for further peculiarities of human sexual anatomy and behavior, including concealed ovulation, visual sexual signals, frequent intercourse, and frontal position.

Many criticisms have been leveled at the Lovejoy model. The low birth rate of apes (based on observations of a single population) may not be typical of the species (Allen et al. 1982). The higher birth rate cited for hominins is not necessary applicable to modern hunter-gatherers (Isaac 1982; Wood 1982; Wolfe et al. 1982). Monogamy is not a fundamental human characteristic at present, and there is no reason to believe it ever was (Allen et al. 1982; Hrdy and Bennett 1981). Nor is monogamy the only social structure that could provide the economic support for females required by this model. There are independent reasons to believe that monogamy and nuclear families were not typical of early hominins. Finally, in no other monogamous animal is sexual activity used to maintain a pair bond.

Much of the critique of the food-carrying hypothesis applies here. Lovejoy inverted the feminist perspective by giving males the active role of carrying food home to feed the family and once again left the female passive. Not surprisingly, much of the resulting criticism of Lovejoy's hypothesis came from feminist views.

Time and more data have lent support to the demographic hypothesis (Key 2000; Kaplan et al. 2000). Human females do indeed appear to be able to maintain a higher fertility rate than chimpanzees with less investment from their own foraging. However, ethnographic evidence from hunter-gatherers does not support the idea of a pair bond at the basis of this economic unit. Instead, food is exchanged with or provided by other relatives and nonrelatives of the social group. Lovejoy's insights have been further developed in a new series of studies of human life history theory (see Chapter 21), but his model probably cannot explain bipedalism.

Energetics. Rodman and McHenry (1980) observed that chimpanzees (as models for our ancestors) are not particularly good quadrupeds and asked whether bipedalism is actually an improvement in locomotor efficiency (Figure 9.2). Studies of cebus monkeys and chimpanzees, each trained to run on a treadmill both bipedally and quadrupedally, showed that energy expenditure was the same on two legs as on

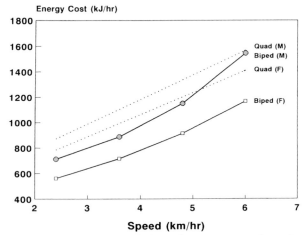

Figure 9.2 Energy cost of bipedalism versus quadruped-alism. This plot compares the relative energy costs of travel in similar-sized mammalian quadrupeds and bipedal humans. Bipedalism is more efficient, particularly at lower speeds. (From W.R. Leonard and M.L. Robertson. 1997, Rethinking the energetics of bipedalism. *Curr. Anthropol.* 38(2):304–309. Copyright 1997. Reprinted with permission from the University of Chicago Press.)

four (Taylor and Rowntree 1973). These results implied that the hominin ancestor would not have become less efficient when shifting to bipedalism. Previous studies had shown that although humans were less efficient than many quadrupedal mammals (Fedak and Seeherman 1979; Heglund 1985), they were more efficient than others, and there were no consistent differences in energy requirements of bipedal birds and mammals compared with quadrupedal lizards and mammals when adjustments were made for size and speed. This observation has been supported by more recent studies (Roberts et al. 1998a, 1998b).

Again, we must ask why, if bipedalism has little or no energetic cost, have chimpanzees not made this switch as well? Why did they specialize as knuckle-walkers? A different habitat and the continuing commitment to arborealism may explain this.

It has been suggested by others that bipedalism might be favored in running or jogging, rather than walking (Washburn 1960; Shipman 1984). However, as speed increases there appears to be an increasing discrepancy in energy use. Obviously humans cannot match the maximum speeds of many quadrupeds of comparable body size, although human running appears to occur with greater endurance than that of many other mammals and is faster than that of other primates running bipedally (Alexander 1991). It is at slower walking speeds that humans become especially efficient (Leonard and Robertson 1995, 1997; Steudel-Numbers 2003).

Shipman (1984) proposed, based on archeological evidence, that early hominins were scavengers and omnivores. Effective scavengers have to cover large areas of ground quickly to find enough to eat. Bipedal jogging is a relatively efficient means of doing just that. Sinclair et al. (1986) noted that scavenging also means following migrating herds over long distances; again, bipedalism might be an adaptation for distance travel. The scavenging hypothesis has yet to be confirmed. More important (as Shipman recognized), bipedalism long preceded the shift to tool-based exploitation of carcasses. The transition from austral-opithecine to modern patterns of bipedalism does correspond to exploitation of more open environments and to increasing evidence of meat consumption, whether by hunting or scavenging. It is likely that this change in ecological niche was the impetus for selection toward greater efficiency (Foley and Elton 1998; Leonard and Robertson 1997).

Thermoregulation. Another recent model is that proposed by Wheeler (1984, 1991a, 1991b), that upright posture exposes less of the body surface area to the tropical sun at midday. Bipedalism might have reduced heat load, exposed the body better to cooling breezes, and enhanced early hominins' ability to exploit the diurnal niche (Figure 9.3). Although this advantage can be quantified, the scenario is entirely speculative. Thermoregulation is a significant consideration, and there are many other physiological adaptations for it; but, again, why only hominins? This scenario is also rooted in the assumption that a savanna environment had a key role in hominin origins. In fact, the fossil evidence for earliest bipedalism occurs in a forest context and a shift to more open country occurs 2 million years later.

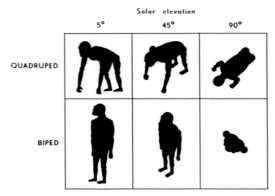

Figure 9.3 Effect of posture on exposure to sun. These silhouettes demonstrate the ability of bipedal posture to reduce exposure to the sun at various angles. A knuckle-walking ape on the top row is compared with an upright hominin on the bottom. (From P.E. Wheeler. 1984. The evolution of bipedality and loss of functional body hair in hominids. *J. Hum. Evol.* 13:91–98. Copyright 1984. Reprinted with permission from Elsevier Science.)

THE PROBLEM WITH CULTURAL MODELS

Humans take advantage of upright posture by using hands for a wide variety of tasks, especially wielding tools and carrying objects. Despite the obvious advantages these functions have for our species, there are problems for any specific evolutionary story that uses them to explain bipedalism. Two questionable assumptions are commonly made.

First, most cultural hypotheses build upon the premise that the reasons for bipedalism can be seen in modern-day advantages. It is thus necessary to assume that the behavior was established in the species and that the species was dependent on that behavior when the transition occurred. Given the realities of the fossil record, it will be difficult to identify the emergence of any specific cultural behavior. Tool use is the easiest behavior to detect, but stone tools do not appear until well over I million years after signs of bipedal locomotion. Therefore, these models are effectively untestable.

Second, most cultural hypotheses assume that the shift to bipedalism meant giving up adapted quadrupedalism. If this were the case, the difficulties of the transition may have been prohibitive. An intermediate stage would hardly have been adaptively preferable to quadrupedalism. Explaining the shift to bipedalism depends upon our first understanding the prebipedal posture of our ancestors.

RECONSTRUCTING THE TRANSITION

How an organism evolves and what selective forces act on it depend completely on its immediate past and present, not on the future. The constraints that the past state of the species places on its later evolution have been termed historical contingencies.

What was the prebipedal ancestor like? The fossil hominoid *Proconsul* serves as a good model for an early hominoid, although not a direct ancestor, primarily because it is well known from a number of remarkably complete specimens (Figure 9.4). Several species of this ape flourished in East Africa during the early and middle Miocene, about 22 to 15 Mya. *Proconsul* had a generalized arboreal quadrupedal mode of locomotion, capable of varying degrees of walking and climbing, but few anatomical specializations for any one specific mode (Walker and Teaford 1989; Ward 1993). Later monkeys and hominoids show many possible types of specializations, for terrestrial running (some monkeys), for brachiation (gibbons), for large-bodied slow suspensory locomotion (orangutans), or for active arboreal climbing (chimpanzees). We can speculatively derive each of these from an unspecialized *Proconsul*-like ancestor (Figure 9.5).

Quadrupedal walking places weight-bearing and locomotor forces on all the limbs. The joints must be constructed with enough stability to withstand such forces. As body size increases or as forces increase with running and jumping, the adaptations for stability generally include greater joint congruity, more precisely fitted articular surfaces, and more narrowly defined motions.

Vertical climbing, which predisposes an animal to bipedal postures, requires longer upper limbs, orthograde trunk posture, grasping hands and feet with long digits, suspensory rather than supportive role for upper limbs, and lower limbs supportive of the trunk. Vertical climbing is incompatible with terrestrial cursorial quadrupedalism. Climbing requires mobile

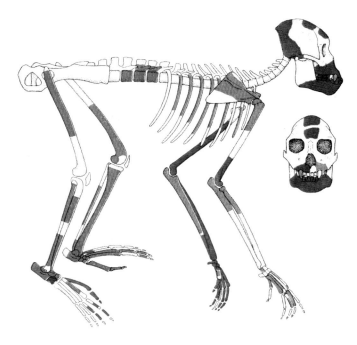

Figure 9.4 Reconstruction of the Miocene hominoid _Proconsul_. This generalized quadruped serves as a good model for the ancestral body form and locomotor pattern of Old World apes and monkeys. (From A. Walker and M. Teaford. 1989. The hunt for _Proconsul_. _Sci. Am._ 260[Jan]:76–82. Copyright 1989 by Tom Prentiss. Reproduced by permisson of the artist.)

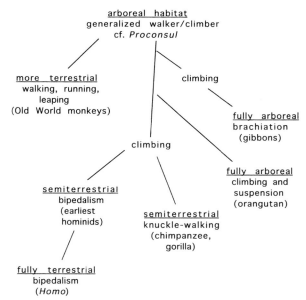

Figure 9.5 Divergence of locomotor patterns from _Proconsul_-like ancestor. _Proconsul_ itself was not an ancestor to the primates on this chart, but it represents a body form and locomotor pattern that was probably similar to those of the ancestor of later hominoids.

joints acting in many dimensions, a laterally facing shoulder that is destabilized, and slow powerful movements. Climbing certainly characterized the common ancestor of humans and African apes and helped to facilitate some degree of bipedalism.

A poorly known contemporary of _Proconsul_ had a body form more similar to that of the modern great apes. _Morotopithecus_ is known from a fragmentary scapula, lumbar vertebrae, and femur but appears to be adapted for suspensory behaviors and more climbing (Gebo et al. 1997; MacLatchy 2004). Thus it represents the evolutionary pathway that our hominoid ancestor eventually took.

The next stage in locomotor evolution involved a shift toward more terrestrialism and the assumption of bipedalism before 4.0 Mya. How do climbing hominoids move on the ground? Gibbons are clumsily bipedal, compromising for the sake of long upper limbs and long grasping toes. Orangutans fist-walk inefficiently on hypermobile joints of the hands and feet. Neither of these species spends time on the ground unless absolutely necessary. Chimpanzees and gorillas knuckle-walk to protect their long digits and unstable joints. In all of these apes, compromises are made in terrestrial travel to preserve climbing abilities. When these species do walk bipedally, their basic movement pattern is not dissimilar to that of humans in kinematic and electromyographic studies of muscle use, when one takes into account anatomical differences (Aerts et al. 2000; Ishida et al. 1975; Okada et al. 1976; Shapiro and Jungers 1988).

Our arboreal ancestors could certainly not have avoided spending some time on the ground. In such circumstances, they may well have walked on fists or knuckles. Early hominin fossils show some upper limb characteristics that might represent weight-bearing by the forelimbs (see Chapter 7). However, the ambiguity of anatomical evidence makes it likely that they were never as specialized in that respect as are modern African apes. The divergence of ape and human lineages calculated by the molecular clock, the appearance of hominins in the fossil record, and the first development of bipedalism occur closely enough together in time (none has been pinpointed more precisely than 6 to 8 Mya) that human ancestors may never have gone far down the knuckle-walking path. As the hominin ancestors adapted to increased terrestrialism even in a forest setting, bipedalism might have been their best option, given their existing climbing specializations.

If this is true, bipedalism should not be seen as a unique event demanding an extraordinary explanation. We should expect many climbing hominoid species to have experimented with it, and that is what we see. Living primate species are all bipedal occasionally. A Miocene hominoid from Italy, *Oreopithecus*, apparently was habitually bipedal when not climbing in the trees (Köhler and Moyà-Solà 1997). Recent discoveries of different patterns of adaptation to bipedalism in early hominins or near relatives, such as *Orrorin* and *Ardipithecus*, raise the possibility that bipedalism evolved more than once in our family (Gibbons 2002). *Orrorin tugenensis*, about 6 million years old, is represented by several femoral fragments, one of which bears a large human-like head and a distinctively human scar for the obturator externus. This small but suggestive bit of anatomy may indicate an advanced degree of bipedalism (Senut et al. 2001). The distantly related *Ardipithecus ramidus* (found at about 5.0 and 4.4 Mya) possessed robust humeri suitable for arboreality. However, a proximal toe phalanx shows human-like hyperextension ability, an indicator of bipedalism (Haile-Selassie 2001).

Subsequent selection on the early hominins improved the efficiency of bipedal posture and locomotion. Some of these changes include reduced climbing abilities of the foot—loss of opposition, fixation of joints, shortening of toes, and lengthening of limb segments. The upper limbs remain relatively unmodified. These changes are observed in varying degrees among the australopithecines.

In a final stage, the australopithecine postcrania evolved into that of *Homo*. We do not fully understand the differences, but it appears (or is assumed) that *Homo* represents a more efficient pattern that would be more compatible with full terrestrialism and possibly long-distance travel. Anatomical changes include lengthening the lower limb, shortening the toes, redesigning the pelvis, and reducing the robusticity of the peroneal musculature.

This scenario permits a gradual shift through a period of perhaps limited arboreal activity (perhaps using trees for sleeping). It recognizes at least two evolutionary events in the hominin lineage that may signal completely different selective pressures, before 4.0 and about 2.0 Mya. Carrying and other cultural adaptations almost certainly came after the first assumption of bipedal posture and walking. Although culture may have had a small role in shaping bipedalism in the final phase, such models as have been proposed are poorly supported by evidence.

10

ORGANIZATION OF THE BRAIN

Although the human brain achieved its current size rather late in our evolutionary history, our brain and its intellectual functions clearly are foremost in distinguishing us from all other life forms (Table 10.1). Nonetheless, despite the enormous cultural gap between ourselves and our nearest neighbors, the gross anatomical plan of our brain is equally clearly the same as that of other vertebrates. Moreover, even the functional uniqueness of our brain is largely a difference of degree rather than of kind from the abilities of our nearest relatives.

The central nervous system derives evolutionarily and embryologically from a simple nerve cord. For each body segment, that cord sends out motor signals to the muscles and receives sensory information. This information is integrated and shared with other parts of the cord to coordinate responses. The vertebrate spinal cord still maintains this functional role, while the brain is a specialization of its cranial end. Although its functions may also be conceptually reduced to input, processing, and output, the brain has assumed the role of initiating and imposing multisegment and body-wide responses over the more localized reflexes.

Table 10.1 Characteristics of the Human Brain

Mammalian characteristics
  Brain larger relative to body size
  Clear development of the cerebrum beyond an olfactory function
  Cerebrum larger in proportion to the rest of the brain
  Neocortex
  Complex limbic system
  Visual processing centered in cerebral cortex
Human characteristics
  Cerebrum extremely large relative to body size
  Lateralization of the cerebrum
  Areas of the left hemisphere specialized for language
  Expanded limbic cortex

Whether we can ever learn exactly how the brain functions and behaves has yet to be demonstrated. Can an organ understand itself? Presently we are pecking at the edges of our ignorance from two directions. Studying the physiology and connections of individual neurons, we can grasp some of their potential for processing complex information. Yet it is apparent that the immense populations of interacting neurons can accomplish far more than models of small numbers are able predict. Examining whole brains, their gross functions, and their disabilities helps us to comprehend what the brain and its parts do; but this method only allows us to theorize about how they work. The brain remains the great frontier in the study of anatomical sciences.

INITIAL DEVELOPMENT OF NERVOUS SYSTEM

The central nervous system first appears developmentally as a hollow cylinder of cells called the *neural tube*. The neural tube develops from the inside out. Actively dividing cells along its inner wall produce neurons and glial cells that are pushed outward. These cells will send out fibers to claim still more space. Thus is formed the basic configuration of the spinal cord, with a small central canal surrounded immediately by gray matter (cell bodies), with nerve fiber tracts, the white matter, running longitudinally along the outside (Figure 10.1). Neurons of the more ventral portion of the gray matter, called the basal plate, will become the motor neurons. Sensory information will enter and be processed by the dorsal gray, the alar plate.

Muscles of the body wall first appear as parts of segments of undifferentiated connective tissues called somites, which line up along the neural tube. At each segment, motor axons from the basal plate project out

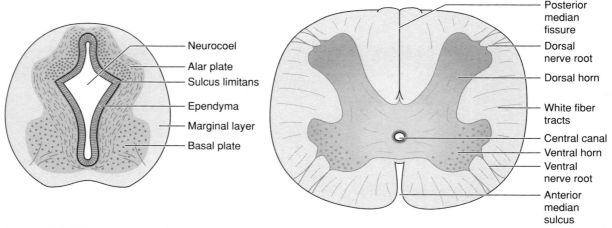

Figure 10.1 Differentiation of the spinal cord from the neural tube. In the early stages (*left*), neurons and glial cells are generated by division of cells in the ependymal layer. The neurons form the basal and alar plates. Cell processes define an outer marginal layer. In the mature spinal cord (*right*), the basal and alar plates become the ventral and dorsal horns, containing motor and sensory relay neurons, respectively. The marginal layer fills with nerve fibers of the spinal tracts.

of the spinal cord to innervate the adjacent somite. These form the ventral root of the future spinal nerve. As mesodermal cells of the somite migrate to appropriate parts of the body wall and mature, axons of the spinal neurons elongate to follow these migrations.

The brain develops from the rostral end of the neural tube. At the end of the fourth week, the three primary vesicles are apparent, distinguished by flexures, or bends in the tube (Table 10.2). The cephalic flexure separates the prosencephalon (forebrain) from the mesencephalon (midbrain). The pontine flexure, which is temporary and appears by the sixth week, divides the rhombencephalon between the future pons and medulla. Caudal to the rhombencephalon, the cervical flexure marks the beginning of the spinal cord. By the fifth week, divisions within the primary vesicles have further developed into the five secondary vesicles: telencephalon, diencephalon, mesencephalon, metencephalon, and myelencephalon.

GENERAL ANATOMY OF THE BRAIN

The brain probably evolved as special sensory organs concentrated at the front of the organism (Figure 10.2). These organs, for smell, vision, sensing vibrations, and equilibrium, enable the organism to scout out the environment ahead, to identify and track down food, and to orient itself. The early brain serves to process and integrate such information.

On a gross functional level, the brain is divided into the brainstem and the forebrain. In lower vertebrates, the brainstem handles immediate processing and coordination of the special senses except for smell and regulates motor activity in the spinal cord and cranial nerves. The forebrain offers a somewhat higher level of integration of somatic senses and handles behaviors associated with learning and emotional drives. Although the major parts of the brain are recognizable across vertebrates, they differ in relative size, corresponding to the importance of their

Table 10.2 Divisions of the Brain

Primary Vesicles	Secondary Vesicles
Prosencephalon (forebrain)	Telencephalon (cerebrum)
	Diencephalon
Mesencephalon (midbrain)	Mesencephalon (midbrain)
Rhombencephalon (hindbrain)	Metencephalon (pons and cerebellum)
	Myelencephalon (medulla oblongata)

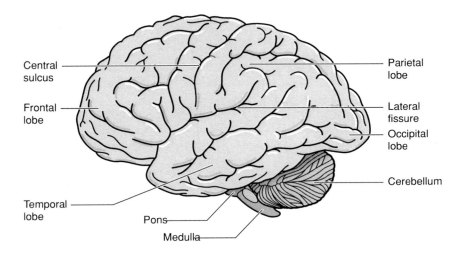

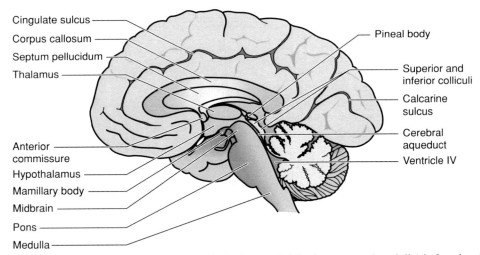

Figure 10.2 The human brain and its major parts. The brainstem (midbrain, pons, and medulla) is functionally and structurally continuous with the spinal cord.

function. Historically there has been a tendency to emphasize continuity and similarity, but the more profound are subtle and require much more effort to discern (Northcutt 2002; Preuss 2001).

THE BRAINSTEM

Structurally, the brainstem represents a gradual transformation of the spinal cord. The caudal part is named for the positioning of the spinal tracts on the outside:

myelencephalon means "myelinated brain" because the myelinated fiber tracts are conspicuously on the surface. (Myelin is a white protein used to insulate many of the long fibers of the nervous system.) Likewise, *medulla* refers to the positioning of the gray matter internally, in the medulla or middle of the structure. The pons, just rostral to the medulla, is distinct only in external appearance. The pons bears the cerebellum on its back, and the massive bundles of fibers entering and leaving the cerebellum give the pons a distinctive bulge. The major tracts and columns of nuclei present in the medulla continue through the pons and into the midbrain.

The midbrain bears two pairs of nuclei visibly protruding on its dorsal surface. These are the colliculi ("little hills"). The more rostral ones (superior colliculi in humans) have a visual function. They are well developed in prominent optic lobes in most vertebrates and are centers for visual processing and mapping of visual space. They are greatly reduced in both function and size in mammals. The cerebrum has taken over most of the sensory function, and the mammalian colliculi are mostly concerned with movements of the eyes and posture of the head. The more caudal nuclei form auditory centers. These are small in fish and much larger in tetrapods, reflecting the relative significance of hearing in those groups. Again, in mammals, much of the function of sound has been relocated in the cerebrum, and the inferior colliculi are relatively small.

Of the gray matter, or nuclei, of the brainstem, the most important components are the columns of cranial nerve nuclei and the reticular formation. The cranial nerves handle the somatic senses of the head, the special senses, eye movements, chewing and swallowing, and speech. The reticular formation is a collection of nuclei extending throughout the brainstem in which neurons and fibers are intermingled. This region of grossly uniform appearance contains nuclei of very different functions. Overall, the reticular formation handles both sensory and motor information, both somatic and visceral activities. Some of its more important functions include the regulation of sleep and consciousness, the regulation of autonomic functions, and the regulation of muscle tone and reflex sensitivity.

The cerebellum, riding on the back of the pons, monitors sensory input from all over the body. The best-understood function of the cerebellum relates to the motor system, for which it coordinates muscle activity with the changing state of the body. The cerebellum is connected to the pons and thence to the rest of the body by the cerebellar peduncles, containing large numbers of incoming and outgoing nerve fibers. In mammals, the cerebellum also assists the forebrain in sensory processing and other tasks. Evidence of the

past decade indicates that the human cerebellum participates in cognition and language (e.g., Aksoomoff and Courchesne 1992; Bower and Parsons 2003; Daum et al. 1993; Gao et al. 1996; Leiner et al. 1993; Middleton and Strick 1994).

Like the rest of the nervous system, the brainstem is a hollow structure. The small central canal of the spinal cord expands between the pons and cerebellum to form the fourth ventricle. More superior ventricles in the cerebrum and diencephalon drain inferiorly through the cerebral aqueduct into the fourth ventricle. Cerebrospinal fluid is secreted by choroid plexus in the ventricles and by lining tissues. It drains caudally through the ventricles and then from the fourth ventricle out into a space between layers of meninges surrounding the central nervous system before it joins the bloodstream.

THE FOREBRAIN

The forebrain includes the cerebrum, which structurally dominates the rest of the human brain, and the diencephalon, which is nearly entirely hidden under the cerebrum.

Diencephalon

The diencephalon ("twin brain") consists of a series of mostly paired structures, or regions, including the thalamus, hypothalamus, epithalamus, and the eye itself (Figure 10.3). The thalamus is a gateway to the cerebrum, receiving all incoming sensory information except the sense of smell. It performs some processing of that information and probably is involved (in humans, at least) with rudimentary consciousness. The hypothalamus is the lower portion of the diencephalon. It contains numerous small paired nuclei that regulate the autonomic functions of the body. Through the limbic system, it mediates the visceral components of emotions. The output of the hypothalamus is both neuronal (via the autonomic nervous system) and humoral (via hormones).

The epithalamus is a region superior to the thalamus that contains a variety of small structures, including pathways of the limbic system. The habenula is a nucleus involved with the sense of smell. In laboratory rodents, it appears to mediate parenting behaviors. The

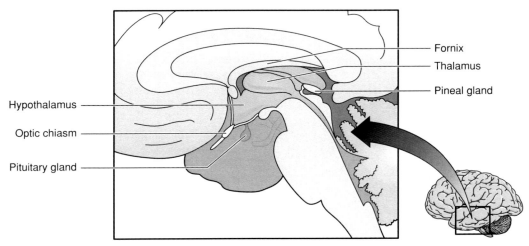

Figure 10.3 The diencephalon.

association between smell and behavior is fundamental and provides the basis for social communication with pheromones (see Chapter 12). The pineal gland is another, unpaired, structure of the epithalamus. It secretes the hormone melatonin in a rhythm tied to the daily cycle of light and dark (see further discussion later). In many lower vertebrates, the pineal is a "third eye" containing light-sensitive neurons. In mammals, it simply receives indirect connections from the retina of the eye.

Cerebrum

The cerebrum is the most variable division of the brain among vertebrate groups. In fish, it is present as an olfactory lobe, processing the sense of smell. Smell is a crucial factor determining behavior, making this region of the brain as reasonable as any other for the eventual development of more complex functions. The cerebrum reaches its greatest development in mammals. Structurally, it consists of a series of nuclei called the striatum, or basal ganglia, that lie ventral to the ventricles and multiple layers of neurons on the surface comprising the cortex. Deep to the cortex are fibers interconnecting neurons of the cortex and carrying signals in and out of the cerebrum.

The expansion of the cerebrum in mammals is not merely an increase in size. New parts of both the striatum and the cortex evolve. The cortical neurons

in fish and amphibians primarily receive olfactory fibers. These form regions called the *paleocortex* and *archicortex*, which visually consist of three layers of neurons. Homologies of these layers persist in mammals but are much reduced in proportion and are greatly overwhelmed by the *neocortex*, containing six layers of neurons. (Reptiles and birds have also added an area of cortex, but it is relatively simple and not homologous to the mammalian neocortex.) The neocortex provides a basic plan on which further evolution could occur as different areas expanded or were elaborated in different groups of mammals (Krubitzer 1995; Northcutt and Kaas 1995).

The basal ganglia also underwent changes. The paleostriatum (globus pallidus in mammals) in fish is the major processing area of the cerebrum. It receives input from the paleocortex and a smaller number of other sensory signals from the thalamus. The output descends to motor areas of the brainstem. As the neocortex expands on the mammalian cerebrum, corresponding new basal ganglia are present. These are the caudate and putamen, also called the neostriatum. While they are most commonly associated with the motor system, the basal ganglia are better understood as parallel processors assisting the all parts and all major functions of the cortex.

Although general trends in relative brain size increase can be identified throughout the fossil record, the greatest leaps occur in the contrast between reptiles either mammals or birds. What does this expansion and reorganization in the mammalian brain signify? Jerison (1973) postulated that a major correlate of this trend was increasing integration of the senses in the cerebral cortex and a mapping of the environment as a "perceptual world." In lower vertebrates, the senses are used primarily to scan the environment and its immediate state. Mammals and birds are more likely to plot the environment and its resources across space and time. Sensory information can be compared and analyzed according to this mental map, and there is a greater capacity for integrative memory. Certainly this integration is consistent with the shift of visual and auditory processing to the mammalian neocortex. Perhaps more important, the higher vertebrates can project beyond the present. They can direct behavior according to a knowledge of resources not immediately tangible or even according to future resources. Such a perceptual world in some sense permits an animal to explore alternate hypothetical possibilities for action or behavior and to make decisions. These are the foundations of consciousness and abstract thought.

The cerebral cortex is divided into lobes named according to the adjacent bones—frontal, parietal, occipital, and temporal. The outer surface or cortex of the human cerebrum consists of gray matter that is highly convoluted to increase its surface area and the volume available for the cell bodies of the neurons. Ridges in the cortex are called gyri, while the grooves are sulci. The contours of the cortex permit identification of specific regions of the cerebrum that are to be associated with specific functions, as discussed later.

The human cerebrum handles conscious mental activity, including perception, learning and memory, decision-making, personality, language, and aesthetics. It is divided into right and left hemispheres, which communicate by means of a large bundle of fibers called the corpus callosum. Although the forebrain is often considered the evolutionary culmination of the brain, performing its most sophisticated functions, it does not regulate the rest of the nervous system. The regulation of sleep and states of consciousness appears to be a function of the brainstem. The onset of sleep, periods of dreaming, and awakening all correlate with projections from the reticular formation.

FUNCTIONAL ORGANIZATION OF THE BRAIN

Just as the brain can be divided by superficial anatomy into cerebrum, diencephalon, midbrain, pons, and medulla, so can it also be divided longitudinally into a number of functional systems. These systems interact and overlap anatomically. The sensory system is responsible for monitoring and interpreting the environment of the brain, both within and outside of the body. The motor system initiates and coordinates the muscular response. The autonomic nervous system regulates the behavior of visceral organs. The limbic system handles learning and memory and coordinates consciousness with the autonomic functions.

These systems share certain principles of organization. (1) Projections and functional systems in the nervous system may be identified as either specific or nonspecific in their targets. (2) Many brain functions occur in specific locations within the brain. (3) Sensory and motor connections with the body are mapped spatially onto the brain in a way can be described by outlining images of the body itself on the cortex and nuclei. (4) Functional systems were constructed through evolutionary time, sometimes by adding new layers of organization and control. (5) Longitudinal systems, such as sensory and motor pathways, cross the midline, so that each hemisphere of the forebrain relates to the opposite side of the body. (6) Many higher-level functions are handled by specific areas of the cortex that are not symmetrical, so these functions are preferentially lateralized to one hemisphere or the other. (7) Nervous system function is closely integrated with other body systems, especially the endocrine and immune systems. These principles are discussed in greater detail later.

SPECIFIC AND NONSPECIFIC PROJECTIONS

Neuronal projections within the brain may be described as targeted purposefully on individual neurons (specific projections) or as distributing diffusely over functional classes of neurons (nonspecific projections).

Specific projections are responsible for the detailed operations of the brain, such as sensory interpretations, motor commands, and learning and memory. The functions we are commonly most interested in when relating the human brain to experience are these specific connections. They are shaped by a combinations of genetic and experiential factors. Fetal axons establish connections by following molecular tags within the developing brain. Once they have found their target tissues, the brain nucleus and target may migrate far apart from one another as the body continues to increase in size or targets, such as muscles, move to another region. The synapses formed in this way are in great excess of what the brain can use or support. However, those synaptic connections that are used are reinforced. Subsequently, unused connections and neurons that have failed to establish connections are weeded from the system. Thus early experience plays a significant role in determining the mature form and function of the brain.

Nonspecific projections determine how the brain will function, in a general sense. For example, the wide broadcast of a certain neurotransmitter or combination of neurotransmitters from the brainstem and diencephalon may change the sensitivity of parts of the forebrain to incoming stimuli. These contribute to changing states of consciousness. In other parts of the cerebrum, nonspecific projections may adjust responsiveness of the brain, or mood. Although nonspecific functions are easier to study chemically than to appreciate at a functional level, they are the only components of the brain that we are able to manipulate by pharmacological medical intervention.

CONSCIOUSNESS

Consciousness, defined physiologically rather than psychologically, is a description of the general level of activity of the brain and its responsiveness to external stimuli. Although the acts of consciousness mostly lie in the cerebral cortex, its regulation appears to be a more primitive function and may be sought in the brainstem. Centers in the reticular formation appear to be responsible for nonspecific suppression of cortical activity during sleep and equally nonspecific arousal of the forebrain to wake us up. The observation that destruction of a nucleus within the reticular formation in a cat prevents the animal from sleeping shows us that sleep is an active brain function and not merely the absence of activity. The diencephalon is now known to also have an active role in regulating sleep. The levels of sleep, of which there are several, are part of a wider spectrum of states of consciousness. While we can shift sometimes rapidly from one level of alertness to another, the brain also follows daily cycles of sleep and wake called circadian rhythms.

During normal sleep, the body is mostly paralyzed by a general inhibition of the motor system from the brainstem on down. Although the senses may operate, the reticular formation and thalamus apparently screen them out. Only strong and unusual stimuli trigger the reticular activating system and wake us up. Both onset and termination of sleep periods appear to depend on sites in the brainstem. A different phase of sleep is characterized by the present of rapid eye movements (REM). REM sleep correlates with activity in a nucleus of the midbrain called locus ceruleus, which projects directly to the thalamus and cerebral cortex. The locus ceruleus projects the excitatory neurotransmitter norepinephrine and stimulates a general increase in the level of cortical activity similar to that of wakefulness. The characteristic eye movements may be motor commands that have escaped the braistem suppression because they did not need to descend beyond the midbrain. REM periods often correspond with dreams.

An explanation of the necessity of sleep and dreams has been elusive. It is commonly asserted that humans sleep more deeply, if not longer, than other animals. Is this related to the larger size and higher metabolic demands of our brain? Hypotheses about rest and repair functions must confront the fact that simple rest is not an adequate substitute for it. A correlation has been argued between the need for sleep and glucose depletion in the brain (Gillis 1996). Certainly our large brain has a tremendous appetite for glucose and cannot function without it. However, sleep is not merely the passive absence of conscious activity. It appears to play an important role in both memory and insight (Huber et al. 2004; Mednick et al. 2004; Wagner et al. 2004). It incorporates another more active, essential, and even more curious state of mind—dreaming. Individuals deprived of sleep for several days are generally able to function adequately at necessary tasks but lack motivation for anything additional. Individuals deprived of REM sleep may begin to hallucinate (dream?) during their waking hours.

All other mammals sleep, and at least many of them have REM periods, which are generally associated with dreaming. The mammal with the highest frequency of REM periods is, surprisingly, the platypus, a mammal possessing many primitive traits, including a relatively small brain (Siegel et al. 1998, 1999). Evidence is accumulating that dreaming is a part of memory consolidation (Fischer et al. 2002; Maquet 2001; Siegel 2001; Stickgold et al. 2001). Kavanau (2002) described an intriguing model for the evolution of sleep and dreams as a mammalian adaptation to allow synapse recovery and memory consolidation from a day's activity. Compared with reptiles, mammals have a much higher level of continuous activity of motor and other neural circuits, so that refreshing of the neurons cannot occur without turning the syste off. In addition, memory consolidation and maintenance (i.e., learning) are also more important and more effectively performed among mammals. Such a hypothesis is consistent with much of what we know about the brain but, like many such evolutionary scenarios, difficult to evaluate and to distinguish which aspects of sleep were the original target of selection.

FUNCTIONAL LOCALIZATION

Brain functions reside in specific regions of the brain. There will be predictable relationships between the locations of injuries and the functional losses experienced by the patients. Such functions may be mapped on the cerebral cortex (Figure 10.4). While this concept was a significant conclusion from research that culminated earlier in this century, we are now appreciating the fact that not all functions are discretely localized.

The primary motor cortex lies along the posterior border of the frontal lobe. This contributes many of the descending motor commands on the corticospinal tract. The secondary motor cortex, just anterior to it, assists in organizing and planning motor activity. Broca's area at the base of this region on the left side of the brain deals with complex motor activities such as speech and manipulation. Other anterior and inferior regions of the frontal lobe, the frontal association area, are responsible for the expression of emotions, motivation, and planning.

The parietal lobe has a primary somatosensory cortex alongside the primary motor cortex. This receives ascending information from the body. More posteriorly are the secondary sensory cortex and parietal association cortex for the interpretation

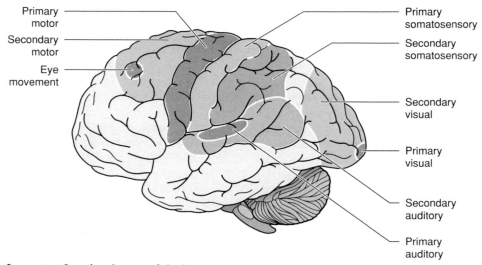

Figure 10.4 Important functional areas of the human cortex.

of sensory input. Taste is processed in the mouth area of the primary sensory cortex and equilibrium may have its own region nearby. Much of what we consider rational thought, including language processing, calculations, and recognition of information gained from sensory pathways, also occurs in the parietal association area.

The occipital lobe is mostly concerned with vision. A primary visual cortex lies on the medial side of the lobe, while higher levels of visual processing occur in secondary and association areas. The visual cortex of primates is huge and spills out of the occipital cortex into adjacent parts of the parietal and temporal lobes.

The temporal lobe contains the primary and secondary auditory cortex. On the left hemisphere of the brain, the secondary auditory cortex appears to be specialized for the processing of the sounds of speech. This is sometimes known as Wernicke's area. The medial side of the temporal lobe contains an infolding of cortex called the hippocampus, which is an important center for learning. Other cortex medial to the temporal lobe processes olfactory information.

Recently, it has been possible to use real-time imaging to extend localization into much more detail, identifying regions of the cortex responsible for remembering and interpreting faces, for example, or for interpreting emotions.

SOMATOTOPIC ORGANIZATION

Most pathways in the brain maintain a somatotopic organization. The arrangement of fibers within these pathways and especially at the termination of the pathways reflects the spatial arrangement of the neurons or receptors from which that pathway arose. For example, the touch/pressure fibers of the dorsal columns of the spinal cord are arranged within those columns in a specific manner. The more medial fibers come from the lower spinal levels, while the lateral fibers represent progressively higher spinal levels. This permits signals from the foot to ascend close together and next to those of the leg, which are next to those of the knee, etc. As the pathways ascend and synapse first in the brainstem, a second time in the

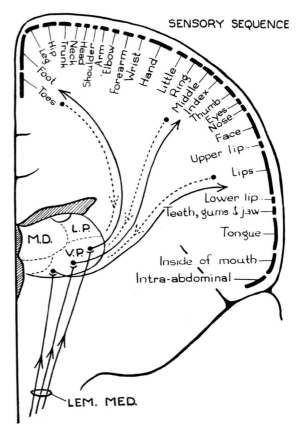

Figure 10.5 The somatotopic arrangement of touch/pressure sensations. (From W. Penfield and L. Roberts 1959. *Speech and Brain Mechanisms.* Copyright 1959 by Princeton University Press. Reprinted by permission of Princeton University Press.)

thalamus, and again in the cerebral cortex, that spatial arrangement is preserved. The body can be mapped on the surface of the cortex according to the origin of signals arriving in that region (Figure 10.5). It can also be mapped on the cortex of the cerebellum. Other pathways also preserve spatial arrangement, including those for motor commands, hearing, and vision. In general, all sensory systems have not one, but multiple maps of the sensory surface in the appropriate area of the cortex. It would appear that this organization helps the brain to make sense out of the sensory input and to put bits of information into larger pictures.

PHYLOGENETIC LAYERING

The functional organization of the brain to some extent reflects its evolutionary history. Newer levels of control may be represented by additional specialized brain structures superimposed upon older systems. For example, the brainstem of lower vertebrates handles most motor and somatic sensory functions. The integration of those functions into mammalian consciousness required that the cerebral cortex assume some access to both sensory perception and movement. This was accomplished by adding these controls rather than replacing existing ones or by expanding existing structures and adding new functions.

THE MOTOR SYSTEM

The motor system is a good example of the complexities resulting from phylogenetic layering. It consists of those centers of the brain and spinal cord that determine the behavior of skeletal muscle plus the spinal motor neurons and their peripheral fibers (Figure 10.6). All of this encompasses a diverse array of functions, including background muscle tone, involuntary reflexes, conscious and unconscious adjustments of posture and equilibrium, programmed motor patterns such as locomotion, novel or unique voluntary movement, and learned motor patterns.

The organization of the motor system appears to reflect the successive rostral placement of higher controls. The segmental nervous system, wholly contained in the spinal cord, produces organized motor responses (reflexes) to sensory stimuli (Poppele and Bosco 2003). The primary motor neurons lie in the ventral horn of the spinal cord, surrounded by interneurons that receive and integrate signals from other parts of the nervous system. Stimuli from sensory receptors and other parts of the nervous system thus may reach the motor neurons either by a single synapse or indirectly via the spinal interneurons. If the signals arrive from outside the central nervous system, the responses are considered to be reflexes. Such responses may be complex, integrating a variety of types of sensory information. A human example is the withdrawal reflex. A sudden painful or otherwise unexpected touch stimulus to one of the limbs may initiate a quick involuntary jerk away from the stimulus. The muscles

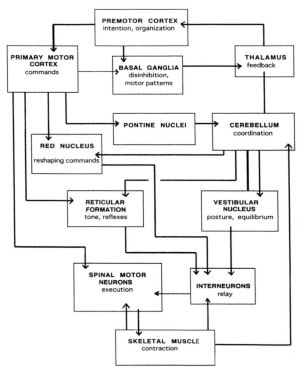

Figure 10.6 Major components of the motor system and their principal connections.

involved are usually flexors but may represent whichever motions are needed to move away from the stimulus. A more elaborate expression of the response may include extensors on the opposite side that help to stabilize the body. Because multiple muscles and spinal levels are involved, this is a complex motor behavior initiated within the spinal cord.

In fish, a locomotor center lies in the brainstem reticular formation and initiates muscular contractions that sweep alternately down the two sides of the body (ten Donkelaar 2001). Both the vestibular system and the cerebellum are active participants in motor control. Because of the importance of the vestibular system in determining the orientation of the organism and in righting reflexes, the vestibulospinal tract is also one of the more primitive descending pathways. The cerebellum is an important center for the convergence of sensory information, including vestibular input, so that motor control can take into account the state of the body.

The motor cortex appears in mammals with the elaboration of the cerebrum but does not necessarily assume motor control. Cats were the subjects of classic experiments that deduced the roles of different parts of the motor system. It was found that destruction of the cerebral cortex did not destroy the ability of the cat to stand, maintain its balance, and even walk on a treadmill. These are functions of the brainstem, including the vestibulospinal and reticulospinal systems, which gives a preferential stimulation to "antigravity" muscles that keep the animal in a stable standing position. Since animals are not selected to function in a decerebrate state, we can best understand these as stabilizing background motor patterns on which the cerebral cortex can overlay more complex behaviors.

The cats' walking pattern may be anatomically located in a nucleus of the midbrain reticular formation called the mesencephalic locomotor region. This center receives input from the cerebrum and also projects more inferiorly. It appears that rather than taking motor functions away from the brainstem, the motor cortex supersedes it by handling higher-level activities relating to behavioral drives and memory—the cortex decides what to do; the brainstem still determines how to do it.

Later laboratory studies of monkeys found that the primate motor system functions in a significantly different manner from that of a cat or other nonprimate mammal (Capaday 2002; Eidelberg et al. 1981; Vilensky 1989). A primate that has experimentally lost its cerebrum cannot stand or walk. "Antigravity" muscles appear to be organized differently from those of other quadrupeds and are less consistently defined for the forelimb. This may reflect a different fundamental posture for primates, perhaps a sitting one, in which the forelimbs are not weight-bearing. Alternatively, it may be a secondary effect of other changes in the motor system.

Primates also have no mesencephalic locomotor region. It appears that primates have shifted a greater proportion of control over motor behavior to the cerebrum, so that the brainstem centers are less effective at independent operation (Heffner and Masterson 1983). Surprisingly, the motor cortex and corticospinal tract in monkeys and humans are not necessarily critical components of gross voluntary motor behavior. Loss of small areas of the motor cortex may not result in permanent loss of function. Precise experimental lesions in the medulla of just that tract in monkeys (or similar rare lesions of human) interfere with fine finger manipulation and little else (Hepp-Reymond 1988). Pathways from the cortex connecting through brainstem centers apparently are also important for gross movement. A simultaneous loss of the corticospinal and other descending tracts is more commonly seen in spinal cord injuries and results in paralysis.

We can attempt to understand these unique primate patterns, first, in terms of different uses of the limbs in locomotion and, second, with reference to the greater importance of manual manipulation in primates. The upper limb is no longer merely a part of the locomotor/support system. Its increased independence proved to be useful in the later bipedalism of hominins and the acquisition of tools and material culture. However, the presence of an underlying quadrupedal motor control may be identified in humans (Dietz 2002). Upper limb movements are coordinated with those of the lower by a circuitry within the spinal cord. This may be observed in both infants and adults crawling on a treadmill. Again, this reflects a superposition of a primate pattern on top of a more fundamental mammalian pattern of control.

Yet another set of players in motor control are the basal ganglia. These nuclei within the cerebrum receive input from all parts of the cortex and help to shape its activity, including motor functions. To prevent spontaneous activity of the motor cortex from triggering muscular contractions, the basal ganglia appear to have a role in general inhibition of motor activity. Signals of voluntary activity from the cortex produce a selective disinhibition (cancellation of the inhibition) pertaining to selected muscles and free them for use. This explains why disorders of the basal ganglia may express themselves either as spontaneous movement (failure of inhibition) or as difficulty in initiating movement (failure of disinhibition). Basal ganglia also appear to store some species-specific motor programs, such as some locomotor controls and grooming and mounting behaviors. Such stored programs increase efficiency by reducing the need for the cortex to plan often repeated activities each time they are needed.

DECUSSATION

All pathways reaching or leaving the forebrain and midbrain decussate, or cross the midline. This includes descending and ascending commands to or from the cerebrum, thalamus, red nucleus, and superior colliculus. However, individual tracts do not necessarily cross at the same level. The corticobulbar fibers, from the cerebral cortex to the brainstem, cross within the brainstem. Corticospinal fibers, carrying motor commands for the spinal motor neurons, cross together just below the pyramids in the medulla.

The sensory tracts ascending from the spinal cord to the thalamus also cross at different places (Figure 10.7). Pain fibers on the spinothalamic tract cross at levels throughout the spinal cord. Touch fibers of the dorsal columns cross in the medulla. The right half of the cerebrum thus receives sensations from and sends commands to the left side of the body and visa versa. Vision and hearing are also treated preferentially by the opposite hemispheres. The adaptive meaning of this pattern, if any, is unknown.

Clinical consequences of these patterns are quite clear. A sensory or motor tract or nucleus may be injured on one side above its decussation and display a functional loss on the opposite, or contralateral side of the body. Stroke in the motor cortex, caused by interruption of blood flow and death of neurons, commonly results in weakness or paralysis of muscles of the opposite side. An injury that destroys one half of the spinal cord at a given level will cause loss of pain and temperature sensation of the contralateral side below the lesion, because those fibers have already crossed, but also paralysis and loss of touch and pressure sensations on the same (ipsilateral) side, because these fibers have not yet crossed.

FUNCTIONAL LATERALIZATION

Structurally and functionally, the left and right sides of the nervous system are nearly symmetrical. The outstanding exception to this occurs in the cerebrum, where there is a tendency for many higher functions to be better expressed on one side or the other. For example, the left hemisphere is most commonly dominant in language; the right hemisphere is usually dominant

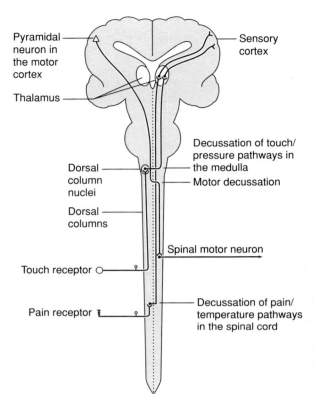

Figure 10.7 Decussation of long spinal pathways. Longitudinal pathways connecting the forebrain with the spinal cord and brainstem cross the midline. Three of the major systems are shown. The pyramidal tract for voluntary motor activity (shown descending from the right side of the brain) crosses in the lower medulla. Touch/pressure sensations (shown ascending from the right side of the body) cross in the medulla before climbing to the thalamus. Pain and temperature climb only briefly in the spinal cord before they synapse, cross the midline, and continue to the thalamus.

in spatial manipulation and musical ability. This does not mean that the nondominant hemisphere is silent or cannot perform the given function. Rather, both generally work together. The dominant hemisphere is more active, and damage to that hemisphere is more likely to result in a functional loss. Lateralization of function has proved to be more incomplete, individualized, and less consistent than early studies suggested and popular ideas continue to suggest.

LANGUAGE AND LATERALIZATION

Language is one of the better studied functions of the human brain, in part because it is easier to isolate and observe a patient's language abilities than any other cortical function. Language is also significant as it is one of the few traits that draws a clear boundary between humans and other living species. Aspects of language are processed on both sides of the brain, but the left side is dominant in language functions. Specialized language areas appear only on the left side. A brain injury to the left hemisphere is more likely to interfere with language than is an injury to the right side. A patient whose corpus callosum has been cut, severing communication between the two hemispheres, is more likely to respond verbally to stimuli addressed to the left hemisphere.

On the basis of brain injury patterns observed since the nineteenth century, electrode stimulation studies of patients undergoing open-brain surgery, and more recent studies of metabolic activity in regions of the cortex, two areas on the left side of the brain have been identified that have important roles in language (Figure 10.8). These were named after the anatomists who described them, Wernicke's area and Broca's area.

Wernicke's area, part of the supplementary auditory cortex in the left temporal lobe, is responsible for auditory imaging of sounds, including words. It enables a person to retain and examine a sound in short-term memory or mentally to "sound out" a written word. Memories for individual words appear to be localized in the temporal cortex. Damage to Wernicke's area interferes with the ability to recognize linguistic meaning in sounds, including the sounds on one's own speech. This condition has been known as Wernicke's aphasia but is better called receptive aphasia. Such patients can hear, but not understand, spoken language. To the extent that a reader may silently sound out words as he or she reads, receptive aphasia may also interfere with reading comprehension.

Broca's area, in the lower lateral frontal lobe, has been associated with the production of speech. The most recent studies have distinguished a language area just anterior to Broca's area that helps to formulate motor patterns of verbal expression. Broca's area is thus reassigned to the role of executing complex motor sequences (Bookheimer 2002; Fox 1989; Montgomery 1989), including but not limited to those of speech. It also seems to play a role in processing musical syntax (Maess et al. 2001). Damage to the general region of Broca's area may cause expressive aphasia, or "Broca's aphasia," involving difficulty in speaking but not in comprehending language.

These two functional regions, Wernicke's and Broca's, create physical as well as functional asymmetries in the human brain. Many other parts of the brain are essential for language and communication, even though they are not visibly asymmetrical (Bookheimer 2002; Lieberman 2002; Scott and Johnsrude 2003; Taylor and Regard 2003). Parts of the accessory visual cortex process written language.

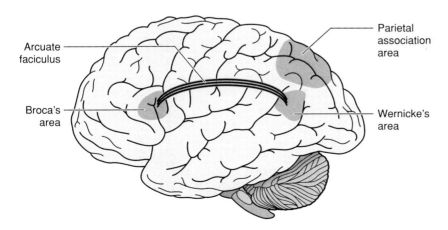

Arcuate faciculus

Broca's area

Parietal association area

Wernicke's area

Figure 10.8 Major language areas of the brain. Damage to Broca's or Wernicke's area was long ago associated with disruptions of language processing. However, numerous other areas of the brain are also involved with both language use and language pathologies, including the arcuate fasciculus connecting these two areas and the parietal association area.

The parietal, frontal, and temporal cortices are involved in perceiving communication in humans and other primates. The parietal association cortex interprets language signs, spoken or written, as one of its many functions. The inferior frontal cortex participates in semantic processing. Regions of the motor and sensory cortex for the face, tongue, larynx, and hand participate in expressing language. Basal ganglia may participate both in motor sequencing, in syntax, and in learning. The right hemisphere is preferentially responsible for expressing and interpreting prosody, the nonlinguistic aspects of communication, such as facial expression, intonation, and hand gestures. It also participates in processing metaphors and nonliteral speech.

Lesions and disorders elsewhere in the cerebral cortex reveal some of the processes by which language is handled. Patients with aphasias (linguistic deficits) may have general problems of comprehension or expression, but sometimes their disabilities are remarkably specific (Bower 1994; Caramazza and Hillis 1991; Cubelli 1991; Damasio and Tranel 1993; Hinton et al. 1993; McCarthy and Warrington 1988; Sutherland 1990; Vigliocco 2000). Individual subjects have been described with an inability to process vowels, but not consonants; an inability to see the last half of a written word, whether written normally or even backwards; an inability to process language relating to a specific subject, such as color or people; an impairment in retrieving verbs; or an inability to match plural and singular nouns and verbs. Each of these disabilities occurring in isolation gives us a clue about the mechanisms the brain uses to process language. Imaging studies on normal patients are also beginning to reveal a remarkable level of specificity. Recordings of the behavior of individual neurons may show increased activity when the individual is shown a particular face or a particular word but not a different one. Different areas of the cortex are responsible for processing word meaning from those responsible for handling word syntax.

The specialization of the human brain for language is evident in several ways (Cheour et al. 2002; Dehaene-Lambertz et al. 2002; Jackendorff 1994; Kuhl et al. 1992, 2003; Moskowitz 1988; Peterson et al. 1990; Petitto et al. 2001; Pinker 1998; Saffran et al. 2001; Zatorre et al. 1992). The left auditory cortex responds preferentially to speech sounds over non-language sounds. The secondary visual cortex reacts differently to verbal and nonverbal stimuli, including pronounceable versus unpronounceable strings of letters. The left hemisphere shows specialization for manipulating the symbolic properties of language. The brain appears to have predetermined patterns for organizing language (i.e., grammar). Human infants show a specialized facility for perceiving and learning language, whether spoken or gestured, and even while sleeping, that begins in the first few months of life and proceeds not only by imitation but also by reconstructing language within a preexisting pattern of grammatical rules. What can we say about the evolutionary roots of such specializations?

THE EVOLUTION OF LANGUAGE AND LATERALIZATION

There has been extensive speculation on the origin of speech and language. Most of that discussion remains speculation. Language origins have been linked to the lateralization of the brain because language was the first function to be recognized as lateralized and because physical asymmetries in the cortex are associated with language areas. There is not yet complete agreement on whether lateralization is a human trait or occurs earlier in primate ancestry. Nor is it clear why functions need to be lateralized or whether a single explanation will suffice for all lateralized functions.

Lateralization in Primates

There have been several attempts to document lateralization in nonhuman primates, mostly involving animals performing very specific and unnatural manual tasks under unnatural conditions. The resulting observations, plus field studies, have enough inconsistencies to leave the subject open to debate (Boesch 1991; Falk 1987; Hopkins and Morris 1993; Marchant and McGrew 1991; Mittra et al. 1997; Palmer 2002; Panger 1998). Lateralization has generally been examined through preferential hand use, although that is only a single dimension of lateralization in humans. Nonhuman primates often show individual preferences for hand use for a given task but that preference may change for different tasks. Complex tasks are more likely to be lateralized than simple tasks. Finally, as with humans, it is likely that a focus on a simple dichotomy of left hand versus right hand may obscure a more complex underlying pattern (McGrew and Merchant 1996). While it appears likely that lateralization and functional asymmetry have deep phylogenetic roots, we can say little about their origin or adaptive significance.

Structurally, Broca's area and Wernicke's area have long been thought to be uniquely human features. Although there is some asymmetry in size and patterning of the fissures of the cortex among some anthropoids, none match the human pattern. In aspects of gross dimensions, humans show small but consistent differences between the sides—the occipital lobe is a bit longer on the left and the frontal is longer on the right. Such asymmetries are absent in monkeys and lower primates. Among the great apes they may be expressed to a lesser degree and are more variable (Gilissen 2001). Only recently have homologues of Wernicke's area and Broca's area been identified in the brains of chimpanzees and other great apes (Cantalupo and Hopkins 2001; Gannon et al. 1998, 2001; Sherwood et al. 2003). It appears that even the structural correlates of language in humans show continuity with the brains of nonhuman primates.

Evidence of Language in Nonhuman Primates

A search for the antecedents of language in nonhuman primates leads to two questions: Are there any behaviors under natural conditions that foreshadow the development of language? Do any other primates have a capacity for the use of human language?

Many primates have fairly complex systems of vocal communication (Mitani et al. 1992; Seyfarth 1986; Snowdon 1990; de Waal 1997a). These systems include a small but distinct vocabulary, may show differences among regional populations, and can be used in sophisticated behaviors, such as deceit. Primate communication gives us a baseline for reconstructing the evolution of human language, but there still is a significant gap separating our species in this regard (e.g., Arcadi 2000; Burling 1993; Ploog 2002). Furthermore, animal calls differ from human language in the lack of abstract content, the absence of grammar, and a general inflexibility in incorporating new concepts. This has led some to question the relevance of calls

to language. Burling (1993) argues that primate communications are more closely related to our gestural and nonverbal signals and that human language evolved from a different set of cognitive functions. However, different, but parallel circuits may be involved in auditory and gestural aspects of speech (Scott and Johnsrude 2003).

Experiments in teaching human language to chimpanzees and gorillas in captivity has a long history. Reports of their use of sign language since the 1960s claimed remarkable success, as measured by the size of vocabulary (hundreds of words) and the spontaneous use of signs in interactions with humans (e.g., Linden 1974; Patterson and Linden 1981). Critics claimed the researchers were overly generous in their interpretations and that the apes were responding to cues (Terrace et al. 1979). Rigorous criteria for evaluating language that included grammar and syntax, spontaneity, ability to express new concepts, and make abstract references were met, but such rules are poor measures of natural language use and fluency (Wallman 1992). Complicating the studies were a diversity of methods and testing standards and the natural differences in the abilities of the subjects themselves. Some animals, such as the chimpanzee Kanzi (Savage-Rumbaugh 1994; Savage-Rumbaugh and Lewin 1994) and even the grey parrot Alex (Caldwell, 2000; Pepperberg 1999), are clearly superior to their conspecifics in their ability to learn and use human language. Certainly a rudimentary ability to communicate with abstract symbols is present in chimpanzees and gorillas, and this has revealed some fascinating insights into the minds of apes. However, apes differ substantially from humans in the effort needed to master language, and the level of language usage, at best, may be surpassed by that of human children. Language studies are moving beyond the question of whether apes can use language to better understand the nature of the differences in how they can learn it and use it (e.g., Parker and Gibson 1990; Savage-Rumbaugh et al. 1998).

Fossil Evidence for Language

When did language evolve? There is little direct evidence in the fossil record. Asymmetry in the manufacture of stone tools may indicate predominant right-handedness as far back as 2.0 million years (Toth

1985). Some hominin skulls and natural endocasts (imprints of the internal surface of the braincase) may bear information about patterns of sulci in key areas of the brain. For years a debate has raged concerning whether the australopithecine brain contours more closely resembled those of apes or humans or indeed whether any relevant details can be deciphered on them. Some specimens suggest a reorganization of the visual cortex had already occurred in *Australopithecus afarensis* (Holloway et al. 2001), but there is no evidence of language-related asymmetries on them. Modern patterns of the brain, including Broca's area, are more convincingly present in early species of *Homo* (Tobias 1987, 1995*)*. Because comparable asymmetries and probably homologies of both Broca's and Wernicke's areas have now been observed in chimpanzees, such observations are no longer convincing evidence for language. An independent line of investigation relates to the anatomical design of the respiratory tract that makes speech possible (see Chapter 15). While all of this evidence suggests that the brain could have been reorganized for language at any phase of the genus *Homo*, it is difficult to imagine what evidence we might reasonably turn up that would tell us definitively whether language was present in a fossil population.

Other clues, equally ambiguous, may come from signs of symbolic behavior in the archaeological record. The use of mineral pigments, manufacture of crudely shaped figurines, and carving marks on stone or bone all appear several hundred thousand years ago. Ochre, possibly used for color symbolism, has been found in a Middle Stone Age site from Africa (Barham 2002) and later from Qafzeh Cave 92,000 years ago (Hovers et al. 2003). A lump of rock from Israel may have represented a female figure about 233,000 years ago (Bahn 1996); a more convincing figurine from the Acheulean culture of Morocco is not well dated (Bednarik 2003). Incised marks on stone or bone appear at Bilzungsleben and other Lower Paleolithic sites in Europe (Bednarik 1995). These may or may not represent complex communication and information processing before anatomically modern humans. While our ability to recognize symbolic behavior in the archeological record has been debated (e.g., Bednarik 1995; Chase and Dibble 1987; Lindly

and Clark 1990; Marshack 1989), some researchers believe that the acceleration of technological innovation and the worldwide appearance of anatomically modern humans within the past 100,000 years are significantly linked with the invention of language in that time period.

INTEGRATION WITH OTHER SYSTEMS

The nervous system has all too often been studied and taught in isolation from the rest of the body. However, it is more than a command center sending signals on long slender axons to other organs. The boundaries of the anatomical system and its regulatory activities blur into other systems. These connections become especially apparent in the neuroendocrine system. The fact that in our brains these unconscious regulatory functions are themselves interacting with consciousness suggests a more complex gradation between human and nonhuman brains.

THE NEUROENDOCRINE SYSTEM

The somatic nervous system, producing our sensations and muscular actions, encompasses most of our conscious awareness. However, there is much that the body must do beyond this. The visceral functions such as digestion, respiration, movement of fluids around the body, adjustment of temperature, elimination of wastes, and regulation of growth and repair need to be coordinated for the entire body and also come under the umbrella of operations of the nervous system. Because these functions are carried on largely without conscious involvement, they are termed "autonomic."

The brain has two pathways of action for autonomic functions (Figure 10.9). Signals may be sent out to body organs on nerves. This network forms the peripheral autonomic nervous system. Alternatively, the brain can release chemical signals, called hormones, into the bloodstream to broadcast instructions. These humoral pathways constitute part of the "endocrine system." Although usually anatomically distinct, the autonomic nervous system and the endocrine system overlap significantly in function. Furthermore, the more we learn about chemical signals within the body, the

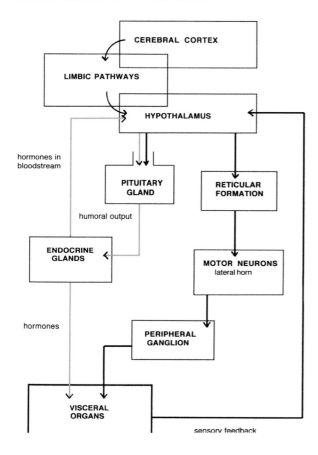

Figure 10.9 Functional connections of the neuroendocrine system.

and this precision is enhanced by the presence of enzymes and reuptake channels in neighboring cells to remove transmitters nearly as fast as they are released. Hormones in the bloodstream are broadcast broadly and potentially interact with large numbers of cells. While hormonal signals act more slowly, the hormones themselves may persist in the blood for much longer periods of time. Clearly hormonal communication is much more expensive in terms of material and energy than activity at a synapse. On the other hand, neuronal networks require much prior investment in constructing the axonal pathways.

The brain center most responsible for regulating autonomic functions is the hypothalamus, a cluster of nuclei in the diencephalon forming the floor of the third ventricle. It receives information from a number of sources. Ascending fibers from the brainstem carry sensory information directly from the visceral organs. The limbic system reports on the emotional and cognitive state of the cerebrum. Through the bloodstream, the hypothalamus can monitor other aspects of the internal state of the body, including blood chemistry, osmotic pressure, and temperature.

The output of the hypothalamus is both neuronal and humoral (hormonal). Ascending limbic tracts of neurons may influence conscious states, producing emotions. Descending nerve tracts mostly regulate operations of the reticular formation, but some pass directly to the spinal cord. These signals oversee sympathetic and parasympathetic activities. The hormones released by the hypothalamus may either regulate the activity of the pituitary gland or pass directly into the bloodstream to be broadcast throughout the body.

The Autonomic Nerves

The autonomic nervous system has two complementary divisions: the sympathetic nervous system and the parasympathetic nervous system.

The sympathetic nervous system is a set of peripheral nerves that are activated together (in "sympathy" with one another) to assist the body in mobilizing for activity. This sympathetic arousal is commonly described as occurring in stressful situations to prepare the body for fear, flight, fight, or sex. The specific effects of sympathetic stimulation are many and varied but mostly relate to the increased distribution of nutrients

less distinct the endocrine system becomes. For this reason, the notion of an "endocrine system" is here discarded in favor of a more functionally defined neuroendocrine system that includes both pathways.

Neurotransmitters and hormones work in very similar ways. They are chemicals released by one cell that bind to receptors on or within a second cell, carrying instructions or information. The primary difference is that hormones are released into the bloodstream to seek out target cells at great distances in the body, while neurotransmitters cross only tiny synaptic clefts between neurons. This difference in action has important implications. Neurotransmitters can be delivered very precisely in time and space;

to the skeletal muscles. Sympathetic functions include an increase in the rate and force of heart contraction, redirecting of arterial blood away from the viscera and to the limbs and body wall, dilation of the bronchi, inhibition of glands and muscles of the digestive tract, production of sweat, erection of the body hairs, and release of hormones from the suprarenal gland. The senses are generally sensitized and put on alert. It is also probable that sympathetic fibers inhibit tissues of the immune system. Recent discoveries that autonomic fibers penetrate lymphatic tissues—including tonsils, salivary glands, and thymus—and bone marrow, where red and white blood cells are produced, are consistent with observations that immune functions are depressed during sympathetic arousal.

The parasympathetic nervous system consists of autonomic efferent fibers that, unlike sympathetic functions, are not necessarily related to one another in objective and that may be elicited independently of one another. Although these functions often are directly opposite those of the sympathetic system, it is incorrect to think of the parasympathetics as merely an opposition to the sympathetics. Parasympathetic functions are regulated independently of sympathetic ones, and sometimes are complementary to them. They include slowing the rate of heartbeat, constricting the respiratory passages in the lungs, stimulating motility and glandular secretions of the digestive system, and emptying the bladder. Sexual arousal and climax represent an interplay between both sympathetic and parasympathetic systems.

The Neuroendocrine Axis

The neuroendocrine axis describes a pathway for signals to the body that begins in the hypothalamus where it projects to the pituitary gland, or hypophysis, regulating its release of hormones. These hormones are varied in number and function. The pituitary gland has two functional parts. The posterior part, or neurohypophysis, consists of axons projected from the hypothalamus along a slender connecting stalk called the infundibulum. Hormones secreted by cells in the hypothalamus are carried down the infundibulum to the pituitary, where they may be released into the bloodstream.

The anterior portion, the adenohypophysis, consists of glandular cells that manufacture different hormones. The hypothalamus is connected to the adenohypophysis by several portal blood vessels, which drain venous blood from the hypothalamus and carry it into small channels in the pituitary called sinusoids. This vascular connection permits the hypothalamus to secrete hormones into the blood to give instructions to the glandular cells. The hypothalamic hormones are generally named releasing factors or inhibiting factors for their actions on pituitary cells.

The pituitary has been called the "master gland" of the endocrine system because it regulates the activity of some, but not all, other endocrine glands. Among these glands are the thyroid, which regulates body metabolism; the suprarenal gland, which regulates blood levels of minerals and other nutrients; and the gonads, which release sex hormones. Additional pituitary hormones regulate growth, milk production, urine concentration, and some smooth muscle contraction.

Neuroendocrine Integration:
The Stress Response

Stress is any perceived threat to the well-being of the body. It may arise from sensory observations or from the imagination. It may reflect a physical threat or an emotional one. The hypothalamus is the center for coordinating the body's response to such dangers, and it does so through several channels. The limbic system alerts the conscious mind to danger and enhances sensory alertness. The sympathetic nervous system is switched on to prepare for action. A cascade of hormones is released to further support activity.

The sympathetic arousal was described earlier and accounts both for cardiovascular and respiratory facilitation and for the temporary suppression of maintenance functions, such as digestion. Sympathetic signals also cause the release of norepinephrine and epinephrine, hormones of the suprarenal medulla. Norepinephrine and epinephrine work together to stimulate cardiac output further. While they generally exert the same effects on the visceral organ systems as the norepinephrine released by the sympathetic neurons, these hormones respond more slowly but have longer lasting effects.

Simultaneously, the pituitary is stimulated to release adrenocorticotropic hormone (ACTH) and its byproduct, beta-endorphin. The endorphin acts in the brain to suppress pain and elevate mood. ACTH stimulates the release of hormones by the cortex of the suprarenal gland. These include cortisol, which reduces inflammation and immune activity, and other corticoids that cause the release of nutrients into the bloodstream. Other effects in the pituitary include increased output of thyrotropin to stimulate metabolism and interference with normal reproductive cycling.

This synergy between sympathetic outflow and hormonal pathways is clearly adaptive in producing an intense and immediate arousal to enable us to meet whatever dangers present themselves. In the environment of the past, such dangers were more often life-or-death crises—facing predators, prey, or enemies, for example. An occasional rush of adrenalin (epinephrine) is actually beneficial to keep the body in shape.

In modern urban society, stress is more likely to be caused by economic, relational, or occupational uncertainties. The physiological stress response is an ineffective weapon for meeting these challenges. Moreover, because these are usually chronic sources of anxiety, the stress response works overtime. The result is bad for our health. Elevated blood pressure, depression of the digestive system, immune suppression, and interruption of reproductive function take their toll when maintained for long periods and account for many contemporary disease patterns.

INTEGRATION WITH THE IMMUNE SYSTEM

Although there has long been thought to be an influence of mood and mind on health, only gradually is the physiological connection between brain and the immune system beginning to emerge (e.g., Bateman et al. 1989; Blalock 1989; Cotman 1987; Pert 1986; Pert et al. 1985; Plata-Salaman 1991; Sternberg 2000). Autonomic fibers have been traced to nearly every immune tissue, including bone marrow, thymus, spleen, and lymph nodes. The detail and complexity of the interaction, however, is bewildering. Knowledge is still piecemeal.

One of the major sources of the interaction involves the corticoids (suprarenal cortical hormones) and their controls. ACTH and the related beta-endorphin have direct enhancing effects on macrophages, lymphocytes, natural killer cells, and other immune components. However, the release of corticoids (such as cortisol) that ACTH causes has an inhibiting effect and can seriously depress immune responses. Medically administered cortisone is an important means of reducing inflammation.

The immune system, in turn, can act on these hormone centers. Macrophages, when challenged by antigens, release substances that trigger secretion of corticotropin-releasing hormone in the hypothalamus and ACTH in the pituitary and simultaneously inhibit the response of the adrenal cortex to ACTH. Considerable fine-tuning is involved, but the system can be upset by external influences, such as stress, lack of social support systems, and, presumably emotional state. Such interactions will certainly play important roles in the future of medicine.

The wall of the intestine has been offered as a small-scale model of what neuroimmune cooperation can accomplish (Wood 1991). The neurons there form a "mini-brain" with sensory and effector pathways plus an integration of information. Large numbers of mast cells in the gut monitor for antigens. At the appropriate stimulus, the mast cells release histamines, which directly excite the neurons and cause a rapid emptying of the intestine to rid the body of invading pathogens. The result is "sudden-onset diarrhea."

The exchange of information among these systems —conscious thought, autonomic nervous pathways, endocrine glands, immune tissues—appears to be so extensive that it is no longer possible to appreciate any of them in isolation from the others. As we learn more about the digestive, excretory, reproductive, and other systems, it is clear that they participate in mutual influence with the nervous system.

INTEGRATION WITH CONSCIOUSNESS: THE LIMBIC SYSTEM

The limbic system is conceptualized as a physical and functional "boundary" that separates and joins the cerebrum and lower levels of the brain. It specifically includes the more primitive elements of the cerebral cortex—the paleocortex and archicortex. Functionally, it deals with three topics: olfaction, emotion, and memory. Through these it provides the basis for cerebral influence on the autonomic system. While these three functions may not appear to have any intrinsic overlap, aside from their all being necessary brain functions in even the simplest of vertebrates, it should be obvious that they are closely interrelated. For example, we can draw upon our own experience to attest that odors have the potential for evoking strong memories and emotions. Similarly, strong emotions may affect, for better or worse, our ability to etch an experience in memory.

The limbic system is poorly defined anatomically. It consists of a series of structures interconnected by fiber tracts and implicated in limbic disorders. Their individual roles are not clearly understood. As usually described, the limbic system involves parts of the cerebral cortex, basal ganglia, and much of the diencephalon, along with numerous tracts and connections among these. Most important are its direct two-way links to the neocortex and consciousness on the one hand, and to the hypothalamus and neuroendocrine system on the other.

Emotions

The concept of emotion is not well defined. Emotions themselves are not discrete, and it is impossible to make a comprehensive listing of the different emotions to everyone's satisfaction. We could include such obvious ones as rage, lust, hunger, fear, anxiety, and pleasure. We might note that their origin may be visceral (hunger), external (a threatening animal), or entirely cerebral (erotic daydreaming). As a working definition, we can consider emotions as forebrain activities that involve two critical components—a subjective feeling or perception and an autonomic response. This corresponds nicely with our concept of the limbic system as the interface between the cerebrum and the hypothalamus. This anatomical localization of emotions is supported by a list of clinical and experimental observations on individual components of the limbic system. Studies of emotions, especially in laboratory animals, naturally concentrate on the behavioral and physiological expression of emotions rather than on

subjective feelings. These give ample evidence of the relationship between the limbic structures and controls of autonomic functions (e.g., Konner 2002).

Some emotions provide behavioral drives, such as hunger and lust, and to that degree can be understood as an important basis of behavior for vertebrates. However, more abstract emotional states and moods are commonly observed in nonhuman animals (Bekoff 2000; Plutchik 2001). Emotions are very important in the complex social interactions of many mammals and even negative or painful emotions may be considered adaptive by directing our efforts away from unsuccessful pursuits (Fredrickson 2003; Nesse 1990, 1991, 1999). Emotions may enhance memory and, indeed, become a part of the memory, as discussed below (e.g., Dolan 2002).

The Reward System

Experiments with implanted electrodes in rats in the late 1950s revealed certain centers of the brain that the rats would stimulate voluntarily if given the opportunity (Routtenberg 1978; Schultz 2000). Offered a choice between pressing a bar to deliver a stimulus to these "pleasure centers" or a bar to deliver food, the rats would opt for the electrode up to 100 times per minute, even as they starved. These studies gave rise to numerous hypotheses about such questions as the nature of pleasure, emotion, memory mechanisms, and drug addiction. The structures and pathways implicated have been termed the "reward system" of the brain. They include, with varying degrees of self-stimulating behavior, regions of the hypothalamus and brainstem. A tract between the frontal cortex and the hypothalamus known as the median forebrain bundle appears to be critical to this functional system. The transmitter dopamine appears to be central to this pathway, but it may relate more to identifying rewarding stimuli than to the sensation of pleasure itself (Spanagel and Weiss 1999).

A simple interpretation of this system is that it maintains vitally important drives, such as those for food and sex, by providing positive feedback (Schultz 2000). This hypothesis gives an adaptive foundation to emotions, particularly to pleasure (Cabanac 1992). The reward system is also implicated in memory encoding, in that continued stimulation blocks learning. Probably a pleasurable stimulus accompanying an experience is conducive to remembering that stimulus (and thus continuous stimulation does not permit any new experience to stand out).

Memory

Learning and memory represent a special type of function for the brain that causes permanent changes in it. The addition of information must correspond to some physical process that constitutes learning and the long-term changes that we call memory. Investigations of what those changes are and how they are produced are beginning to produce answers.

There are many types of memory and possibly many different memory mechanisms in the brain. For example, memory can broadly include our knowledge of events and experiences, conditioned responses, spatial familiarity with our surroundings, recognition of faces, recognition and use of words, data storage, learned motor functions, memory of body movements, and procedural knowledge (such as how to speak, tie a shoe, turn a key, and tell time). We cannot say how many of these are manifestations of significantly different mechanisms. At least two have been distinguished: declarative (recollection of facts or events) and implicit (including skills and classical conditioning).

Two parallel anatomical pathways have been identified as encoding sensory experience into memory (LeDoux 1994; Mishkin and Appenzeller 1987; Squire and Zola-Morgan 1991; Zola-Morgan and Squire 1993). The first involves a circuit from the neocortex to the hippocampus (part of the paleocortex), anterior nucleus of the thalamus, and prefrontal cortex and the basal forebrain (parts of the frontal lobe). The second uses information directly from the sensory relay nuclei of the thalamus to the amygdala (a basal ganglion) and then to different regions of the diencephalon before communicating with the basal forebrain. Complete impairment of learning requires disruption of both of these circuits—both the hippocampus and amygdala, for example. The structures comprising these circuits are necessary for encoding but not for recall once the memory is consolidated. The memories themselves appear to be "stored" in disparate but discreet locations across the cortex, especially in the frontal lobe (Smith and Jonides 1999).

These two circuits may be differentiated according to the role of the amygdala in emotion (Cahill and McGaugh 1998; LeDoux 1994). Bilateral loss of the amygdala prevents a subject from learning the appropriate emotional (i.e., autonomic) response to a stimulus, even though that subject may learn a factual association between the stimulus and its emotionally meaningful content. (Experimenters paired a color with a loud startling sound. Subjects understood the pairing but did not display a startle response to the color alone.) On the other hand, subjects with a bilateral loss of the hippocampus did not learn the factual association, but did display the autonomic response to the conditioning stimulus. A subject with bilateral loss of both amygdala and hippocampus showed no learning.

The close association of memory encoding with the limbic system is apparent from these circuits. Emotional intensity of an experience or emotional stress very definitely affects the likelihood of encoding a memory. Memories of emotion do not require that the experience register in the cortex or in declarative to be encoded. The presumed pathway involves direct projections from the thalamus to the amygdala.

11

THE HUMAN STRATEGY

Brain Size

Brain volume tends to increase through evolutionary time. This is a pattern that is revealed in the fossil record within a number of mammalian groups, but humans appear to have benefitted the most by this trend. Superficially, larger brains would seem to imply more complex behavior and greater intelligence. At the levels of species, where we generally assess brain size, and of individuals, where natural selection operates, the significance of bigger brains is not clear; the correlation with intelligence is weak, at best. Multiple factors contribute to determining brain size and only some of them relate to function, yet constructing and maintaining a large brain is expensive. It has been estimated that the active human brain can consume as much energy as the resting metabolism of the rest of the body. What adaptive significance does greater intelligence—much less a larger brain—possess that could justify such a drain on resources? Is our uniquely inflated cerebrum a cause or consequence of our complex behavior and culture?

COMPARING BRAINS

THE RELATION OF BRAIN AND BODY SIZE

Larger animals have absolutely larger brains than do smaller animals of similar design and functional ability. Presumably the increase in body tissues requires additional neurons to use or maintain them. Thus the largest living animals, whales, have the largest brains. To make meaningful comparisons of functional differences of brains among species, it is necessary to make allowance for body size. The most obvious standard would be to compare brain size as a percentage of body mass. What would this mean? Would we expect "equivalent" brains in different-sized animals to be in the same percentage range?

If brain function simply depended on the ratio of neurons to body cells and if cell size did not vary between species, we would expect the ratio of brain weight to body weight to be constant. This sort of ratio might make sense for motor output, for example, where an increase in the number of muscle fibers would have to be accompanied by an increase in the number of motor neurons. Cutaneous receptors should be proportionate in number to the surface area of the skin, but surface area increases less quickly in proportion to mass. For some neural functions, however, body size is not important. For example, sensory receptors monitoring blood pressure or water balance in the body would not need to multiply with body size. The actual relationship of brain to body mass reflects the sum of the needs of the different physiological systems and other functional circuits.

Prediction and observation agree that increasing body size should necessitate an increase in brain size, but not in a constant ratio. A predictive equation for brain size has been worked out empirically for placental mammals by Martin (1990):

$$\log_{10} B = 0.755\log_{10} b + 1.774$$

where B is brain weight and b is body weight, both measured in grams (Figure 11.1). The constants in this equation vary slightly within different taxonomic groups.

Various other factors that determine brain and body size also affect the relationship between them. For example, body mass is quite responsive to nutritional levels and activity levels during the growth periods. Brain size is also sensitive to nutrition but generally only during briefer critical periods. Although both

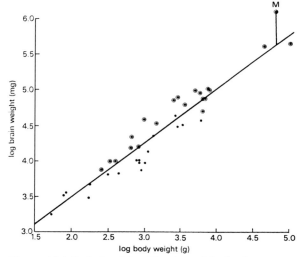

Figure 11.1 Relationship of brain and body size among primates. This logarithmic plot shows a regular relationship between them. Humans (M) have a brain size significantly greater than predicted for our body size. (From *The Human Primate* by Richard Passingham 1982. Copyright 1982 by W.H. Freeman and Company. Used with permission.)

BRAIN SIZE AND FUNCTION

Many aspects of primate life appear to be correlated with brain size in some way (Aboitiz 1996; Armstrong 1990; Barton 1996; Clutton-Brock and Harvey 1980; Dunbar 1998; Martin 1988; Sawaguchi 1988, 1989). Comparisons of relative brain size among primate families show a correlation between brain size and diet, home range size, group size, and social structure. Brain size appears to relate to overall metabolic level and to gut development (Armstrong 1985; Hofman 1983a; McNab and Eisenberg 1989), but these are also a reflection of diet, and the causal relationships among diet, metabolism, gut, and brain size have not been untangled. The brain also scales with life history parameters, including longer gestations, periods of immaturity, reproductive life spans, and longevity (Harvey and Krebs 1990; Hofman 1983a, 1984). What can these correlations tell us about the evolution of human brain size and function?

Which regions of the brain account for changes in brain size? Finlay and her colleagues reported a survey of primate, bat, and insectivore species showing that the size of different parts of the brain were highly correlated (Finlay and Darlington 1995; Finlay et al. 2001; Kaskan and Finlay 2001). They argue that nonolfactory parts of the brain are under similar developmental controls and thus enlarge together. Selection for enhancement of one brain area would affect others. If this is true, it may be very difficult to identify specific factors leading to brain size increase. Other researchers find linkages between functionally related structures, particularly within mammalian orders (Barton and Harvey 2000; de Winter and Oxnard 2001). Thus, for example, the size of the vestibular nuclei, monitoring equilibrium, correlates well with the size of the cerebellum but not with that of the amygdala, part of the limbic system. Such a finding implies that the functional systems have evolved independently of one another.

Passingham (1982) reviewed data on the sizes of different parts of the brain in attempt to identify those that have shown differential growth specifically in humans. He found the medulla is close to the predicted size for anthropoid primates, but other areas, especially the cerebellum and neocortex, are disproportionately expanded (also see MacLeod et al. 2003). The

have some genetic determination, brain size has less variance than body size. Because of the different influences on brain and body size, we might expect to see a different pattern of ratios within species than between closely related species.

The relationship between brain and body size varies within different groups of vertebrates or even within different groups of mammals, although certain patterns are evident, as Figure 11.1 indicates. Compared with other mammals, primates as an order have a slightly greater relative brain weight, or encephalization quotient. Within primates, monkeys and apes have higher values than prosimians. The largest relative brain sizes for nonhuman mammals (i.e., dolphins) overlaps with the figures for the monkeys and apes. Human brains are significantly larger than this equation would predict. They exceed the quotients of monkeys and apes by a factor of 3 and those of primitive mammals by a factor of more than 20. This result is not surprising, but we must attempt to understand its significance.

neocortex and lateral cerebellum appear to be evolutionarily linked in size and function (Whiting and Barton 2003). Historically, it has been a commonplace to assert that expansion of the frontal lobes, believed to be the center of intellect, was crucial in human evolution. Careful measurements of the frontal lobe with magnetic resonance scans have now refuted that idea (Semendeferi et al. 2002). The percentage of cortex given to the frontal lobe is approximately the same in humans and great apes.

The key questions concerning brain evolution revolve around the selective pressures that brought about the capabilities we now observe. More fundamental is the problem simply of discerning the components of intelligence. When did these abilities appear in our history, and what immediate benefits did they confer? Several models for the evolution of mammalian and primate brains are discussed later. They attempt to describe or explain different dimensions of the brain and therefore are not mutually exclusive or even easily compared. Each, however, represents an interesting perspective on what the foundations of intelligent behavior might be.

THE RATE OF HOMININ BRAIN EVOLUTION

We have still to account for the extraordinary development of the human brain size beyond that of other primates. The degree of relative expansion observed in the human brain appears to be a unique characteristic of our lineage. *Australopithecus* has a cranial capacity absolutely similar to that of the great apes (Table 11.1, Figure 11.2). However, because *Australopithecus* had a lesser body size, this indicates a small relative increase in brain size. Brain expansion takes another step with the genus *Homo* and continues at least until about 100,000 years ago.

Understanding the cause of brain expansion requires an appreciation of the pattern of brain growth. Given the spottiness of the fossil record, it is difficult to draw precise species boundaries around populations to compare brain size. The groupings of fossils in Table 11.1 are somewhat arbitrary. It is even more difficult to assign precise dates to each of the fossils. It is impossible to determine both body and brain sizes for more than a handful of individuals. Either one must use arbitrary species designations to which body size estimates may be added (Figure 11.3), or one must assume an insignificant role of body size in hominin brain evolution and use the data in Figure 11.2 at face value. Therefore, the chronology of brain enlargement is still a matter of dispute. The basic question is whether brain increase has been gradual or episodic. Several authors have interpreted the fossils to indicate discrete events of brain expansion followed by longer periods of stasis or slower increase (Cronin et al. 1981; Hofman 1983b). Others find the evidence consistent with a complex pattern of more continuous expansion, accelerating in the last 700,000 years (Lee and Wolpoff 2003; Trinkaus and Wolpoff 1992).

Table 11.1 Brain Size of Fossil Hominins and Living Hominoids

Observed ranges and means of brain size (cm³)

	Mean	Range
Chimpanzee	400	282–500
Orangutan	397	276–540
Gorilla	469	340–752
Modern *H. sapiens*, male	1450	
Modern *H. sapiens*, female	1350	

Cranial capacities of fossil hominins (cm³)

	N	Mean	Range	Date of Specimens
Australopithecus	29	487	387–650	3.2–1.2 Mya
Early *Homo* species	9	651	510–870	1.9–1.5 Mya
H. erectus group	59	1052	600–1450	1.75–0.05 Mya
Neanderthals	23	1435	1200–1740	160,000–40,000BP
Early *H. sapiens*	50	1469	1120–1748	500,000–38,000BP

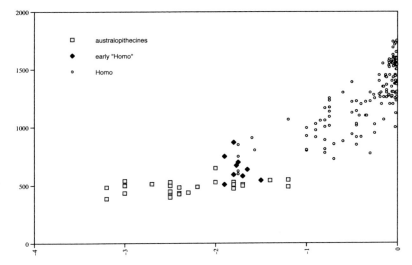

Figure 11.2 Increase in cranial capacity over time. Capacities of 180 fossil hominin crania are plotted over time (millions of years before the present). Australopithecine brain size changes only slightly over time, while that of *Homo* shows a steady increase. However, the sample cannot resolve the question of whether the increase was continual or episodic. Cranial capacity is reported in cubic centimeters, unadjusted for body size.

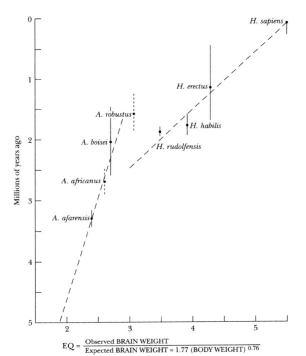

$$EQ = \frac{\text{Observed BRAIN WEIGHT}}{\text{Expected BRAIN WEIGHT} = 1.77 \text{ (BODY WEIGHT)}^{0.76}}$$

Figure 11.3 Increase in encephalization over time. A plot of brain weight relative to body size (encephalization quotient [EQ]) over time reveals a clear pattern of increase. However, this is based on single average values of body size for each species and cannot reflect variation, overlap of species and the arbitrary nature of classification of fossils. (From G.C. Conroy 1997. *Reconstructing Human Origins: A Modern Synthesis.* Copyright 1996 W.W. Norton.)

An even more serious problem with this conventional discussion of brain evolution is the current trend in classification that splits hominins into multiple lineages, most of whom are not ancestral to later humans. If Asian *H. erectus* and Neanderthals are independent species from our African lineage, for example, by definition they are not our ancestors. Including them in an analysis of brain size such as that in Figure 11.2 is valid only if we assume that the brains of our actual contemporary ancestors were subject to the same selection and other evolutionary factors as these species and are otherwise similar in size through time. Bruner et al. (2003) tested and rejected that model. They observe that humans outside of Africa (*H. erectus, H. heidelbergensis, H. neanderthalensis*) have followed a different pathway to achieve brain expansion than modern humans, in which the latter showed an expansion expecially of the parietal region. However, if our analysis were to be limited to certain ancestors of humans, we would be left with a uselessly small data set.

What is the significance of this controversy? Gradual and steady growth of the brain supports a hypothesis that larger brain size is good for its own sake. There thus is a steady directional selection to increase brain size, tempered by other constraints such as energy budget, delayed maturation, and obstetric complications. If size increases steadily, then the qualitative differences observed in brain function between apes and humans either emerged gradually or are not tied to episodes of sudden change in brain size.

The punctuated equilibrium or episodic model suggests a series of specific evolutionary stages, each one demanding an adaptive explanation. An episode of brain expansion either represents the conquering of a given constraint to brain size or it represents a functional breakthrough within the brain itself. One interpretation of brain evolution argues the human brain is qualitatively different from that of apes and that it has undergone rewiring as well as expansion. Each episode of brain enlargement in the fossils might then correspond to a reorganization of the cortex and a significant adaptive advance.

Both of these models have significant implications and raise interesting possibilities. Each is consistent with current observations of the fossils and with some models for brain evolution.

SELECTING FOR BRAIN SIZE

Do brain size and function correlate well? Relationships are not obvious within species, but Gibson et al. (2001) argue that incremental increases of absolute size can be important. If so, adaptive models need to explain not only how incremental increases in size and function are advantageous but also why other primate and mammalian lineages have not taken the same evolutionary pathway. What circumstances common to anthropoids, for example, have led to greater encephalization than that exhibited by prosimian primates? Why do apes have relatively larger brains than monkeys, and humans the largest of all? Three models of long standing, involving society, diet, and the environment, respectively, probably all have some validity. However, we are not yet able to tease out the precise variables to answer all of our questions.

THE SOCIAL MODEL

Living in a complex society can be very challenging mentally. A politically successful individual must be able to distinguish and recognize all individual members of the troop, to remember the history of individual interactions with them, and to anticipate their behaviors based on individual personalities. Success in competition, alliance, manipulation, and deceit depends on those capacities. Because primates as an order tend to have more complex societies and social relationships than those of other mammalian groups, many researchers have argued that social demands are closely related to the general advance of primate intelligence and the increase in brain size (Cheney et al. 1986; Dunbar 1998; Essock-Vitale and Seyfarth 1987; Humphrey 1976; Jolly 1966; Steele 1989). Furthermore, a large brain provides more capacity for social learning, including imitation of such skills as tool use, transmission of culture, and creative innovation. Such behaviors have been shown to correlate with brain size (Reader and Laland 2002).

This model is also consistent with observations cited above that brain size correlates with troop size and breeding system. Dunbar (1992, 1995) suggests that the correlation with group size is causal: The size of the cortex places an upper limit on troop size. Beyond that limit, increasing conflict and instability cause the troop to fission.

Such a model for brain size does not readily explain the superior intelligence possessed by great apes compared with other primates. Is chimpanzee society, for example, more intellectually challenging by its complexity, or is the complexity a secondary effect of the greater intelligence of its members? Gorillas have a different social organization, the single male troop, that is observed in many species of monkeys. Orangutans are usually solitary, although they may congregate if resources permit. These descriptions of social structure do not explain why apes might live more challenging social lives than monkeys, for example, but it is apparent that behavioral complexity could provide positive feedback for intellectual growth. It is more difficult to compete successfully in a society of intelligent animals than among less intelligent ones. Thus, competition should favor the evolution of increasingly intelligent brains up to the point where other constraints intervene.

DIETARY MODELS

An ecological model suggests that challenges of foraging represent the stimulus for brain development. Grasses and leaves are relatively easy to find and provide neither challenge nor a particularly rich diet. Many primates forage over large ranges and are very selective about their diet. Being able to concentrate on rich foods gives a greater energetic return to their efforts. Nutritionally rich foods tend to come in small scattered parcels that are seasonally available, yet chimpanzees, in particular, seem to be able to anticipate food sources throughout their range. Mapping resources in space and time may be a critical capacity leading to more general intellectual development (Goodall 1986; Milton 1981). However, the foraging model also falters in explaining the greater intelligence of all the great apes. While both chimpanzees and orangutans focus on fruit and other rich scattered foods, gorillas also consume large quantities of leaves that present little challenge to find.

Metabolic Needs

The energy budget of a pregnant female correlates with brain development (Martin 1983b). If her infant is to have a large brain, she must devote a greater amount of dietary energy into its development and must lengthen its gestation. Brain size of a neonate correlates closely with the mother's metabolic rate and with the quality of her diet. Put another way, to support an infant with a larger brain, the mother must consume a high-energy diet.

Diet quality is generally inversely correlated with body size (Chapter 14). The larger an animal, the lower is its energy consumption (per kilogram of body weight) and the poorer is its diet. Small mammals burn energy rapidly and consume small quantities of energy-rich foods, such as insects, nectar, gum, and carefully selected plant parts. Medium-sized mammals are likely to consume moderately rich diets such as fruits. Large mammals commonly eat energy-poor foods such grasses and leaves. When one animal consumes a diet richer than another of similar body size, the first animal may have extra calories to invest in brain size. Such a pattern has been described among New World monkeys (Radetsky 1995). Significantly, the large hominoids eat unusually rich foods for their body size. Orangutans and chimpanzees eat mostly fruits plus a wide variety of other items. Chimpanzees in particular supplement plant foods with meat of smaller mammals. The leaf-eating gorilla is a notable exception among the apes, although it will eat fruit when that is available. Humans eat very rich foods

—meat, fruits, and other selected animal and plant products. Perhaps the inclusion of larger quantities of meat, providing a concentrated source of calories and protein, was a crucial step in supporting brain expansion (Milton 1999).

An extreme statement of this relationship between diet and brain size is the prediction that gut size and brain size are inversely related. If an animal can take advantage of a rich diet to reduce gut tissue, the energy saved can be invested in the brain (Aiello and Wheeler 1995; Aiello et al. 2001). Such correlations are observed (Fish and Lockwood 2003), but at the present it is not possible to sort out precise causes and effects—whether it is the extra energy provided by the diet or the tissue savings that contribute to the larger brain or whether this is a useful distinction. High-quality diets provide the energy that permits a larger brain size. By the foraging hypothesis, intelligence is needed to achieve a rich diet for a large animal. Intelligence also enables primates to interact in complex societies. We can thus "explain" hominoid intelligence either by reference to its adaptive advantage in ecological or social context or by reference to the metabolic budget necessary to develop and support a large brain.

ENVIRONMENTAL MODELS

Paleoanthropologists have become increasingly interested in the interaction of climate and human evolution. As contemporary climate change has become a topic of intense interest, we gain an appreciation for the significance a past change might have had on evolution.

Past attempts to link climate and habitat change to human evolution were simple and unsophisticated, both in the reconstruction of paleoenvironments and in understanding their significance. Darwin, in his essay *The Descent of Man* (1871), proposed an ancestral move from forest to savanna was sufficient to account for the divergence of humans and apes, starting with bipedalism and continuing through tools and brain expansion. Although his model did not provide a good reason why a savanna ape must become bipedal, variations on Darwin's naive hypothesis continue to play a crucial role in later models (e.g., Ardrey 1961).

There was indeed a significant change in both climate and habitat during the African Pliocene. Temperature and rainfall dropped and forests were eventually replaced by more open grasslands. According to a later interpretation, the crucial ape–human divergence occurred because the ancestors of chimpanzees persisted in the western forests while hominins stayed in the east, where the change were most pronounced (Coppens 1994).

Unfortunately, this convenient story is not correct (e.g., Shreeve 1996). The earliest evidence of the hominin lineage, discovered in the past decade, occurs in a forested habitat. The shift to bipedalism, for example, began before our ancestors "left the trees." In fact, the breakup of forests in East Africa and their replacement by grasslands occurred over only a period of millions of years, beginning before and continuing well after the appearance of *Australopithecus*. The time span of *Australopithecus* was one of continuing fluctuation of climate and habitat, with patches of forest breaking up into woodland and grassland, reforming, and breaking up again. If we wanted to tie the occupation of the savanna with a specific event in human evolution, it now fits with the origin of *Homo*.

Changes in climate and habitat are traumatic for species that have adapted closely to a deteriorating environment. They must change or go extinct. The Pliocene and Pleistocene have been periods of continuing climatic upheaval rather than a mere change from forest to savanna. The habitat in any given part of East Africa changed repeatedly as the climate fluctuated between warmer and cooler, wetter and drier extremes. There have been waves of extinction in the wake of such changes. Conroy (1997) calls this the "climate forcing model," by which we understand climate change to drive, or force, evolutionary adaptation. Potts (1996, 1998a, 1998b) uses the term "variability selection" to describe the pressure to become adaptable to a wide range of circumstances.

The great expansion of the brain in *Homo* came later, with growth accelerating over the past million years. At that time our ancestors were experiencing the more intense climatic shifts we know as the Ice Ages. The expansion and contraction of habitable areas of the Old World might have caused repeated squeezes on human populations. Such bottlenecks may cause intense selection and speed along evolution. A few authors have argued that the Ice Ages were important agents pushing brain evolution (e.g., Calvin 1991; Stanley 1996).

Hominins did survive by becoming generalists. A broad omnivorous and opportunistic diet enabled them to exploit a wide range of habitats and eventually to occupy an immensely diverse geographic range. Culture was important in making such adaptability possible. Humans have emerged as the most versatile of all organisms, not in the least because of our intelligence. It is plausible that environmental challenges played a major role in selecting for larger and more powerful brains.

OVERCOMING CONSTRAINTS

Another approach to explaining the uniqueness of human brain expansion assumes that larger brain size would be adaptive for any species, but that one or more barriers have limited such growth in nonhuman species. What are the constraints a hominin or any mammal would have encountered that limit brain size? The brain comes with a substantial cost in energy, nutrition, metabolism, reproductive investment, and delayed maturation. Potentially, such a model could be consistent with either continual or episodic brain expansion.

Thermoregulation. Because the brain is extremely sensitive to elevation in temperature, the heat of metabolism from cerebral activity might act as a check on tolerable brain size. Falk (1990) observed that hominin skulls and circulatory systems have a pattern of venous drainage that could contribute to cooling the brain. Venous blood cooled on the face and scalp drain into dural sinuses through foramina in the skull, absorbing heat before flowing back toward the heart. If these pathways represented a significant new mechanism for removing heat, expansion of the brain might have followed quickly.

Energy. The brain demands a continuous flow of energy merely to keep it alive. Neurons will succumb within minutes of being deprived of glucose or oxygen; yet other body systems are demanding energy at the same time. Animals that push their budget too close to the limits of their caloric intake may not be able to maintain a strong immune system or survive a temporary food shortage. Supporting the hominin brain presupposes a dietary surplus relative to the budgets of other animals. It is difficult to identify exactly the source of that surplus, but meat and animal fat from hunting or scavenging (Cordain et al. 2001; Leonard 2002), fish and shellfish (Broadhurst et al. 1998, 2002), and starchy tubers (Hatley and Kappleman 1980; Wrangham et al. 1999) have all been proposed. Most likely it is our dietary flexibility, combined with a food gathering and processing technology such as cooking, that makes new resources available and underlie our extravagant brain.

DHA. Building a brain requires some specific nutrients beyond generic calories. One of these is the long-chain polyunsaturated fatty acid docosahexaenoic acid (DHA). DHA is a critical component of the membrane of neurons. The best dietary source of DHA comes from the marine food chain, including algae, shellfish, and fish. It has been proposed that exploitation of coastal resources was an essential step in developing large brains (Crawford 1992; Chamberlain 1996; Broadhurst et al. 1998). While it is true that the nonhuman mammals with the largest brains (i.e., whales and dolphins) are also part of the marine food chain, not all marine mammals have unusually large brains. As argued in Chapter 14, it is not clear that our ancestors could not have acquired sufficient DHA from a terrestrial diet.)

Furthermore, human milk is not significantly different from that of other primates in DHA content (Robson 2004); thus, DHA supply does not explain human brain size.

Rate of growth. A gross comparison of the rates of human and chimpanzee brain development shows nearly identical trajectories, with human growth simply continuing longer (Rice 2002; Vrba 1998). The brain, regardless of its size, must be assembled from structural molecules. The nutrient transport capacity of the placenta may be a limiting factor, as may the synthesis of crucial components such as the long-chain fatty acids (Ellison 2000). Hominins have not eased that limitation, but our species is unusual in the early occurrence of birth relative to final brain development

(see Chapter 21). Nursing delivers nutrients in greater quantities than placental transport, and the early birth, combined with postnatal growth, extends the time period allowed for brain maturation.

SELECTING FOR REORGANIZATION AND NEW ABILITIES

Is the human brain qualitatively different from those of other species? A comparison of gene expression in human, chimpanzee, and orangutan brains and other organs shows that both species possess similar structural genes and proteins. However, in the brain in particular, humans show a different pattern of gene expression (e.g., Enard 2002). This suggests that at least some of the key changes in human and ape brains occurred in the genes regulating the development and function of the brain rather than in the structure itself.

DEVELOPMENTAL POSSIBILITIES

What sorts of restructuring are possible? The operation of the brain relies on the establishment of proper synaptic connections. The formation of synapses proceeds via a multistep process that involves a combination of genetic programming and a selective elimination of neurons and connections that allows for nongenetic influences (Aoki and Siekevitz 1988; Kalil 1989; Katz and Shatz 1996; Sharma et al. 2000). The genetic determination explains the consistency with which functions are distributed in the brain from one individual to another. The later pruning permits experience, including environmental influences and the degree of utilization, to help shape a brain to individual environments.

Axons grow from their cell bodies and must find their way to their targets. They are encouraged along the way by molecular signals borne by "guidepost cells." If the axon attaches to an appropriate receptor on a guidepost cell, it will hold that position while the tip gropes blindly for the next cue. Properly positioned guidepost cells can lead axons to their targets. An alternative signaling system involves molecular clues secreted by target cells in other tissues. The axons are sensitive to the concentration of such molecules and may follow the concentration gradient toward or away from the source (Tessier-Lavigne and Goodman 1996).

Such mechanisms are liable to errors. Because individual neurons may make thousands of synapses with other cells, they can also produce many inappropriate connections. Neurons may also fail to form meaningful synapses. Consequently, many more neurons and many more synapses are produced than are desired This sets the stage for the second phase of brain development, in which failed neurons and excess connections are eliminated.

A competition springs up among axons to establish a larger number of active synapses and receive appropriate feedback from the target cell. Activity along the nerve fibers is an important factor in the selective process and permits a role for external (environmental and experiential) influences. Active mutual inhibition between competing neurons may further the process. The synapses of losing axons are lost and the axons withdrawn. At the end of the day, neurons left with no active connections die. It is estimated that in different areas of the central nervous system, 15% to 85% of the neurons will be eliminated in this manner. Such a process requires a substantial initial overproduction of neurons but allows for a considerable margin of error in the establishment of connections.

How might evolution play with such as system to reorganize the brain? Extending the period of cell division and axon growth increases brain size and complexity of connections (Rakic and Kornack 2001). Subtle changes in molecular signals might redirect axons. Changes in the pattern of use might lead to changes in the functional allocations of certain brain parts. For example, motor and somatosensory mapping in the cortex represents body parts in proportion to their sensitivity (degree of innervation) rather than their size or surface area. The boundaries of these maps are influenced in part by the activity of a small number of genes (Rakic 2001). This allocation of cortex shows different patterns between species (e.g., Allman 1999) or between humans (Elbert et al. 1995). Furthermore, it can shift within the lifetime of one person (Bao et al. 2001; Florence et al. 2001; Nudo et al. 2001). On a larger scale, the human prefrontal areas, used for planning functions, may have increased their functional area by annexing neighboring regions of cortex (Deacon 1997a). Other integrative areas originally served in visual processing. Such functional changes

are not apparent at the gross anatomical level but may have evolutionary importance.

QUALITATIVE DIFFERENCES IN FUNCTION

Arguments for qualitative differences point primarily to language and less well defined aspects of behavior and cognition. Although humans have distinct patterns of cortical folding, there is not yet good evidence that gross differences above the cellular level determine functional differences (Deacon 1997b). Functional differences must occur at the level of synaptic connections and gene expression. What can hominin brains do, or what do they do that others cannot?

Culture. Since Darwin, tool use and other cultural attributes have long been thought to provide the positive feedback necessary to drive the expansion of the brain. For a number of reasons, the archeological record of stone tools is not sufficient to evaluate this hypothesis. Although the first known appearance of stone tools in the archeological record coincides roughly with that of genus *Homo* about 2.5 Mya, it is not certain that only *Homo* made the tools, because *Australopithecus* was also present. Certainly perishable materials would have been used as well, possibly much earlier (Panger et al. 2002). Several other species have also been observed to make and use tools in the wild, most notably chimpanzees but also monkeys and birds. We now recognize in other species rapidly accumulating evidence of learned traditions that suggest culture evolved gradually and that it is not unique to the human brain (Boesch 2003; Boesch and Tomasello 1998; Fragaszy 2003; Humle and Matsuzawa 2001; Perry and Manson 2003; Rendell and Whitehead 2001; van Shaik et al. 2003; de Waal 2001; Whiten and Boesch 2001; Whiten et al. 1999, 2003). The interplay of material culture and tools is not sufficient to explain why humans and not other hominoids developed large brains, but certainly the complexity of human material culture paces the expansion of the brain in the archaeological record.

Hunting. Hunting has been considered a central theme for hominin evolution to explain much of human behavior and physical evolution (Washburn and Lancaster 1968; Laughlin 1968; see also Chapter 14).

It dropped from favor when fossils were reinterpreted to argue that *Australopithecus* was not substantially carnivorous. However, we now recognize that the archeological evidence for hunting approximately coincides with the appearance of *Homo*, and it is difficult to rule out some degree of linkage, however indirect. Jerison (1973) argues that carnivores in an ecosystem consistently show slightly larger brains than their prey. Although Order Carnivora includes many of the more intelligent nonprimate mammals (such as dogs and cats), none of these display either brain size or behavioral complexity comparable to that of higher primates. Meat consumption is consistent with the hominin brain. Meat and other body parts, such as marrow, provide high protein and calorie content necessary to develop it and to support the metabolic needs of a large brain. Human hunting solves the question posed by Martin of how we can afford cerebralization; but whether it has a causal relationship with brain expansion remains to be demonstrated.

Social cooperation. A wide array of other species exhibit cooperation within a social group. Again, humans have developed this behavior to an unparalleled level through language. Language facilitates cooperation within society. Hunters may work together to enhance their success. Band members may divide labor for the more efficient performance of necessary duties. Resources can be shared with greater confidence of reciprocity.

Humans are better able to assemble information about the present. Communication permits individuals to pool their information so that each can benefit from the experiences of the others. For example, if hunters or gatherers forage in different directions from camp, an even discussion is sufficient to update each member of the band about their shared range.

Humans can transmit information across generations. While primate infants learn from their mothers which foods to eat and what strategies to use in obtaining them, such learning and overt teaching are limited to concrete facts of the immediate environment. Humans can teach one another abstract concepts, rules, histories, and facts. Culture becomes a cumulative experience extending well beyond the lifetime of any one individual.

Humans have a more complex sense of time. Humans can communicate about the past and the future, to remember and to plan. The universe acquires a new dimension—time—when the brain is equipped to handle it. We can anticipate and we can build. It is the accumulation of efforts and of dreams that has transformed human society into something unique on this planet.

Language appears to correlate with the development of other essential human cognitive abilities, including teaching, recursive grammar, creative play, and theory of mind (Premack 2004).

Are these behaviors discrete traits that have been selected? More likely they are emergent properties —abilities of the brain that are achieved when its complexity reaches a critical level. The volume of the cortex and of nuclei reflects numbers of neurons and the density with which they are packed, but the work they do is a property of their synaptic connections and communications within the brain. As new neurons are added, the possibilities for connections increases exponentially, and the functional differences that might emerge are vast and, in our perspective, unpredictable. The complexity of social behavior correlates with innovation, learning, and tools use among primates, and with brain size (Reader and Laland 2002). Such observations make any attempt to single out one trait as evolutionarily most significant very difficult, if not meaningless. The unique consciousness and intellect of the human brain, our facility with language and abstract thought, our creativity with art, and aesthetics may have been selected for because they were adaptive, and some of these may represent evolutionary landmarks. On the other hand, at least some of these functions may be the fortuitous byproducts of a critical number of neurons.

12

THE SPECIAL SENSES

Sensory receptors gather information from the environment and from the body and report it to the brain. Although the sensory system is listed as only one of several functional divisions of the nervous system, a nervous system without sensory input is inconceivable. Unless it can receive some input and feedback from the rest of the world, a brain or an organism is just a machine out of control.

Afferent pathways of the central nervous system are conventionally divided into general and special senses. General senses are distributed throughout the body. Special senses are located within the head and involve a specialized structure to enhance the sensitivity of the receptors. The general senses are many and varied, including touch, pain, and proprioception, each of which has many types of receptors. The special senses can be easily enumerated: taste, olfaction, vision, equilibrium, and hearing.

It is the special senses that show more variability of development among vertebrate species and pose more interesting evolutionary questions. Smell, vision, and hearing are critical for scanning the environment at any distance from the body. These senses command significant portions of the brain and especially the cerebral cortex. Because no animals possess a high level of development of all of the special senses, there probably is competition among them and with other function for neurons and connections within the brain.

OLFACTION

The physiological nature of olfaction has been the most difficult to understand of the major senses because, despite our abilities to discern a large number of scents —as many as 10,000 different odors—much of the brain's response occurs in different parts of the cerebrum and at a different level of awareness than other senses. At the same time, many believe it to play a continuing role of importance in our behavior (e.g., Chiarelli 2001; Stoddart 1990; Watson 2000). The number of odors some humans can detect and distinguish is not only immense but also seemingly open-ended. Newly synthesized chemicals register as completely with the nose as do common foods. Nonetheless, it is well known that humans are relatively insensitive to odors compared with most other mammals. Several models have attempted to explain how this system can possess the wide range of detection that it has.

By analogy with other senses, especially taste, conventional explanations have supposed a finite number of receptors, each responding to a specific class of molecular shapes. These shapes correspond to different smells. Seven have commonly been listed in the past: camphor, ether, musk, pungency, floral scent, putridness, and peppermint. All other smells would represent distinctive combinations of these. Each set of receptors would be determined by a single gene and would register with a particular region of the olfactory cortex. The anatomy of those connections would determine our interpretation of the experience. However, the great variety of smells to which we are sensitive and the ease with which we learn to recognize newly manufactured odors show that this traditional model is inadequate (Axel 1995; Freeman 1991; Freeman 1993).

Modern techniques of genetic mapping and cloning have identified enormous numbers of genes that relate to olfactory receptors (Firestein 1991, 2001; Gilad et al. 2003a, 2003b; Glusman et al. 2001; Menashe et al. 2003; Mombaerts 1999, 2001; Zimmer 2002). Some families of receptor genes have over 100 different members. Current numbers for rodents have been estimated at 1000 to 1500 genes, as many as 1% to 5% of the genes in the entire genome. Human numbers are slightly smaller, about 900 to 1000. Even a greater variety of receptors are designed or cloned from a mix-and-match family of genes as new smells are encountered, in a manner analogous with the synthesis of antibodies by the immune system. Another step in distinguishing odorants is the combined action of different receptors. Even closely related molecules will not stimulate exactly the same receptors but unique combinations of them. The potential number of receptors and permutations of receptors that can be generated by such as system can explain how the nose is prepared to detect a seemingly endless succession of new smells.

Nearly two thirds of the genes appear to be broken or incomplete—"pseudogenes" not producing olfactory receptors in humans but clearly related to other functioning genes in more acute noses of other species (Glusman et al. 2001; Mombaerts 2001). The percentage of pseudogenes for olfaction drops to 50% among the apes and 20% for mice. If human noses have a deficient sense of smell, we have merely extended and accelerated the pattern already present in other primates (Gilad et al. 2003b; Rouquier et al. 2000; Whinnett and Mundy 2003). Yet even the intact genes for odor receptors comprise up to 1% of functioning human genes and may be maintained by selection (Gilad et al. 2003a; Gimelbrant et al. 2004). If a person is heterozygous at these genes, the number of different functioning receptors may be twice this. How is it that so much of our genome is devoted to one sensory modality that admittedly does not function very well for humans? It is possible that this diversity of both genes and pseudogenes has other functions in the body. As membrane proteins, they potentially serve to identify cell classes and receive signals from other cells. The genes may be active early in development, where they may help to determine the interaction and fates of cells and their connections within the brain (Mombaerts 1999; Nef 1998).

CENTRAL PATHWAYS

The olfactory receptors lie in a region of the nasal mucosa known as the olfactory epithelium. Their cell bodies are surrounded by supporting epithelial cells.

Their axons make up the olfactory nerve, which consists of about 20 bundles of fibers on each side projecting through small foramina in the cribriform plate of the ethmoid bone to synapse in the olfactory bulb (Figure 12.1). Within the olfactory bulb, the axons rearrange themselves. Those that have arisen from a specific type of olfactory receptor converge on a cluster of cells called a glomerulus (Mori et al. 1999). Each of approximately 1800 glomeruli, representing a specific smell, refines the signal and relays it to the brain. Within the olfactory cortex, individual smells are be handled by a specific but overlapping clusters of neurons (Zou et al. 2001).

Axons from the bulb continue along the olfactory tract to limbic structures of the frontal lobe of the cerebrum and diencephalon. The information carried on the tract has already been modified according to context and previous experience, so that an observer cannot distinguish smells simply by which neurs report them. Olfactory information to this level appears to be communicated by patterns of relative activity involving a large number of neurons (Bower 1988; Freeman 1991; Laurent 1999; Travis 1999). It is the relative behavior of different neurons, rather than their specific pathways, that contains the information. In fact, the same smell may trigger different patterns of activity in successive encounters because the meaning of each encounter has been altered by the previous ones. In this way olfaction appears to be unique among the conscious senses.

The olfactory cortex is a phylogenetically old part of the cerebrum, the paleocortex, and is not as clearly delineated as the regions for other senses, which are processed by the neocortex. It interacts with other parts of the limbic system, including pathways for emotions, memory, and visceral responses. Olfaction is the only sense that relates directly to the cerebrum without intermediate processing by the thalamus. This fact is sometimes used to waken comatose patients through the application of smelling salts or other olfactory stimulus to induce cerebral activity. The olfactory nerve, bulb, and tract are considered projections of the brain itself, rather than a peripheral nerve, and thus represent the only part of the brain in direct contact with materials outside the body.

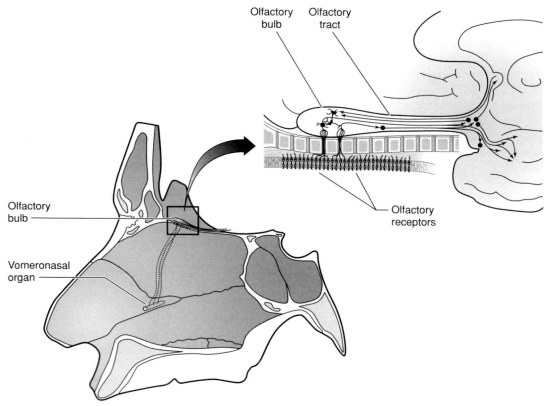

Figure 12.1 The olfactory system. Olfactory receptors lie in the upper reaches of the nasal cavity and send signals into the olfactory bulb and parts of the brain immediately superior. The receptors of the vomeronasal organ lie more inferiorly on the septum but use the same pathways.

THE EVOLUTION OF SMELL

Smell may be thought of as the simplest of senses in terms of perception. It is independent of distance, light, spatial relationships, and direct physical contact. Olfaction is especially critical because nearly all naturally occurring odiferous compounds are derived from other organisms and organic material, including food, potential mates, social interactions, and predators. At least three different olfactory systems have evolved in vertebrates, of which two—our olfactory epithelium and the vomeronasal organ—are found in humans (Eisthen 1997). The paleocortex, containing the olfactory areas, is also considered the most primitive part of the cerebral cortex. It has fewer layers of

neurons (three rather than six) and is more extensive in lower vertebrates than other regions.

In a phylogenetic context, smell is the best developed sense of early mammals, but it has conspicuously declined in relative importance in primates and interacts with behavior often at unconscious levels in humans. This decline is anatomically evident in the reduction of the size and complexity of the nasal passages and in the absolute area of olfactory membrane (Table 12.1). Because animals in a variety of habitats, including birds and fish, may retain acute senses of smell, we cannot explain its decline directly on our arboreal heritage or on any other single dimension of our environment. Instead, the decline parallels the emergence of vision as the more important special

Table 12.1 Comparison of Olfactory Membranes in Mammals

	Area of Olfactory Membrane (mm^2)	Density of Receptors (per mm^2)
European deer	900	3×10^8
Domestic dog (German shepherd)	1500	2.5×10^8
Rabbit		10^8
Tarsier (*Tarsius bancanus*)	39	
Human	25–50	3×10^7

Source: Ankel-Simons, 2000.

sense in primates. Perhaps we are observing the outcome of competition for brain resources.

THE VOMERONASAL ORGAN

The vomeronasal organ (VNO) is a parallel olfactory system well developed in many vertebrates, such as snakes. The characteristic tongue-flick of a snake conveys molecular traces of prey from the ground to the VNO in its palate and enables the snake to follow a trail. In mammals, the VNO is believed to respond especially to pheromones, odorants emitted as chemical signals by other individuals of the same species (Døving and Trotier 1998; Holy et al. 2000; Leinders-Zurfall et al. 2000; Luo et al. 2003; Sam et al. 2001; Stowers et al. 2002; Thorne and Amrein 2003; Watson 2000). Pheromones convey information about physiological and emotional states and are very important in sexual attraction and mating, in parent–offspring recognition and imprinting, and in other social interactions. For example, male mice lacking an ion channel expressed in the VNO failed to distinguish the gender of other mice or to behave appropriately toward other males.

When present, sensory neurons from the VNO project to the accessory olfactory region of the cortex and directly into the limbic system without registering in consciousness, although some of the stimulus molecules may also evoke responses on the olfactory membrane. Such signals are in an excellent position to influence behavior and, at the same time, to escape our conscious experience (Mori et al. 2000; Watson 2000).

The VNO and pheromonal communication are present in prosimian primates and in at least some New World monkeys. The VNO has not been found in Old World monkeys. In recent years the presence of a much diminished VNO has been confirmed to be present in adult chimpanzees and humans as a small concentration of receptors in a pit on the nasal septum (Bhatnagar and Smith 2001; Døving and Trotier 1998; Jacob et al. 2000; Smith et al. 1998, 2001a, 2001b; Takami 2002; Trotier et al. 2000). However, questions remain as to whether these structures maintain innervation and function in the adult.

Does the VNO play a significant role in human behavior? A number of authors argue in the affirmative, scarcely making any distinction between human and nonhuman behaviors (Kohl and Francoeur 1995; Watson 2000). Monti-Bloch et al. (1998) described neural responses of the organ following olfactory stimulation. Certain observations appear to confirm some level of olfactory communication (Meredith 2001). For example, it has been observed that women living in close quarters tend to synchronize their menstrual cycles (Graham 1991; Graham and McGrew 1980; McClintock 1971; Quadagno et al. 1981; Russell et al. 1980; Stern and McClintock 1998; Weller and Weller 1993; Weller et al. 1999; Wilson 1992). Body odors presented experimentally may influence the perception of sexual attractiveness. However, the sensory pathways for these signals have yet to be confirmed, and aside from menstrual synchrony, the evidence for human responsiveness to pheromones under natural conditions is very weak. Nor is there much evidence for any significant function of the primate or human VNO. Mice contain approximately 100 genes for VNO pheromone receptors. Of the homologous genes in humans, only one is known to be functional. Most, if not all, of the others are pseudogenes (Keverne 1999; Kouros-Mehr et al. 2001; Liman and Innan 2003; Mombaerts 2001; Rodriquez and Mombaerts 2002; Sullivan 2002; Zhang and Webb 2003). Apparently the sensory neurons and the gross structure are largely vestigial.

TASTE

Taste and smell are closely related special chemo-senses. Both detect molecules from the environment that arrive at receptors dissolved in saliva or mucus. Potentially, smell provides information about a source of the odor that is still at a distance, while taste requires direct contact with the tongue. In practice, the odor of food detected through smell is a major aspect of the experience of taste. One of the chief complaints of patients who have lost the sense of smell is that their food has lost its flavor. However, the two senses are neurologically quite distinct in their receptors and brain pathways.

Human taste receptors are clustered in taste buds, which are scattered across the surface on the tongue as well as on our soft palate and adjacent regions of the pharynx. Within a taste bud, receptor cells are supported by interspersed epithelial cells. The complex is sunken beneath the surface of the tongue but opens to the surface via a small pore. Molecules from the food dissolved in saliva may enter the pore and contact processes of the receptor cells.

Tastes are distinguished according to the ability of molecules to bind to receptor molecules on the receptor cell membranes. Different receptors are sensitive to sweetness, sourness, bitterness, saltiness, and umami (glutamic acid), although we now understand these categories to be oversimplified (Lindemann 2001; McLaughlin and Margoleskee 1994; Smith and Margolskee 2001). For example, there may be more than one receptor distinguishing different "bitter" tastes (Caicedo and Roper 2001). The discrete number of five (previously four) basic tastes likely represents an arbitrary imposition of order based on limited knowledge. Although textbooks continue to map the basic tastes to different parts of the tongue, in actuality the receptor types are equally distributed across it (Smith and Margolskee 2001). The modalities of taste are understood to have adaptive value (e.g., Hladik and Simmon 1996; Hladik et al. 2002).

Sweetness is a response to glucose and a class of molecules of similar shape and properties. Sweet-tasting foods indicate a relatively high content of carbohydrates and easily available energy. Both the ability to assess sweetness and a dietary preference for it probably evolved in primates for rapid discrimination of edibly ripe and more easily processed plant foods. Carnivores, who do not need such discrimination, lack the sense of sweet. Interestingly, the sensitivity of primates to sweetness is inversely proportional to body size, roughly indicating the decreasing proportion of sugars in the diets of large animals (Hladik and Simmen 1996).

Sourness is a sign of acidity. Together with sweetness, it helps to determine the ripeness of fruit and other vegetable foods. Hydrogen ions produced as acids dissolve in water act directly on receptor membranes to trigger neurotransmitter release.

Bitterness is usually an indication of toxic alkaline secondary compounds produced by plants as a defense against herbivores. Because these toxins may have other negative effects on the digestive or nervous system, it is adaptive for animals to avoid bitter tastes (Hladik and Simmon 1996). Perversely, humans have often pursued such plant defenses for their striking tastes or side effects. Caffeine, for example, is a bitter-tasting neurotoxin whose stimulating effects are much sought after. Pepper and other "hot" spices have similar origins.

Salt, of course, represents a vital nutrient for the body not commonly encountered in a concentrated form in nature. Sodium ions of dissolved salt again act directly across the cell membrane to trigger a response. Many animals are known to seek out salt sources in response to a nutritional need for a sodium. The human response to a temporary deficiency is more subtle, but may be reflected in a preference for salty foods (Denton 1982; Schulkin 1991). A taste for salt may also be learned and may exceed dietary need to the point of endangering health.

Umami, or savoriness, is the taste of compounds related to glutamic acid, an amino acid. This is common in protein-rich foods such as meat and fish. It is also used as a flavor enhancement in the form of monosodium glutamate (MSG). Because umami has only recently been recognized as a distinct taste modality (Lindemann 2000), it is less well understood than the others. Perhaps the ability to detect protein-rich foods is an important adaptation for the human diet. On the other hand, glutamate is an important neurotransmitter in the brain. Some individuals have an adverse reaction—headache—when they consume excess amounts of it. They are considered "allergic" to MSG.

VISION

The eye converts radiant light into neural impulses. Photoreceptors are arranged in a layer within the retina, where light falls upon them. The retina is composed of several layers of neurons, of which the outermost layer consists of the photoreceptors. The inner layer contains ganglion cells whose fibers make up the optic nerve. In between these neurons is a variety of interneurons that begin the processing of visual information (Hubel and Wiesel 1979; Kolb 2003; Masland 2001). The remainder of the structure of the eye is concerned with focusing that light so that it forms a sharp image. The various tissues through which light must travel are designed either to bend light in a precise manner or to permit light to pass through with minimal distortion.

Primates place a greater emphasis on the visual sense than do other mammals. Although most mammals rely primarily on smell or hearing to monitor their environment, even nocturnal primates continue to

depend on vision, enlarging the eyes dramatically to maintain sensitivity. Primates have several specializations for vision that indicate a relatively greater dependence on visually oriented behavior than that of other mammals (e.g., Martin 1990; Noback 1975). These specializations include frontally directed vision with a macula and stereoscopic vision, enhanced color vision, and an elaboration of the visual cortex. When these traits are combined with specializations for cortical processing, the result is perhaps the most sensitive and sophisticated visual system of any living mammal (Table 12.2).

FRONTALLY DIRECTED VISION

Frontally directed vision enables an animal to perceive and center attention on objects immediately in front of it. Herbivores generally position their eyes more laterally to scan a wider angle of the environment, while predators have directed vision anteriorly to facilitate stalking and catching prey. The high degree of convergence in primate vision is probably related to hand–eye coordination and predation on insects. To accomplish frontal vision, the skull had to be reshaped. The orbits shifted anteriorly and were reinforced by

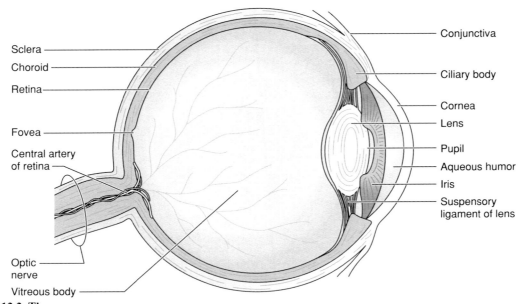

Figure 12.2 The eye.

Table 12.2 Characteristics of the Primate Visual System

Vertebrate characteristics
   Midbrain processes information from opposite eye
   Six muscles coordinate the movement of each eye
Tetrapod characteristics
   Eyelids
   Lacrimal gland
Mammalian characteristics
   Eyelashes
   Domination of retina by rods
   Visual cortex processes information from opposite eye
Primate characteristics
   Frontally directed vision
   Visual cortex processes opposite visual field
   Midbrain processes information from opposite visual field
   Stereoscopic vision
   Macula
   Elaboration of the visual cortex
Anthropoid characteristics
   Domination of retina by cones
   Three-color vision (catarrhines)

a postorbital bar and the snout fell below the line of vision and was reduced in size (see Chapter 4).

Stereoscopic vision requires that the visual fields of the two eyes overlap and that retinal axons from each eye are shared between the two sides of the cerebrum. The overlap in visual field is a consequence of frontally directed vision and has appeared to varying degrees in many groups of mammals but reaches its best expression in primates. The optic nerves leaving the retina carry visual information that is already partly processed. Information from clusters of receptors has been pooled to indicate patterns of light— shapes, movements, colors, and contrasts. The two optic nerves cross and mingle with one another at the optic chiasm, exchanging approximately half of their fibers with one another (Figure 12.3). The optic tract, which continues from the chiasm to the thalamus, contains a mixture of fibers from the optic nerve on the same side of the body (47%) and from the nerve

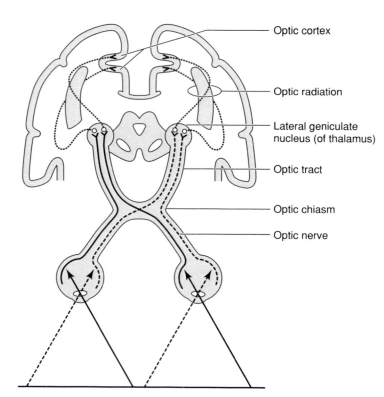

Optic cortex

Optic radiation

Lateral geniculate
nucleus (of thalamus)

Optic tract

Optic chiasm

Optic nerve

Figure 12.3 Visual pathways. About half of the fibers of the two optic nerves cross the midline at the optic chiasm and recombine to form the optic tracts. Thus, each tract carries information to the brain from the opposite visual field of both eyes.

on the opposite side (53%). Among nonmammalian vertebrates, 100% of the fibers cross the midline and there is no sharing of visual information from the two eyes. The figure varies among mammalian orders. In tree shrews, for example, thought to be closely related to primates, 90–97% of the fibers cross. This pattern of crossing fibers also extends to the midbrain tracts. Primate retinal fibers to the superior colliculus divide so that each colliculus handles only the opposite visual field just as the visual cortex does. This is not typical of other mammals.

This exchange of fibers is not random. In a given optic tract, all of the fibers carry information relating to the opposite visual field only. Light from the left side of the visual field strikes the receptors on the right side of each retina. The right part of each retina projects axons to the right optic tract and thence to the right thalamus and visual cortex. Each hemisphere of the cerebrum processes visual information relating to the visual field of the opposite side, just as it processes sensory information from the opposite hand and sends motor commands to that hand. Moreover, since the information arriving from the two eyes concerns the same portion of the visual field, the visual cortex can compare the information from the two eyes to determine relative distance of objects from the body.

Primates have also evolved the macula for enhanced acuity in the center of the visual field. The term *macula* refers to any dense cluster of receptors, but in the retina it describes a closely packed array of photoreceptors that correspond to the center of the visual field. Acuity is a function not only of the density of receptors but also of the amount of cortical area on which the macula maps. This region receives a disproportionate part of the retinal map on the primary and all secondary visual areas of the cerebrum. Thus, we are able to see objects in the center of our visual field with much greater resolution than those more peripheral.

PHOTORECEPTORS AND COLOR VISION

The sensitivity of photoreceptors depends on pigment molecules that undergo chemical change in the presence of light (Schnapf and Baylor 1987; Stryer 1987).

For example, vertebrates use the pigment rhodopsin, which converts temporarily to a form of vitamin A when struck by a photon of light of certain wavelengths. Rhodopsin-containing receptor cells are called rods because of the shape of their light-sensitive processes. Excitation of rhodopsin molecules changes the membrane properties of the rods and signals the presence of light to the next layer of neurons.

The human eye contains three other visual pigments, called opsins, each of which is selectively sensitive to particular wavelengths and occupies a different type of receptor called a cone (Merbs and Nathans 1992a; Nathans 1989). The three types of cones permit us to discern colors (Figure 12.4). The similarities of the cone pigments to rhodopsin and to one another indicate that they evolved through gene duplication and subsequent modification. The genes for red and green pigments, for example, lie adjacent to one another on the X chromosome and differ by only 2% to 4%, with substitutions in only 2 of the 364 amino acids responsible for most of the functional difference. Other copies of these genes have also been discovered on the chromosome and may account for variations in human color vision (Merbs and Nathans 1992b; Nathans 1994; Nietz and Nietz 1995).

Rods and cones have different properties and distributions. Cones require relatively high levels of light and are most useful during the day. Rods are

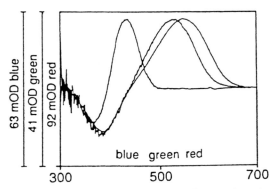

Figure 12.4 Absorption frequencies of the color visual pigments. (From S.L. Merbs and J. Nathans, 1992. Absorption spectra of human cone pigments. *Nature* 356:433–435. Copyright 1992 by Nature Publishing Group. Reprinted with permission of Nature Publishing Group and the authors.)

more sensitive to low levels of light and thus are responsible for relatively colorless night vision. The macula is populated entirely with cones, permitting us to use color vision to enhance the visual discrimination in the center of our visual field but only in relatively good light. Toward the periphery of the retina, the proportion of rods increases and the overall density of receptors decreases. Because the rods, unlike cones, are sensitive enough to detect subtle or very brief changes in light intensity, they are good at detecting motion. The shadow cast by a fly passing through our visual field may register in the periphery but not on the cones of the macula. While the brain commonly directs visual attention to a specific object in the center of the visual field, it also continuously monitors motion and activity in the broader environment.

Color vision has evolved independently in dozens of different animal groups, including arthropods and vertebrates, although not to the same extent (Jacobs 1993; Kelber et al. 2003; Pichaud et al. 1999). While higher primates distinguish three primary colors, some other vertebrates may perceive two or four. Birds have excellent three-color (trichromatic) discrimination, while mammals generally have poor dichromatic vision. It has been suggested that ancestral mammals lost an opsin gene during an early nocturnal phase. Color adds an important dimension to visual analysis of the environment and identification of food items. The colors of flowers and fruits appears to have coevolved with color vision in animals—specifically with potential pollinators and seed dispersers. Thus, color vision is best developed in frugivores and in animals that consume pollen, nectar, or flowers. These animals include primates.

The ancestral vertebrates may have possessed only cones, although without some diversity of pigments, a single type of cone provides only monochromatic vision. Rods were selected to provide greater acuity of vision and greater sensitivity in low light. Rods presumably increased in frequency and proportion in early mammals in accord with their nocturnal habits. Nocturnal behavior and rod-dominated retinas characterized the ancestral primates as well. Higher primates reacquired a cone-dominated retina along with a diurnal niche (Heesy and Ross 2001; Mollon

1989; Rowe 2002; Surridge et al. 1993; Weiss 2002). Old World monkeys and apes and one South American monkey, the howler, are unique among mammals because they have evolved three-color vision by diversifying the cone pigments. (The nocturnal owl monkey, a subject of much vision research, is a notable exception among higher primates because it secondarily returned to the rod-dominated night vision.)

Color vision is not necessarily an important adaptation for predation, nor is it for nocturnal habits. Its evolution in the ancestors of anthropoids indicates that this group made a significant ecological shift away from nocturnal insectivory toward frugivory and selective browsing and diurnal habits. The dichromatic (two-color) vision of New World monkeys appears to be well adapted for discerning fruit. The highest development of trichromatic (three-color) vision may relate to fruit or to the discernment of young edible leaves (Dominy and Lucas 2001; Dominy et al. 2003; Lucas et al. 1998, 2003; Regan et al. 2001; Shyue et al. 1995; Wolf 2002).

EVOLUTION OF VISION IN THE BRAIN

Nonmammalian vertebrates have a simpler cerebrum that handles primarily the sense of smell and limbic functions. Visual information is processed by the midbrain in the homologue of the superior colliculus. This region is called the optic lobe in amphibians and birds, in which it may be especially well developed. Mammals have shifted the processing of all somatic senses to the cerebrum where they are integrated for a more complex view of the environment. The visual cortex in the occipital lobe is a supplement to the midbrain visual area. The latter is maintained mostly for control of eye movements while the cerebrum has taken over the analysis and interpretation of visual information.

The elaboration of mammalian vision corresponds to increasing specialization of the secondary visual cortex. In addition to the primary cortex bearing a retinotopic map, there are numerous additional maps of the retina on nearby regions of the cortex (Allman 1977, 1999; Tootell et al. 1996; Van Essen et al. 1992). More than 10 are reported in the human cortex, and about 30, in the macaque. Because of the difficulties of identifying such areas, these are likely conservative figures. Each of these may handle a specific function, such as calculating distance, processing color, and integrating the right and left halves of the visual field. These secondary maps are especially numerous and well developed in primates, where vision assumes critical importance in social interactions as well as in foraging and exploiting the environment.

EQUILIBRIUM

The sense of movement is a critical and widespread sense among most multicellular organisms and some unicellular ones. For humans it is more subtle and less conscious than vision or hearing, generally not more than a vague awareness in our minds unless called to our attention. Nonetheless, it is an essential component of all motor activity, and the vestibular system initiates its own involuntary movements to help us maintain balance.

The vestibular organ detects changes in body position or momentum. More specifically, it detects both linear and rotational acceleration. The vestibular organ is a complex membrane encased in the petrous portion of the temporal bone in a cavity called the inner ear (Figure 12.5). The chambers of the membrane contain a fluid, endolymph, and are surrounded by another fluid, perilymph. Because of its complex shape and multiple passages, the organ is called the membranous labyrinth.

The mechanoreceptor of the vestibular organ is called a *hair cell* because it lies in an epithelium with a number of hairlike stereocilia projecting into the endolymph (Figure 12.6). When the cilia are bent, they open or close ion channels and excite or inhibit the receptor, depending on the direction of bending. Hair cells are clustered within a chamber of the labyrinth with the tips of their cilia embedded in a gelatinous structure called a *cupula*. Also on the cupula are calcite crystals called *otoliths* ("ear stones"). The otoliths possess a density and an inertia greater than that of the endolymph in which they lie. Consequently, when the head or body is accelerated (or decelerated, which is merely an acceleration in the opposite direction of travel), the otoliths respond more slowly than the fluid. If acceleration occurs in the plane of the cupula, the hesitation in the movement of the otoliths causes the stereocilia to bend and the acceleration to register on the vestibular nerve. One macula lies in a roughly horizontal plane in a chamber called the utricle. This is responsible for recording linear, or straight-line, acceleration in a horizontal direction. A second macula lies in a more vertical plane in a chamber called the saccule and may report linear acceleration in the third dimension.

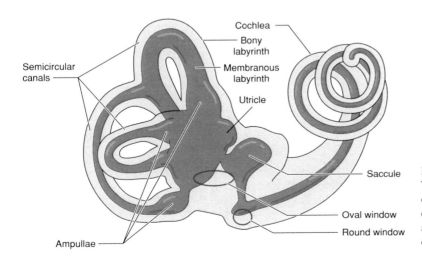

Semicircular canals

Cochlea

Bony labyrinth

Membranous labyrinth

Utricle

Saccule

Oval window

Round window

Ampullae

Figure 12.5 The vestibular organ. The various changes of the vestibular organ contain sensors for acceleration of the head. The membranous labyrinth appears shaded within the convoluted cavity of the bony labyrinth.

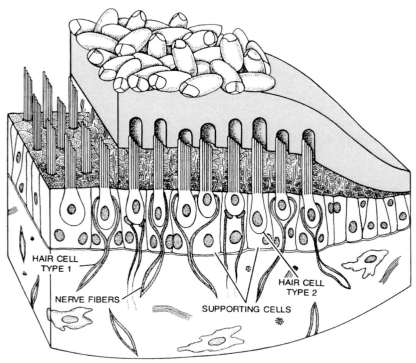

Figure 12.6 Hair cells in the vestibular organ. The hair cells are concentrated in a macula within the chambers of the membranous labyrinth. Their cilia project into a cupula weighted with otoliths. When the head moves, the inertia of the cupula bends the cilia and allows the macula to signal the brain. (From E. Parker, "The vestibular apparatus," *Scientific American* 243(5): 118–135. Reproduced by permission of Patricia J. Wynne.)

Angular, or rotational, acceleration is reported by the semicircular canals. The canals are curved membranous passages connected at both ends to the utricle. When the body begins to rotate in the plane of one of the canals, the endolymph lags behind and effectively moves along the canal in the opposite direction. A similar effect can be observed by turning a glass of water: the water does not begin to turn as quickly as the glass, but it continues to swirl after the glass is made to stop. This current creates a pressure against the cupula to which stereocilia of hair cells in the canal are attached and enables neurons to report the movement to the brain. By orienting the three semicircular canals in mutually perpendicular planes, the vestibular organ can detect angular as well as linear acceleration in any orientation or direction.

The vestibular organ is large and well developed among fish. Because their movement occurs in three dimensions to a far greater degree than that of land animals, and because a body suspended in water is easily moved by currents, a good sense of equilibrium and direct motor connections to vestibular centers are very important.

The delicate stability of human bipedalism also places unique demands on the vestibular system (Spoor and Zonneveld 1998; Spoor et al. 1994; for an opposing view, see Graf and Vidal 1996). The anterior and posterior canals, which would first detect a loss of balance, are enlarged to increase sensitivity. Such changes can be detected within fossil skulls using computed tomography. The human pattern is not present among australopithecines or early *Homo*, but it is found in *Homo erectus*.

Table 12.3 Characteristics of the Auditory System

Primitive tetrapod characteristics
   Inner ear sensitive to bone conduction
   Stapes employed in palatal, spiracular manipulations
Early therapsid characteristics
   Bone conduction included articular, quadrate, and stapes
Most primitive mammalian characteristics
   Dentary–squamosal joint established
   Articular and quadrate specialize for transmitting vibrations
Advanced mammalian characteristics
   Quadrate reassigned as incus
   Articular reassigned as malleus
   Angular reassigned a tympanic bone
   Middle ear sealed off behind tympanic membrane
   Development of pinna
   Mechanisms for sound localization
   Sensitivity to high frequencies, 4 kHz to 100+ kHz
Human characteristics
   Specialization for perception of frequencies corresponding to
     human voice
   Neural specialization for pitch, volume, direction, and language
     discrimination

HEARING

The cochlea, the site of receptors for hearing, is a derivation of the vestibular organ. Although its receptors operate on the same mechanical principles as those of the utricle and semicircular canals, the cochlea is clearly specialized for an altogether different function. The ear has undergone more gross evolutionary transformation in vertebrate history than have other sense organs and provides a particularly satisfying example of evolutionary adaptation (Table 12.3).

The human cochlea ("snail") is a long spiral-shaped basilar membrane that bears rows of hair cells. The cilia of the hair cells are embedded in a second membrane, the tectorial membrane. This complex, including the two membranes, the receptors, and related structures, is called the *organ of Corti*. Sound vibrations are transmitted to the perilymph outside the organ of Corti and cause the basilar membrane to vibrate against the relatively stationary tectorial membrane. This vibration causes the cilia to bend and initiate action potentials on the cochlear nerve. Different regions of the basilar membrane are "tuned" to vibrate at different preferred frequencies of sound.

A long membrane enables the ear to detect a greater range of pitches. The spiral shape of the cochlea is an arrangement to permit the mammalian ear to fit a long basilar membrane into a relatively small space.

THE OUTER AND MIDDLE EARS

One of the problems facing terrestrial animals is that airborne sound waves contain relatively little energy. The function of the outer and middle ear cavities is to make certain that sound vibrations in air reach the inner ear with enough energy to excite the organ of Corti. The outer ear consists of the pinna and the external auditory meatus. The middle ear is an enclosed air-filled cavity beyond the eardrum where sound vibrations are amplified.

The pinna, the fleshy visible portion of the ear, acts as a receiver to funnel sound waves to the meatus. Dogs and other species with mobile pinnae can direction their listening in a specific direction. The facial muscles surrounding the human ear are only vestiges of that ability in distant ancestors. The meatus is the short passage leading to the eardrum, or tympanic membrane. It is present so that the tympanic membrane can be protectively recessed into the skull.

The tympanic membrane is the boundary between the outer and middle ear. It is the site where air vibrations are transformed into mechanical vibrations of body tissues. The vibration of the eardrum is transmitted along a chain of three tiny ossicles: the malleus, which is attached to the eardrum; the incus; and the stapes, which is attached to another tympanum covering the oval window of the inner ear. When the vibrations reach the stapes, they are transmitted across the oval window to the perilymph of the cochlea.

The tympanic membrane is much larger than the membrane over the oval window, having about 17 times the area. The force of air waves striking the eardrum is thus concentrated on a much smaller area before it is conveyed to the inner ear. Pressure (measured in force per unit area) on the second membrane is about 22 times that of the first. This intermediate stage of conduction of sound vibrations is important in ensuring that they possess enough energy to register in the cochlea. The use of rigid ossicles to convey the vibrations minimizes energy loss through elastic deformation.

Concentrating sound energy increases the sensitivity of the ear but leaves it vulnerable to damage from excessively loud sounds. High-amplitude vibrations are dampened by a pair of muscles that contract reflexively to loud sounds. The tensor tympani inserts on the malleus where it attaches to the eardrum. Contracting this muscle pulls the malleus and membrane inward, tensing the eardrum. It has the same effect on the amplitude of vibration as does placing a hand on the head of a drum. The stapedius muscle performs a similar job on the stapes and the oval window. Because our own voices are among the loudest sounds consistently assaulting our ears, these muscles are automatically activated when we speak.

Because the tympanic membrane is sensitive to pressure on it, sensitive hearing requires that there be no pressure differential of the air in the outer and middle ears. To avoid this, the middle air cavity has an open passage to the pharynx, the eustachian tube. Air exchange will rapidly balance any change in atmospheric pressure. Similarly, fluid accumulation in the middle ear normally will drain to the pharynx.

Perception of sound depends on the physical and dynamic characteristics of all of these conducting elements of the middle and inner ear and on the behavior of receptors in the organ of Corti. Slight changes in their properties can significantly affect relative sensitivity to different pitches. Comparative studies indicate that species vary their ranges of hearing adaptively. It is not a coincidence that the frequencies to which the human ear is most sensitive are those of the human voice.

THE EVOLUTION OF THE EAR

The design of the ear has changed dramatically during the course of vertebrate evolution. High-frequency sound vibrations do not travel far or well in water, so hearing as we know it is not of importance to most fish. (A few have independently evolved an apparatus for hearing.) Short-distance vibrations of very low frequency in water may be extremely important for the detection of swimming animals and even stationary objects nearby. These are detected by a different sensory system, the lateral line system, which also uses hair cells but is not homologous to our ear.

Perception of airborne vibrations required the evolution of a new and much more sensitive apparatus. The modern arrangement of a tympanic membrane with one or more ossicles supporting it appears to have evolved independently but in parallel three times in the amphibian, mammalian, and reptilian/avian lineages (Clack 1989; Panchen 1989). The opening spanned by the eardrum was the spiracular opening of bony fish, one of the pharyngeal slits. The ancestral arrangement is still in evidence in the eustachian tube and middle ear cavity, which maintain their communication with the pharynx. Amphibians and reptiles used the stapes to transmit vibrations from the eardrum to the inner ear, but the stapes had a respiratory function in the gill in fish and early tetrapods.

The independent development of the mammalian system can be traced through numerous intermediate stages represented in the fossil record (Allin 1975; Gould 1990; Kermack 1989; Rowe 1996; Wang and Li 2001). Therapsid reptiles, the ancestors of mammals, used a very different strategy. The ear perceived vibrations transmitted through the bones more effectively than through the air. Sound vibrations would have been picked up by the mandible in direct contact with the ground. Analogous behaviors can be observed in some modern reptiles. (Human ears are also more sensitive to vibrations when they are transmitted through the temporal bone. This fact is used in standard clinical tests for diagnosing hearing loss.) The conduction chain for such vibrations included smaller bones in the posterior mandible, the angular and articular. From the articular, vibrations crossed the jaw joint to the quadrate bone in the cranium and then to the stapes and the inner ear (see Chapter 4, Figure 4.6). Such a hearing apparatus would have been relatively cumbersome and restricted to higher energy and lower frequency vibrations. Obviously, chewing would have interfered with hearing as well.

Later stages of therapsid evolution witnessed significant redesign of the mandible and the joint with the cranium. The posterior mandibular bones were substantially reduced in size as the dentary expanded. The dentary established its own articulation with the cranium, moving on the squamosal. The earliest mammals thus had two jaw articulations which specialized for different functions. The dentary–squamosal joint,

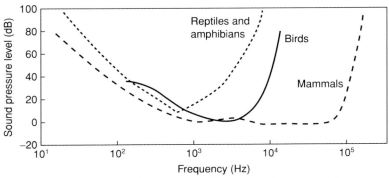

Figure 12.7 Ranges of hearing of different vertebrate classes. The mammalian ear provides a greater sensitivity than other designs, particularly in the potential range of frequencies detected. The audiograms indicate the threshold (lowest volume detected) for a range of animals in each class. Approximately 90% of the animals tested fall above the line indicated. (From J.M. Allman. *Evolving Brains.* Copyright 1999 by W.H. Freeman. Reproduced by permission of Henry Holt, Inc.)

which defines the origin of mammals, became the primary joint for mastication and is homologous with the human TMJ. The mandible was mechanically strengthened by the removal of all articulations between the dentition and the functional jaw joint.

The quadrate–articular joint continued to serve in the conduction of vibrations. As these two bones became smaller and lighter, they also became more sensitive. Eventually the articular and angular bones completely separated from the mandible. The quadrate and articular are homologous with the modern incus and malleus, respectively, while the former angular supports the tympanic membrane. This transformation of bones serves the purposes both of mastication and of hearing. Both adaptations probably contributed to the evolutionary process.

The mammalian design has resulted in a more flexible hearing system than that of other vertebrates (Fray and Popper 1985; Kermack 1989). In particular, mammalian ears are much more sensitive to higher-frequency vibrations—those beyond 2 kHz (2000 cycles per second; see Figure 12.7). The upper limit of hearing in birds and reptiles is about 10 kHz (and, so, apparently, was that of dinosaurs), while many mammals maintain sensitivity to 50 kHz. The extremes of the normal human range are considered to be 20 and 20,000 Hz. Although many animals may outperform us at one end of the range or the other, humans have an extraordinary ability to discriminate intensity, frequency, and direction in the preferred middle range. This is part of our neural adaptation for processing spoken language.

13

SKIN

The integument, or the skin, is the largest organ of the body, covering an average of 18,000 cm². It accounts for 14–18% of body weight. The skin incorporates structures of many of the body's organ systems, including nerves, blood vessels, and muscles, and serves many different functions. Skin is the boundary of the body, defining us as individuals and separating us from the outside world. Most of our interactions with the environment are mediated through the skin. The skin physically encounters that environment, for which it has an immense array of sensory receptors to monitor our surroundings and to facilitate appropriate responses. Passively, the skin serves as a barrier to keep foreign organisms and materials out and to keep water and nutrients inside. However, skin is more than a passive barrier. It actively contributes to regulation of body temperature, through fat insulation, changes in blood flow, and perspiration. Incidental to the latter, it adds to the capacity of the kidney to filter waste products

Table 13.1 Characteristics of Human Skin

Vertebrate characteristics
  Skin has two layers: epidermis and dermis
Mammalian characteristics
  Hair
  Sebaceous and sweat glands
  Increase in the thickness of the subcutaneous fat layer
Primate characteristics
  Nails replace claws on at least some digits
  Hands, feet, and digits possess friction pads with dermatoglyphs
  High density of Meissner's corpuscles in hands and feet
  Eccrine glands exist in hairy skin
Distinctively human characteristics
  Reduction of body hair
  Eccrine sweat glands are thermally sensitive
  Apocrine glands have reduced distribution and importance
  Subcutaneous fat layer is increased in thickness

from the bloodstream. These functions contribute to a unique homeostatic strategy, helping humans to maintain a relatively uniform internal environment even as the external environment undergoes drastic changes (Table 13.1).

THE ANATOMY OF SKIN

The skin is composed of several layers of tissue (Figure 13.1). The outermost layer is an epithelium, called the epidermis. Beneath the epidermis is a layer of connective tissue, the dermis. Deeper still is loose connective tissue called the superficial fascia or hypodermis. Each of these layers makes its own functional contributions to the integument.

THE EPIDERMIS

An epithelium is a sheet of cells arranged on a noncellular membrane to form a boundary. The cells may be tightly or loosely packed in one or many layers. The epidermis consists of many layers, or strata, of tightly packed cells. The cells of the epidermis are flat, or squamous. The deepest layer of epidermal cells is actively dividing, whereas the outermost layers are dead or dying.

The limiting resource for these cells is nourishment and oxygen from capillaries in the dermis that do not cross the basement membrane. The cells closest to the membrane receive a plentiful supply and are able to maintain normal growth. As these cells grow and divide, they are pushed increasingly farther away from the basement membrane and thus from their source of nourishment. Their activity slows down and they fill up with the protein keratin, a waxy water-resistant material that limits water loss from the body. The outermost layer of the epidermis consists of dead

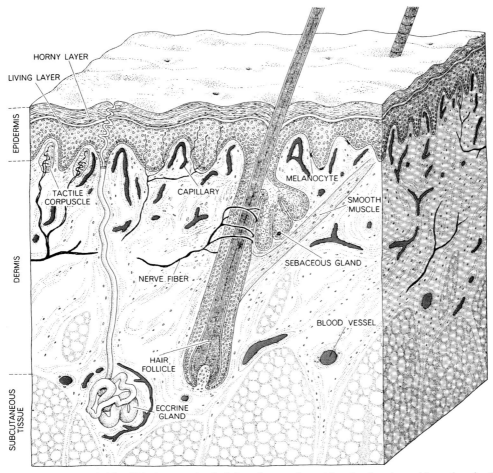

HORNY LAYER
LIVING LAYER
EPIDERMIS
TACTILE CORPUSCLE
CAPILLARY
MELANOCYTE
SMOOTH MUSCLE
SEBACEOUS GLAND
DERMIS
NERVE FIBER
BLOOD VESSEL
HAIR FOLLICLE
SUBCUTANEOUS TISSUE
ECCRINE GLAND

Figure 13.1 The structure of skin. Skin is comprised of three layers: the epidermis, dermis, and hypodermis. (Reproduced from *Scientific American* by permission of the estate of Bunji Tagawa.)

and dying keratinized cells and serves as mechanical protection for the living skin. These cells are disposable and are constantly sloughed off and lost from the body. While it might be possible to redesign the skin in order to keep the outer cells living, the same mechanism that allows continual replacement of cells also permits more rapid repair of the epidermis when it is injured. If the skin did not need to replace epidermal cells so frequently, wounds would heal that much more slowly.

The human epidermis is generally thin, about 0.1 mm thick. However, it is considerably thicker (about 1.0 mm) in the palms of the hands and soles of the feet, where frictional wear is greater. Mechanical stimulation of the skin accelerates the production of cells and causes a thickening of the epidermis called a callus.

In the deepest layer of the epidermis is a set of specialized cells called melanocytes. The function of melanocytes is to produce a black pigment called melanin. Melanin is exported from the melanocyte and taken up by neighboring cells. This is the most conspicuous of several pigments that contribute the color of skin and hair.

THE DERMIS

The dermis is a connective tissue; that is, it consists of loosely associated cells suspended in a matrix of extracellular fibers and small molecules. It contains several different types of cells performing different functions, including synthesis of the protein fibers, fat storage, insulation, and scavenging debris. Within the dermis are blood vessels, nerve endings, glands, and hair follicles. The variety of cells contained in the dermis carry out a number of maintenance functions, including synthesis of protein fibers, immune surveillance, and fat storage.

The fibers of protein that give strength to the dermis are primarily of two types, collagen and elastin. Collagen fibers have great tensile strength and tend to be arranged parallel to one another. This gives skin a grain at any given part of the body. Elastin fibers are able to stretch and recoil, thus permitting the skin flexibility and resiliency. They are present in human skin in greater quantities than are seen in other animals.

The boundary between the dermis and epidermis is a point of potential weakness, where the two tissues might be shorn away from one another. Consequently it is highly interdigitated, especially in regions of skin likely to experience shear forces, such as the palm and the sole.

THE HYPODERMIS

The hypodermis is a loose connective tissue that contains some of the larger vessels and nerves supplying the skin. Throughout the surface of the body, most of the cutaneous fat is stored in the fat cells of the hypodermis. The extent and thickness of this layer are unusually developed in the human lineage. The deepest layer of the hypodermis is more fibrous, blending into the tougher fascias that define muscles (epimysium) and surround bones (periosteum). The thickness of the hypodermis and its binding to deeper structures help to determine how easily skin can be made to glide across the surface of the underlying organs.

Subcutaneous fat is preferentially stored in a series of deposits that are mostly in the same locations in different mammals, even though the relative size of the deposits may vary considerably within and between species (Pond 1998). Deposits are found on the back,

the chest, the abdominal wall, and the base of the tail. Excess fat is also stored within the abdominal cavity. Humans tend to enlarge the size of those deposits nearly to the point of creating a continual layer of fat under the skin.

The distribution of subcutaneous fat within the human body reflects both individual variation and strong sexual dimorphism. As rule, subcutaneous fat is preferentially deposited in women in the thighs, buttocks, and breasts. A minimum amount of energy stored in subcutaneous fat is essential for reproduction. In men, excess subcutaneous fat is more likely to be placed on the abdominal wall. There has been considerable speculation and discussion about the origin of such a pattern. It is likely that the sex differences in part play a role in gender identification and attraction.

DIFFERENCES AMONG VERTEBRATES

The arrangement of an epithelium lying atop a specialized connective tissue is characteristic of the skin of all vertebrates. However, the relative thickness of the different layers and the development of specialized structures, such as scales, hair, feathers, claws, and glands, varies among groups of vertebrates according to the demands of the environment and habits of individual taxa.

The integument of a fish is concerned with limiting osmotic exchange of water and with reducing drag as the animal swims. For this purpose, skin glands secrete mucus, making fish feel "slimy." Because a fish's skin is more concerned with providing defense against predators and parasites than with interactions with objects in the environment, it is commonly lined with scales.

Modern amphibians have reduced or eliminated the protective armor. Because dehydration is not a concern for species living in or near water, amphibians may retain a thin skin that is physiologically active in the exchange of gases and able to supplement the respiratory system. Some amphibians out of water may secrete a protective layer of mucus that reduces water loss.

For fully terrestrial species, the need to conserve water results in a cutaneous barrier that prohibits respiration but can be made to withstand the abrasions of life on the ground. Hence, reptiles, mammals, and birds have a thicker epidermis. Mammals and birds have the additional need for thermal insulation and thus have acquired hair or feathers and a thicker layer of fat.

GETTING A GRIP: FINGERPRINTS AND NAILS

FRICTION PADS AND FINGERPRINTS

The skin over most of the body is relatively smooth. On the palms of the hands, the soles of the feet, and the digits, the skin is thicker and is marked by a series of fine ridges. These ridges represent more substantial foldings of the epidermis and the epidermal–dermal boundary. On the tips of the fingers and toes, they form unique patterns of dermatoglyphs or fingerprints. Dermatoglyphs serve two important functions: they increase the frictional properties of the skin and they enhance tactile sensitivity.

The skin of the hands and feet of primates, including humans, is supported and strengthened by special pads containing thicker dermal layers of both fiber and fat. These friction pads lie under the base and tip of each finger. Two more pads lie near the wrist on either side of the heel of the hand (Figure 13.2). Pads on the foot have a similar distribution under the toes, the ball of the foot, and the heel. These pads provide cushioning and distribute forces to protect the skin.

Numerous very tough collagen fibers are oriented perpendicular to the surface to counter traction forces that tend to shear the skin away from deeper structures. The interdigitation of the dermis and epidermis also prevents the two layers from shearing apart. The further ability of the ridges of the skin to interdigitate with irregularities of the substrate enhances the ability of the skin to embrace small or rough objects and increases the friction between skin and substrate (Cartmill, 1974b, 1979). Dermatoglyphs also appear on the prehensile tails of some New World monkeys and on the knuckles of chimpanzees and gorillas. These surfaces are also used to grip the substrate during locomotory behaviors.

The penetration of the epidermis into the dermis along the ridges also increases the sensitivity of the skin to movement. Under each ridge on the surface of

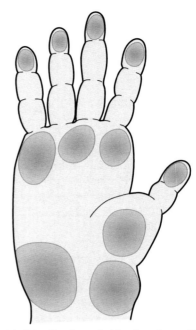

Figure 13.2 Distribution of ridged pads on the primate hand. Primate hands and feet are protected by a series of fat pads on the tip of each digit, at the base of the digit, and next to the wrist or on the heel.

the finger are two parallel papillary ridges of dermis separated by a small infolded ridge of epidermis. The papillary ridges are lined with touch receptors, specifically Meissner's corpuscles. These are especially sensitive to motion of the intermediate ridge and thus more sensitive than endings in other regions of the skin (Cauna 1954). The high concentration of Meissner's corpuscles is a characteristic of primates and coincides with the emphasis on the hands as tools for manipulation and exploration.

The development of the ridges is determined by patterns in the dermis. Dermatoglyphs, or fingerprints, are under genetic controls and appear as early as the third month of gestation, but there are an undetermined number of genes influencing them. That dermatoglyph patterns are not mere accidents of development is shown by the fact that fingertips that have been severed and regenerated will display the same fingerprint. The seemingly infinite variety of prints that makes each person unique has well known forensic implications. Of more clinical interest is the correlation of dermatoglyph pattern with certain prenatal factors and genetic disorders.

NAILS

The nail is a claw modified by the loss of the one or two of the original three tissue layers and by making it broad and flat. The nail overlies a layer of epithelium called the nail bed (Figure 13.3). The proximal end of the nail bed, distinguished by a white partial circle, or lunula, is the region of active growth. The process of growth is similar to that of the epidermis itself—keratinized cells die and are added to the nail as tiny scalelike plates.

Nails, as opposed to the claws of other tetrapods, are characteristic of primates and were probably present in the common ancestor (Soligo and Muller 1999). At least one nail, on the first digit, appears early in the primate fossil record, as shown by the

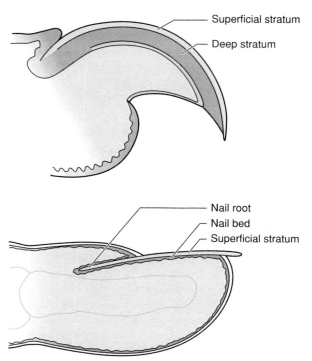

Figure 13.3 Structure of nails and claws. The nail has greatly reduced or eliminated the deeper layer tissue.

Paleocene fossil *Carpolestes* (Bloch and Boyer 2002). All living primates have at least one nail on each hand and foot, aside from some South American monkeys that have secondarily lost them. A nail gives a broad firm support to the pad of the digit to help dissipate compressive forces on the pad.

The relative adaptive advantage of friction pads and nails versus claws presents an interesting evolutionary question. A traditional view held primates to be first and foremost arboreal animals, and interpreted primate characteristics as adaptations for life in the trees (Le Gros Clark 1970). A grasping thumb and first toe seem to fit this expectation, giving individuals a more secure grip, and the transformation of claws into nails better accommodates that digital grasp. However, superb arboreal acrobats such as squirrels show none of these primate characteristics. On large vertical tree trunks, the claws of a squirrel that can dig into the bark have a clear advantage over grasping digits in a small-sized animal. What, then, is the adaptive advantage of nails?

Opposable digits appear to correspond more with the ability to grasp and manipulate objects, including food items, than to mere arboreality (Figure 13.4). They might be of advantage in a small-branch environment, although squirrels are obviously comfortable there as well. Cartmill (1974a) challenged the arboreal model with an alternative hypothesis that early primates were arboreal insectivores, stalking individual arthropods and catching them by hand. His model is consistent with other primitive primate characters, including visual development and hand-eye coordination. Not all early primates were insectivorous; however, eating fruit and other small objects in the terminal branches of the trees would also exploit the advantages of grasping hands and feet, skilled visual identification of food, and dexterity of hands to obtain and manipulate it (Rasmussen 1990; Sussman 1991).

Most modern primates are medium-sized animals or larger and continue to function well with nails, even though their niches have diversified. It is worth observing that some of the small squirrel-like South American monkeys that travel, climb, and feed on large tree trunks have converted the nails back into claws and no longer have an opposable thumb (Hamrick 1998). This reconvergence to a squirrel-like form is consistent with

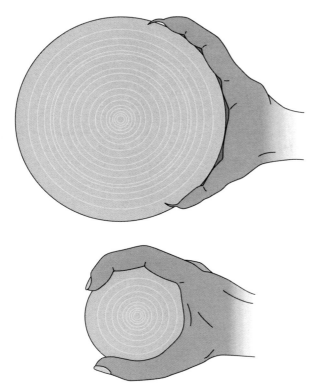

Figure 13.4 Climbing with nails and claws. Claws can dig into bark to get a better hold on a trunk that is large relative to the hand. Finger pads supported by nails are more secure for grippping small branches. It is likely that primate hands with nails evolved in small-branch settings.

a close relationship between the diameter of the substrate and the development of nails, showing a full range of development from nails to claws and intermediate forms within one group of primates.

This particular evolutionary pathway has been a fortunate exaptation for humans. Our legacy of hands capable of holding and manipulating objects encouraged the later development of tool use by several different species of monkeys and apes. Clearly it was an important facilitator of hominin culture.

HAIR

Hair follicles are complex structures (Figure 13.5). Although the cells and glands of the follicles are derivatives of the epidermis, hairs are rooted more deeply in the dermis. Like nails and the epidermis, the hair itself is composed of dead keratinized cells that have been molded about the base, or root, of the hair before being sloughed away from their epithelium. In this way the hair grows from its root outward. Within the root is a papilla composed of dermal cells, including blood vessels. These nourish the cells of the root to permit continued growth.

Hairs grow to a predetermined length before they are shed and replaced by new hairs of the same length. The determinants of hair length are not known, but they do account for differences between human body hair and that of other mammals. There are sex differences in hair length as well. Under the influence of hormones, male body hair is longer than female body hair. Conversely, female scalp hair may be longer than male scalp hair. These patterns suggest the action of sexual selection. Cultural practices often highlight this difference.

Surrounding the epidermal cells of the hair follicle is a dermal layer containing numerous collagen fibers, nourishing blood vessels, and sensory nerve endings.

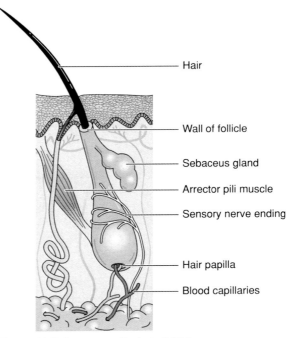

Hair
Wall of follicle
Sebaceus gland
Arrector pili muscle
Sensory nerve ending
Hair papilla
Blood capillaries

Figure 13.5 Structure of a hair follicle.

A small smooth muscle called the arrector pili inserts into this fibrous coat. This muscle causes the hair to stand upright away from the skin. Large numbers of these contracting together incidentally create small dimples in the skin over their origins and slight elevations around the follicles ("goose bumps"). The arrectores pilorum are under control of the sympathetic nervous system and are stimulated to contract during times of fear or emotional arousal or when one is simply cold.

Hair has two primary functions in mammals. The first and more obvious is insulation. By trapping a blanket of "dead" air around the body, heat transfer by convection is sharply reduced. Erection of the hairs increases the thickness of this air space. Insulation considerably increases the efficiency of maintaining a constant body temperature. The layer acts both to keep heat in the body when surrounding air is cool and to keep heat out of the body under the hot sun.

The second function, but probably the first to evolve, is sensory. At the base of each hair follicle are sensitive peritricial nerve endings that report any signs of movement by the hair. These make it possible for us, for example, to detect and chase away an insect attempting to land on our skin. Nearly all mammals except humans have specialized hairs on the face called vibrissae or, more commonly, whiskers. Vibrissae are anchored in especially sensitive beds of nerve endings and serve as the most important organs of touch for many mammals. It is speculated that hair originally evolved as just such a specialized sensory organ and later proliferated to become significant for insulation. Although humans have entirely lost vibrissae, all human hairs are richly innervated, while many of those on other mammals are not. It is probably this sensory function, rather than any residual insulation, that accounts for the evolutionary retention of human body hair.

Human hair is visibly reduced compared with that of most other mammals; but it has not been eliminated. Varying degrees in the reduction of hair density occurs among higher primates. Schultz (1931) counted an average of 170 hair follicles per square centimeter of chest skin in Old World monkeys versus 90 in apes and 1 in humans. Taking into account the difference in body size and area of skin, humans may have a similar absolute number of body hairs (Montagna

1985). The most conspicuous difference is a reduction in the length and thickness of individual human hairs. The adaptive explanation for this lies with a unique strategy for thermoregulation (see later).

Humans have retained hair in the axilla (armpit), in the pubic area, and on the scalp. In addition, mature males have varying amounts of facial hair. Although these patterns have been interpreted as evolved visual signals of sexual maturity or of threat and intimidation, it is difficult to test such speculation. Axillary and pubic hair may likewise serve a useful function of reducing frictional irritation between two moving surfaces of skin (Montagu 1962). One can speculate that they also play a role in dispersing pheromones from apocrine glands in those places (see later). Scalp and facial hair represent a different puzzle. Scalp hair is useful in insulating the head against solar heating, because the brain is particularly sensitive. The sexual dimorphism of head hair length has triggered a number of hypotheses about evolutionary role of sexual attractiveness of hair that can neither be confirmed nor dismissed.

HAIR REDUCTION

Because fur represents insulation to keep an animal warm, it is most reasonable to suppose that the elimination of it enables a body to cool down. Extremely large mammals have more difficulty dumping body heat. Thus, all tropical mammals greater than 1000 kg are naked, with the exception of the giraffe, whose long lines probably disperse heat effectively. Human ancestors have always been below this threshold size, so further explanation is needed.

The most widely accepted explanation for reduction of hair, addition of subcutaneous fat, and changes in the sweat glands relates to a more efficient thermoregulation that enables humans to maintain high activity levels for long periods of time without overheating (Ebling 1985; Porter 1993; Wheeler 1984, 1985, 1992b, 1994; Zihlman and Cohn 1988). In the 1960s, most models for human evolution focused heavily on hunting. Montagu (1964) and later others suggested specifically that hair reduction permitted long-distance chasing of prey animals without overheating. The emphasis on hunting has had many reinterpretations in the past two decades, but this hypothesis coincides with other aspects of our anatomy and is discussed in more detail later. Here are some other proposed answers.

Body size. Some scholars believe that the reduction of body hair occurred earlier than the shift in behavioral strategies. Newman (1970) suggests that our forest-dwelling ancestors lost their hair while still living an apelike existence. Schwartz and Rosenblum (1981) argued that it was increasing size in the ape lineage that led to hair reduction to allow a larger body to lose heat more effectively. A size-related, or allometric, change may or may not explain the relatively reduced density of hair in the living great apes, but not the extreme condition found in modern humans.

Aquatic habitat. The feminist response to the hunting hypothesis was the aquatic hypothesis. An aquatic mammal, such as a whale or hippopotamus, faces very different problems of maintaining body temperature than do terrestrial mammals. Flowing water carries heat away from the skin more quickly than does air. Evaporative cooling through perspiring is impossible in water, and wet hair is useless as insulation; therefore it has been discarded in those mammals that never leave the water, although not in most mammals that spend at least some time on land (Hardy 1960; Morgan 1972, 1982, 1990). As current models of the aquatic hypothesis do not envision a wholly aquatic phase for human ancestors, this is an unlikely explanation for our extreme reduction of hair.

Sexual attraction. Darwin (1871) and Morris (1967) understood the reduction of hair as an adaptation to heighten sensitivity and thereby sexual attractiveness and arousal. This view fails to acknowledge that the nerve endings associated with hair can be just as sensitive, if not more so, as the endings on naked skin. In fact, touch through grooming is an extremely important vehicle of social interaction among primates that humans have had to forgo, even though we still derive pleasure from pawing through our pets' hair and sometimes that of other people.

Clothing and culture. Because there is no fossil evidence bearing directly on the presence of absence of hair, it is not impossible that hair reduction was a very recent event. Hamilton (1973) and Kushland (1985) suggested that the use of clothing and fire made hair redundant and led to its reduction. This hypothesis would ignore the physiological implications and other changes in the integument in the hominin lineage.

Parasites. Fur is a common site for parasites to lodge, including fleas, ticks, and lice. It has been suggested that reduction of our body hair would make it more difficult for them to cling and hide, and humans would bear less disease (Pagel and Bodmer 2003). However, if hair loss offered such an advantage to humans, why have so few other species lost their body hair? Many animals already do have adaptations to discourage biting insects and other parasites. Primates, including humans, use grooming for the dual purpose of hygiene and social bonding.

This speculation might be more informed if we better understood when hair was reduced. It has been common for artists to place substantial amounts of body hair in reconstructions of *Homo erectus* and even Neanderthals, despite the lack of direct evidence for this. However, a recent analysis of the melanocortin I receptor gene (*MC1R*) has addressed this question. The *MC1R* gene is responsible for some of the population differences in pigmentation among modern humans. Before hair was lost, there would have been little advantage for pigmented skin, even in the African savanna. According to the genetic diversity present in this gene in Africa and to its selection constraints, Rogers et al. (2004) calculate that human ancestors have been hairless for more than 1 million years. Such a date would argue against the role of recent cultural developments such as clothing and places the loss of hair among our ancestors in Africa.

GLANDS OF THE SKIN

There are three different classes of glands embedded in the dermis that differ in structure, function, and distribution. Hair follicles are accompanied by sebaceous glands, consisting of a collection of fat cells clustered around a duct. As individual cells in the gland deteriorate, their oily product, called sebum, empties along the duct into a hair follicle and thence to the skin. The oils appear to keep the skin from drying out and to add to the chemical barriers protecting the body. In a few sensitive regions of the skin, such as on the lips, sebaceous glands may be found independently of hair follicles. Glands that secrete in this manner are called holocrine glands. In addition to the holocrine sebaceous glands, mammals possess two types of sweat glands. Apocrine glands secrete a viscous fluid in hairy regions of the body. Eccrine glands produce the watery fluid in nonhairy regions that is more familiar to humans as perspiration.

APOCRINE GLANDS

Apocrine glands secrete their products by pinching off a portion of the cell itself to pass through its duct to the surface of the skin. The fluid contained in these fragments can evaporate, removing heat from the body. In most mammals, these are the most important glands for evaporative cooling. However, apocrine secretions are wasteful in that they shed nutrients along with the fluids. Those same nutrients provide a favorable environment for bacterial growth.

Human apocrine glands, located particularly among the axillary and pubic tufts of hair, contribute little to evaporative cooling. It has been hypothesized that they have a more important role in olfactory communication. Surface bacteria break down certain compounds in the apocrine secretions to produce aromatic compounds. Because apocrine glands are stimulated in response to emotional arousal (e.g., fear or pain), the resulting body odors may signal emotional state. While such olfactory cues constitute a vital part of the sensory world of other animals, such as dogs, it is not clear whether humans respond to them to a significant degree.

If body odors are being produced as an adaptive communication with other individuals, these compounds would qualify as pheromones. Pheromones are known to be used by other vertebrate and invertebrate animals for chemical communication regarding sexual state, but evidence for their importance in humans is slight (see Chapter 12). In American culture, apocrine odors are considered offensive, and many people go to great lengths to eliminate them. This is not a practice in most world cultures, but the use of perfumes to enhance or disguise body scent is common. In fact, humans often select perfumes that imitate animal sex pheromones (e.g., musk and civet) or plant analogs of them (floral scents).

ECCRINE GLANDS

Eccrine glands produce perspiration by a process of filtration. Plasma is filtered to remove larger molecules and is then subjected to an active extraction of desirable solutes. The remaining filtrate, a dilute solution containing waste products, is discharged through a duct to the surface of the skin. This process is metabolically expensive but has some clear benefits. Because the resulting perspiration is primarily water, it more effectively removes body heat. The recovery of metabolites makes eccrine sweating less wasteful than apocrine secretions. Furthermore, the elimination of waste materials, such as urea, supplements the function of the kidney in cleansing the bloodstream.

The skin possesses an estimated 2 to 5 million eccrine glands, whose density varies considerably in different parts of the body. These are under a complex control involving both central and local mechanisms. One set of eccrine glands that are physiologically closer to those in nonprimate mammals is found in the soles of the feet and palms of the hands, where they keep the skin moist. These are stimulated by the sympathetic nervous system using the neurotransmitter norepinephrine and are activated in response to stress and emotional state.

Eccrine glands elsewhere in the body relate more directly to thermal stress. They are stimulated by the sympathetic nervous system using the neurotransmitter acetylcholine. A slight rise in body temperature will register in the hypothalamus of the brain, which will send commands to these sweat glands. Alternatively, warming of a specific region of the skin may stimulate local eccrine activity without involving the rest of the body. The responsiveness of these pathways to change in temperature indicates that thermoregulation is the primary function of human eccrine glands. The use of a different neurotransmitter from that of other sympathetic pathways makes functional sense. Because active sweat glands must process large quantities of fluids, they should not be activated by the same sympathetic signals that cause vasoconstriction.

SWEAT AND HAIR REDUCTION AS A HUMAN STRATEGY FOR THERMOREGULATION

Humans are unusual in the number and functioning of the eccrine glands. In most other mammals, apocrine glands vastly outnumber eccrine glands, and the latter are restricted in distribution to nonhairy regions, such as the pads of the feet. Among higher Old World primates, eccrine glands do appear in hairy skin, but these may not be sensitive to temperature state. Even animals noted for their ability to cool themselves by sweating, such as horses, rely on apocrine glands. These glands are responsive to systemic epinephrine levels, reflecting activity level or emotional state rather than temperature.

Most mammals with a coat of hair would find sweating of limited effectiveness. Fur traps a layer of dead air against the skin that is very effective for insulation. Perspiration in this area would quickly saturate the air, inhibiting further evaporation. Unless air can circulate freely over the moisture, perspiration cannot evaporate; without evaporation, there is no cooling. Thus, an animal that relies primarily on evaporative cooling from the body surface must reduce its coat of hair. This explains one of the most conspicuous of human adaptations.

Terrestrial mammals that cannot sweat may use other means for cooling the body. One common strategy is evaporation from the mouth and pharynx through panting, commonly observed in dogs. Water buffalo, elephants, and pigs apply water or mud to their bodies. Most medium-size or large mammals have a limited capacity to shed excess heat and thus modify their behaviors accordingly. In the tropics, they avoid activity during the heat of the day and strenuous activity is limited absolutely. Few mammals have the endurance for long-distance running or sustained high metabolic rate that humans have. Predators such as lions or cheetahs can maintain a sprint for only short distances, and their prey need run only a little longer to be safe. Overheating is a primary consideration for them.

The human ability to shed excess heat permits us to sustain high levels of activity for long periods of time in a warm, tropical habitat. With the emphasis on eccrine glands and the reduction of hair, our perspiration is a most effective mechanism. However, there are other consequences that demand compensatory adaptations. One is the need to replenish body fluids. Humans drink more and are more dependent on standing water resources than are most other mammals. Another consequence is the need to ingest and consume more energy. Thus humans have a diet that is relatively rich in meat for primates. A third problem is the need to replace the layer of hair with another form of insulation to prevent hypothermia between periods of activity. Modern humans may have two sources of assistance, one physiological (fat) and the other cultural (fire and clothing).

A PROBLEM OF FAT

Humans possess an unusually great quantity of fat, a fact that deserves explanation in an evolutionary context. It has been noted that we have about 10 times the number of adipocytes predicted for our body mass, suggesting that humans have a unique strategic use for fat. Our species has been compared with other fat species, such as marine mammals, with the suggestion that we are adapted to an aquatic environment (Hardy 1960; Morgan 1997). Fat sometimes serves as thermal insulation and it decreases the body's specific gravity, two functions that have been offered as adaptive strategies in the framework of an aquatic hypothesis (Cunnane 1980; Morgan 1986, 1997; Verhaegen 1985). However, superficial observations may be misleading. It is likely that adult human fatness is merely a by-product of the infant's need to store energy observed in a culture that commonly oversupplies adults with more energy to be stored.

Mammalian adipocyte size and number follow an allometric relationship with body size: larger animals possess proportionately fewer but larger fat cells (Pond 1987, 1998). Thus, it is possible to predict expected values for a given species. Humans are found to have approximately 10 times as many fat cells as expected, but they are smaller, comparable to the size of rat adipocytes. Humans and macaques (and presumably other primates) increase adiposity by increasing the number of adipocytes (Pond and Mattacks 1987). This is unlike the mechanism of other mammals, including rats, which increases the size of the cells rather than their number. The number of human adipocytes is a reflection of individual fatness rather than a definitive species characteristic.

Human adults vary greatly in the proportion and distribution of body fat. However, that distribution within the body follows the same pattern observed other mammals. Anatomical homologies of specific depots may be readily identified by physiological properties (Pond and Mattacks 1989; Pond 1991, 1998). Depots are also found to differ in fatty acid composition, implying functional differences as well (Calder et al. 1992; Malcom et al. 1989; Pond 1999).

Fat tissues and lipid reservoirs potentially serve several functions, including thermal insulation, thermogenesis, energy reservoirs for local tissues or for the body as a whole, buoyancy, storage of hormones, and mechanical insulation. Because adipose tissues may perform more than one of these functions simultaneously, it may be difficult to determine which function, if any, is the primary adaptation. The variability of size of human fat deposits and other aspects of body structure must be considered in speculating on any species adaptation.

FAT AS INSULATION

Although the locations of fat deposits are the same in humans as in other mammals, the increased size of individual deposits permits them to expand under a greater proportion of the skin. Moreover, fat has the advantage of lying deep to the outer layers of the body and thus can be bypassed when necessary. Blood vessels, which are responsible for moving heat within the body, mostly run deep to the fat, where they are potentially insulated from the cooling environment of the skin. If it is desirable to loose heat, the arterioles of the dermis dilate to carry a volume of blood greater than the actual metabolic needs of the skin. Much of this extra flow passes through shunts that connect directly into venules. Thus, the blood bypasses the capillaries but is still exposed to the cooler surface.

If it is desirable to retain body heat, the vessels constrict and minimize dermal blood flow. The outer layers of the skin are highly tolerant of such a temporary reduction of oxygen supply, although they may loose sensation. As a protection against actual frostbite (freezing of tissues), prolonged exposure to extremely cold conditions may produce a secondary vasodilation due to cold-induced paralysis of the vasoconstricting muscles. This mechanism brings a flush of warm blood to restore the tissues.

Vascular dilation is under both local and central control. Warming a region of the skin causes a relaxation in the smooth muscles of the vessel walls and valves. The result is flushing, or redness in the skin. Changes in the blood flow also have significance for the body as a whole. Expanding or closing cutaneous channels lowers or raises blood pressure accordingly and can be manipulated by the neuroendocrine system for that purpose. Vessels to the head do not undergo vasoconstriction in response to cold. Rather, they maintain a flow of warm blood. The obvious advantage to this in terms of maintaining a constant temperature for the brain comes at a cost of the additional heat lost to a cold environment. This suggests that scalp hair has been retained as functional insulation.

Subcutaneous fat is in an ideal location to work with this circulatory pattern as a layer of insulation that can be switched on or off; and it has been assumed that that is its primary role (Pawlowski 1998; Thong 1982). However, the pattern of human fat is not consistent with this premise (Pond 1991). It is not evenly distributed over the body, and leaves some areas less protected than others. Superficial fat is not preferentially deposited or conserved when total fat content changes, as would be expected if it had a static function. Finally, the rate of heat loss is not primarily related to the amount of fat. Differences in body build between warm-adapted and cold-adapted human populations reflect changes in muscle and bone mass more than fat quantity; and the rate of heat loss in cold water also reflects body build. Likewise, heat conservation in infants correlates better with lean body mass than with fat volume. Even in obese individuals, resistance to heat loss is as much a function of cutaneous circulation as it is of fat (Jéquier et al. 1974).

This is in agreement with observations in other terrestrial mammals. In polar bears, for example, heat conservation is promoted by increasing body size rather than fat alone and by adaptations of fur (Pond and Ramsey 1990). While there may be no independent increase in fat levels, there may be a differential distribution of the fat to better insulate the body in cold-adapted mammals (Young 1976). We would do better to look to other explanations to understand the evolution of our own adiposity.

FAT IS AN ADAPTATION FOR THE BRAIN

Fat is an optimal form of energy storage. Because of its chemical properties, adipose tissue contains reduced water content and thus concentrates energy while minimizing the weight burden for the animal (Young 1976). Because fat stores scale with absolute body size in most mammals, large species such as bears, but not smaller ones, are able to hibernate and support themselves through the winter on fat stores alone.

The distribution of fat is best explained as an energy reserve, either locally or systemically. Characteristic fat deposits near lymph nodes probably service immune tissues; fat in yellow bone marrow supports blood cell formation (Pond 1999). The percentage of body weight represented by fat declines after age 6 months but in girls the percentage rises again in later childhood approaching puberty and reproductive age (Fomon, et al. 1982). A minimum level of nutritional health and energy balance, partly reflected in body fat, is one of several essential conditions for normal ovulatory cycling (Frisch 2002; Ellison 1990, 2001). These fat reserves will enable a mother to maintain the delivery of calories to her child, both before and after birth. Sexual dimorphism in fat quantity is typical across mammalian species (Young 1976).

Higher levels of body fat appear to increase reproductive success. Food restrictions can have a severe impact on fertility across mammalian species (Wade and Schneider 1992). However, humans are remarkable among mammals in their ability to ignore nutritional insults during pregnancy (Peacock 1990). While some species terminate pregnancies or resorb fetuses, human

pregnancies are likely to continue, although the infant may suffer from extreme deprivation. Perhaps this reflects an extraordinary investment that a pregnant woman is making and cannot afford to abandon. Unusually large maternal fat reserves would be a logical dimension of this strategy.

The fat represents not only reserves through gestation but also energy supply to support lactation, during which they are typically depleted in many mammals. Women need an additional 670 kcal per day during lactation if they are not to lose weight (Dewey 1997), but this need was probably rarely achieved in the past (Merchant and Martorell 1988). Studies of modern women indicate that the increased caloric needs of lactation are likely to be met with increased food intake and reduced activity level (Lunn 1994), although fat may serve as a critical reserve. Both general weight loss and fat loss, in particular, during this period are common (Dewey 1997; Merchant and Martorell 1988; Prentice and Prentice 1988). Marine mammals, also noted for fat reserves, transfer enormous quantities of energy to their offspring through lactation. Their nursing periods tend to be short and intense, corresponding to fasting by the mother and rapid growth in body weight by the infant (Gittleman and Thompson 1988).

Because of the close association between adipose tissue and energy use, it is likely that human fatness evolved as an energy reservoir specifically for the newborn infant and its developing brain (Kuzawa 1998; Correia et al. 2004). The developing brain is sensitive enough to energy supply that its final size is closely correlated among mammals with maternal metabolism (Martin 1983a). At the same time it is very demanding. The human brain uses 20–25% of the total resting metabolism, compared with the 8–9% cost for other primates (Cordain et al. 2001). Because of its brain, the developing human infant requires 9% more energy investment from the mother than a chimpanzee would. By age 5, the human brain is calculated to have consumed three times as much maintenance energy as the chimpanzee brain (Foley and Lee 1991).

As human reproductive investment has been transferred from the prenatal to the postnatal period, the energy stores of lipid likewise are transferred. The most consistent development of subcutaneous fat

appears in human fetuses and infants, where it accumulates at a greater rate than in other species. Fetal
fat is established almost entirely in the last trimester.
Figures calculated from infants born prematurely show
that in the same period in which the body weight
increases 269%, the brain increases by 220%, and total
fatty acids increase by 739% in white adipose tissue
and 240% in brown adipose tissue (Clandinin et al.
1981). Fetal fat development is so consistent that is
has been used to calculate gestational age of preterm
neonates (Ellis et al. 1994): its deposition is so rapid
that adipose weight correlates to the seventh power
of age. Likewise, paucity of fat may be diagnostic for
prematurity (Petersen et al. 1988). Birth weight is a
prime indicator of chances of infant survival and of
later performance of intelligence tests (Hrdy 1999).

Infant fat requires the establishment of large numbers of adipocytes. As long as individuals in later life
consume excess calories for storage, those adipocytes
will persist as subcutaneous fat. This critical need of
the infant for a continuous supply of calories therefore
appears to be the best explanation for the unusually
large fat reserves in not only infants and women but
also all members of the species.

Perhaps because of the relationship between fat
levels and reproductive cycling, human fatness is
also unusual in its employment in sexually dimorphic
features (see Chapter 19). Different subcutaneous
depots are preferentially expanded in females and males,
emphasizing breasts and figure development. The
importance of these sexually differentiating features
is suggested by the fact that they are retained even
in malnourished women who are unable to ovulate.
Kirchengast and Huber (2001) argue that the figure
provides continuing sexual attractiveness so that a
woman can attract and retain a mate when conditions
improve.

EXPLAINING HUMAN SKIN PIGMENTATION

The pattern of pigmentation, or coloration, of the human
skin remains socially one of the more troublesome
aspects of human anatomy. Even as we learn more
about the physiology of skin pigments, the subject
is surrounded by superstition and controversy, both

sociological and biological. Sociological problems arise
because our natural tendencies to divide ourselves into
exclusive populations have focused excessively on skin
color as a means of classification. Biological problems
arise because we do not fully understand why pigments
are distributed as they are among populations.

There are several chemicals contributing to the
color of the skin. These include hemoglobin in the
bloodstream, carotenes from dietary nutrients, and,
most important, the black pigment melanin. Melanin
is synthesized by specialized epidermal cells called
melanocytes, which lie in the germinative layer.
These package melanin into small bodies called
melanosomes, which are exported into neighboring
epidermal cells and thus distributed widely in the skin
and hair. All human beings, except for those with
certain genetic errors, produce significant amounts of
melanin by roughly the same number of melanocytes.
Individuals with "black" and "white" skin tend to
produce slightly different forms of the molecule. Dark
skin has eumelanin, which is dark brown to black
in color. Lighter-skinned individuals produce a more
yellowish pheomelanin. In addition, lighter-skinned
people may produce less pigment and tend not to
disperse is as widely among the cells.

MELANIN AND PHOTOPROTECTION

The best explanations available for the adaptive significance of melanin are not entirely satisfactory and many
others have been attempted. The most conspicuous
function of melanin is the absorption of sunlight and
protection of the dermis from solar radiation. Most
attempts to understand the adaptive value of skin
color focus on this function (Coon 1982; Frisancho
1981; Krantz 1980b; Roberts 1978).

Shorter-wave radiation is potentially damaging to
cells because it destabilizes molecules and produces
free oxygen radicals, which are highly reactive chemically. Free radicals have been implicated as causative
factors in cancer. Fortunately, the atmosphere screens
us from most radiation of shorter wavelength than
visible light, except for part of the ultraviolet range.
It is the ultraviolet radiation that is responsible for
sunburn and genetic damage that might cause skin
cancer.

By definition, pigments "absorb" radiant energy of specific wavelengths of visible light. Melanin apparently does this by neutralizing the free radicals produced by light and by ultraviolet rays and releasing their energy as heat. Within epidermal cells of lighter-skinned populations, melanosomes are positioned around the nucleus, where they are in the best position to protect DNA, although in more heavily pigmented individuals, melanosomes disperse throughout the cell. It thus appears that melanin functions primarily to protect the skin cells against excessive solar radiation.

Light-skinned individuals usually have the ability to synthesize increased quantities of melanin after exposure to sunlight. This is the basis of a suntan and may be interpreted as a seasonal adaptation in temperate climates to the more direct summer sunlight. However, pigments may produce oxygen radicals even as they absorb the damage from UV radiation (Wu 1999). A tan probably provides less effective protection from intense sunlight.

Within the human species, the skin color in different populations roughly correlates inversely with latitude and with potential exposure to ultraviolet radiation (see Figure 13.6). Because melanin reduces reflection of light as well as penetration into the dermis, the amount of melanin is commonly estimated by the percentage of sunlight reflected from the skin. The most pigmented populations are in the tropics of Africa, Asia, and Australia, where they receive more direct and intense solar radiation. Europe and northern Asia are populated by lighter-skinned peoples. One survey found that the skin of Europeans reflected 30–45% of the sunlight; skin of Asian Indians, 22%; and that of sub-Saharan Africans, 16–19% (Blum 1945).

Frisancho (1981) has attempted to estimate the dangers of sunlight for light-skinned people living in the tropics. European descendants in tropical Australia have an annual death rate of 677 per 100,000 from malignant melanoma and other skin cancers. In the absence of medical care, this rate would probably double to an estimated 1.4%. Other researchers have observed high rates of premalignant cells or cancers in the skin of albino (unpigmented) and white children in tropical America and New Guinea (Daniels et al. 1972). To the degree that melanization prevents such

cancers and that the victims are of reproductive age or younger, this represents a strong selective force.

Melanized skin would have been an essential adaptation for the hominins who drastically reduced their body hair in tropical habitats. This conclusion raises additional questions. Why would populations in higher latitudes have reduced pigmentation? Sunlight in the higher latitudes, north and south, strikes the atmosphere at a lower angle than it does in the tropics and passes through more of it before it strikes the ground. The ozone in the atmosphere has a greater opportunity to filter more of the harmful ultraviolet radiation. Therefore, heavily pigmented skin is unnecessary for high-latitude populations. Is this sufficient reason to evolve light skin, or is there some adaptive advantage to reduce melanin?

THE VITAMIN D HYPOTHESIS

Ultraviolet light penetrating into the dermis has several physiological effects, of which the most interesting is the synthesis of vitamin D. The quantities of vitamin D taken in through food is supplemented by chemical synthesis within the body from cholesterol. This synthesis is a pathway of chemical reactions. In the last reaction, 7-dehydrocholesterol in the skin interacts with sunlight to produce an active form of vitamin D. Vitamin D participates in calcium transport systems to absorb calcium in the intestine and to manufacture bone. A deficiency of the vitamin causes abnormalities of bone development (rickets) and metabolism. Children who do not have sufficient exposure to sunlight may have this disease. It has been hypothesized that people migrating from lower to higher latitudes may sustain a shortage of vitamin D unless the amount of melanin in the skin is reduced. Over evolutionary time, this hypothesis predicts that skin color will lighten in proportion to the distance from the equator (Murray 1934; Loomis 1967; Jablonski and Chaplin 2000, 2002).

Is the impact of pigmentation on vitamin D production great enough to direct the evolution of skin color? Differential production in dark versus light skin has been observed, but some authors believe the differences are not nutritionally significant (Clemens et al. 1982; Matsuoka et al. 1991). Frisancho (1981;

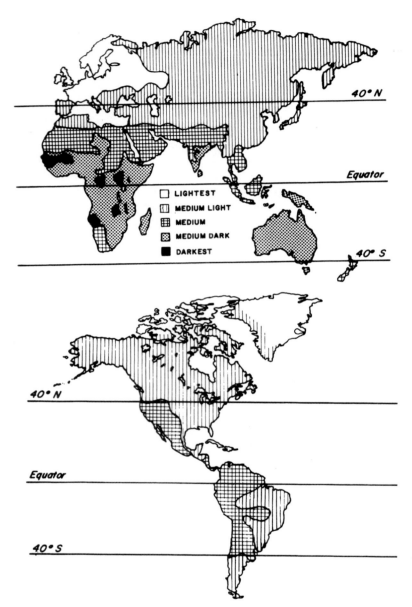

Figure 13.6 The world distribution of human skin pigmentation. (Reprinted from A. Roberto Frisancho, *Human Adaptation: A Functional Interpretation.* Copyright 1981 by A. Roberto Frisancho. Reproduced with permission from the University of Michigan Press.)

also Robins 1991) reports that in the early twentieth century, pelvic deformities occurred in 15% of African American women and only 2% of white American women; other surveys found the ratio of incidence between the two groups to be 3:1 or 4:1. Pelvic malformation might have serious consequences for child bearing and therefore would exert natural selection. However, it is not possible to separate the effects of vitamin deficiency on pelvic formation from those of differential nutrition and medical care.

Critics of the vitamin D model question whether the observed difference in synthesis between European and African peoples is great enough to be clinically significant (Blum 1961; Neer 1975; Robins 1991). The vitamin is readily stored by the body, so that constant exposure to the sun is not necessary. The frequency of rickets greatly increased in the modern industrial and urban environment of the nineteenth century, when children sometimes spent excessive hours laboring indoors. Its significance in the evolutionary past has been disputed. In the absence of a viable alternative model, scholars are likely to continue to debate the vitamin D hypothesis for some time to come.

With or without the vitamin D hypothesis, the solar protection model goes a long way toward explaining the diversity of skin color observed in the human species. Habitat and climate variables account for 80% of the observed variation in skin color (Roberts 1978). There are some populations, however, that do not fit the predictions. Indigenous peoples of the New World, for example, show less variation in moderately pigmented skin color from the Arctic to the Amazonian basin than do Old World peoples. It is possible that the populating of the Western Hemisphere was so recent (25,000 to 15,000 years ago) that there has not been time for local adaptations. However, the Americas were populated by migrations from northern Asia and were culturally adapted for Ice Age conditions. It is unlikely that the Asian ancestors were recent arrivals from tropical lands, yet the skin colors are darker than one would expect from high-latitude adaptation. Other smaller groups, in such regions as equatorial Africa and Tasmania, also have lighter or darker skin tone than predicted and are not likely to represent recent immigration.

Are these exceptions so great as to invalidate the radiation hypothesis? Some authors think so (Blum 1961; Diamond 1992b; Hamilton 1973). Others believe that recent migrations and secondary fine-tuning (e.g., lighter skin color among people living in the shade of dense tropical forests) can account for the discrepancies. It is also possible that the diversity of skin color in higher latitudes is simply a reflection of random changes occurring under reduced selection. In summary, the radiation model goes far toward explaining skin pigmentation, but it leaves much room for argument.

OTHER HYPOTHESES OF HUMAN PIGMENTATION

Camouflage. Dark coloration might confer protective coloration for hunters and human prey in tropical forests and light skin color might be less obtrusive in the tundra (Brues 1977; Cowles 1959). This may explain a general tendency, named Gloger's rule, for mammals to have fur color that gets progressively lighter in higher latitudes. However, there is a greater variation of habitat and local lighting conditions in latitudinal zones than this hypothesis acknowledges.

Protection of nutrients. Solar radiation may break down certain essential nutrients, such as folates (members of the vitamin B group), as they circulate in cutaneous blood vessels. Folates are particularly critical for normal fetal development. Melanin may protect these, but that link has not been established (Branda and Eaton 1978; Jablonski and Chaplin 2000, 2002).

Protection against cold stress. Observations of soldiers in the Korean War showed that Americans of African descent had a higher frequency of frostbite and other cold-stress syndromes (Post et al. 1975). Intriguingly, differential cold damage has been observed in pigmented versus nonpigmented skin of the same laboratory animal. It is suggested that reduction of pigmentation in northern populations is an adaptation toward greater cold tolerance, but the role of pigmentation itself has not been isolated from that of other aspects of climatic adaptation, such as control of cutaneous blood flow.

Protection against tropical diseases. Tropical habitats have other medical hazards, especially parasites. There is some evidence that melanin has a role in the reticuloendothelial system, which helps to protect against germs in the skin and elsewhere. Wasserman (1965, 1969) suggests that higher levels of melanin reflect a more potent line of defense against tropical disease, but his evidence is largely circumstantial.

Thermoregulation. Hamilton (1973) observed that because pigments absorb radiation, dark skin is more effectively warmed in sunlight than light skin. Thus, individuals with dark skin can take advantage of solar heating and conserve on calories otherwise spent warming their bodies. Other authors have observed the same effect but considered this a liability in the tropics, where heat dissipation was more important that heat gain. In fact, black hair and feathers have been shown to be superior to white in radiating heat to the environment as well as in absorbing it. Some desert animals have dark pigmentation to pursue this strategy. Attempts to examine these factors in humans have been inconclusive, but it appears that other physiological variables are more important than pigmentation (Robins 1991).

Sexual selection. Several studies have shown that humans choose as mates individuals most like themselves in countless subtle traits, including skin, hair, and eye color. In small populations, this pattern of positive assortative mating would result in exaggeration of differences from one isolated group to the next. Over time and geographical space, one would expect to see increasing diversity in traits that otherwise had little selective value. Diamond (1992b) believes that this pattern, an aspect of sexual selection, explains the apparent exceptions to the adaptive hypotheses.

MORE PIGMENTS

Hair and Eye Color

Hair color and eye color pose far more difficult questions. Because hair is not a living tissue, its intrinsic properties such as color and shape have little functional value. Most explanations that have been offered, such as camouflage, fail to explain observed distribution. Hypotheses that eye color relates to visual acuity have also been rejected (Hoffman 1975). Melanin in the iris of the eye may serve a similar photoprotective function as it does in the skin (Short 1975). Light-colored eyes appear only in European populations and a few small groups where local mutations have occurred. Perhaps it is better to regard the European case as genetic drift where selection pressure has been relaxed rather than an adaptation in itself.

Melanin in Other Tissues

The picture is greatly complicated by the fact that melanin appears elsewhere in the body in addition to the skin. It appears in the choroid layer of the eye, where it has a clear function in absorbing light, and also in the cochlea and the meninges, where it does not. Melanin gives the name to a cluster of cells deep in the brain, the substantia nigra ("black substance")

and inhabits other brain nuclei. Melanin thus appears to have some functions not directly related to sunlight.

The existence of melanin in the brain is independent of epidermal pigmentation. Brain melanin is manufactured by a different metabolic pathway, as a byproduct of such neurotransmitters as dopamine and norepinephrine. For this reason, it has been dismissed as a waste product. Brain melanin also has the ability to bind with certain drugs and to scavenge free radicals. However, these functions may lead to the creation of more free radicals and thus contribute to other problems (possibly including cell destruction and such disorders as parkinsonism). Consequently, one hesitates to accept them as adaptive functions.

It is safe to say that our understanding of melanin under the skin is in its infancy. Whatever its purpose in deep tissues, there is no evidence at the present that the production of melanin by epidermal cells plays a role in those functions.

GENETICS OF SKIN PIGMENTATION

Skin color is often cited as a well-understood example of a polygenic trait—that is, one that is determined by the actions of several different genes. The truth is that skin pigmentation is complex and involves many more genes than we can understand. Earlier models estimated that three or four genes could account for known variation, but this number is too few.

In addition to the several genes determining the quantity of melanin produced and its distribution within and among cells, we must include the variations in cellular metabolism and function that might affect pigment expression. For example, the metabolic pathway for the synthesis of melanin can be interrupted at various stages by a variety of different mutations that may cause conditions, such as albinism (congenital lack of skin melanin). There also are interactions between genes and the environment that may affect skin color, such as the distribution of dietary carotenes or the ability of a person to tan. More than 70 genes have been found to influence pigmentation in mice (Montagna and Parakkal 1974). We have no reason to think that the human pattern is any simpler.

14
DIGESTION AND DIET

It is easy to think of the body as a complicated black box into which food is placed and from which waste is eliminated. This image ignores the fact that the food is not a part of the body until individual molecules are extracted and absorbed across the cells lining the gut. It might be more instructive to view the body as a hollow cylinder that is open at both ends. The

digestive tract is the hole through the middle. Food passes through the hole, nutrients are extracted, and what is left of it emerges from the other side.

Digestion is therefore a temporary interaction with food as it passes through the body in order to extract as much of the nutrient content as possible. Larger animals, and active mammals, in particular, have an enormous need for energy intake. To increase their efficiency for digesting larger quantities of food, the digestive system is designed to retain food for a longer time and to increase the surface area that interacts with the food. The simplest modification toward these goals is to make the canal very long. The human digestive tract is approximately 10 meters in length (Figure 14.1). Within that tract, organs and tissues specialize to carry out ingestion, propulsion of the food, mechanical digestion, chemical digestion, absorption, and elimination of the remainder. To certain degree, we expect these specializations to reflect the "natural diet" of our species. The questions for this chapter are "What is the natural human diet?" and "How is that diet reflected in our anatomy?"

A NATURAL DIET

A natural diet is one to which our bodies are best adapted, presumably because it describes what our ancestors ate. Our present society is very much aware of nutritional advice and natural foods; but our concept of "natural" has more to do with marketing rhetoric than biological history. Thus, our pursuit of a natural diet takes us into the past as much as it involves a study of our anatomy.

Why is diet important in understanding human evolution? Food, an animal's source of energy from the environment, is the centerpiece of a niche. It relates directly to the species' habitat, activity and ranging

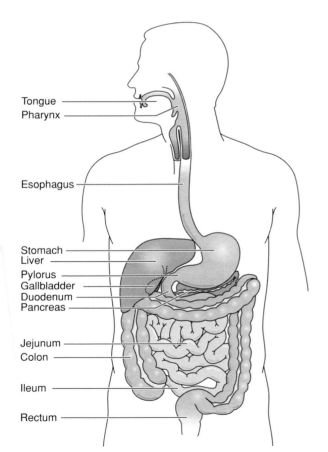

Figure 14.1 The digestive system.

Tongue
Pharynx

Esophagus

Stomach
Liver
Pylorus
Gallbladder
Duodenum
Pancreas

Jejunum
Colon

Ileum

Rectum

Table 14.1 Characteristics of the Digestive System

Animal characteristics
  Heterotrophy
  Extracellular digestion
Characteristics of more complex animals
  Mouth and anus separate
  True coelom
Chordate characteristics
  Pharynx present, originally for filter feeding
Vertebrate characteristics
  Pancreas, liver, and gallbladder developed from the wall of the
    duodenum
Mammalian characteristics
  Secondary palate isolates nasal cavity and respiration from
    mouth and food intake
  Mastication of food
  Salivary glands well developed
Primate characteristics
  Mobile lips participating in facial expression
Human characteristics
  Food and air passages overlap in the pharynx
  Relatively simple stomach and small colon

patterns, positional behavior, social structure, inter-specific relations, tool use, and culture. It influences where an individual sleeps, with whom it competes, and what it fears. It is reflected in the design of an animal's teeth, masticatory apparatus, digestive tract, metabolism, brain size, senses, limb capabilities, and health issues. Many hypotheses of the critical stages of human evolution revolve around diet. Different researchers have hypothesized meat, bones, shellfish, tubers, seeds, and fruit as the dietary staple of our ancestors. If a determination of paleodiet does not prove any of these models right, it is likely to prove some of them wrong. While we can bring many lines of evidence to bear on this question—anatomy, dental microwear, stable isotope analysis, archaeology, ecology, comparative nutrition, and paleopathology (Table 14.1)—the problem of defining a diet becomes increasingly complex the closer we examine it.

For what diet did the hominin pattern evolve? The certain answer is that they were like ourselves, and like most primates—omnivorous. Humans represent a generalized mammalian digestive system in many ways. Among primates, it correlates with an adaptation for a somewhat higher-quality diet that at some stage began to include meat (Table 14.2).

THE DIGESTIVE SYSTEM

All animals practice extracellular digestion, which means that they must define a space in which food is to be held as it is broken down and from which nutrient molecules are absorbed. For most animals, that space is a gut cavity. Simple invertebrates such as a flatworm possess a single opening to the gut cavity that serves as both mouth and anus. Food is absorbed directly in body tissues surrounding the cavity and then may diffuse to other parts. Indigestible materials can be expelled through the mouth. Such an arrangement implies that one meal must be fully processed before a second can be taken. The development of a true anus distinct from the mouth was a significant addition to the design of the animal body (Figure 14.2). Such animals can eat continuously and move food in a single direction as it is digested.

In the digestive tract, new incoming food displaces the previously consumed food. If digestion has not been completed, nutrients will be lost. However, the presence of a single-direction digestive tract opens the possibility for specialization of different regions of it. Food is broken down in a predictable sequence. Like workers on an assembly line, organs along the tract can specialize for faster and more thorough extraction of nutrients. The design of complex digestive systems for larger animals focuses on increasing the efficiency of digestion in a limited span of time.

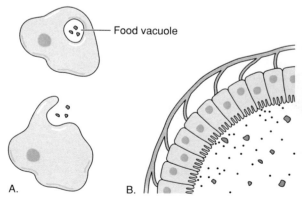

Figure 14.2 Intracellular and extracellular digestion compared. *A*, Intracellular digestion by an ameboid cell. *B*, Extracellular digestion by secreted enzymes in an animal intestine.

Table 14.2 Evidence for the Past Hominin Diet

Evidence	Indicated Diet	Species	Date
Analogy with chimpanzees	Omnivory; predominantly fruit with some meat	Early hominins	
Dentition	Omnivory; incorporating fibrous or gritty plant foods	Australopithecines	3.0–1.5 Mya
Tooth microwear	Herbivory, fruit	*A. robustus* and *A. africanus*	3.0–1.5 Mya
Carbon isotopes	Omnivory	*A. robustus*	2.5–1.5 Mya
Carbon isotopes	Significant inclusion of meat	*A. africanus*	3.0–2.5 Mya
Strontium	Some meat	*A. robustus*	2.5–1.5 Mya
Strontium	Roots, tubers, other plant foods	Early *Homo*	2.5–1.5 Mya
Cut marks, fractures of animal bones	Meat or animal products	Hominins	From 2.4 Mya (Bourhi, Ethiopia)
Association of stone tools and animal bones	Meat or animal products	Hominins	From 2.4 Mya (Kanjera, Kenya)
Cut marks, fractures of animal bones	Primary access to carcasses	Hominins	1.75 Mya (Olduvai Gorge, Tanzania)
Pathological skeleton	Carnivore liver(?)	*H. ergaster*	1.6 Mya (Koobi Fora, Kenya)
Nut shells and pitted stones	Nuts	*Homo* sp.	Early-Middle Pleistocene (Gesher Benot Ya'aqov, Israel)
Throwing spear	Hunted large animals	*H. heidelbergensis*	400,000 BP (Shöningen, Germany)
Carbon isotopes	Large proportion of meat	*H. neanderthalensis*	200,000–35,000 BP
Strontium	Large proportion of meat	*H. neanderthalensis*	200,000–35,000 BP
Increase in small mammals, birds, fish	Intense local resource exploitation	Modern humans	After 40,000 BP
Carbons, nitrogen isotopes	Varied diet including aquatic resources	Modern humans	28,000–20,000 BP
Molecular evolution of tapeworms	Meat consumption	African hominins	Older than 1.0 Mya
Poor synthesis of taurine and some fatty acids	Dependence on animal foods	Modern humans	
Relatively short gut and rapid passage of food	Rich, less bulky foods such as meat	Modern humans	
Analogy with hunter-gatherers	Omnivory with significant meat or fish consumption	Later hominins	

Mya = million years ago; BP = years before the present.

THE COELOM AND THE PERITONEUM

Vertebrates and complex animals from earthworms to insects illustrate a further alteration of body design to increase efficiency of the visceral organs. The gut is made independent of the body wall by the interposition of a potential space between muscles and other body wall structures on the outside and the visceral organs on the inside. This cavity, the coelom, is lined with a thin serous membrane, the peritoneum, that secretes a lubricating fluid (Figure 14.3). The skeletal muscles and the digestive organs are now able to move and perform their functions independently of one another.

The peritoneum is a continuous unbroken membrane that encloses a peritoneal cavity. This cavity is a potential space, empty and without volume except for a small amount of serous fluid. The peritoneum lines the inner body wall and coats the visceral organs.

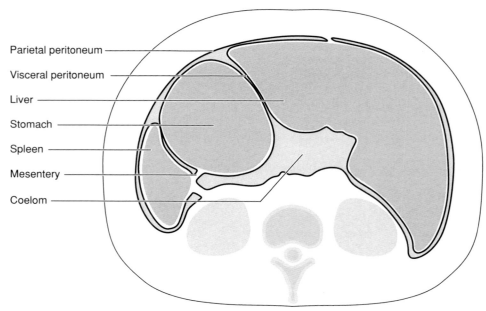

Parietal peritoneum

Visceral peritoneum

Liver

Stomach

Spleen

Mesentery

Coelom

Figure 14.3 The coelom and the peritoneum. This section through the abdomen shows how the peritoneum defines the coelom, or body cavity. Visceral organs are lined by peritoneum and often suspended or held in place by its folds, or mesenteries.

Where the organs are not in contact with the body wall, they are suspended from it by a double layer of peritoneum called a mesentery. The mesenteries conduct vital blood vessels, nerves, and lymph vessels to and from the organs.

TISSUE STRUCTURE OF THE ALIMENTARY CANAL

The vertebrate alimentary canal distal to the pharynx is constructed of four layers of tissue—mucosa, submucosa, muscularis, and serosa. The innermost layer is the mucosa. The mucosa is an epithelium with its supporting connective tissue. The deeper connective tissue provides physical support to the epithelium and contains a thin layer of smooth muscle capable of manipulating its shape. The mucosa is the only layer in contact with unabsorbed food and must handle all interactions with it, including secretion of digestive juices and absorption of nutrients.

The submucosa is connective tissue that incorporates the blood vessels, nerves, and lymphatic tissues supporting the mucosa. The rich blood supply provides the raw materials for digestive secretions and carries away newly absorbed foods. The lymphatic tissues provide an important line of defense against pathogens entering the body from the gut.

The muscularis consists of two layers of smooth muscle. The inner layer contains circular fibers that constrict the canal. The outer layer has longitudinal fibers that can shorten sections of the canal. When the two layers work together, they propel food with rhythmic contractions called peristalsis. The muscle is also responsible for mechanical churning of the food to mix digestive enzymes into it. Sphincter muscles, which can close off the passage of food, are elaborations of the circular layer of fibers. Two of these, the pyloric sphincter and internal anal sphincter, are true sphincters with fibers completely encircling the canal. Others, the cardiac sphincter and ileocolic "sphincters," are narrowings of the passage but not true sphincters.

The outermost layer is the serosa, or the peritoneum. The peritoneum holds the organs in place and maintains their blood and nerve supply as it permits movement of the organs themselves.

This pattern of four tissue layers is maintained from the pharynx to the anus. Within that span, different organs of the tract specialize for different functions by modifying one or more of these layers.

ORGANS OF THE DIGESTIVE TRACT

The digestive tract may be compared with an assembly line, where food is processed in sequence of stages, each carried out at a station specialized for that purpose. The functional specializations are expressed in the structures of the different organs of the tract (Figure 14.4). The organs similar in all vertebrates, but the needs of a given species may lead to differential development.

The Start of the Canal

What we normally think of as the digestive system operates mostly independently of conscious control, regulated through the autonomic nervous system. However, it also includes structures of the mouth, including the tongue and teeth. The long individually innervated skeletal muscle in the pharynx gives way to small smooth muscle tissue in which regional stimulation sets of waves of contractions. The conscious and detailed sensations of the mouth and pharynx stop abruptly to be replaced by the vague sensations of the esophagus and the rest of the gut, limited mostly to the modalities of stretch and pain. The pharynx represents a transition zone under the control of a unique column of neurons in the brain. The muscle wall has circular and longitudinal layers, as in the rest of the gut, but these are reversed in position, with the circular fibers on the outside. Stimulation of the muscles may be voluntary or it may occur through involuntary reflexes, including swallowing and gagging.

Esophagus

The esophagus is a tube about 24 cm long, descending from the pharynx to the stomach. Its sole function is to transmit food to the stomach quickly. It performs no other digestive action. The circular muscle fibers along its course are continuous with the pharyngeal constrictors. There is a gradual transition from skeletal to smooth muscle tissue. The upper fibers are dependent on direct innervation by the vagus nerve to initiate

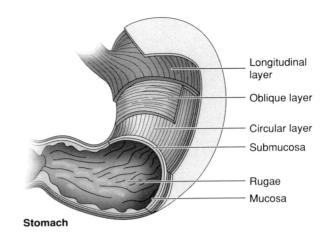

Stomach

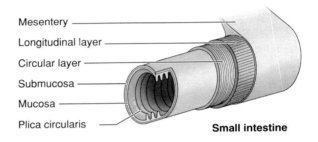

Small intestine

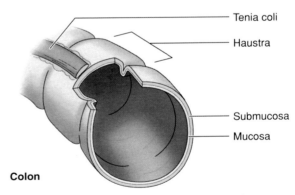

Colon

Figure 14.4 Specialization in the digestive tract. Although the digestive tract has a basic four-layer structure throughout, the regions we recognize as organs modify that plan for the specific tasks they perform. The stomach has added a third layer of smooth muscle, while the colon has reduced its longitudinal layer to three narrow bands. The form and foldings of the mucosa also vary in different regions.

peristalsis. The lower esophagus has a neural network within its walls capable of establishing its own pattern of contraction. It is responsive to the parasympathetics in the vagus but ultimately independent of them. This intrinsic network is typical of innervation in the abdominal parts of the gut.

Stomach

The stomach is a muscular chamber that has specialized for mechanical churning of the food and for some chemical digestion. The muscular layer is strengthened by adding a middle oblique layer of smooth muscle fibers. This enables the stomach forcefully to churn and wring its contents, working digestive juices into the food. The mucosa has long folds, the rugae, that give the stomach the ability to expand when full. It is protected from its own enzyme and acid secretions by a layer of mucus containing a bicarbonate buffer. Although stomach contents are highly acidic to activate digestive enzymes, the walls are protected by this neutral layer. Nonetheless, there is a high turnover of epithelial cells lining the stomach (replacement every 3 to 6 days) to recover from normal wear.

Small Intestine

The small intestine is the most critical region of the digestive tract. It is here that most of the enzyme secretion and chemical digestion and nearly all the absorption of nutrients occur. Time for digestion and surface area for absorption are critical. Thus, it is in the small intestine that the more dramatic adaptations to increase surface area occur. The small intestine is extended to well over 4 meters long. For our convenience, anatomists divide it along its length into three regions—the duodenum, jejunum, and ileum.

Digestive enzymes are secreted from glands at the ends of intestinal crypts in the walls of the intestine, primarily in the duodenum and jejunum. These are holocrine glands, shedding entire cells with their enzymes—as many as 17 billion cells per day. A wide variety of enzymes are produced that break macromolecules into their smallest units. The intestine gains time for chemical breakdown and absorption by several mechanisms. The gut can control the rate of passage of food by the speed of peristalsis. Closure of the pyloric sphincter, as described earlier, can limit the flow of food from the stomach into the duodenum. Plicae circulares, circular foldings of the mucosa that lie perpendicular to the direction of flow, slow the food down by partially obstructing the proximal intestine. These plicae diminish in number toward the ileum.

The great length of the intestine and the plicae circulares both increase the surface area for secretion and absorption. The plicae approximately triple the surface area to about 1.8 m². In addition, the surface of the mucosa is covered with minute finger-like processes called villi, which increase the surface area by an additional 8 times (Figure 14.5). Individual villi are scarcely visible to the naked eye; collectively they give the mucosa the visual texture of velvet. The exposed membranes of the individual epithelial cells of the villi repeat this strategy with even smaller projections, microvilli. The microvilli increase surface area by another factor of 20, bringing the total membrane area to approximately 300 m².

Accessory Glands

Three glands have developed from the duodenal lining to become major organs. These are the pancreas, liver, and gallbladder. Embryologically and in many adult vertebrates, the pancreas is a paired organ, the two parts lying anterior and posterior to the duodenum. The anterior pancreas in humans rotates to the posterior side and fuses with the dorsal pancreas to form a single organ. This origin explains the irregular shape of the gland and the variable presence of two or only one duct.

The liver is one of the defining characteristics of the vertebrates. The largest organ of the body after the skin, the liver has many vital functions in the body relating to the digestive, circulatory, and endocrine systems. As a digestive gland, the liver secretes bile, which helps to break up fat droplets and expose the lipid molecules to enzymes. Normally the lipids do not dissolve into the water-based secretions of the digestive tract and with most foods but separate themselves into inert pockets. Emulsifiers such as bile salts break the larger pockets of lipid into much smaller ones, speeding digestion.

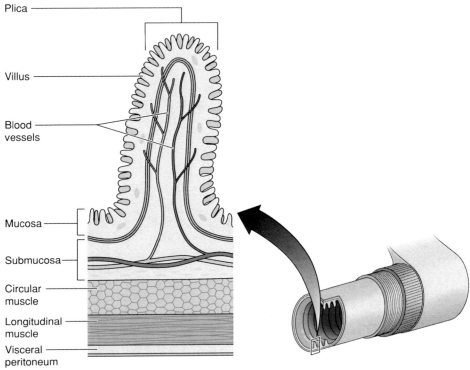

Plica

Villus

Blood vessels

Mucosa

Submucosa

Circular muscle

Longitudinal muscle

Visceral peritoneum

Figure 14.5 The wall of the small intestine. The small intestine shows numerous strategies to increase surface area for effective absorption of food. The plicae circulares are folds of the mucosal lining and submucosa that also help to slow the passage of food. Villi on the surface of the mucosa and microvilli of the individual cells lining the villi also greatly multiply the exposed membrane surface.

The liver plays a major role in processing nutrients that have been absorbed by the intestine. Many vitamins and minerals are stored in liver tissue (hence liver is a good dietary source for vitamins and minerals). Monosaccharides are converted to glucose by the liver and stored in hepatocytes, among other places, in the form of glycogen. Amino acids are processed and synthesized in the liver and used in the construction of numerous proteins and enzymes for the body, including the major blood proteins. The liver plays a major role in processing lipids, as well, including synthesizing and storing cholesterol and packaging lipids for transport throughout the body.

The liver has other functions less closely related to the process of digestion. It filters blood and removes worn-out red blood cells. As these are broken down, vital parts such as the iron in hemoglobin are extracted and recycled. Toxins in the blood are removed and neutralized by the liver. Ammonia is converted to urea, to be excreted by the kidneys. Alcohol is metabolized largely by the liver, which bears the brunt of its negative effects on the body. The liver also participates in hormone synthesis, including somatomedins, intermediaries in the actions of pituitary growth hormones on other body tissues.

Bile is created continuously by the cells of the liver and secreted into the hepatic duct system. These ducts lead to the duodenum, but their exit is normally closed by a sphincter. Accumulating bile backs up into the gallbladder, whose sole function is to serve as a reservoir for bile. The walls of the gallbladder are muscular and are stimulated to contract by the hormone cholecystokinin from the small intestine. Thus, bile is released only as it is needed for digestion.

Colon

The large intestine, or colon, is shorter than the small intestine, but it has a greater diameter and considerable potential for distension. The human colon is approximately 1.5 meters in length. It begins at the ileocecal junction, describes three quarters of a circle about the abdominal cavity and descends into the pelvic cavity to exit through the perineum. Although its different segments have different names based on their relations to the peritoneum, functionally there is very little differentiation within the colon.

The cecum is a "blind" pouch at the beginning of the colon. It is separated from the ileum by a sphincter-like valve. In some herbivores, this pouch is enlarged and can be used for storing food undergoing bacterial fermentation to break down cellulose. The human cecum is comparatively small and has no special role.

At the inferior end of the cecum is the vermiform appendix, a process about the length of the fifth digit and slightly more slender. The appendix has no digestive function and is notorious for becoming infected and inflamed; however, it is not as completely vestigial as has been commonly stated. The walls of the appendix contain lobules of lymphatic tissue that supplement similar tissues elsewhere in the colon and in the mesenteries (Fisher 2000; Hill 1948). For this role in immune surveillance, the appendix has been referred to as the "abdominal tonsil."

Most of the nutrients should have been absorbed by the ileum before the feces enter the colon; but some water and electrolytes remain to be recovered here. The cavity of the colon also houses a considerable population of symbiotic bacteria. These bacteria obtain energy from certain indigestible or undigested foods such as cellulose and excess fats, and they synthesize certain vitamins, including vitamins K and B_{12}, thiamin, and riboflavin, that become available to the host. Resident bacteria also form a benign protective coating of the wall that inhibits the establishment of pathogenic colonies.

The primary function of the colon is storage of feces as they are compacted and water is removed from them. The glands of the colon are much reduced, producing only mucus. The walls are made thinner and expandable; and the outer layer of longitudinal muscle is reduced to three narrow bands that each run the length of the colon. The final composition of feces varies considerably with the diet but is typically about 75% water. Of the solids, as much as a third is composed of dead bacteria. Another third is indigestible dietary fiber and dead epithelial cells. The remainder is mostly undigested fat and inorganic matter.

RECONSTRUCTING THE DIET OF EARLY HOMININS

The human body should reflect adaptations appropriate to the diet of our ancestors. We may regard it as important for our health to understand what that natural diet is, but our proclivity for sampling almost anything edible along with our ability to colonize almost any earthly habitat makes this a difficult question to answer. Ironically, it may be more useful to interpret the diet of our fossil ancestors than it is to interpret our own anatomy and physiology.

THE EVIDENCE OF DENTAL ANATOMY

Although the later australopithecines are conspicuous in the relatively greater mass of chewing muscles and in the increased size of their teeth, these adaptations do not necessarily indicate increased bite force. Rather, they tend to cancel out one another (McHenry 1984). Stronger muscles are required just to maintain the same pressure over larger teeth. The presence of larger molars makes more sense as a strategy to increase the life of the dentition. Adaptations that support greater molar wear and longevity of the teeth are most consistent with a diet that is high in fiber or in grit, particularly small or hard objects (Teaford and Ungar 2000). Such foods require a greater amount of oral processing and cause more abrasion. Among the candidates proposed for this diet are seeds (Jolly 1970), underground tubers (Hatley and Kappleman 1980), hard objects such as nuts (Kay 1981), or simply a variety of tough, gritty savanna foods (Ryan and Johanson 1989; Picq 1990). These hypotheses extrapolate beyond dental and facial adaptations to locomotion, manipulative skills, tools, and habitat. Unfortunately, on the basis of gross anatomy alone, these models are difficult to differentiate and test.

THE EVIDENCE OF DENTAL MICROWEAR

Microscopic examination of the dentition reveals additional details of wear patterns. Different food categories produce distinct types of damage and abrasion lines in the teeth; counting and measuring the sizes of pits and scratches distinguish meat-eating and bone-crunching and herbivory. Among herbivores, grazing on grasses, browsing on leaves and branches, and eating fruit produce different patterns of scratching and polish (Figure 14.6).

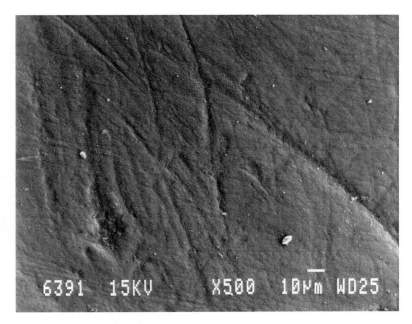

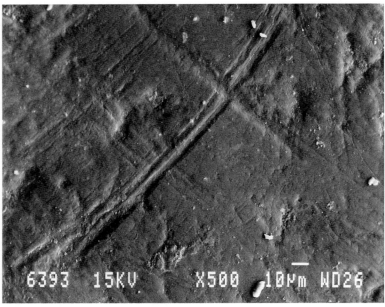

Figure 14.6 Different foods create microscopic scratches on the surface of teeth. These microphotographs of prehistoric Native American teeth show different consequences of an Archaic foraging diet (*above*) and a Woodland diet (*below*) that is partially agricultural. More abrasive foods in the latter produced increased scratching and pitting. Examination of these markings provides clues to the diet of an extinct species. (Courtesy of Christopher W. Schmidt)

Images generated with a scanning electron microscope indicate that the wear on the molars of robust australopithecines most closely resembles that of frugivores and is consistent with a diet of whole fruits, including the rinds, husks, and seeds (Grine 1981; Walker 1981). Microwear studies of the anterior teeth of *Australopithecus afarensis* also support a largely herbivorous diet (Puech and Albertini 1984; Ryan and Johanson 1989). The patterns exhibited on robust and gracile australopithecine teeth are similar but suggest *A. robustus* included more hard food items in its diet (Grine and Kay 1988). However, there are limitations to such studies. Daily eating erases the wear patterns and replaces them rapidly, in 1 day or less. Hence, the fossil teeth reveal only the animal's last few meals, which probably varied with the season.

BONE COMPOSITION AND DIET

The elemental composition of the minerals in bone is related to the dietary intake of the individual during life. Because the composition is preserved for a long time after death, mineral composition possesses clues to diets in prehistoric populations. The basis for dietary reconstruction depends on different concentrations of minerals and isotopes in different classes of foods (Katzenberg 1992; Schwarcz and Schoeninger 1991).

Different stable carbon isotopes, ^{12}C and ^{13}C, are taken up differentially by plants through alternate pathways of photosynthesis. The C_4 photosynthetic pathway (referring to a particular metabolic sequence) is an adaptation of plants in hot, dry environments, including some grasses and grains and, particularly, maize. The more common C_3 pathway is found in most other plants. The C_4 pathway takes up a higher proportion of the ^{13}C, and that higher proportion is passed on and incorporated in the bodies of the animals that feed on such plants. Analysis of carbon isotope ratios in bones of prehistoric populations has helped us document the transition from a hunting-gathering lifestyle to one that depended on maize agriculture.

Nitrogen also has stable isotopes with dietary implications. Atmospheric nitrogen is nearly entirely ^{14}N, whereas nitrogen found in the soil is about 10% ^{15}N. The legumes, which fix atmospheric nitrogen with the aid of symbiotic bacteria, consequently have a lower proportion of ^{15}N than other plants that depend on soil nitrogen. That difference also ascends the food chain to herbivores and carnivores. Analysis of nitrogen ratios helps distinguish populations that ate primarily marine plants and animals from those that ate terrestrial foods.

Strontium is an element that can be incorporated into bones in place of calcium. Strontium occurs in soils in quantities that vary in different localities. Within a habitat, higher concentrations of strontium are present in the bones of herbivores, whereas strontium will be rejected in favor of calcium at higher trophic levels. Thus, the ratio of strontium to calcium in bones can be a guide to an animal's position in the food chain (Sillen and Kavanaugh 1982). Analysis is complicated, however, by certain variables, including the nature of the local soils, gender (Thackeray 1995), and the presence of mineral-rich foods in the diet (Burton and Wright 1995).

These analyses have more commonly been performed on bones of relatively recent human and nonhuman populations, but it may be possible to apply them to human ancestors. Carbon isotope analysis supports the idea that *Australopithecus robustus* from Swartkrans Cave in South Africa was omnivorous (Lee-Thorp 1994); strontium analysis confirms a diet that included some meat (Sillen 1992). Strontium isotopes in the remains of contemporary *Homo* suggest it was more vegetarian and may have intensively exploited roots and tubers (Sillen et al. 1995), but carbon isotope ratios could not differentiate *Homo* from *A. robustus* (Lee-Thorp et al. 2000). Elevated levels of ^{13}C have also been identified in *Australopithecus africanus* from Makapansgat Cave (Sponheimer and Lee-Thorp 1999) and from Sterkfontein Cave in South Africa (van der Merwe et al. 2003), with the latter showing a high degree of variability. It is unlikely that australopithecines were eating grasses (no primate is capable of digesting a grass diet). It is more likely that they were consuming grass-eating animals. The observed variation is consistent with an eclectic and opportunistic diet.

ARCHAEOLOGICAL EVIDENCE OF MEAT CONSUMPTION

Historically, the greatest amount of interest, speculation, and controversy regarding early hominin diet has centered on the role of meat. The "killer ape" model developed by Dart (1949) and popularized by Ardrey (1961) built on the idea that the gracile australopithecines, *A. africanus*, were omnivorous, incorporating significant amounts of meat in their diet. Robinson (1962) proposed that the difference between gracile and robust australopithecines was one of carnivory versus herbivory. A partly carnivorous diet has significant potential implications for behavior, social organization, culture, and language, as well as anatomical evolution (Cartmill 1993; Laughlin 1968; Stanford 1999; Washburn and Lancaster 1968). By implication, robust australopithecines were evolutionarily conservative peaceful gorilla-like herbivores. Gracile australopithecines had made a departure in the direction of humanity by adopting a violent carnivorous behavior and diet. This dichotomy represented the separation of the ape and human lineages.

This "hunting hypothesis"—the assertion that the adaptations of a hunting niche were formative for humans—dominated paleoanthropology through the 1960s. Other models partly displaced it after a better understanding of the fossil evidence showed australopithecines were probably not significantly carnivorous and when cultural influences on anthropology shifted. The differences between gracile and robust australopithecines, in particular, probably represent similar diets in different habitats, in which robust australopithecines were consuming drier and tougher fruits. In many ways they resemble one another more closely than either resembles humans or apes.

From the perspective of the modern human diet, the question is not whether hominins ate meat but when. Chimpanzees (Boesch 2002; Boesch and Boesch 1989; Stanford 1995, 1996, 1999; Stanford et al. 1994; Watts and Mitani 2002), bonobos (Hohmann and Fruth 1993), and many other primate species (e.g., Butynski 1981; Fairgrieve 1997; Fedigan 1990; Harding and Strum 1976; Rhine et al. 1986) have been observed hunting smaller animals for meat on many occasions at different study sites. Hunting is a cooperative male activity, and often the meat is shared. We should assume that the earliest hominins killed and ate small animals when opportunities arose. Clearly, however, recent humans perform hunting as a more sophisticated endeavor, seeking out much larger and more dangerous game animals.

There is overwhelming archaeological evidence of hunting weapons and butchered animals bones from the recent prehistoric past. More circumstantial evidence of hunting and/or scavenging goes back more than 2 million years. At Bourhi, Ethiopia, stone tools are associated with animal bones that bear cutmarks and have been broken open to expose the marrow (de Heinzelin et al. 1999). The 2.4-million-year-old hominin that may be associated with these activities has been named *A. garhi*. At about the same time period, a hominin was accumulating small animals and probably scavenging larger ones on a lake shore at Kanjera, Kenya (Plummer et al. 2001). At later East African sites, such as those at Olduvai Gorge and Koobi Fora, stone tools and animals bones with cut marks abound. Carcass processing is beyond dispute. While it is likely that hominins then, as now, were not reluctant to scavenge freshly killed animals to supplement their hunting, there remains discussion about the relative importance of hunting versus scavenging. However, comparison of butchering at Olduvai with that of recent hunters suggests early hominins at least sometimes had first access to larger game (Bunn 2001). The debate is probably a moot point: modern mammalian carnivores and scavengers generally partake in both behaviors when opportunities present themselves.

There is a clear overlap in time of the presence of australopithecines and *Homo* and of stone tools in both South and East Africa, but there is no definitive evidence that butchery or even tool-making can be associated with *Australopithecus*. A conservative approach cannot accept as well supported the idea that australopithecines consumed any more meat in their diet than chimpanzees do today (which is still significant). Nor can we identify with certainty any fundamental adaptations in the hominin fossils that can be associated with hunting or carnivory.

Tools do not always imply carnivory. Chimpanzees have been observed making and using a variety of tools but not for hunting or eating their prey. We should expect the same of hominins, from the earliest time to the present. The earliest bone tools, from Swartkrans Cave in South Africa, appear to have been used to dig into termite mounds (Backwell and d'Errico 2001). This discovery should remind us both of the variety of foods we expect our ancestors and relatives to have consumed and of the fact that we should not limit our speculation of diet to those food categories appealing to the Western palate (see also McGrew 2001).

AND OTHER EVIDENCE . . .

In 1973 a partial skeleton assigned to *H. ergaster* was unearthed at Koobi Fora, Kenya. Pathological developments of the bones suggest the individual suffered a painful death caused by an excess of vitamin A (Walker et al. 1982). Vitamin A is found in many naturally occurring foods but becomes particularly concentrated at the top of the food chain in the liver of carnivores. Hypervitaminosis A is rare but has been recorded among polar explorers who consumed polar bears and among health food faddists who overdosed on vitamin pills. One interpretation of the Koobi Fora skeletons is that the individual was consuming meat and ate a fatal dose of liver from a dangerous animal. An alternative suggestion is that the victim fed extensively on honey from wild bee nests and consumed a lethal dose of vitamin A in bee larvae (Skinner 1991).

Another unexpected contribution to this investigation comes from studies of the tapeworms that some humans harbor in their intestine. Typically, tapeworms of the genus *Taenia* have a life cycle that alternates between two hosts, a herbivore and a carnivore, but each parasite species co-evolves with a specific pair of host animals. The three species of human tapeworm are presently adapted to domestic animals as their intermediate hosts, either cattle or swine. The closest relatives of these parasites are tapeworms that infest bovids (antelope) and felids (cats) living in sub-Saharan Africa. Molecular comparisons suggest that these parasites have been co-evolving with humans for well over 1 million years (Hoberg et al. 2000), although the confidence interval for this calculation is very broad. The most likely scenario is that humans began their association with the tapeworms early in the history of genus *Homo* as they competed with leopards and lions for game. Only much later, less than 10,000 years ago, did the parasites switch to farm animals.

Several indications from our digestive tract also suggest a reliance on meat. The human gut is relatively

short, as is passage time through it. Unlike herbivores, humans share with cats a poor ability to synthesize the amino acid taurine or longer chain fatty acids such as arachidonic acid and docosahexaenoic acid (Bloch 1995; Cordain et al. 2002). All of these are essential— that is, our bodies cannot manufacture them in sufficient quantities. They can be obtained from animal foods, but not as well from a vegetarian diet.

LATER HOMININS

In the Middle Pleistocene, the time of *H. erectus* (*H. heidelbergensis* in Europe), we have occasional detailed glimpses of hominin activities in the archaeological record. At Gesher Benot Ya'aqov, Israel, shells of seven edible species of nuts—including pistachio, almonds, acorn, and water chestnut—were found alongside pitted stones used to crack them open (Goren-Inbar et al. 2002).

Sites with stone tools and animal bones continue to be numerous up to the present. Early excavations were crude by modern standards and merely reinforced assumptions that hominins have been hunters and carnivores. Meticulous recoveries in recent years, such as the 500,000-year-old site at Boxgrove, England, have reconstructed tool manufacture and butchery in great detail (Pitts and Roberts 2000). While the association of tools with the bones of large animals becomes increasingly prevalent and convincingly indicate meat consumption, doubts have persisted among some archaeologists as to whether these animals were hunted or merely scavenged. Perhaps the earliest conclusive evidence for hunting appears in the form of four spears discovered in a coal mine at Schöningen, Germany (Thieme 1997). Four hundred thousand years ago, these were deposited in a bog. The skilled shaping of the shafts provides aerodynamic stability and resembles javelins used by modern athletes.

After 200,000 years ago, Neanderthals in Europe consistently turn up alongside the remains of large game animals. Carbon and nitrogen isotope tests on Neanderthal specimens indicate a highly carnivorous diet (Bocherens et al. 1991, 2001; Dorozynski and Anderson 1991). Studies of strontium content that indicated a diet of up to 97% meat provide independent confirmation (Balter et al. 2001). Neanderthals lived through the glacial cycles of Europe and are often compared with later hunters of the Arctic. High latitude and cold forest environments have relatively low productivity, but open tundra and grassland can support herds of larger game animals. Northern hunter-gatherers today have the highest proportion of meat and fish in their diet, and it is not surprising that Neanderthals would have had similar tendencies.

Modern humans appear in Africa after 100,000 years ago and in Europe after 40,000 years ago. At these times, some changes in behavior and foraging strategies are apparent. In Italy and Israel, sites show a sudden greater emphasis on fast and fast-reproducing animals such as birds and hares rather than slower game such as shellfish and tortoise. This suggests a rapid increase in population density that is placing stress on certain types of resources (Stiner et al. 1999, 2000). Carbon and nitrogen isotope samples from several European skeletons 28,000 to 20,000 years old also revealed a change of diet (Richards et al. 2001). Compared with the Neanderthals, these people were consuming a wider variety of food, including significant quantities of fresh water animals and waterfowl. Mannino and Thomas have argued that overexploitation of coastal resources would present pressures on foragers to migrate and disperse (Mannino and Thomas 2002); and anatomically modern peoples certainly did spread out across the globe.

A similar pattern appears in Africa. In southern Africa, there is an increase in coastal sites exploiting marine resources and creating shell middens (refuse deposits), as well as accumulations of mammalian bones, in the Middle Stone Age, 200,000 to 40,000 years ago (Grine et al. 1991; Henshilwood et al. 2001; Klein 1983, 1999). There is another transition from the Middle to the Later Stone Age observed in numerous sites, such as Klasies River Mouth, Die Kelders, and Blombos. From the age profiles of animals, it is evident that humans were effective hunters throughout both periods. However, there is evidence of more effective exploitation of resources in the Later Stone Age. Remains of fish and flying shore birds, previously rare, are now common. Tortoise shells and shell fish decline in size, indicating that collection by humans had become much more intense. Isotope studies of Later Stone Age skeletons showed that coastal peoples obtained up to 50% of their diet from marine resources, while a few inland individuals consumed an almost entirely terrestrial diet (Sealy and van der Merwe 1985). The general pattern is one of increasing diversity and adaptability of culture, enabling people more efficiently to exploit whatever resources were available in their neighborhood.

A BACKWARD LOOK FROM THE PRESENT

MODERN HUNTER-GATHERER DIET

The modern Western diet is quite removed from that available to hunter-gatherers. Plants and animals consumed differ not only by name, but also by quality. Traditional agricultural societies rely heavily on one or a few starchy grains or tubers as the primary supply of calories. Gathered plants have neither the concentration of starch nor the density in the field to provide such an easy harvest. Foragers therefore must collect a greater variety of plants over a wider area. The variety, however, contributes to a better balance of nutrients and a more consistent supply food across the seasons. Animal foods are similarly different. Domestic animals are fattened on grains, while wild animals are much more lean and have a healthier, higher ratio of polyunsaturated to saturated fats. Eaton and Konner (1985) estimated such a ratio at 1.41 compared with 0.44 of the contemporary American diet.

By all estimates, hunter-gatherer and ancestral diets are healthier and more nutritionally balanced than our own (e.g., Eaton and Konner 1985; Haenel 1989; Milton 1991; O'Dea 1991; Southgate 1991). In comparison to the average modern American, a survey of several groups show they consume relatively more protein, less fat and carbohydrates, twice as much fiber, a fraction of the sodium, twice as much calcium, and more than four times the amount of vitamin C (Cordain et al. 2000, 2002; Eaton and Konner 1985). While hunter-gatherers may suffer from periodic famine, Americans contend with diet-related obesity, diabetes, cardiovascular disease, cancers, and other forms of affluence malnutrition. Although these are generalized statements, recent hunter-gatherer societies are widely scattered geographically, occupying a wide range of ecological zones from tropical rain forest to desert to tundra. Their diet and the proportions of meat, seafood, and plants they consume vary accordingly. The contrasts with the Western diet are important from a health care perspective, but it is clear that we all follow the species pattern of an eclectic omnivory.

FEEDING THE BRAIN

The most important variable interacting with the evolution of hominin diet has been our brain. The modern human brain consumes a great deal of energy, accounting for approximately 25% of our metabolism, as well as quantities of protein, lipids, and other nutrients. Whatever dietary patterns our ancestors had must have been consistent with a steady increase in brain size and nutritional demands over the past 2 million years (see Chapter 12).

The increased protein and calorie needs were most likely met through hunting or scavenging (e.g., Leonard and Robertson 1997). Modern hunter-gatherers vary greatly in their meat consumption, but at the lower end there is a 10-fold increase over the amount of meat available to chimpanzees (Wrangham et al. 1999). The lean meat typical of undomesticated animals delivers large quantities of protein and little fat. Processing protein alone consumes approximately one third of its caloric value; thus, an additional source of calories is needed. Without a supply of carbohydrates or fat, hominins consuming meat will suffer from a disorder sometimes called "protein poisoning" (Harris 1986; Speth 1991). Symptoms include weight loss (because body tissues are consumed to fuel protein metabolism), impaired liver and kidney function (as the body attempts to handle excess urea), and even death. Pregnant women and their offspring are especially sensitive to excess protein in the diet. For this reason regional cuisines around the world typically pair a starchy carbohydrate (e.g., bread, rice, yams, potatoes) with meat. Hunter-gatherers have less access to such domesticated plants. They highly prize fat on an animal and readily consume fatty organs such as the liver, brain, and bone marrow—and that taste is all too apparent in our own society. Beyond this, there may be limits to the amount of lean meat

hunters or fishers can consume, even though it is highly valued (Malainey et al. 2001; Speth 1991).

Wrangham et al. (1999) proposed that starchy tubers represent a source of calories underexploited by other African species. The technology (including cooking) needed to extract and prepare them would have stimulated social evolution as they supported brain development. Tubers would have been only a part of the diet, since they offer little in the way of protein, fat, or other necessities. The social implications are highly speculative. On the other hand, cooked tubers would have represented a decrease in dietary fiber, permitting a more nutritious dietary balance. Brittain-Conklin et al. (2002) suggest tubers might have represented an adaptive advance for australopithecines. Perhaps such a calorie source helped early *Homo* to balance a diet in which meat assumed an increasingly larger role.

Other nutrients specific to the needs of the brain are long chain polyunsaturated fatty acids. Two families of unsaturated fatty acids, referred to as omega-6 and omega-3, are responsible for about a quarter of the dry weight of the nervous system in all mammals. Nutritionists recognize medium chain members of these two families, linolenic acid and linoleic acid, respectively, as essential nutrients because we cannot synthesize them. However, it is the longer derivatives of them, docosahexaenoic acid (DHA) and arachidonic acid (AA), that are essential in brain structure and function. Synthesis of DHA and AA by the human body is very slow, and some nutritionists argue that these are themselves essential in our diet. They propose that the key to brain evolution was a diet sufficiently rich in these fatty acids—specifically from an aquatic or marine food chain (Broadhurst et al. 1998, 2002; Cunnane et al. 1993). Long-chain polyunsaturated fatty acids are synthesized by algae and are present in the food chain of shellfish and fish that depend on them. Moreover, such a diet provides these fatty acids to humans in approximately the same proportions needed by the brain.

Both linolenic acid and linoleic acid are also synthesized by terrestrial plants and are metabolized to longer chain fatty acids by the animals that consume them. The omega-6 fatty acids, including AA, are available in adequate quantities in meat, but DHA is scarce in the contemporary non-marine diet. It is concentrated in the liver of some animals and in brains. Would a prehistoric terrestrial hunting or scavenging diet provide sufficient quantities of these essential fatty acids?

Today's domesticated plants often have fewer unsaturated fatty acids than wild plants (Simopoulos 1990). Our dependence on agricultural grains and grain-fed animals skew fatty acids in favor of the omega-6 rather than the omega-3 series. Domestic animals fed on grains and domesticated plants pass this skewed distribution up the food chain to us (Chamberlain 1996; Eaton and Konner 1985; Leaf and Weber 1987). Deer, for example, can concentrate omega-3 fatty acids from browsing on ferns and mosses, and chicken eggs may vary the ratio of omega-6 : omega-3 fatty acids between 1.3 and 19.3 depending on whether they were free range or factory raised (Simopoulos 1990). Our ancestors' diet probably contained a rough balance between omega-3 and omega-6 fatty acids (Eaton 1992; Simopoulos 1990, 1991). Both a decrease in saturated fatty acids and a relative increase in the proportion of omega-3 fatty acids to omega-6 fatty acids would increase the rate of DHA synthesis (Cordain et al. 2001). Additional DHA is available from brains and possibly other organs of game animals that most Westerners would not consider eating. Cordain et al. argue that this would be a sufficient supply.

Surplus DHA is stored in a variety of body tissues, including fat, and can be released to the bloodstream as needed for the brain to take it up. The developing infant receives nearly all of its DHA from the stores of its mother via the placenta or traditional nursing practice or by synthesis within its own tissues. Therefore, dietary needs should be more accurately defined by lifetime intake rather than daily intake. In that time frame, the ability to synthesize DHA, even at a slow rate, becomes much more important. In order for the argument for the necessity of aquatic resources to stand, we would have to establish that a terrestrial diet and the body's synthesis of DHA are insufficient sources. Neither have been established, and there is no evidence that diets low in DHA have caused nutritional problems, either in modern Western society or among hunter-gatherers.

Many hunting-gathering societies do not consume significant quantities of aquatic resources. In a survey of the literature pertaining to 66 modern hunter-gatherer groups, Hayden (1981) found that 17 did not produce a measurable part of their diet from fishing. Cordain et al. (2000) published a similar survey for 229 hunter-gatherer cultures. Of these, 36 obtained 0% to 5% of their diet by fishing. Such lists are hardly representative of humanity, since they include disproportionate representation from both the arctic and the desert. However, they indicate that aquatic foods are commonly incorporated into hunting and gathering but are not an essential part of the human diet.

The hypothesis of an evolutionary dependency on aquatic or marine resources or on any other single food source is therefore unsupported. Current data do not suggest that an ancestral population modeled after modern terrestrial hunter-gathers would have encountered fatty-acid deficiencies that would limit the evolution of the brain. Eaton et al. (2002) suggest that the declining quality of the modern diet, specifically regarding DHA, may have been responsible for the 11% decline in brain size observed in the past 30,000 years. However, it is not clear to what extent that decline may be related to the overall decrease in body mass and to more general malnutrition that accompanied agriculture until quite recently.

ANATOMICAL ADAPTATIONS

Among the primates, three trends for specialization of the alimentary canal have been identified, corresponding to different diets (Chivers and Hladik 1980; Martin 1990; Martin et al. 1985; Milton 1993). The same patterns are present in other mammals, often with a greater variety and degree of specialization.

Faunivorous primates, those eating a significant quantity of animal foods, have a relatively simple stomach and colon, as measured by size, surface area, weight, and structural complexity. Although many primates are opportunistic enough to eat smaller animals as they are available, relatively few specialize in predation. Those are mostly insectivorous prosimians, representing possibly the original adaptation of the order. Animal foods tend to have a higher concentration of proteins and calories per weight and are poor in fiber. They are thus nutritionally rich foods with less bulk. The nutritional advantages offered by an animal diet are balanced by relative scarcity or difficulty in obtaining it. Insectivorous primates are all of small size but support a relatively high metabolism.

Folivorous primates specialize in eating leaves. Leaves are poor nutritionally but relatively easy to obtain in large quantities. Folivorous animals must therefore process a large volume of food to extract sufficient nutrients. Primates depending on leaves have an enlarged complex stomach or an expanded cecum and colon capable of retaining more food for a longer processing time, or both (Lambert 1998). In some cases, digestion is aided by bacterial fermentation to break down complex fiber that otherwise cannot be digested by mammalian enzymes. Folivory tends to correlate with larger body size and lower metabolic rate in mammals, as observed in several species of monkeys.

Frugivorous primates consume relatively large amounts of fruit. They are selective feeders on vegetation, eating foods more nutritious and less bulky than leaves but not so rich as insects and meat. Fruit provides a higher yield of carbohydrates but may be limited in protein and fats. Fruits may be

supplemented by tender shoots and young leaves, flowers, and nuts. Just as frugivorous primates have a diet intermediate in value, they display a gut intermediate in its size and complexity. Frugivores tend to be of medium body size, but frugivorous primates overlap the folivores in size. Examples may be found among the lemurs and monkeys and also include the orangutan. While chimpanzees have been commonly listed as frugivores, they also consume more meat than most other primates.

In the context of the patterns observed above, humans lie in the direction of faunivory (Figure 14.7). That is, our species has a relatively simple and small stomach and colon (Leonard 2000; MacLarnon et al. 1986; Martin 1990; Milton 1986). Approximately 56% of our gut volume is small intestine, compared with 14–29% in the great apes. Twenty percent of the volume is represented by the human colon, compared with more than 50% among the great apes. Our position is shared only by the capuchin monkey, *Cebus capucinus*. Outside of the primates, we tend to sort with more carnivorous mammals. As a cautious note, let us observe that a simple stomach and small colon are less precisely indicative of faunivory than of a rich and easily digested diet that is relatively

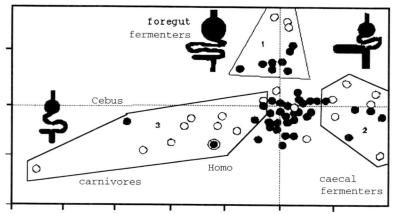

Figure 14.7 Relative development of the organs of the digestive tract in primates and other mammals. This plot combines multiple measurements of the tract to compare nonprimates (open circles), nonhuman primates (solid circles), and humans (double circle). The proportions of the gut relate generally to diet. Humans and the cebus monkey have proportions similar to those of carnivores. (From R.D. Martin, 1990. *Primate Origins and Evolution: A Phylogenetic Reconstruction.* Copyright 1990 by Princeton University Press. Reprinted with permission of Princeton University Press.)

low in vegetable fiber. With the mastery of fire, domestication of crops, reliance on starchy grains, and extensive use of cooking and food preparation, even the vegetable aspect of the human diet has converged in that direction.

Both *Homo* and *Cebus* have diets best described as omnivorous. We opportunistically consume any nutritious food available, whether plant or animal. The result is a diet containing much more meat and less plant fiber than that of other monkeys or apes. The capuchin consumes approximately 50–60% rich vegetable foods and 40–50% small vertebrates and invertebrates (Milton 1987). It is impossible to offer a similarly quantified account of the modern human diet, since it varies so dramatically in different cultures. Nonetheless, both anatomy and ethnographic observation support the conclusion that humans consume a high quality diet (e.g., Leonard 2000).

Various lines of study of australopithecine dentition and jaw structure have indicated that the earliest hominins were adapted to a diet of tough vegetable material, probably high in fiber. Such a diet would be more consistent with the observed diet and the gut morphology of chimpanzees. Chimpanzees are now known to consume meat frequently, obtained by hunting and killing small mammals such as monkeys. Nonetheless, fruit and vegetable foods remain the mainstay of their diet. The colon of chimpanzees and other great apes exceeds half of the volume of the gut. Considering what has been deduced about *Australopithecus*, it is likely that the relative reduction of the hominin colon, along with dental reduction and an increase in the proportion of meat in the diet, occurred in genus *Homo*. It is worth emphasizing that at no time in the evolution of the hominin lineage was our diet characterized by exclusive frugivory or exclusive carnivory. The trend has been a shift in emphasis from predominantly plant foods toward a more balanced omnivory.

FRUGIVORY, CARNIVORY, OMNIVORY

What confounds all of these investigations is the difficulty of defining dietary categories. It is clear that modern humans across world cultures eat a tremendous variety of animals, plants, and fungi and that food preferences vary markedly in different places. We write of dietary niches such as insectivory (insect eating), frugivory (fruit eating), folivory (leaf eating), and carnivory (meat eating) as though they represented real and mutually exclusive alternatives. Most certainly they do not. Although there some general patterns that relate to body size, the majority of primates sample foods from many food categories on a regular and opportunistic basis and commonly consume both plant and animal foods. Humans are the prime example of this strategy.

What question are we really asking when we attempt to reconstruct early hominin diet? Usually we are looking for specializations of diet that may help explain adaptive behaviors and anatomical features. According to the data described above, later australopithecines may have evolved such specializations not by changing broad dietary categories, but by shifting to the tougher foods of a drier habitat. To the extent they may have incorporated additional amounts of meat in the diet, more recent hominins would have become more general in their diet, rather than more specialized. As omnivores, hominins may define dietary generalization as a reduction of specialization. Generalists are, by definition, less likely to display anatomical structures and behavior patterns associated with any particular dietary category; it is less likely that "human nature" will be explained by a specific diet. Nonetheless, this omnivory does not mean a lack of discrimination. The human diet is opportunistic and highly selective to assure a higher return for foraging effort.

RECENT CHANGES IN DIET AND HEALTH

Diets in modern affluent nations have greatly changed from this ancestral pattern. Affluent societies such as our own have increased their consumption of meat. Since domesticated animals have long been bred to increase their fatness and tenderness, fat consumption has increased and makes up a greater percentage of caloric intake than in the diets of hunters. At the same time, the variety and intake of vegetable foods have declined except for starchy staples. Indigestible dietary fibers has been reduced.

Such tastes have sound evolutionary roots. Meat is one of the richest sources of balanced protein and calories available. It is celebrated across modern cultures as a festive food to be shared. A craving for it is natural. One aspect that has changed is the ease with which it can be obtained in Western society. A second is the quality of the meat available to us. Domesticated animals have been selected for centuries for higher fat content. In the process, there has also been a great increase in the proportion of the less healthy saturated fats.

The quality of vegetable foods has also declined for a number of reasons. Hunter-gatherers, as well as nonhuman animals, typically consume a wide variety of species. This results in part from the limited availability of any one food source for very long, but also because wild plants have defensive toxins that may cause sickness in high doses. Observations of folivorous monkeys, for example, reveal that they voluntarily switch foods without exhausting any one source (Milton 1993). The consequence of a varied diet is a healthy balance of nutrients.

Intensive agriculture is associated with a decrease in the variety of fruits and vegetables consumed. Again artificial selection has favored those crops that can provide a greater bulk of calories from starch and sugars. Reliance on fewer different foods contributes to nutritional imbalance. The adoption of agriculture in the fossil record can be identified in part by a decline in the size and nutritional health across the population.

These dietary changes have continued to exert a great impact on patterns of health (Eaton et al. 1988). Malnutrition is widespread even in affluent societies, caused less by a lack of food than by an imbalance of nutrients or an excess of calories. Even as modern medicine has reduced mortality from infectious disease, new disease patterns have arisen that can be tied to changing diets. High levels of fat intake and a decreasing ratio of unsaturated to saturated fats are directly related to cardiovascular disease, including arteriosclerotic placque. Digestive tract disorders, including many types of cancer, can be linked to dietary composition.

15

RESPIRATION

Breathing is synonymous with human life as the most fundamental interaction with the environment. The body must supply itself with a constant flow of oxygen, but making oxygen move in the direction we want it to is tricky—oxygen will diffuse along a gradient only toward a lesser concentration. The most important adaptations of the vertebrate respiratory systems relate to increasing the quantity of oxygen available in the bloodstream. Terrestrial life adds a second problem: once a doorway into the body cavity is open for oxygen to enter, water and heat begin to flow out. Human speech adds a third functional consideration, manipulating the flow of air for complex communication. The design of the human respiratory tract represents a compromise to address all of these needs (Figure 15.1; Table 15.1).

Table 15.1 Characteristics of the Human Respiratory System

Vertebrate characteristics
    Development of the pharynx as an organ of respiration
    Transport of oxygen in a closed circulatory system
    Lungs or swim bladder (bony fish and tetrapods)
Tetrapod characteristics
    Loss of gills and dependence on lungs for respiration
    Respiration through rib movements
Mammalian characteristics
    Elaboration of turbinals in nasal cavity
    Secondary palate
    Larynx
    Development of alveoli in the lungs
    Thoracoabdominal diaphragm
    Sagittal flexion of the spinal column
Primate characteristics
    Simplification of the nasal passages and conchae
Human characteristics
    Decoupling of breathing from locomotion
    Elimination of thermoregulatory panting
    Enlargement of the pharynx and its use in speech
    Enlargement of the spinal canal for increased motor control of breathing
    Protruding nose

EVOLUTION OF VERTEBRATE BREATHING

THE PROBLEM OF GAS EXCHANGE AND TRANSPORT

Respiration involves the exchange of oxygen molecules taken into the organism for carbon dioxide molecules shed by the organism. Molecules move within the body by active transport or passive diffusion. Active transport can be very efficient, but it is energetically expensive and therefore not practical for moving small numerous molecules such as water and oxygen. Therefore these molecules move by passive diffusion.

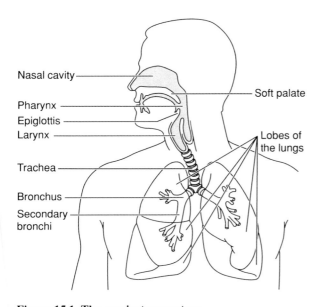

Figure 15.1 The respiratory system.

Nasal cavity

Pharynx
Epiglottis
Larynx

Trachea

Bronchus

Secondary bronchi

Soft palate

Lobes of the lungs

Respiration in early chordates may well have been accomplished by diffusion directly through the walls of the body, as it is in many aquatic invertebrates. However, this strategy places upper limits on body size. Oxygen diffuses across a surface area but must supply a volume of tissues. Surface area is calculated as a second power of a linear dimension (length × width), but volume is calculated as a third power (length × width × height). Therefore, an increase in body dimensions without distorting the shape causes the body mass, and the demand for oxygen, to increase faster than its surface area, or the ability to supply that oxygen (Figure 15.2). Moreover, as the size of an organism increases, the distance individual molecules have to travel to reach the innermost tissues increases, but the ability of the oxygen to penetrate does not increase. Therefore, the larger body sizes that are typical of vertebrates require a more efficient system of acquiring and delivering oxygen to the body tissues.

Diffusion involves the net shift of molecules from a region of high concentration to one of low concentration due to random movement of individual molecules. The rate of diffusion across a membrane (number of molecules moved in a given time period) is a function of the surface area of that membrane; the gradient, or difference in concentration between the two sides; and the thickness or resistance of that membrane. Diffusion rate may be increased by increasing the surface area, increasing the gradient, or decreasing the resistance. All three of these strategies are exploited in the vertebrate respiratory systems.

The problem of delivering oxygen was solved much earlier in animal evolution with the carrier molecule hemoglobin and a circulatory system. By temporarily binding with oxygen and flowing with the bloodstream, hemoglobin can retard the uptake of oxygen by local body cells and ensure a more even distribution of oxygen throughout the body.

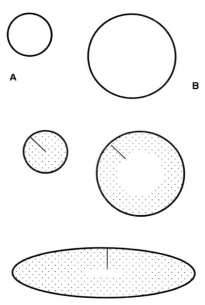

Figure 15.2 The relationship between surface area, volume, and gas exchange. As linear dimensions of an organism increase, surface area, volume, and physiological needs increase at a much faster rate. "Cell" A above has twice the diameter and circumference of cell B. It also has four times the area. If the ability of a substance to diffuse into the cell is dependent on surface area and distance, molecules will reach the center of the smaller cell in the time they penetrate only part way into the larger one (below). The larger cell can function more efficiently if it changes its shape and increases the ratio of its surface area to volume. In the ellipse below, much more of the cell contents lie reasonably close to the membrane.

GILLS

A **slitted pharynx** is one of the definitive characteristics of the chordates. This was originally an adaptation

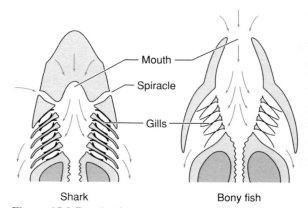

Shark   Bony fish

Figure 15.3 Respiration through gills. Gills provide surfaces at which to exchange gases with a constant flow of water. They are specialized adaptations of the feeding pharyngeal slits of ancestral chordates.

for filter-feeding of small food particles. Water was drawn into the mouth by expanding the pharynx or elevating its arches or by facing into a current. Water then flows between the slits of the pharynx while food particles are trapped and swallowed (Figure 15.3).

As vertebrates increased in body size and began actively pursuing larger food items, the slits were no longer relevant to feeding and were available to specialize for a second function. Filter-feeding creates a situation in which a great volume of water is flowing over a large surface area; and these are the properties needed for respiration, as well. Respiration through the pharyngeal arches became more effective as their

surface area and vasculature increased to match body needs. This arrangement still describes the gills of all three living classes of fish.

LUNGS AND SWIM BLADDERS

The ancestors of bony fish elaborated the respiratory function of the pharynx by extending a pouch from it caudally into the thorax (Figure 15.4). This pouch offered a further increase in surface area and incorporated a bubble of air into the middle of the body. Depending on the current function of this pouch, modern bony fish have either a lung or a swim bladder.

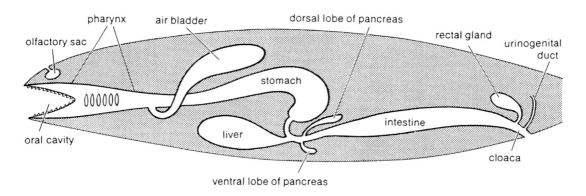

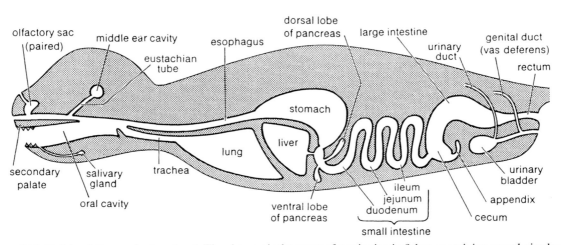

Figure 15.4 Origin of the respiratory tract. The pharynx is the organ of respiration in fishes, containing vascularized membranes for gas exchange. The lungs and swim bladder originated as outgrowths of the pharynx as well. (From M.H. Wake, 1979. *Hyman's Comparative Anatomy*, 3rd ed. Copyright 1979 University of Chicago Press. Reproduced with permission from the University of Chicago Press.)

The structure serves as a lung if it maintains a connection with the pharynx through which air may pass. Some fish ventilate the lung by gulping air to supplement gill respiration. This is an adaptive strategy in oxygen-poor ponds, for example, and it enables several types of fish to survive temporarily out of the water or to estivate in the mud of a dried-up pond. A swim bladder has lost its functional connection with the pharynx. Its gaseous contents are maintained instead by countercurrent exchange with the circulatory system. Fish may use the swim bladder to regulate buoyancy in the water, control swimming depth, or maintain an upright posture. The swim bladder is more useful than a lung for fish that rarely come to the surface.

It is not clear from comparative anatomy whether the lung first evolved for respiration or buoyancy control, although evidence now leans toward respiration. The ancestors of the tetrapods clearly had functioning lungs that made the transition to land possible.

BREATHING IN SIMPLE TETRAPODS

Modern amphibians and reptiles utilize diverse strategies for obtaining oxygen. All have functional lungs, but there are many mechanisms to force air into them. Aquatic amphibians also maintain skin and/or mucous membranes that are vascularized and thin enough to supplement their oxygen supply by diffusion from the air. These membranes would dry out in terrestrial species. Larval stages and some adult amphibians possess external gills, which are highly vascularized folds of skin around the neck with large surface areas capable of carrying on respiration.

Frogs and some other amphibians and reptiles "swallow" air. This action, called pulse ventilation or gular pumping, use muscles of the pharynx to force air into the lungs. Various reptiles create suction by expanding the ribcage or, as in the case of crocodilians, using a muscle analogous to the mammalian diaphragm. The rib suction pump was probably the primitive design used by the first tetrapods. The use of pulse ventilation in frogs is related to their secondary loss of ribs.

Salamanders and most reptiles use lateral flexion of the trunk during locomotion (Figure 15.5). This motion interferes with breathing as it does not permit free movement of the ribs, and it causes an alternating asymmetric compression of the lungs (Carrier 1987; Owerkowicz et al. 1999). Therefore, lizards commonly alternate bursts of locomotor activity with pauses for breathing, although a few may use pulse ventilation. In addition, many reptiles have independently acquired muscles for active movement of air into the lungs, operating in a way more analogous to our diaphragm (Gans 1970).

BREATHING IN MAMMALS

Several of the evolutionary innovations of mammals had direct implications for respiration. Homeothermy and an elevated metabolic rate require a dramatic increase in the rate of oxygen consumption, as does the increased body size of many mammals. The lung, which is little more than an empty membrane in fish and lower vertebrates, is filled with spongy alveoli in mammals. Alveoli are tiny membranous chambers whose walls add up to a very large surface area where gases may be exchanged. A further mammalian innovation was the secondary palate, which separates the respiratory and digestive pathways and permits mammals to feed and breath at the same time. The bony secondary palate is extended by the muscular soft palate. The soft palate contacts a cartilage called the epiglottis and further isolates the nasal-laryngeal airway from the oral pathway in nearly all mammals except adult humans.

Ventilation, or the flow of air in and out of the lungs, ultimately determines the amount of oxygen available to the animal. Many of the other mammalian adaptations relate to increasing the volume of air taken in with each breath. Mammals have evolved a true thoracoabdominal diaphragm. The diaphragm forms a muscular posterior wall or floor to the ribcage. It is shaped like a dome with the most cranial point in the center. The contractile fibers are arranged radially about a central tendon so that they can pull the center of the diaphragm caudally, increasing thoracic volume and creating a partial vacuum into which air flows. The diaphragm develops in the neck and descends during fetal development to assume its position in the trunk. This follows the origin of the respiratory tract as an outgrowth of the pharynx. Even in the adult,

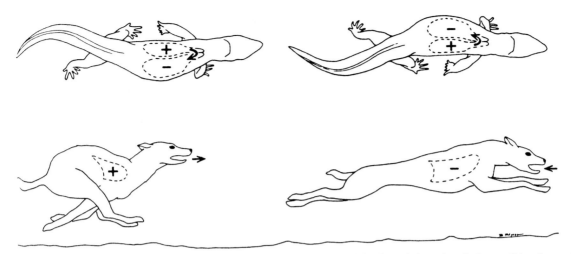

Figure 15.5 Breathing constraints in reptiles and mammals. The lateral flexion of the spine during walking in reptiles interferes with respiration. Each bend of the body compresses the lung on one side, forcing air into the other lung, and reducing the exchange of air outside the body. Sagittal flexion of the spine during locomotion in mammals, especially galloping, constrains respiration. Alternate flexion and extension compresses and expands the body cavity, driving the flow of air. Furthermore, a heavy gut, such as that of leaf or grass eaters, acts as a piston driving into the diaphragm to further compress the lungs. Breathing therefore is synchronized with gait. (From D.R. Carrier. *Paleobiology* 13(3): 326–41. Copyright the Paleontological Society, Reproduced with permission of the Paleontological Society.)

the diaphragm is innervated by nerve fibers off the cervical spinal cord that form the phrenic nerve rather than by the much closer thoracic spinal nerves.

The diaphragm creates an active rhythmic flow of air into the lungs. This action is more effective than the rib movements of early tetrapods, but it supplements rib movements rather than replaces them. Mammals redesigned the ribcage, eliminating ribs from the abdominal region. Thoracic ribs were retained because rigid walls must surround the lungs in order to support a partial vacuum. The ribcage was strengthened by adding a sternum to anchor them anteriorly. Nevertheless, the design of the individual ribs and their joints permits movements to expand thoracic volume in all three dimensions. When a quadrupedal mammal is standing still, the action of the diaphragm is responsible for much of the air flow. In addition, the intercostal muscles cause the ribs to flare laterally and shift cranially, further increasing the volume of the thorax. Expiration depends mostly on the elastic recoil of the ribs.

Eliminating abdominal ribs allows a significant degree of spinal flexion during locomotion. Flexion brings the cranial and caudal ends of the trunk closer together and also influences respiration. Compression of the trunk as the spine flexes forces air out of the lungs. Relaxing it as the spine extends draws in air. When the animal is running (especially galloping), breathing movements must work with the movements of the spinal column and the visceral organs (Figure 15.5). Spinal flexion occurs as the hind limbs are brought forward together and placed on the ground. This action compresses the abdomen, places pressure on the diaphragm, and forces air from the lungs. Conversely, the spine extends as the forelimbs reach forward, and the thorax is able to expand. Since breathing itself is aided by these locomotor movements, efficient breathing must synchronize with them (Lee and Banzett 1997). Running mammals, especially large ones, breathe exactly once during each stride (Bramble and Carrier 1983; Young et al. 1992). Possibly as a consequence of this limitation, respiratory efficiency drops in mammals as level of activity increases (Wagner 1987). Furthermore, panting, an important strategy of many mammals for dumping excess heat, cannot occur while the animal is running.

A second constraint linking breathing with running is the movement of the visceral organs. The rhythmic shifting of weight by the intestine, liver, etc., en masse acts like a pendulum with its own resonant frequency. (That is, the rhythm of shifting organs is determined as much by their own properties as by the stride of the animal.) To avoid interference with breathing, the diaphragm must help to stabilize these organs by contracting against them (Bramble 1989). Thus, in quadrupeds, unlike humans, the diaphragm may not be a primary muscle of respiration. Other muscles of the trunk, such as those of the abdominal wall, may also switch behavior from respiratory to locomotor when the animal moves (Deban and Carrier 2002). The intercostals appear to participate in respiration and locomotion at the same time (de Troyer and Kelly 1982).

Other gaits also constrain breathing. Trotting in dogs is synchronized with gait (Bramble and Jenkins 1993); and a complex relationship has been observed among the asymmetric limb movements of trotting, thoracic wall shape, and air movements within the lungs.

In summary, mammals display a number of structural adaptations that increase the efficiency of ventilation of the lungs. Locomotion involving spinal flexion assists in ventilation, but gait places serious constraints on breathing frequency on all but very small animals. The adoption of an upright stance by humans largely isolates breathing from locomotion.

MECHANICS OF HUMAN RESPIRATION

As in quadrupedal mammals, the human diaphragm is anchored to the lower margin of the ribcage and to the spinal column. Contraction of the diaphragm places a downward pressure on the liver, stomach, and other organs and increases thoracic volume. Since the muscles of the abdominal wall generally maintain a firm tone, this action increases intra-abdominal pressure. The effects on the abdomen and pelvis are significant. Venous blood in the abdominal cavity and

walls is forced upwards toward the heart, and lymph circulation is likewise stimulated. The lumbar spine is reinforced along its curvature for greater stability. Diaphragmatic pressure can be applied adaptively in vomiting, defecation, and delivery of a baby.

Rib movements also occur in humans to enlarge thoracic volume. Rotation at the costovertebral joints produces an upward movement of the sternum and anterior ends of the ribs and, simultaneously, an outward movement of the lateral parts of the ribs. Thus the diameter of the thorax is increased in all dimensions. The muscles that contribute to inspiration are those that help to move the ribs apart.

Expiration again is largely a passive movement, using the force of gravity to bring the ribcage back down to its resting position. Pressure maintained in the abdomen pushes the diaphragm back up so that air is expelled. The rate of expiration is increased by the active contraction of abdominal wall muscles, especially the rectus abdominis. This is useful in such actions as blowing, coughing, sneezing, and laughing.

The rate of human breathing is roughly independent of locomotion, since neither spinal flexion nor cranial–caudal visceral movements are significant. The jarring effect of foot contact with the ground does tend to drop the ribcage downward, but this is cushioned by numerous mechanisms in the joints and muscles. Experienced runners tend to synchronize breathing and stride, but may do this in a variety of ways. Most common is a breath-to-stride ratio of 1:2, but other ratios are observed—5:2, 3:1, 3:2, 1:1 (Bramble and Carrier 1983). The result is that humans can adjust stride rate (and thus speed) and breathing rate (and thus oxygen supply) independently.

Maintaining an appropriate delivery of oxygen is important. If oxygen supply cannot keep up with demand, the oxygen that is stored in myoglobin in the muscles is quickly depleted and the body cells shift to anaerobic respiration. Glucose, the fuel of muscles cells, is consumed incompletely and therefore inefficiently. The partially oxidized byproduct lactate accumulates and impairs performance. These processes contribute to exhaustion. At the same time, the person has accumulated an "oxygen debt" such that oxygen taken in later will be needed to complete the oxidation of lactate and to replenish stores in the muscle. An ability to handle exercise obviously requires a reserve capacity for respiration.

On the other hand, normal respiration should not be too efficient. If the respiratory rate exceeds demand (hyperventilation), blood levels of carbon dioxide are depressed, creating other problems. Carbon dioxide dissolved in the blood interacts with water molecules to form carbonic acid. A sudden decrease in carbon dioxide raises the pH, with implications for other aspects of body chemistry. Furthermore, the nervous system used monitors the bloodstream for carbon dioxide and acidity levels to adjust the rate of breathing, Hyperventilation may be followed by a dangerous suppression of breathing.

The ability to fine-tune respiratory rate independently of locomotion permits the body to maintain an appropriate balance of oxygen and carbon dioxide for a longer period of time and through a greater variety of activities. Therefore these consequences of bipedalism may be considered an important contribution to the human ability to sustain activity over long periods of time. Respiratory changes correspond to adaptations of the skin that also facilitate sustained activity. Efficient sweating sheds the excess heat and makes panting unnecessary. The elimination of the panting mechanism was an adaptation that later facilitated the development of speech.

FUNCTIONAL SPECIALIZATIONS OF THE HUMAN RESPIRATORY TRACT

The respiratory tract extends from the nose to the lungs. It includes the nasal passages, pharynx, larynx, trachea, bronchi, and the lungs themselves. Apart from being an air pathway, it actively participates in regulating the temperature and humidity of the air in the body and in the neutralization and removal of germs and dirt.

The entire passage as far as the lungs is lined with a ciliated mucous membrane. The secretions of the mucous glands keep the membrane moist. The cilia beat rhythmically to create a constant flow of mucus that converges on the pharynx. As the mucus entraps foreign particles, these are swept back through the nose and mouth and upwards from the lungs and trachea. Once in the pharynx, the mucus and debris are swallowed and destroyed in the digestive system. If mucus and other fluids build up in the lungs due to enhanced secretion or simply to overnight accumulation, their expulsion can be aided though coughing.

THE NASAL PASSAGES

The nasal passages represent a filter through which the air must pass en route to the lungs. Beginning in the nasal cavity, the air is modified to make it more tolerable to the body. To accomplish this, a large surface area is necessary. The walls of the nasal cavity are expanded by the conchae, bony scrolls anchored to the lateral wall. Incoming air is forced to pass over these formations. Heat is drawn from the membranes so that the incoming air is close to body temperature by the time it reaches the larynx. Sufficient moisture evaporates to bring the air to 90–100% of saturation before it reaches the lungs.

This function represents an expenditure of energy and loss of water. Even that loss can be adaptive. Many animals use panting as a means of ridding the body of excess heat and lowering body temperature. Although humans developed the more efficient mechanism of bodywide perspiration to accomplish the same thing, the loss of heat from the upper respiratory tract does have an important function in cooling the temperature-sensitive brain (see Chapter 16).

Under other circumstances, the heat and water loss is an undesirable burden. When the outside air is very cold or very dry, the expenditure increases. It is therefore significant that the mucous membranes, which have lost heat to the incoming air, are now cooler than the outgoing air. Even under very cold conditions, air that has been warmed to body temperature (37°C) leaves the body at about 32°C, the balance of heat having been recovered. As the temperature of the expired air drops, a certain amount of the moisture condenses and is recovered by the body (Figure 15.6). Overall, approximately 20–25% of the heat energy and water may be conserved (Courtliss et al. 1984). So important is this function, that the presence of a complex set of conchae has been cited as an indicator of homeothermy among fossil animals (Ruben et al. 1996; Zimmer 1994).

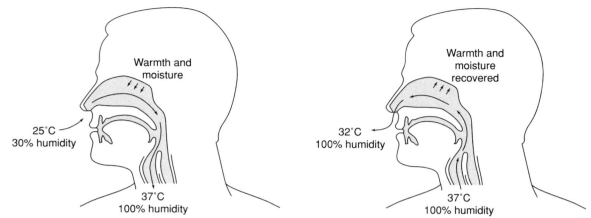

Figure 15.6 Heat and water exchange in the upper respiratory tract. Cool, dry inhaled air receives warmth and moisture from the membranes of the nasal cavity, so that its state has less impact on the lungs. During exhalation, the air from the lungs, now warmer and nearly saturated, again passes through the nose. The relatively cooler membranes there (cooled by the previous inhalation) cause condensation on the membranes, recovering much of the warmth and moisture lost earlier.

The air is cleansed by a number of mechanisms. The hairs at the entrance to the nostril represent the first barrier to particles. The natural stickiness of mucus traps particles which come into contact with it, and the probability of contact is enhanced by the narrow passages and sharp turns in the airways. Further, the membrane maintains a positive electrostatic charge which attracts negatively charged particles. Thus an estimated 85–90% of particles 6 μm (0.006 mm) or larger are removed.

The nasal passages are also active in regulating airflow by several valves. The nostrils themselves widen (flare) by the actions of facial muscles during inspiration. Septal cartilages of the external nose also open the passage during inspiration and narrow it during expiration. This action alters both volume and dynamics of airflow through the nose. The inferior concha performs a similar action, constricting the passage by engorging its vessels with blood.

The superior concha and corresponding region of the septum are covered with olfactory epithelium, in which the receptors of smell are embedded. Compared with that of other mammals, this is a small area and corresponds with the reduced emphasis on the sense of smell. Both number and complexity of the conchae (called turbinals in other mammals) are greater in nonprimates. The horse, for example, has 37 turbinals.

Primates in general and the anthropoids in particular reduce the number of turbinals along with the size of the snout.

The nasal cavity also connects with the paranasal sinuses. The sinuses are air spaces of uncertain function in several of the facial bones: frontal, ethmoid, sphenoid, and maxilla. Each of the sinuses is lined with mucous membrane and drains to a opening in the lateral wall of the nasal cavity under cover of one of the conchae. Although several functions have been proposed (e.g., reduction of weight, thermal insulation for the brain, cooling of the brain), the factors that determine sinus size and position are not clear (Blanton and Biggs 1968). Because they represent "dead ends" in air circulation, sinuses are frequently sites of infection and congestion.

THE PHARYNX

The air passages of the nasal cavity lead through the choanae into the pharynx. The soft palate provides a mobile valve to separate the nasal cavity from the oral cavity and pharynx and prevent the upward movement of food during swallowing. The pathways for air and food diverge in the lower pharynx, continuing as the larynx and esophagus, respectively. The epiglottis forms a barrier that deflects food away

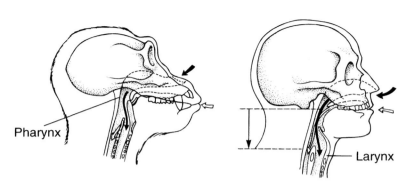

Figure 15.7 The position of larynx in humans. In most mammals (*left*), the palate overlaps the epiglottis to form an effective separation of the air and food passages. In adult humans (*right*), the larynx has descended to a lower position, enlarging the pharynx but making it possible for us to choke on food. (From *Vertebrates: Comparative Anatomy, Function, Evolution* by Kenneth V. Kardong. Copyright 1998 by the McGraw-Hill Companies. Reproduced with permission of McGraw-Hill Education.)

from the entrance to the larynx during swallowing to avoid choking.

In nonmammalian vertebrates, breathing and eating overlapped even more. It was the mammalian need for extensive chewing that led to the development of the secondary palate, soft palate, and epiglottis and to the conversion of internal nares to the choanae. By soft tissue arrangements in the pharynx, particularly an epiglottis high at the back of the mouth and overlapping with the palate, most mammals are able to continue to breathe even as they chew, drink, and swallow (Laitman et al. 1977).

Adult humans have departed from this plan. The pharynx descends during infancy and childhood, to a lower position in the neck, so that the epiglottis no longer contacts the palate (Figure 15.7). Human infants still have a high larynx and can quite happily nurse and breathe simultaneously for long periods at a time. It is adults who are at greater risk of choking and drowning.

THE LARYNX

The larynx is a triangular chamber enclosed on two sides by the thyroid cartilage and in back by soft tissues and smaller cartilages. It contains the vocal folds, which constitute a valve guarding the entrance of the trachea. The vocal folds also regulate the flow of air out of the lungs and are responsible for voice. A vocal fold consists of a vocal ligament stretching from the thyroid cartilage anteriorly to the arytenoid cartilage posteriorly, a vocalis muscle alongside the

ligament, and a membranous cover continuous with the lateral wall. By movements of the cartilages, the free edges of the vocal folds can be shifted medially and laterally to open and close the air passage.

When the larynx is closed, an attempt to exhale builds pressure below the vocal folds. A sudden release results in a cough or sneeze and can be used to expel food, mucus, or other objects from the respiratory tract. More subtle manipulation can fine-tune the flow of air during normal breathing (Bartlett 1989). If the vocal folds are tensed, air passing between them causes them to vibrate. The vibrations are translated into sound waves as voice. The pitch of the voice corresponds to the degree of tension in the vocal folds. Voice is further modified by resonance in the pharynx and nasal cavities and by manipulations of the tongue, lips, and other soft tissues of the mouth.

The low position of the larynx in adult humans enlarges the pharynx, enhances its resonant properties, and makes the current range of human speech possible. Changes in the shape of the larynx, brought about by its muscular walls, determine vowel sounds. Consonants are produced in the mouth. Although a few other mammals, such as the red deer, have lowered the pharynx to make deeper and louder calls (Fitch and Reby 2001), complex speech in other mammals is not possible in part because of the lack of specialized neurological control over the necessary muscles.

It has been commonly asserted that the descended human larynx is an adaptation to speech. However, recent observations show that the larynx of an infant chimpanzee descends in parallel with that of humans (Nishimura et al. 2003). Although the human pharynx continues to elongate downward, at least the initial phase of laryngeal descent can no longer be linked to a distinctively human trait such as speech. Nonetheless, because of the importance of the human form of the pharynx and larynx for modern speech, it would be of great interest if we could identify from the fossil record exactly when the modern morphology was attained. Although several attempts to extrapolate this from skulls have been made, the evidence is controversial and must be regarded as inconclusive.

THE EVOLUTION OF SPEECH

Language is one of the fundamental characteristics defining humanity. Unfortunately, the capacity for language, depending as it does on neuronal circuitry and soft tissue anatomy, is very difficult to identify in the fossil record. Therefore we are very uncertain when and why language evolved. As discussed in the text, knowledge of the morphology of the pharynx and position of the larynx might give us a good idea of the complexity of speech of which an ancestral hominin was capable. Several attempts have been made to infer these features from the skeletal record (e.g., Frayer and Nicolay 2000), although we must recognize that the capacity for speech is neither necessary nor sufficient for true language.

The debate over the relationship between basicranial flexion and the position of the pharynx was discussed in Chapter 4. Other researchers have sought functional correlations with speech elsewhere. Duchin (1990) argues that the mouth is more important than the pharynx, and that the reduced size of the mouth and mobile tongue in humans, as opposed to chimpanzees, facilitates articulate speech. Lieberman et al. (1992) dispute this by observing that it is not possible to separate functionally the different parts of the respiratory tract for speech. Krantz (1980a) proposes that most of the changes that occurred in the skull between *Homo erectus* grade hominins and modern humans relate to the evolution of language capacity and that, aside from cranial capacity itself, these traits place Neanderthals with the linguistically primitive

H. erectus. These include both reduction of the mouth and changes in the basicranium.

The discovery of an intact and very modern looking Neanderthal hyoid bone led to another round of controversy (Ahrensburg et al. 1989, 1990). The hyoid bone is positioned superior to the larynx and anchors muscles important to tongue and laryngeal movements. Critics point out that no good correlation between hyoid form and its position in the neck (reflecting pharyngeal length) has been established. Certainly the correlation with speech is unproved (Lieberman 1993, 1994).

Kay and colleagues noted a difference in the hypoglossal canal size of modern humans and Neanderthals versus chimpanzees (Cartmill 1998; Kay et al. 1998). The hypoglossal canal conducts the hypoglossal nerve, which supplies the muscles of the tongue. In order to supply the fine motor control needed for speech, it is expected that the human nerve would be much larger than that of nonspeaking animals. Despite the earlier observations, however, it was found that the canal size is a poor indicator of nerve diameter across primate groups (DeGusta et al. 1999). Therefore, although the relative size of the canal shows an increase in more recent hominins, its relationship with speech cannot be considered proved.

What is the evolutionary relationship between cranial form, respiratory tract, and language? Has the acquisition of speech been a sufficient adaptive advantage to reshape the anatomy? When did the capacity for speech evolve? These are unanswerable questions. We can assume that either some rudimentary speech had to be present to encourage the reshaping of the anatomy; or that the anatomical changes were in response to other functions. Cranial flexion has been interpreted as facilitating head balance in a bipedal posture. The upper respiratory tract may reflect climatic adaptations. Speech capability cannot be considered independently of these physiological functions. The use of language, as opposed to the capability for it, may or may not have appeared at the same time.

The present anatomy of the upper respiratory tract certainly represents a compromise to meet diverse functional demands. For example, the human infant condition is not merely a recapitulation of a previous evolutionary state. The high position of the larynx is adaptive in permitting prolonged suckling as the infant breathes. The later descent of the pharynx occurs when the infant has better motor control of its head, is more alert, and can alternate between feeding and breathing as necessary. Furthermore, nonlinguistic considerations alone would probably not have given humans a pharynx that makes us unusually susceptible to choking and drowning.

More recent evidence has come from a 1.5-million-year-old *H. ergaster* skeleton discovered in 1984. Analysis of the bones revealed many startling findings, including the fact that the neural canal, the passage in the vertebral canal for the spinal cord, was remarkably narrow. After considering alternative interpretations, Walker and his colleagues (Walker and Shipman 1996; MacLarnon and Hewitt 1999) concluded that the increased diameter of the modern spinal cord can be understood as an adaptation for increased motor control of the diaphragm and intercostal muscles for the control of breathing during speech. Their *H. ergaster* specimen had not yet acquired this adaptation.

CLIMATIC ADAPTATIONS

The respiratory tract directly encounters the environment with each breath. The properties of the tract will adapt to local conditions, such as temperature and humidity. These adaptations are most conspicuous in the form of the nose, which is unusual in its protrusion in humans and is progressively more protruding in populations of higher latitude (Carey and Steegman

1981). A protruding, rather than flat, nose increases the area of the mucosal membrane that is in contact with the air. It also may increase the turbulence of air flow and hence the ability of the nose to warm and humidify otherwise cold and dry air and then to recover condensing moisture of the exhalation. Although such a relationship with climate is not observed in other mammalian groups, the additional length of the nasal passages in humans may be more significant because of their already reduced size. The breadth of the nose appears to be more a function of the breadth of the palate, which reflects the influence of nonclimatic factors.

A protruding nose is first suggested by the facial bones of *H. erectus* crania. *H. erectus* was also the first hominin population to leave the tropics for colder and drier habitats. Trinkaus (Franciscus and Trinkaus 1988; Trinkaus 1986) suggests that the nose form was specifically an adaptation for the strenuous exercise of hunting in that climate.

A different habitat shift with consequences for the respiratory system involves high altitude. The decreased atmospheric pressure requires compensations to increase the intake and transport of oxygen by the body. Individuals who move into high altitude environments for long periods of time increase their hemoglobin levels and red blood count. Populations who have adapted to such environments in the Andes and Himalayas over generations also show an enlarged thorax and are capable of moving a greater volume of air in and out of the lungs.

16
CIRCULATION

The circulatory system is more elaborate in vertebrates than in other animals (Table 16.1). This is a direct consequence of body size. The simple mechanical demand of moving greater quantities of blood over greater distances has required the evolution of a more complex and stronger heart with multiple chambers. Larger masses of tissues reduce the effectiveness of diffusion and have required the development of closed circulation within an extensive system of blood vessels and capillaries. The efficiency of the blood in its ability to carry and deliver oxygen has been finely adjusted by the controls on blood pressure and the molecular mechanisms of oxygen transport.

The vertebrate system was able to make suitable adjustments for the greater demands of the high metabolic levels of mammals, including their need for an even higher rate of oxygen delivery and the new requirement of finely tuned thermoregulation. This

Table 16.1 Characteristics of the Human Circulatory System

Vertebrate characteristics
  Hemoglobin contained in blood cells
  Chambered heart
  Closed circulation
  Aorta ventral to the spinal column, dorsal to the viscera
Tetrapod characteristics
  Reduction of aortic arches
  Pulmonary circulation
Mammalian characteristics
  Complete separation of pulmonary and systemic paths through
    the heart
  Aortic arch passes to the left
  Elaboration of countercurrent exchange to regulate body
    temperature
Human characteristics
  Increased efficiency of emissary connections to cool the brain

chapter will explore some of the contributions that anatomical design can make to circulatory function.

THE HEART

The blood vessels in the body form a closed circulation, in which blood does not leave the vessels. Although some white blood cells do actively roam outside of the capillaries and plasma routinely escapes from them, the erythrocytes (red blood cells) and the bulk of the plasma continue to follow an endless circuit away from and back to the heart. Many invertebrates possess open circulation, in which blood is pumped initially through an aorta but then flows freely among the body cells before finding its way back into the heart. Closed circulation is a vertebrate characteristic.

Blood within the vessels must be put into motion to circulate throughout the body. In the digestive system, the body uses peristaltic contractions to move food along the alimentary canal. A rhythmic peristalsis works for arteries in some small invertebrates but could not generate the force necessary to supply the capillary beds of a vertebrate. Thus, a larger pump, the heart, is needed. The introduction of valves into the heart controls the direction in which the blood will move. The complexity of the heart is increased by its division into chambers that initially served to increase the effective force of contraction.

It is easy to imagine a heart created by strengthening the muscle already present in the walls of the vessels. However, the muscle of the heart has diverged into a unique tissue, different from both skeletal and visceral muscle. Cardiac muscle is striated—that is, its contractile proteins are aligned in bands within the fibers for more powerful, coordinated action. Individual cells are stacked end to end to form long branching muscle fibers. Cardiac cells do not required individual innervation to stimulate contraction. Instead, excitation is initiated rhythmically by a small cluster of specialized cardiac cells and then spreads directly from one cell to another. In its striation and formation of long fibers, cardiac tissue resembles skeletal muscle. By its composition of individual cells and its ability to spread rhythmic waves of excitation across the tissue, it more resembles smooth muscle.

THE CHAMBERED HEART

The force with which a muscular chamber can contract is proportional to the initial stretch of its walls. If sufficient blood enters a chamber of the heart under pressure, it will distend the walls, stretching the fibers, to permit the fibers to generate a stronger contraction. This relationship of stretch and force of contraction has been formalized as Starling's Law.

How does the body create an initial pressure to fill the heart? The solution is to use a series of chambers. The first chamber fills with blood returning from the veins that has very little pressure. When it contracts, it fills a second chamber with blood under a greater pressure. The second chamber is thus able to produce a stronger force to push blood into the third chamber. The third chamber contracts more strongly yet, and so on with each successive chamber.

Primitive vertebrates, if we are to make inferences from living fish, had hearts with four successive chambers—the sinus venosus, atrium, ventricle, and conus arteriosus (Figure 16.1). The sinus venosus is simply the confluence of the major veins of the body. It pumps blood through a valve into the atrium. Pressure builds as the blood successively passes through the other chambers. From the conus arteriosus, blood is sent under the greatest pressure into the arterial system. The same pattern is seen in early developmental stages of other vertebrate hearts.

STRUCTURE AND FUNCTION OF THE HUMAN HEART

The human (and, more generally, mammalian) heart is an elaboration of this primitive vertebrate pattern. It has been reduced to a series of two pumping chambers, the atrium and the ventricle. The primitive

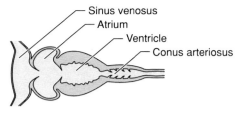

A. Primitive heart

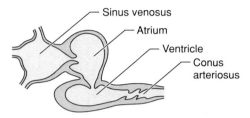

B. Fish heart

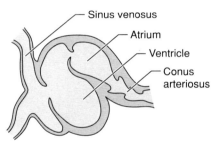

C. Human embryo heart

Figure 16.1 The primitive vertebrate heart. The heart of fish and of the human embryo each has four chambers, but they are not the same divisions found in the adult mammalian heart. *A*, In the primitive heart, the chambers are lined up in a single series. *B*, Fish heart. *C*, Human embryo heart.

sequence of four chambers has reduced to two. The sinus venosus has joined with the atrium to make a single chamber. One side of it still retains the smooth inner walls of the sinus. The conus arteriosus forms the base of the aorta and pulmonary trunk, splitting longitudinally to do so.

On the other hand, both the atrium and the ventricle have divided into right and left chambers. The outline of human circulation is illustrated in Figure 16.2.

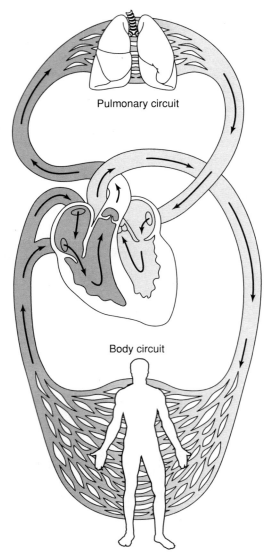

Figure 16.2 The circulation of blood through the heart.
The circulation is divided into two systems. The right side
of the heart sends blood to the lungs, while the left side
receives that blood and pumps it to the rest of the body.

Blood from the veins of the body is collected into the
superior and inferior venae cavae, which empty into
the right atrium. The right atrium pumps into the right
ventricle, enabling the right ventricle to pump blood
more forcefully out of the heart. From the right ventricle,
blood travels through the pulmonary trunk and arteries

to the lungs to be oxygenated. Blood returns from the
lungs via the pulmonary veins into the left atrium. From
the left atrium it passes through the left ventricle and
out the aorta to be distributed to body tissues.

The ventricles have much thicker and stronger
walls than the atrium, while the left ventricle is thicker
and more rounded than the right. These proportions
indicate the force of contraction needed to move the
blood to its immediate destination. The left ventricle
must work harder to propel blood throughout the
body than the right does to move it to the lungs.
Nonetheless, all of the chambers must move the same
volume of blood in the same number of strokes. Thus
their capacities are of comparable size, the human
chambers expelling approximately 70 to 90 mL (2 to
3 ounces) of blood with each contraction.

EVOLUTION OF VERTEBRATE CIRCULATION

The evolutionary transition from the primitive vertebrate four-chambered heart to the modern mammalian
four-chambered heart is elegantly described by comparative anatomy and reiterated in human embryology.
So lucid is this model that we must remind ourselves
that living vertebrates do not themselves form an
evolutionary sequence but represent only the end
products of many evolutionary lineages. The intermediate stages in this transformation were adaptive
for different habitats and physiological contexts, so
that lineages have survived and continued each of
these stages.

The developmental recapitulation of these stages
tells us how the evolutionary changes came about.
If at some phase the aortic arches of the mammalian
embryo resembles those of a fish, it is because the lineage leading from the common ancestor to mammals
changed the developmental process by extending it in
a new direction (peramorphosis).

CIRCULATION THROUGH GILLS

The four-chambered heart of a fish described earlier
may be understood as a primitive vertebrate form.
Blood leaving the conus arteriosus enters a ventral
aorta and runs anteriorly into the branchial arches
(Figure 16.3). These arches, containing skeleton,

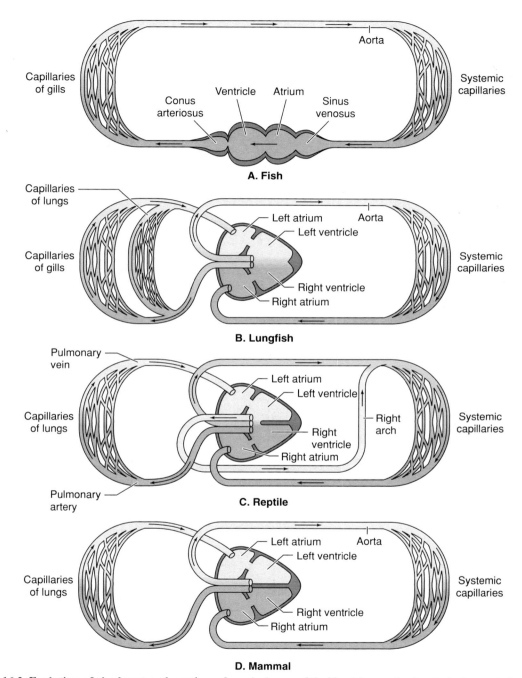

Figure 16.3 Evolution of the heart and aortic arches. *A*, Among fish, blood leaves the heart via the ventral aorta and then passes through capillaries of the branchial arches in the gills. *B*, A transitional state is seen in lungfish, which maintains both functioning gills and lungs. *C*, Lower tetrapods have altered the arches to direct blood to the lungs. While a separate circuit returns the blood to the heart after it has been oxygenated, this blood is not completely separate from the stale blood returning from the rest of the body. *D*, Mammals maintain distinct circuits for the lungs and the rest of the body.

muscle, nerve, and other tissues, are specialized for respiration as internal gills in modern fishes. The artery supplying each of the branchial arches is known as an aortic arch. Each of the aortic arches produces a rich capillary bed from which gases are exchanged with the passing water. The aortic arches then reform, collecting oxygenated blood into single dorsal aorta that runs posteriorly. Branches of the dorsal aorta supply the body and its visceral organs, while other branches from the anterior arches run forward to the brain.

This circulatory system loses most of its blood pressure at the capillaries of the aortic arches. Blood flows slowly through other body arteries because of fluid pressure and skeletal muscular contractions. Obviously the oxygen supply is adequate for the needs of the fish, but the flow of arterial blood must be described as sluggish by mammalian standards. The evolution of the lung as a respiratory organ provided an opportunity to improve on this pattern.

CIRCULATION THROUGH LUNGS

The lung/swim bladder organ system that appears in bony fish is, like the branchial arches themselves, a derivative of the wall of the pharynx. Its blood supply, the pulmonary artery, is a branch of the sixth aortic arch. Blood draining from the primitive lung joins other venous blood in returning to the heart. As the lungs became increasingly important in respiration for fishes and amphibians, the pulmonary veins became more important as carriers of oxygen. Any alterations of circulation or heart structure that might keep this oxygenated blood from being diluted before it reaches other body tissues would increase the efficiency of oxygen supply to those tissues and would be evolutionarily favored. Changes in this direction appear in modern lungfish and amphibians.

Air-breathing terrestrial vertebrates should be able to bypass the capillary beds of the aortic arches with much of their blood flow. This is accomplished in amphibians and all derived tetrapods by directing that flow primarily through an enlarged fourth aortic arch, as a direct link between the ventral and dorsal aortas (see Figure 16.3). This shunt thus elevates the pressure in the dorsal aorta.

At the same time, lower tetrapods display modifications of the heart that reduce blending of oxygenated blood from the pulmonary veins with deoxygenated blood from the other veins of the body. Incomplete septa within the atria and ventricles and a flap of tissue within the conus arteriosus called the spiral fold direct parallel streams of venous blood through these chambers. Mixing of the streams is incomplete. More important, the two streams of blood are targeted separately. The more oxygenated blood is directed to the more anterior aortic arches, supplying the head and body via carotid arteries and dorsal aorta. Less oxygenated blood enters the fifth and sixth aortic arches to the lungs.

This amphibian design is clearly a successful evolutionary strategy, as it has been maintained for hundreds of millions of years. It has the advantage, among others, of varying the proportion of blood entering the lungs or passing directly to the body. For a lungfish that can carry on respiration alternately with gills or lungs, or for an aquatic amphibian or vertebrate that ceases to breathe for periods of time, this option is essential.

Complete separation of the two circuits occurs in birds and mammals. The mammalian pattern, of which humans are typical, reduces the aortic arches to one (the left fourth arch) and converts the sixth arch into the pulmonary artery (see Figure 16.4). The septa within the atria and ventricles are now complete and divide each into right and left chambers. The spiral fold of the conus arteriosus now makes a complete division into the entwined vessels, the pulmonary trunk and the ascending aorta. Birds follow an almost identical strategy except that the right fourth aortic arch is retained and the left one lost.

HUMAN ONTOGENY
RECAPITULATING PHYLOGENY

The evolutionary sequence inferred from comparative anatomy is replayed in every human fetus. In the human embryo six pairs of branchial and aortic arches are created. The fate of each one can be traced into the adult form (see Figure 16.4). The first two arches are lost. The most rostral extension of the

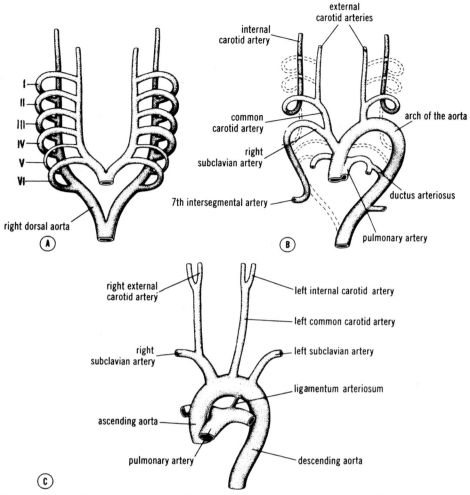

Figure 16.4 Development of the major arteries. The adult pattern of the aortic arch and its major branches can be traced developmentally to the paired branchial arches in the fetus. These in turn show an evolutionary homology with the pharyngeal arches of ancestral chordates and the gills of fish. (From J. Langman, 1975. *Medical Embryology*, 3rd ed. Copyright 1975 Williams & Wilkins Co. Reprinted with permission from Lippincott Williams & Wilkins.)

ventral aorta persists as the internal carotid artery to the brain, while the third arch becomes the external carotid artery to the neck and face. Of the two fourth arches, the right one is lost (except as the base of the right subclavian artery) while the left arch persists. The fifth aortic arch is lost and the sixth becomes the pulmonary artery. The connection of the sixth arch with the dorsal aorta is retained until birth.

The Fetal Circulation Pattern

This developmental sequence accommodates the two different patterns of circulation that must serve mammals before and after birth. A fetus obtains its oxygen from the maternal bloodstream at the placenta. Its lungs are nonfunctional. Worse, the lungs are filled with fluid, which maintains the pressure around pulmonary vessels and greatly increases the resistance

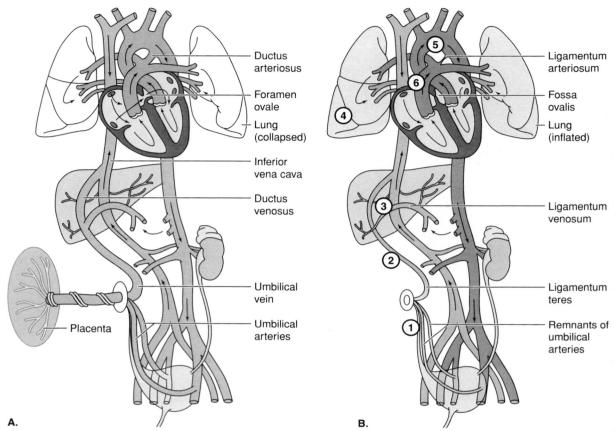

A.   **B.**

Figure 16.5 Changes in the human circulatory system around birth. The demands on the circulatory system are quite different before (*A*) and after (*B*) birth, and the fetus has only moments to make the transition from one state to the other. The transition requires a number of nearly simultaneous changes: 1, the umbilical arteries constrict; 2, the umbilical veins constrict; 3, the ductus venosus collapses; 4, the lungs inflate and reduce resistance to blood flow; 5, ductus arteriosus closes; 6, foramen ovale closes.

to blood flow there. If the right ventricle of the heart were strong enough to pump a full stream of blood through the lungs before birth, that chamber would be hypertrophied and critically overpowered for its role after birth.

Fetal circulation resolves these problems with several structural modifications from the adult pattern (Figure 16.5). Oxygenated blood from the placenta enters the body via the umbilical vein. It passes through the liver via a shunt called the ductus venosus and enters the inferior vena cava. The inferior vena cava and superior vena cava both drain into the right atrium

of the heart. However, the two streams do not completely mix. The parts of the atrial septum do not fuse and leave a passage, the foramen ovale, between the right and left atria. Oxygenated blood from the inferior vena cava is preferentially directed through the foramen ovale, while deoxygenated blood from the superior vena cava preferentially passes into the right ventricle. The transfer of blood from the right to the left atrium reduces the volume and pressure of blood pumped by the right ventricle.

The fetal pulmonary trunk remains connected to the aortic arch by the ductus arteriosus. This passage represents the primitive connection of the sixth aortic arch with the dorsal aorta. Because the lungs still have a high resistance to blood flow, the ductus arteriosus functions as a relief valve, permitting even more blood to be diverted away from the lungs and into the aorta. Relatively little blood trickles through the lungs and returns to the left atrium; thus, blood in the left side of the heart is predominantly that which was received from the inferior vena cava through the foramen ovale.

Note that the most highly oxygenated blood flows to the coronary arteries (to the heart tissue) and the carotid arteries (to the head), while other parts of the body are supplied with blood that has been diluted by the ductus arteriosus. Since other visceral systems of the fetus are not performing vital functions and the skeletal muscles do little work in the nearly weightless environment of the womb, the heart and brain are the only tissues that demand a large supply of oxygen.

Changes at Birth

At birth, the newborn circulation must immediately switch to an adult circulation pattern. The groundwork for this transformation has already been laid, and only the stimulus of birth itself is required to trigger the changes. The following changes happen almost simultaneously.

The vessels of the umbilical cord are emptied of their contents and collapse. The umbilical arteries constrict, permanently shutting off the flow of blood out of the fetus to the placenta. The now unneeded umbilical vein also collapses. Its remnants persist as the ligamentum teres, a strand of connective tissue in the adult running from the umbilicus to the liver on the inside of the abdominal wall.

Ductus venosus in the liver is constricted and closed. Its remnants persist as the ligamentum venosum between the right and left lobes of the liver. Before birth, the ductus venosus carried blood from both the umbilical vein and the hepatic portal vein to the inferior vena cava. The hepatic portal vein drained the nonfunctioning digestive tract. After birth, the portal vein carries newly absorbed nutrients. When the ductus closes, these are forced to pass through sinusoids of the liver for reprocessing and storage.

The lungs empty of fluid and resistance to blood flow drops. Muscle fibers around the ductus arteriosus

constricts and all of the blood in the pulmonary arteries is now directed into the lungs. Ductus arteriosus persists in the adult as a nonfunctioning connection between the pulmonary trunk and the aortic arch, the ligamentum arteriosum.

The increased flow of blood through the lungs causes an increase in the return of blood to the left atrium. As pressure rises in the left atrium, it equals the pressure in the right. The overlapping flaps of the septum are now pushed together to effectively close the foramen ovale. The final shunt of blood away from the lungs is now eliminated and the full volume of blood passes through the right ventricle and pulmonary trunk. A depression in the wall of the septum, the fossa ovalis, marks the site of the foramen ovale. In a substantial number of individuals, the parts of the septum never fuse completely, but are held together by the balanced pressure.

CIRCULATION AND THERMOREGULATION

As the circulatory system transports blood, oxygen, nutrients, and wastes within the body, it also distributes heat. Blood is warmed as it passes through metabolically active tissues, and it transfers that heat to superficial structures cooled by contact with the environment. Several specific mechanisms have evolved to increase or decrease the efficiency of this heat exchange. The circulatory system thus plays an important role in thermoregulation of the body.

The effectiveness of heat exchange depends on the rate at which blood passes through a tissue and the surface area of the vessels in contact with that tissue. Exchange is therefore more complete in a capillary bed, where fluid moves slowly and the surface to volume ratio of the vessels is greatest.

COUNTERCURRENT EXCHANGES: MAINTAINING A TEMPERATURE DIFFERENTIAL

A countercurrent exchange may occur where venous and arterial pathways pass one another in opposite directions. Generally the arterial blood carries warmth from the core of the body, while venous blood is

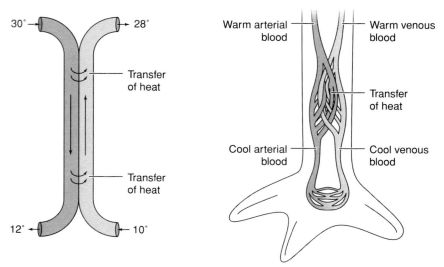

Figure 16.6 A countercurrent exchange. The countercurrent exchange is the principle used in a household heat exchanger and is found in a number of anatomical contexts. As shown schematically here, a countercurrent exchange is used to reduce heat loss in the feet of a wading bird. Warm blood descending from the core of the body is brought in close contact with cool venous blood returning from the foot. When the arterial blood transfers heat to the venous blood, that heat is conserved by returning it to the body. The cooled arterial blood continues to the foot, but now there is less heat to be lost. The consequence is that different parts of the body can efficiently maintain different concentrations of heat (or oxygen or salt, etc.).

returning from the cooler periphery. In such a case, venous blood may be warmed so as not to chill the center of the body. Arterial blood may be cooled to reduce heat loss from the skin. In this way a countercurrent exchange helps to conserve body heat (Figure 16.6). Countercurrent exchanges commonly occur in the limbs. They are of vital importance to such animals as wading birds, whose feet would lose heat in the water, and arctic foxes living in the snow.

Humans have a potential for vascular countercurrent exchanges. Deep veins lie in contact with the arteries in the limbs. Often two small veins exist for each artery; this facilitates heat transfer by increasing the area of contact. In the spermaticord, an exchange occurs between the testicular artery and a plexus of veins returning to the body. This mechanism maintains the testis and epididymis below body temperature.

COOLING THE BODY: REGULATING CUTANEOUS BLOOD FLOW

A second vascular mechanism for thermoregulation is an adjustment of the volume of blood flow to the skin.

Major cutaneous vessels lie deep in the superficial fascia, insulated by adipose tissue from the surface. Smaller vessels carry blood through the fascia closer to the surface. From the more superficial capillaries, heat is more readily lost and fluids are more readily available to sweat glands. By dilating or constricting the vessels to the skin, the body can influence the rate of heat loss. When the body is hot or is producing excessive heat, as by exercise, the arteries open and blood comes more to the surface. This accounts for the reddening or flushing of the skin. If the body is cold, vessels are closed down and the skin becomes visibly more pale.

COOLING THE BRAIN

The brain is extremely sensitive to temperature fluctuations. Thermoregulation in the head may be expected to be more elaborate than elsewhere in the body. The most important factor determining brain temperature is the temperature of blood ascending in the internal carotid arteries (Baker 1982). Additionally, an active brain may produce a significant amount of waste energy.

Many mammals, including carnivores and ungulates, keep the brain cool by an elaborate countercurrent exchange mechanism. The internal carotid artery typically passes through the wall of the cavernous sinus, one of the dural sinuses surrounding the brain. In these animals, the artery briefly divides into a network of small arterioles, called a rete mirabile ("marvelous network"), as it passes the sinus and then reforms into a large artery once more. This network increases the effectiveness of heat exchange. As the cavernous sinus receives blood from the face via the ophthalmic vein, its ability to cool the carotid artery is that much greater.

Primates do not possess a rete at the carotid sinus, although their increased brain size requires effective cooling. Other mechanisms contribute to this function. Blood from the upper respiratory tract may drain to the sinus. Inspiration especially cools this blood in order to warm air going to the lungs (Cabanac 1986; White and Cabanac 1995a, 1995b, 1995c, 1996). Experimental evidence shows that this brain cooling function may be actively manipulated by adjusting the flow of blood in this area. The evaporation of human perspiration from the face and scalp provides another mechanism for cooling blood. This surface blood may be redirected to enter the cranial fossa to mingle with blood in the cavernous sinus and other dural sinuses instead of draining inferiorly toward the neck (Baker 1982; Dean 1988). Such a pathway is possible because of the absence of valves in these connections. Additional small and variable communications between the dural sinuses and external veins are called emissary veins. One of the more consistent pairs of these pass through the parietal foramina on either side of the sagittal suture. Other conspicuous canals for emissary veins are sometimes found at the base of the occipital bone.

Human upright posture may use gravity to facilitate the flow of blood from the brain. In addition to the internal jugular vein, a plexus of veins in the vertebral canal drains the cranial fossa (Eckenhoff 1970). The combination of connections from the perspiration-cooled skin with the gravity-driven siphon of the vertebral plexus appears to be highly effective in cooling the human brain. Falk (1990) suggested that the achievement of this "radiator" mechanism overcame thermal barriers that limited brain size and permitted its further expansion in hominins. As critics have observed, however, the logic may be reversed. The evolution of a large brain may have required the secondary development of a mechanism to cool it.

17
EXCRETION

The excretory system evolved for the vital function of regulating water and electrolyte balance (Figure 17.1; Table 17.1). The osmotic pressure within the body, pH, and the delicate balance of ions are constantly fine-tuned by the kidneys. The excretory system is also the body's means of eliminating wastes generated by the tissues. Carbon dioxide may be eliminated through the respiratory system, and the digestive tract passes undigested and unabsorbed items from food; with few exceptions, all other unrecycled wastes, metabolic byproducts, and debris drifting in the body fluids must be removed by filtering them from the bloodstream.

Table 17.1 Characteristics of the Human Excretory System

Vertebrate characteristics
   Nephrons develop segmentally but drain via a common duct
Amniote characteristics
   Kidney derived from only a small region of embryonic tissue
   Corpuscle reduced and tubule lengthened to conserve water
   Elaboration of the urinary bladder from the urinary tract
Mammalian characteristics
   Loop of Henle enhances the conservation of water
   Ureter separates from the female vagina; common cloaca lost
Human characteristics
   Urine concentration ability is slightly less than that predicted
     for body size
   Eccrine sweat glands supplement the nephrons in filtration and
     excretion of wastes

The production and elimination of urine operate in effectively the same way in all vertebrates. The most conspicuous differences have arisen in response to different environmental demands and concern for the balance of water and salt. A primary concern for a freshwater animal is the elimination of excess water. Marine animals must rid themselves of excess salt. In contrast, terrestrial vertebrates are in danger of dehydration and must conserve water.

EVOLUTION OF THE VERTEBRATE KIDNEY

The evolution of a kidney and the more efficient elimination of metabolic wastes was an essential step in the origin of vertebrates (Gans 1989; Ruben 1989). Among the ecological shifts in that step were a switch from filter feeding to active predation. Filter feeders can lead a relatively sedentary and stable lifestyle. Wastes are generated and discharged at a constant low level. The faunivorous diet of the earliest vertebrates meant periodic feeding and the intake and processing

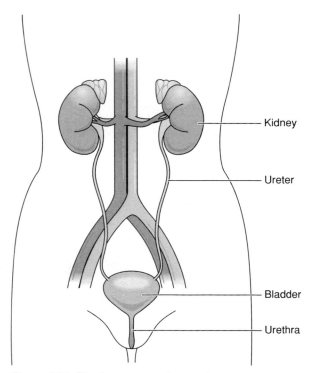

Figure 17.1 The human excretory system.

Kidney

Ureter

Bladder

Urethra

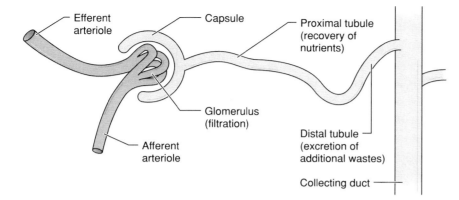

Efferent arteriole

Capsule

Proximal tubule (recovery of nutrients)

Glomerulus (filtration)

Afferent arteriole

Distal tubule (excretion of additional wastes)

Collecting duct

Figure 17.2 The vertebrate nephron. The nephron consists of two parts, a glomerulus, which is usually within a capsule, and a tubule.

of a much greater quantity of protein. Urea, one of the primary waste products in the bloodstream, is a nitrogenous compound generated during the oxidation of amino acids, the building blocks of protein. Active swimming and pursuit of prey require an increase in energy flow in the body, again generating metabolic wastes. Vertebrates often use glycolysis (the inefficient anaerobic initial breakdown of glucose) to provide energy during spurts of activity, which lowers the pH of the tissues. These physological changes imply episodic fluctuations in the metabolic load and require a substantial increase in the peak capacity of excretion. The assembly of a larger and more efficient excretory system was the adaptive solution.

The functional unit of the kidney is the nephron. Millions of nephrons in each individual vertebrate are collected in a single pair of organs, the kidneys. In invertebrates, nephritic structures commonly are arranged in each body segment and excrete independently to the outside. In the embryonic stage of some vertebrates and in the presumed ancestor, nephrons also arise within each segment but are still drained by a common duct. The trend in amniotes is to derive the kidney from a smaller number of embryonic segments so that it appears as a nonsegmental paired organ.

THE VERTEBRATE NEPHRON

The blood plasma consists primarily of water. In addition, it contains large blood proteins; dissolved nutrients circulating to the tissues; electrolytes such as sodium, potassium, calcium, and chlorine that regulate the osmotic pressure across cell membranes; and waste products. Of the wastes, perhaps the most critical are the nitrogen compounds urea and uric acid produced by protein metabolism. The kidneys have the responsibility of determining how much water to retain or eliminate. By retaining large quantities of dissolved electrolytes, the kidneys retain water in the system. If they excrete more dissolved material and the urine is hypertonic, water tends to follow passively. A second task of the kidneys is to sort desirable molecules for retention in the body from waste materials for excretion. This juggling of water, electrolyte, and waste balances is complicated by their interdependence.

The functional unit of the kidney is the nephron (Figure 17.2). The nephron has two parts, a corpuscle and a tubule. Blood is filtrated in the corpuscle, removing most of the water and smaller solutes. Within the tubule, the filtrate is sorted and desirable materials recovered by the body.

Filtration is filtering under pressure. Blood enters the corpuscle via an afferent arteriole. The arteriole divides into a network of smaller capillaries called the glomerulus. As the diameter of the passages decreases, the resistance to blood flow increases and the pressure within the vessels increases. The walls of the capillaries are loosely constructed and relieve the pressure by permitting serum from the blood to escape. Blood cells and proteins are left behind to be carried back into the body via an efferent arteriole. The water and solutes that have left the capillaries are the filtrate.

Filtrate that has been forced from the glomerulus is contained within a membranous capsule. In many embryonic forms of the nephron and perhaps in the ancestral type, that capsule is the peritoneum itself. Filtrate collecting in the coelom is drained by small tubules to the outside of the body. In nephrons of mature vertebrates, the tubule is independent of the coelom. Filtrate is captured in a thin-walled capsule that is continuous with the tubule. Cells lining the tubule are able to interact with the filtrate, recovering useful molecules and secreting additional wastes. As ions are recovered by the tubule, water is recovered as well.

VERTEBRATE STRATEGIES

The probable ancestral form of the vertebrate nephron appears designed for eliminating large quantities of water. This observation has given rise to an unresolved debate suggesting that vertebrates might have evolved in fresh water. The opposing argument, for a marine origin, is better supported by the fossil record. Later vertebrates have adapted their ability to regulate waste elimination to a variety of environments, including fresh water, salt water, and dry land (Table 17.2). Fresh water fish and amphibians face the problem of taking in excessive quantities of water that threaten to dilute their body fluids. Such species possess large glomeruli capable of processing large quantities of blood. Their tubules are functionally short and recover very little of the water. Thus large quantities of dilute urine are excreted.

Marine vertebrates live in a salty environment that is hypertonic—that is, having a higher concentration of dissolved material than do body fluids. One strategy used by bony fish to cope with salt water is to reduce the size of the corpuscle and process only small quantities of urine. This prevents the loss of water. Excess salt and urea are excreted by a separate process from the gills. Sharks retain a large glomerulus but have evolved the ability to tolerate high levels of urea in the body. Their body fluids are isotonic with the sea. The urea displaces salt and the kidneys process large quantities of blood to dump both salt and water.

Tetrapods derived from fresh water ancestors, but on land they face a similar problem of water loss to that of marine fish and use similar solutions. Reptiles, for example, have reduced the size of the corpuscle and process less urine. Marine turtles and sea birds excrete salt from special glands on the face to supplement the action of the kidney. Mammals and birds face an additional problem not experienced by exothermic animals. Their elevated metabolism results in a higher production of nitrogenous wastes. Urine production must be elevated, but water has to be conserved. Both groups have evolved a specialized region of the tubule capable of recovering a large proportion of water. This section is referred to as the loop of Henle.

KIDNEYS OF SIMPLE VERTEBRATES

Nephrons are clustered together to form a discrete organ called the *kidney*. There are several forms of kidneys observed in vertebrate groups and in various stages of development. Developmental stages appear to reflect an evolutionary sequence, as well.

Ancestrally, the vertebrate nephrons arose in each segment of the trunk (Figure 17.3). At each segment, they connected directly and individually into the coelom;

Table 17.2 Vertebrate Strategies for Water and Waste Balance

Taxon	Environment	Strategy
Bony fish, amphibians	Fresh water	Large corpuscle, short tubule; high volume of urine processing and excretion
Cartilaginous fish	Salt water	Large corpuscle; high volume of urine processing and excretion; body tolerates high levels of urea
Marine bony fish	Salt water	Corpuscle small; low volume of urine processing and excretion; additional salt excreted via gills
Reptiles	Terrestrial	Corpuscle small; low volume of urine processing and excretion
Mammals, birds	Terrestrial	Large corpuscles, elongated tubule; high volume of urine processing and concentration; low volume of excretion

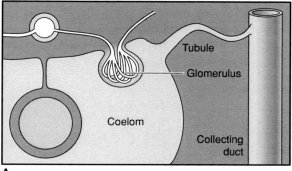

A.

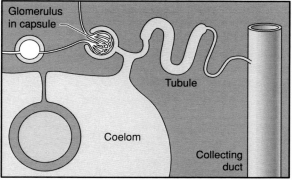

B.

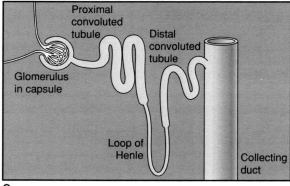

C.

Figure 17.3 Evolution of the vertebrate nephron. The embryonic form of the nephron (*A*) places the glomerulus close to the coelom into which filtrate from the blood may be discharged. The capsule and tubule become more isolated from the coelom (*B*), and eventually the connection is lost in most adult vertebrates. The mammalian nephron (*C*) has a more complex tubule for the concentration of urine and recovery of water.

thus, the kidney did not exist as a discrete organ but as a series of independently functioning glands. In living adult vertebrates, connections between nephrons and the coelom are interrupted by the developmental expansion of the myotome (embryonic trunk musculature) and are replaced by a common passage for urine, the archinephric duct. The result is an idealized kidney called a *holonephros*, having segmented nephrons running the length of the trunk and draining into a cloaca by a duct (Figure 17.4). (A cloaca is a common opening from the body for the urinary and reproductive tracts.) A true holonephros is only observed among living vertebrates in the larvae of hagfishes and one minor groups of amphibians.

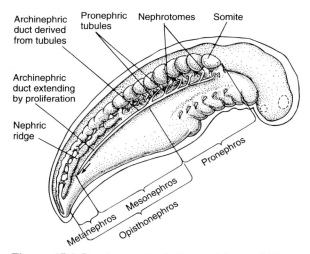

Figure 17.4 Development of the vertebrate kidney. Kidneys evolved and develop from segmented arrangements of nephrons, but different groupings of the segments may function at different times. *A*, A holonephros (pronephros + opistonephros) includes nephrons of all segments functioning at the same time. *B*, The pronephros of the embryo represents only more anterior segments. *C*, The opistonephros, observed in fish and amphibians, consists of the middle and caudal segments. *D*, The middle segments make up the mesonephros, which is the kidney of fetal amniotes. *E*, The metanephros, the most posterior segments, is the adult kidney in amniote vertebrates. (From Vertebrates: Comparative Anatomy, Function, Evolution by Kenneth V. Kardong. Copyright 1998 by the McGraw-Hill Companies. Reproduced with permission of McGraw-Hill Education.)

What is more commonly observed is a separate and sequential development of anterior, middle, and posterior groups of nephrons. The development of nephrons, as with other body tissues, occurs in a distinct anterior to posterior gradient, so that more anterior nephrons begin to function earlier than caudal ones. As the nephrons of the anterior body segments mature, they form a functioning kidney called the *pronephros*. The pronephros exists in the embryo for a limited time and then degenerates to be replaced by a structure formed from more posterior segments. The middle and posterior segments form an opistonephros, which matures in place of the degenerated pronephros. For most fishes and amphibians, this is the functioning adult kidney. Among the amniotes, the middle nephrons form a mesonephros, which is merely the second of three successive developmental versions of the kidney. The final and mature version of the kidney, the metanephros, represents the most caudal portion of the original holonephros. The nephrons themselves have multiplied immensely in number over the original one-per-segment of the ancestral form.

THE MAMMALIAN KIDNEY

The mammalian kidney is a hollow organ. Two layers of tissue, the outer cortex and the inner medulla, wrap around a space containing the renal pelvis (Figure 17.5). The pelvis is a membranous chamber collecting urine from the nephrons. It drains medially into the ureter toward the bladder. The medulla is further subdivided into renal pyramids, which are triangular in a sectional view, and the columns lying between them. The whole organ is surrounded by a tough renal capsule and embedded in a fascial pouch filled with fat. The fat provides mechanical insulation to protect the kidneys.

The corpuscles of the nephrons lie within its cortex. Their tubules run within the cortex and the loops of Henle may descend deep into the pyramids. Mammalian tubules may be divided into several parts, each with its own function. Immediately draining the capsule is the proximal convoluted tubule where nutrients are actively recovered. The loop of Henle descends and returns to become the distal convoluted tubule. The distal tubule empties into a collecting duct.

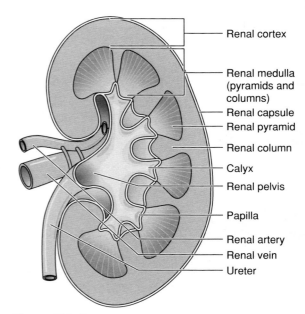

Renal cortex
Renal medulla (pyramids and columns)
Renal capsule
Renal pyramid
Renal column
Calyx
Renal pelvis
Papilla
Renal artery
Renal vein
Ureter

Figure 17.5 Coronal section of a human kidney.

The collecting duct descends once more through the pyramid to its tip and opens into a limb of the pelvis called a calyx. The loop of Henle, the distal tubule, and the collecting duct are all involved in the extraction of water and concentration of the urine.

THE CONCENTRATION OF URINE

Most mammals, including humans, face a danger of dehydration along with a high metabolism that produces a greater quantity of waste materials. It is necessary to dump the wastes on a regular basis while losing as little water as possible. Active recovery of individual water molecules is prohibitively expensive in terms of energy costs. Thus, the body needs a strategy to induce the passive flow of water out of the collecting duct.

The cheapest way to draw water out of the nephron and back into the body is to establish a hypertonic solution around the duct (i.e., a higher concentration of salt outside the duct than inside). Passive diffusion of sodium into the tubule and water out of it will attempt to equalize the concentration. This hypertonic solution is established in the renal pyramids of the medulla. If the walls of the collecting duct passing through the pyramid are made impermeable to sodium, then only water can flow out of the duct into the medulla, where it is taken back into the body.

The challenge is to maintain that very high concentration of sodium against the natural tendency of the sodium to diffuse into neighboring tissues. This is performed by the loop of Henle and a network of vessels called the vasae rectae. Every time urine flows through the loop into the medulla, some sodium is actively pumped out of the ascending tubule into the medulla. More importantly, the loop and the blood vessels set up natural countercurrent exchanges that constantly maintain a high concentration of sodium at the base of the medulla.

A countercurrent exchange occurs between two fluids moving in opposite directions between higher and lower concentrations of a given material (see Figure 16.6). The design of the exchange enables the differential concentration to be maintained against normal diffusion. Around the nephron, the ascending venule of the vasa recta carries sodium away from the deeper part of the pyramid; however, that sodium diffuses out of the venule and into the hypotonic blood of the descending arteriole. In the arteriole it is returned to the pyramid. Similarly, water leaves the arteriole and flows into the venule. The sodium is thus carried back to the base of the medulla in a continual flow and is prevented from escaping, while the osmotic gradient is preserved.

The loop of Henle is slightly more complex (Figure 17.6). The descending limb is permeable to water but not to sodium. Water is drawn out of the tubule by osmotic pressure, and the urine is concentrated. The ascending limb is permeable to sodium but not to water. Sodium leaves (encouraged by active pumping) and water is retained in the tubule. The urine returning to the cortex is much reduced in volume and water content compared to the filtrate in the first part of the nephron. Urine may be concentrated still further as it descends once more through the medulla via the collecting duct. The collecting duct is impermeable to sodium but allows water to escape.

The permeability of the collecting duct to water may be altered (increased) by the action of antidiuretic

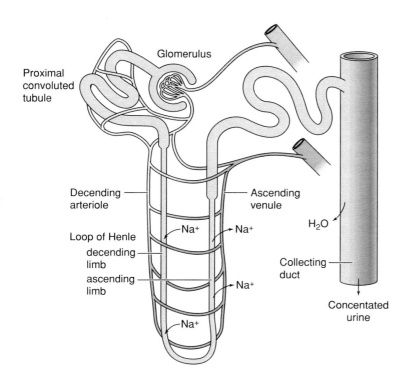

Figure 17.6 Countercurrent exchange in the mammalian nephron. The urine is concentrated at the collecting duct when water is drawn off via osmosis into the hypertonic environment of the pyramid. That hypertonic environment is created by pumping sodium from the loop of Henle. A countercurrent exchange involves the loop of Henle and the vasae rectae (descending arteriole and ascending venule) entering and leaving the pyramids.

hormone. At its most permeable, the collecting duct concentrates urine to nearly the same level as the tissues of the medulla. By adjusting the permeability of the membrane of the collecting duct (i.e., the rate at which water can pass through it), the kidney can control the recovery of water from the urine and thus regulate the overall balance of water in the body. The pituitary's antidiuretic hormone (ADH) has a major role in this regulation. This system not only is elegant but also is the only way we can avoid spending our entire lives running back and forth between a drinking fountain and a latrine.

HUMAN WATER BALANCE

The ability of a mammalian kidney to concentrate urine scales generally with body size (Beuchat 1993). Thus, it is possible to predict how a species of human body size should perform and to compare different species. Human urine is slightly less concentrated than would be predicted. Other animals with low concentrations include those with a moderate to high intake of water, including those spending much time in and around fresh water (e.g., beaver, muskrat, pig, crab-eating macaque). Species with a higher concentration of urine than predicted include many desert and marine animals (e.g., kangaroo rat, camel, whales, and seals). Highly concentrated urine suggests either an adaptation to conserve water, or an adaptation to eliminate excess salt. Humans show neither.

Both in this function and in the use of eccrine sweating for thermoregulation, humans are profligate in elimination of water (Newman 1970). This is not what would be predicted for the savanna environment, where traditional models placed human evolution. It confirms what we can observe of human needs today—that we must have daily access to drinking water. If this level of kidney function was typical of earlier hominins, we can assume reasonably that our ancestors were always tied to a water source, even as they learned to exploit a wider range of habitats. This requirement may have predisposed early hominins toward the modern behavior of orienting to a home base.

18

THE HUMAN STRATEGY

Energy Flow, Homeostasis, and Endurance

In February 2002, American runner Tom Johnson won a 50-mile desert foot race against an Arabian horse. Arabian horses were bred over thousands of years for speed and endurance in the desert. Even the fastest human sprinters could not challenge them in a short race. How can a human defeat a horse? Johnson did not run faster, but he was able to pass up the two 40-minute rest stops the horse made for feeding and watering. For what have humans been bred?

The physiology of our homeostatic systems may answer that question. These systems largely revolve around the flow of energy through the body—its intake and processing and the elimination of waste. The other body systems—musculoskeletal, neuroendocrine, and reproductive—reflect its expenditure. Because they deal with fundamental problems shared by all living organisms, the homeostatic systems differ less among species than do other parts of the body. Significant differences are more likely to correspond to the major

adaptive strategies that define vertebrate classes than to groups within the classes. Humans do show some distinctive features that appear to relate to the quantity of energy processed. In this chapter we review some of those features indicated in the preceding pages and place them in the larger context of a hypothetical human strategy for sustained high energy expenditure (Table 18.1).

LIMITATIONS ON ENERGY PROCESSING

ENDURANCE REQUIRES ENERGY

Our ability to use energy is potentially limited by several factors. The body must be able to acquire and process sufficient nutrients and calories to maintain health, activity, and reproduction. The body must be able to take in and transport sufficient oxygen to sustain metabolism. The body must be able to eliminate the waste products of metabolism, including heat.

On the other hand, few mammals need to sustain high levels of activity for very long at a time. A predator such as a lion stalking a zebra should assess the probability of a successful kill quickly and not waste effort when the chances are poor. Striking from ambush at the shortest range possible, if the lion is unable to overtake the prey within 50 meters or so, it will give up. Therefore neither the lion nor the wildebeest needs to run for more than a couple of minutes at a time. This is the typical strategy of all cats, and most other predatory mammals and reptiles. Wolves are conspicuous exceptions. They may persist and run prey animals to exhaustion or overheating, sometimes taking turns with other members of the pack to prolong the chase. It is therefore not surprising that one mammal most often cited for its endurance, the

Table 18.1 Human Characteristics Promoting Endurance

Adaptations for increased oxygen intake
   Voluntary control over breathing
   Independence of respiratory mechanics from locomotion
Adaptations for increased energy supply
   Increased intake of meat
   Adaptations of gut for rapid processing of high-energy food
   Cultural strategies for increased foraging efficiency
   Omnivorous diet buffering erratic food supply
   Surplus energy efficiently stored in fat
Adaptations for improved thermoregulation
   Thermally sensitive eccrine sweat glands
   Loss of insulating fur
   Increased cutaneous blood flow for dumping heat
   Solar heating reduced by upright posture
   Brain cooled by venous return from facial skin and scalp
Adaptation for increased waste elimination
   Kidney filtration supplemented by eccrine sweat glands

pronghorn, coevolved with wolves. Pronghorns can gallop at high speeds for more than half an hour.

What does sustained running require? Surprisingly, speed is not a major determinant of total energy use—distance is. The metabolic cost of travel per kilometer is roughly constant for a given animal regardless of its speed (Kram and Taylor 1990). An animal running more quickly gains ground at a faster rate and consumes calories and oxygen at a faster rate. The key to endurance is the ability to supply the muscles with calories and oxygen, and secondarily to remove the waste products, including heat, rapidly enough to sustain the effort. Fatigue occurs when the muscles run out of glucose or oxygen or both, or when the body overheats.

The secret of the pronghorn's endurance is a slight superiority in the intake and delivery of oxygen and a higher concentration of mitochondria in which to burn glucose and release energy, allowing it to process energy at a higher rate and thus defer fatigue. The cost is dietary. Although the pronghorn eats grasses of low caloric density, it must still take in enough to support this expensive strategy. Perhaps it lives close to the edge of starvation (Nowak 1992).

ENERGY AND OXYGEN CONSUMPTION

As discussed in Chapter 14, the quality of food tends to vary in inverse proportion to its availability. This problem is offset by the fact that smaller animals maintain a higher metabolic rate and require proportionately more calories for their body size than do large animals. Thus small mammals tend to eat small quantities of highly nutritious foods, such as insects and seeds, while larger mammals are more likely to consume bulkier plant materials, such as fruit or leaves. In order to have endurance and not live on the brink of starvation, a species must have a diet richer than predicted for its size. In this regard, meat is an exceptional food. It may occur in large quantities in a large prey animal. Yet the difficulty and danger of capturing large prey limit the frequency of its availability. The often observed tendency of predator population sizes to depend closely on prey population size indicates that carnivores are more likely to experience limitations from the quantity of their diet than its quality.

Humans appear to use two strategies to circumvent this general caloric limitation. One is our ability to procure and use a wide variety of foods, including both animal and vegetable foods selected for higher nutrition values (Leonard 2000, 2002). A versatile and selective omnivory is the best means to ensure a steady supply of energy. Our second strategy is the ability to store surplus calories as fat (see Chapter 13). Although body fat may have its most critical value in supporting infant brain development, it also allows individuals to take best advantage of temporary windfalls of food. Consumption of large amounts of carbohydrates and fat cause high levels of insulin to be released into the bloodstream. Insulin helps the liver, fat cells, and other tissues to remove glucose and fatty acids from the bloodstream quickly, so that they are placed in storage before they can be oxidized for superfluous energy. The body fat so critical for infants becomes a conspicuous reservoir of calories in adults with an affluent diet.

This ability to cache excess energy within the body is linked to the modern global distribution of adult-onset non–insulin-dependent diabetes. Populations who suddenly have found themselves in an affluent economy with continuous access to a high-carbohydrate grain- or dairy-based diet, coupled with a drop in activity, show a dramatic and deadly surge in the prevalence of diabetes within a generation or two. According to the "thrifty gene" hypothesis, such people are also better able to withstand periods of fasting or starvation by drawing on their stored reserves (Diamond 1992a; Neel et al. 1998; Reitenbaugh and Goodby 1989). In contrast, Western European populations that have existed for many generations on a high-carbohydrate grain- or dairy-based diet respond to high-calorie meals differently by secreting less insulin and permitting more of the nutrients to enter body cells that burn them. Such profligate waste of energy makes one more prone to suffer from food deprivation but also avoids diabetes. Perhaps the resistant genotype has become common in Europeans in the past few centuries because the centralized economy of Western states has buffered the effects of famine and taken away the advantages of the original genotype (Diamond 2003).

The limitations of oxygen consumption are the problems of absorbing enough oxygen at the respiratory membrane. It is necessary for an animal to maintain a reserve capacity to increase respiration above the resting level in times of greater exertion. Although peak oxygen needs may be experienced during such intense activities as fighting, these are generally brief episodes during which an individual can afford to operate with stored oxygen and anaerobic respiration. The most taxing activity is probably running, because of its duration.

For quadrupedal mammals, whose breathing is synchronized with locomotion (see Chapter 15), ventilation is automatically increased with speed by two mechanisms. As the stride rate increases, the breathing rate must also increase. As stride length increases, so does the degree of spinal flexion and therefore tidal volume. Nonetheless, the quadrupeds thus far studied are constrained by their inability to adjust breathing patterns independently of locomotion. This is a liability that may limit endurance.

MANAGING HEAT

Body temperature is important. It determines the rates at which chemical reactions within the body may occur. Animals that are literally cold are more likely to be sluggish and slower reacting. With higher temperatures come faster chemical reactions and more efficient bodily processes. Excessively high temperatures, however, interfere with chemical bonds and may cause the breakdown of some macromolecules.

It is possible for a species to adapt to find optimal efficiency at a particular range of temperature, but this is advantageous only when the species may be ensured of a reasonably constant body temperature. Not surprisingly, strategies for maintaining a stable body temperature (homeothermy) have evolved independently many times. Mammals represent only set of examples, relying for the most part on endothermy, or internally generated heat. Endothermy is expensive. It requires continual combustion of food. Efficiency may be increased and energy conserved through insulation and larger body size.

Chapter 13 discussed the revolutionary approach to the control of body heat represented by human skin. Removal of the insulative layer of fur permits blood vessels to bypass or negate any insulation fat provides to shed excess heat when necessary. Perspiration increases this ability to shed heat. This strategy is superior to that of most other mammals in that it permits us to tolerate the greater head load produced during exercise.

The most sensitive part of the body to overheating is the brain, whose neurons may stop functioning during heat stroke. The enlarged human brain is in even greater jeopardy because of the heat it generates by itself. Mammals require strategies such as the rete mirabile (Chapter 16) to avoid overheating the brain. Humans, as we have seen, use blood cooled on the skin and upper respiratory tract to cool arterial blood arriving at the brain.

Wheeler (1984, 1985, 1991a, 1991b, 1992a, 1992b, 1993, 1994) proposed that our bipedal posture significantly reduces our heat load in a tropical setting. Large quadrupedal mammals expose more of their surface area to the sun and thus require insulating fur to prevent them from passively overheating. Upright hominins, on the other hand, present a minimal profile to the sun, especially at midday. Fur could be shed (except on the head) to expose the skin to cooling breezes. His models also suggest bipedal postures may conserve food and water. Wheeler goes so far as to propose that thermal stress on the savannah was the reason for bipedalism. Current fossil evidence indicates that hominin bipedalism originally evolved in a forest or woodland. However, the modern form of bipedal gait appears to have evolved in early *Homo*, whose fossils are associated with those of savanna animals, and who probably exploited a wide range of habitats, including grasslands. Thus, although Wheeler's model is unlikely to account for the origin of bipedalism, it may have relevance for later hominins.

THE SIGNIFICANCE OF BODY SIZE AND FORM

Body mass is a great determinant of caloric needs. A larger size contains more cells and thus absolutely more metabolic activity to keep them functioning. On the other hand, larger size may result in proportionately fewer calories because some organs (e.g., the heart) are less sensitive to body size and because greater mass conserves heat more efficiently. There have

Table 18.2 Body Size Estimates of Hominin Species

Species	Male (kg)	Female (kg)	Male (cm)	Female (cm)
A. afarensis	45	29	151	105
A. africanus	41	30	138	115
A. robustus	40	32	132	110
A. boisei	49	34	137	124
H. habilis	52	32	157	118
H. erectus	59 (Both sexes)		180	160
H. sapiens	68.2	50.0		

Source: Jungers, 1988; McHenry 1988, 1991, 1992.

been significant changes in body size over hominin evolution, but the reasons for this are not clear.

Australopithecines were much smaller than are modern humans (Table 18.2). The famous partial skeleton of *Australopithecus afarensis*, "Lucy," was tiny, only about 97 cm (3.5 feet) and 27 kg (60 pounds). However, Lucy is one of the smallest adult hominins known and cannot be considered representative. Despite the large size of skulls and teeth of the robust australopithecines, all species are approximately 60% of the body mass of modern humans.

The early species of *Homo* show a definite increase in body height and mass. The few specimens of *Homo ergaster* and *H. erectus* for which size can be estimated suggest individuals equal to or greater than modern people. During the Upper Paleolithic, within the past 30,000 years, body height has decreased (Formicola and Giannecchini 1999), but size changed more abruptly with the origin of agriculture, as the quality of diet and overall health diminished. However, stature in some developed societies has grown dramatically in the present century in response to changing nutritional standards (Harrison et al. 1996).

Skeletal changes of the past million years or so are likely to result at least as much from changes in the environment and behavior as from genetic change. Certainly organic evolution cannot account for the most recent increases that are measurable within a couple of generations. Nonetheless, one may infer from this history that human body size may be quite sensitive to energy availability. The substantial increase in size that occurred in early *Homo*, which almost certainly involve genetic change, may indicate a strategic response to a more efficient diet or foraging pattern.

BODY FORM AND CLIMATE ADAPTATION

Modern human populations differ in more than merely size. There are conspicuous variations in body build, as well. These differences fit into long observed ecological patterns for other mammal species. In general, species or populations living at high latitudes and enduring colder climates tend to increase body mass (Bergman's rule). An increase in mass allows the body to produce and conserve heat more efficiently. Populations living in tropical and warmer climates have longer extremities, increasing surface area and exposure (Allen's rule). Since heat loss or gain from the environment is regulated by the surface area of the skin, a higher ratio of skin to surface area increases the possibility of heat flow. Mammals in cold regions are most concerned with conserving internally generated heat. Those in the tropics are more challenged to shed body heat.

The predictions of these ecological rules are borne out by several exemplary human populations (Ruff 1994), such as the stocky Eskimo and tall, thin Nilotic peoples (Figure 18.1). Although there are some exceptions, major trends in human populations satisfy these predictions. Interestingly, there is some evidence that fossil hominin populations also obey these principles (Ruff 1993). Both Lucy, from Ethiopia, and a *Homo ergaster* skeleton from West Turkana, Kenya, represent individuals from tropical regions. Although differing greatly in absolute size, they have body proportions that yield a similar ratio of body surface area to mass. Lucy is very short, even for an australopithecine. The *H. ergaster* individual is tall, and compensates for

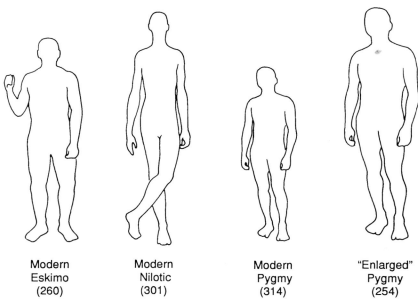

Modern	Modern	Modern	"Enlarged"
Eskimo	Nilotic	Pygmy	Pygmy
(260)	(301)	(314)	(254)

Figure 18.1 Human body form adapts to climate. The proportion of surface area to body mass determines the ease with which body heat is lost or shed. Populations well adapted to very cold climates, as exemplified by the Eskimo on the left, are more likely to be stocky with shorter limbs to conserve heat. The slender build with long extremities in the African figure on the right has a higher surface area and sheds heat more readily—a preferred form for a tropical climate. The numbers represent an approximate ratio of surface area to mass (cm^2/kg). However, absolute body size is also important. A small Pygmy from tropical Africa does not need elongated limbs to maintain an acceptable ratio of surface area to mass, although a tall Pygmy with the same body proportions would have a surface area to mass ratio below that of the Eskimo. (From C.B. Ruff. 1993. Climatic adaptation and hominin evolution: The thermoregulatory imperative. *Evol. Anthropol.* 2[2]:53–60. Copyright John Wiley and Sons, Inc. Reproduced with permission of the author and of John Wiley and Sons, Inc.)

his greater body mass by being slender to increase his surface area (Figure 18.2).

The Neanderthals of glacial Europe are more noteworthy for their robusticity, and this has been interpreted as a cold adaptation (Holliday 1997; Pearson 2000b; Steegmann et al. 2002; Weaver 2003). The relatively short limb bones have a thickened cortex, large joint surfaces, and pronounced muscle attachment that suggest well-developed muscles. Although these muscles probably coincided with a strenuous life, they would also help generate body heat. In contrast, the anatomically modern humans that migrated into Europe to displace the Neanderthals were more lightly built and reflected the body form perhaps of a warmer African climate (Holliday 1999; Holliday and Falsetti 1995).

THE HUMAN STRATEGY

The consequence of the various human mechanisms to sustain aerobic respiration and dump excess heat is that humans have more endurance than do many faster quadrupeds. This comes at a cost, of needing a greater number of calories and a dependence on drinking water. In what evolutionary context was the human strategy adaptive?

An increased tolerance for sustained expenditure of energy offers several adaptive possibilities. Individuals could expand the size of their daily range, covering more ground in a shorter period of time. Shipman (1986) argued that this would be an important and necessary adaptation of scavengers, but there were other reasons for rejecting that model of hominin

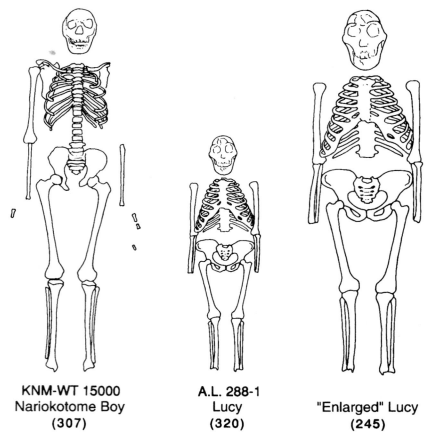

KNM-WT 15000   **A.L. 288-1**
Nariokotome Boy   **Lucy**   **"Enlarged" Lucy**
(307)   **(320)**   **(245)**

Figure 18.2 Fossil hominins from Africa show appropriate body form for a tropical climate. The Nariokotome boy (*Homo ergaster*), an adolescent, would have reached about 185 cm in height. His relatively long limbs and slender build are comparable to those of modern humans from tropical regions. Both have the relatively long slender limbs of modern tropical populations. "Lucy," an *Australopithecus afarensis* female, was a very small individual, about 97 cm. Because of her smaller body mass, she did not need elongated extremities to increase surface area. (From C.B. Ruff. 1993. Climatic adaptation and hominin evolution: The thermoregulatory imperative. *Evol. Anthropol.* 2[2]:53–60. Copyright John Wiley and Sons, Inc. Reproduced with permission of the author and of John Wiley and Sons, Inc.)

adaptation. The adaptations open the possibility for routine long-distance travel for scattered resources, such as suitable materials for stone tools (Leonard and Robertson 1997a, 1997b). Potts (1983) has shown that early stone tool makers often transported materials over many kilometers and cached them for future use.

A third possibility is the conversion of early hominins into effective predators. Superior tolerance for fatigue permits humans to run down and catch faster but less dangerous quadrupedal prey (Carrier

1984). Although the traditional image of "man the hunter" posits well-armed hominins cooperatively combating large mammalian prey, primate models suggest opportunistic exploitation of defenseless small-bodied and juvenile animals that can be trapped or found. "Endurance hunting" offers a reasonable intermediate phase where hominins could incorporate a greater percentage of meat into their diet even if the material technology has not advanced to the level of big-game hunting.

When did this adaptive strategy appear? A major redesign of the body is evident at two points in hominin prehistory (see Chapter 9). The first occurred with the advent of bipedalism at the base of the hominin line. Australopithecines were clearly bipedal and therefore potentially had reorganized respiratory control. The second skeletal reorganization was a more subtle one that appears to signify the transition from *Australopithecus* to *Homo* approximately 2.0 Mya. The differences in postcranial elements between the two enable us to sort skeletal fragments from such sites as East Turkana into two taxa, even when associated cranial remains are missing.

Among the latter changes is the disproportionate elongation of the lower limb. A longer limb increases stride length and reduces the number of steps taken. If we pay metabolically for each step (Alexander and Ker 1990), such changes would yield greater savings in distance travel (Leonard and Robertson 1995, 1997b). However, some studies have produced contrary results for walking. Steudel cites experimental data and Kramer employs a mathematical model to show that shorter limbs need less power and are more efficient with each step, so that limb length does not greatly affect energy expenditure (Kramer 1999; Kramer and Eck 2000; Steudel 1996). Assuming the change to modern limb proportions was adaptive, the advantages most likely represent increased speed over greater distances of travel and the use of running gaits.

There is not yet agreement on the behavioral significance of the last suite of changes to the body. They coincide with the appearance of *Homo* and also with the appearance in the archaeological record of stone tools, of steady increase in brain size (with its caloric needs), and of long-distance migrations outside of Africa. Although the earliest tools, designated the Oldowan Culture, are unspecialized and could have been used for a wide range of tasks, they indicate a sudden broadening of possibilities for subsistence activities. By 2 million years ago, there is increasing evidence for the use of tools in butchering carcasses (see Chapter 14). A increase in dependence on hunting very likely accounts for both a sudden increase in nutrient availability and a selective pressure favoring endurance. It is reasonable to associate the adaptations of soft tissue systems with this phase of evolution.

19

REPRODUCTION

Sexual reproduction is something we often take for granted among complex organisms. The genetic recombination that is a consequence of sex confers obvious advantages of increased genetic diversity and accelerated evolution, although it is not entirely clear

Table 19.1 Characteristics of the Human Reproductive System

Eukaryote characteristics
  Meiosis and sexual reproduction
Animal characteristics
  Heterogamy
Vertebrate characteristics
  Increasing parental investment
Amniote characteristics
  Embryo and fetus protected by an amniotic membrane
  Eggs must be laid on land
Mammalian characteristics
  Viviparity
  Support of the fetus via a placenta (placental mammals only)
  Epitheliochorial placenta
  Uterus and uterine tube
  Nourishment of infants via milk produced by the mother
Primate characteristics
  Simplex uterus (single unpaired structure)
  Testes permanently descended in the scrotum
Human characteristics
  Absence of a baculum
  Enlarged external genitalia (both sexes)
  Copious menstrual flow
  Menopause
  Hemochorial placenta (shared among haplorhine primates)
  Birth canal constricted by reshaping of pelvis for bipedalism
  Close fit between pelvis and fetal head
  Fetal head makes characteristic rotations as it passes the birth canal
  Labor longer and more painful
  Assistance to the mother required during delivery
  Infants in helpless state for first years of life
  Milk thin, high in carbohydrates and low in protein and fat
  Breasts permanently enlarged with adipose deposits

why sex evolved in the first place. Most eukaryotes are not only sexual species, but they produce heterogametes. That is, male and female gametes are structurally and functionally distinct.

In humans, female and male reproductive systems stem from a common anatomical origin. Both sexes begin developing the reproductive tract in the same way, and it is the action of the testosterone during fetal development that redirects the male pattern and prevents it from becoming female. Hormones acting through puberty in both sexes cause further differentiation of the reproductive system; but they also have essential life-long roles in normal body physiology and behavior.

In a Darwinian sense, reproduction is the measure of success for living organisms. How ironic it is, therefore, that reproduction in our species demands repeated compromises in bodily structure and that these compromises represent pain and sometimes life-threatening complications for the mother and the infant. The reproductive responsibility of the mother does not end with birth, however. A dependent infant demands milk; and that involves another curious part of human anatomy, the breast. In this chapter, we will visit some of the anatomical implications of human pregnancy and delivery and address some of the larger questions raised by sexual reproduction itself (Table 19.1).

THE ORIGIN OF SEXUAL REPRODUCTION

Why have sex? Sexual reproduction is expensive. It involves time, energy, and risk. At the end of the process, the individual parent has transmitted only 50% of his or her genes to each offspring. It would appear to be more advantageous for an organism to produce two asexual independently reproducing offspring than to make one male and one female, each with only half the reproductive potential. Furthermore, sexual rearrangement of genetic material may disrupt a well adapted genome. Some significant benefit must outweigh the evolutionary disadvantages. Of the many speculations over this question in the past, most have taken the strategy of judging sex by its consequences, but it may well be that there is not a single answer to the question for all lineages.

What are the consequences of sex? The cells of sexually reproducing organisms are diploid. That is, all chromosomes and genes are present in two copies. The implications of diploidy are far-reaching. The possession of two copies of genes makes it easier for a cell to detect and repair errors occurring during replication (Bernstein et al. 1985). Moreover, it permits mutations to appear alongside the normally functioning genes and cause less disruption. Recessive mutations can hide within the genome for many generations before they are exposed to selection. This greater tolerance for mutation permits greater genetic diversity in the population. Diploidy and meiosis (the production of gametes) permit genetic recombination to occur—a mechanism by which genes are passed on in new combinations, thus further increasing diversity. The same chromosome interactions that allow recombination may also result in gene duplication, another source of diversity. Increased diversity, of course, provides greater opportunity on which selection can act to clean out accumulated deleterious mutations and to promote new advantageous variants (Barton and Charlesworth 1998; Colegrave 2002; Kirkpatrick and Jenkins 1989; Kondrashov 1988; Peters and Otto 2003; Rice 2002; Rice and Chippendale 2001; de Visser et al. 1999). Finally, sexual selection may permit the elimination of more disadvantageous genotypes from a population by stimulating sexual competition (Agrawal 2001; Siller 2001).

Although the accelerated evolution resulting from sexual reproduction enabled life to reach its present diversity, we can not evaluate its benefits on such long-term consequence. How can an increase in the rate of evolution be of benefit to individual lineages? This question has been answered in two ways. Increased diversity of offspring may be adaptive if the environment in which they will compete is heterogeneous, rapidly changing, or unpredictable. In a stable environment, asexual reproduction is to be preferred because it will not disrupt a successful combination of genes. However, when the environment to which the organism must adapt is not stable, the variety produced by sexual reproduction increases the chances that part of the dispersing generation will be successful.

Another aspect of this solution considers the evolutionary "arms race" between parasites, including both predators and disease organisms, and their hosts (Ebert and Hamilton 1996; Hurst and Peck 1996; Ridley 1993). Both parasite and host are under intense selection to outmaneuver their opponents' strategies. As hosts develop structural and molecular defenses, parasites are selected to evade those defenses. This cycle continues indefinitely. The speed with which diversity is generated and with which evolutionary change may occur becomes important and might favor sexual reproduction. Such a model of constant selection and evolution has been called the "Red Queen hypothesis" after the character in Lewis Carroll's *Through the Looking Glass* who exclaimed, "It takes all the running you can do to keep in the same place."

GAMETES AND THE BASIS OF SEXUAL SELECTION

The two gametes in human reproduction are the egg and the sperm. These have different roles and are quite different in form. The egg, or ovum, is larger, engorged with stored nutrients to support developmental stages. It is usually incapable of active locomotion. The ovum retains the capacity for mitosis and for specialization of its progeny into all of the tissues needed in the adult organism. This capacity for specialization into different types of cells is called pluripotency.

In contrast, the sperm is usually smaller and actively able to swim. It is already a specialized cell capable of sensing and seeking the signals given off by the ovum. Organelles not contributing to this function are excluded as the sperm is formed. Thus, the sperm consists of little more than a head, containing genetic material; a small body packed with mitochondria to power the flagellum; and the flagellum, by which it swims. The viability and potential success of the sperm depend greatly on the proper functioning of the flagellum. The head bears a cap of digestive enzymes, the acrosome, that is essential in the fertilization process. Probably as a consequence of this specialization for active locomotion, the sperm has lost the capability for mitosis.

This contrast in the function and potentiality of ovum and sperm help us understand the evolution of heterogamy (Margulis and Sagan 1986). Only one can swim; only one can divide. Motility and dividing are incompatible. The two different needs are resolved in different cells.

The dependence on heterogametes has a long train of consequences in the differentiation of the sexes and their reproductive strategies. On a cell-for-cell basis, ova are more expensive to produce than sperm; thus females are usually the limiting sex in potential reproductive output. A male may manufacture many more sperm for each egg produced. It is to be expected under such conditions that males face more competition among themselves for reproductive opportunity than do females. However, the cost of reproduction, termed "parental investment", must be calculated on a broader basis.

Female investment begins with the energetic cost of producing ova plus variable costs of mating and parenting—protecting the eggs or supporting offspring development before and after the young hatch or are born. Male investment includes both the successful sperm and the millions that failed to fertilize an ovum. To these calculations, one must add all the expense and foregone mating required in the search for a mate, competition with rivals, courtship, copulation, territorial defense, parenting, and other related behaviors. Theory predicts that the total expenses of the two sexes (averaged across the species) will probably equalize. For example, if the direct costs to a male (sperm production, parenting) are low, the indirect costs (e.g., competition) will be that much higher. However, the variance in reproductive success within each sex may remain different. These differences provide the opportunity for sexual selection.

DIMORPHISM AND SECONDARY SEXUAL CHARACTERISTICS

Differences in behavior and reproductive investment between males and females mean that different selective forces are acting on each sex. Not surprisingly, in many species male and female bodies evolve to be conspicuously different, or dimorphic, in ways that do not relate directly to reproductive functions. Such distinguishing traits are called secondary sexual characteristics. Sexual selection may account for many of these differences, but understanding the functional and nonfunctional meanings behind individual traits can be very difficult.

Developmentally, the simplest way to achieve dimorphism is to link the formation of specific traits to the actions of the sex hormones. The sex hormones are families of related chemical signals that are present in both sexes but differentially expressed in them. The primary sex hormones are estrogen in females, and testosterone in men.

Estrogens have wide-ranging effects on the body and brain (Table 19.2). These hormones play essential roles in normal development of the fetus of either sex. Higher levels in a female guide sex determination in the third month of gestation and influence secondary sexual characteristics—throughout life. At puberty there is a dramatic increase in the production of these

Table 19.2 Female Secondary Sex Characteristics

Growth and maturation of the ovaries and accessory reproductive structures at puberty
Growth of the breasts
Increase in osteoblastic activity, including an acceleration of ossification at puberty
Overall increase in quantity of subcutaneous fat
Patterning of subcutaneous fat deposition favoring the hips, thighs, buttocks, and breasts
Skin made relatively soft, smooth, and highly vascular
Increased sodium and water retention (follows ovulatory cycle)
Promotion of reproductive and mothering behaviors (including copulatory positions and nest-building by laboratory animals)
Increased sensitivity to pheromones (including perfumes)
Relative lack of body hair
Wider hips (increased growth of both bone and fat deposits)
Relatively high-pitched (child-like) voice

These secondary sex characteristics represent normal responses of various body tissues to estrogens and the lack of testosterone.

reproductive hormones. The elevated levels of estrogens lead to the immediate growth and maturation of female reproductive organs, beginning with the ovary itself and including the uterus, external genitalia, and the breasts. Numerous secondary effects include the development of secondary sexual characteristics. Estrogens increase osteoblastic activity and thus accelerate ossification of the bones. They regulate the deposition of subcutaneous fat, leading to an overall increase in its quantity relative to males and to preferential placement in the hips, thighs, and breasts. Sex-specific changes in bone growth and fat deposition lead to the characteristic female figure with broader hips and a relatively narrow waist.

Estrogen secretion rises and falls with each ovulatory cycle to manipulate the ovary and uterus, but these changes have other effects on the body. They include such diverse symptoms as increased water retention, altered sensitivity to smells, and release of nutrients to fuel reproduction (Wade and Schneider 1992). There are a number of different estrogen

Table 19.3 Male Secondary Sex Characteristics

Sexual differentiation in the fetus
Descent of the testis at about 8 months' gestation
Growth and maturation of the gonads and accessory sex organs at puberty
Overall growth of the body
Protein synthesis and muscle development
Ossification of long bones and growth of cortical bone
Narrowing of the pelvis to support increased weight
Increase of facial and body hair (Male pattern baldness may result from an interaction of testosterone with other genes.)
Growth of the larynx and deepening of the voice
Thickening of the skin
Increase in the activity of sebaceous glands (This increase in oily secretions on the skin causes commonly causes acne following puberty. The glands later become less sensitive to testosterone.)
Increase in red blood cell production
Initiation and maintenance of spermatogenesis
Stimulation of male reproductive behavior (In other mammal and bird species, testosterone has been specifically linked to sexual arousal, courtship behavior, mounting behavior, birdsong, aggressiveness, and the production of pheromones.)
Altered sensitivity to specific odors; generally a decreased sensitivity relative to women
Inhibition of the release of GnRH in hypothalamus

These secondary sex characteristics represent normal responses of various body tissues to testosterone.

receptors in the brain. Estrogen can act directly or indirectly, by modulating sensitivity to other neurotransmitters. Behavioral effects include, but are not limited to, increased sexual activity (Hedricks et al. 1994b; Udry and Morris 1968).

Testosterone has a similar diversity of effects on the male body. Its production also surges at puberty, and it directs the growth and maturation of the reproductive organs and the appearance of secondary sex characteristics. The latter include an increase in skeletal and muscle mass, broadening of the shoulders, thickening of the skin, growth of facial and body hair, and growth of the larynx and deepening of the voice (Table 19.3).

Testosterone is a necessary determinant of normal male sexual development and function, both physiological and behavioral. Testosterone receptors in the brain mediate its effects in stimulating the sex drive and reproductive behavior and also play a role in enabling aggression. Testosterone plus two other hormones, follicle-stimulating hormone from the pituitary gland and epidermal growth factor, are necessary for spermatogenesis to occur.

VERTEBRATE REPRODUCTIVE STRATEGIES

For all living organisms, successful reproduction is measured by how many offspring survive to adulthood. It is not sufficient merely to produce many children. In fact, there is a rough inverse relationship between the numbers of offspring produced and the number that survive (see Chapter 21). Some animals, such as houseflies and clams, may produce literally millions of eggs, but few of those live to reproduce. Most birds and mammals, on the other hand, are parents to a small number, of whom a higher percentage survive.

Animal strategies are also limited by the complexity of their developmental pattern. If it takes a long period of development to produce an independently functioning offspring, the parent must provide sufficient energy to support that period of maturation and growth. Storing energy in an egg makes the egg bulkier and more expensive and thus places a constraint on the number a female can produce. The eggs of amniote vertebrates—reptiles, birds, and mammals, which can reproduce on dry land—have the additional burden of supplying the embryo's water. Amniote eggs have protective membranes and shells designed to enclose and conserve water while permitting respiration.

Homeothermic vertebrates have the added burden of keeping developing embryos within a certain temperature range. Burying eggs in the sand under the sun is a common reptilian strategy, but such temperature regulation is not reliable. Birds brood over their eggs, using body heat to warm them. A better solution, evolved many times among sharks, bony fish, amphibians, reptiles, and mammals, is for the mother to retain the eggs within her body until they are old enough to hatch. This behavior is termed *viviparity* (bearing live young), in contrast to *oviparity* (egg-laying).

MAMMALIAN STRATEGIES

The three subclasses of living mammals are distinguished by differences in their reproductive strategies. The prototherian mammals (monotremes) lay eggs. Metatherians (marsupials) retain the egg within the uterus and permit it to draw nourishment directly from the mother for an entire ovulatory cycle. At the end of that cycle, the fetus would be flushed out of the uterus with the endometrium in preparation for the next ovulation. To survive, it must emerge from the birth canal and find its way to the mother's nipple. Its development is completed over the next several months using the nourishment of her milk. The helpless young often lies behind the protection of a fold of skin, or pouch.

Eutherians (placental mammals) follow a different strategy of maintaining the endometrium through an extended gestation period. Some trigger causes the pituitary to continue secreting gonadotropins to sustain the corpus luteum. This stimulus may be the tactile stimulation of copulation in a rat or chorionic gonadotropin secreted by the embryo in a human. The result is continued production of progesterone from the ovary to maintain the uterine lining. The placenta is the site of continued attachment to the uterus, where nutrients and gases continue to be supplied to the fetus. The placenta also assumes the role of secreting progesterone as long as the pregnancy lasts. Placenta-

like structures also evolved independently several times in other vertebrates, including sharks and reptiles.

Menstruation

The endometrium of the mammalian uterus is built up and shed with the ovulatory cycle. When implantation of a fertilized egg is possible, the endometrium is thick and rich with blood and nutrients. If there is no implantation, the excess tissue is shed as menses. Menstruation, the periodic flow of blood and remnants of the endometrium, is nutritionally costly. Accompanied as it often is by discomfort, behavioral shifts, and a drop in body defenses, menstruation generally has a negative and even feared connotation. Menstrual blood and menstruating women have been considered taboo in many cultures. It is relevant to ask why menstruation occurs at all.

Human menstrual flow is more conspicuous and copious than that of other primates, although this appears to be the culmination of a trend specific to Old World monkeys and apes (Strassman 1996b). The quantity of flow apparently is proportionate to the thickness of the endometrium, which in turn reflects the potential development of the placental nourishment of a fetus. The condition in monkeys and apes in particular and primates in general may relate the greater support provided by the uterus and placenta in these taxa. Such support would be particularly necessary for relatively larger fetal brains.

A few hypotheses have argued that menstruation is an end in itself—either protection against invading pathogens (Profet 1993) or a signal of ovulatory cycling and thus fertility. It is more likely that menstruation is secondary to the ovulatory cycle. Strassman argues that endometrial tissue is so expensive to maintain that there is a significant energy savings to be gained by shedding it between periods of potential impregnation (Strassman 1996a, 1996b). In contrast, Finn argues that menstruation should not be seen as an independent physiological process but as a secondary consequence of reproductive cycles (Finn 1986, 1987, 1998; also Salamonsen 1998). The development of the endometrium and anticipation of contact with a foreign body (the embryo) induces inflammatory defensive reactions that are held in check by progesterone. When progesterone levels drop, the inflammation follows its normal course and rejects or discards the endometrium.

Women in Western society, where fertility is artificially regulated through contraception, are used to monthly periods. Cessation of cycling (amenorrhoea) before menopause occurs only during and immediately following pregnancy or in a pathological state, such as starvation or extreme levels of exercise. In nonindustrial societies and presumably in our evolutionary past, menstruation was probably relatively rare. A woman past puberty spent most of her reproductive life either pregnant or nursing extensively (May 1978; Short 1976). Menstruation appeared in adolescence and for brief intervals between births —perhaps not more than 100 times in a life span, one quarter of that experienced by normal American women (Strassman 1997). Under such conditions, it is more understandable that the phenomenon was misinterpreted as unusual and perhaps dangerous. Ironically, the Western pattern of years of menstruation may indeed by dangerous, as it correlates with higher levels of disorders such as anemia, endometriosis, and cancer.

The Placenta

The *placenta* refers to the persistent interface between the fetal membranes and the endometrium of the mother (Figure 19.1). Each of these structures contributes a thin layer of tissue, across which oxygen and nutrients must diffuse. As critical as the placenta is for mammalian evolution, the placenta of different mammals may assume different forms in details of its development and structure, including the degree to which it invades the uterine tissue (King 1993; Luckett 1975). The ancestral primate, including modern prosimians, possessed an epitheliochorial placenta, in which fetal and maternal bloodstreams remain separated by tissue layers from each individual. Loss of the epithelium of the uterus permits the walls of the capillaries to come into direct contact with the chorion. This is probably the ancestral condition for all mammals.

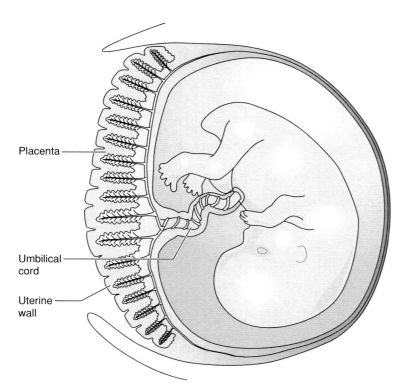

Placenta

Umbilical cord

Uterine wall

Figure 19.1 The placenta. The placenta develops along with the embryo from the fertilized egg. Its processes penetrate the maternal blood supply of the uterine lining.

Haplorhine primates, including tarsiers and all anthropoids, possess a hemochorial placenta. A hemochorial placenta is more invasive of the endometrium. The mother's tissues break down so that her blood is in direct contact with the chorion. The advantage of this arrangement is a more efficient transfer of oxygen and nutrients to the fetus. Among catarrhines, this is enhanced further by increasing invasiveness and complexity of the placental surface. Villi, processes of the chorionic membrane greatly increase the surface area of contact. At birth the human placenta has a diameter of only 15 to 20 cm (6 to 8 in.) but it has a functional surface area of 14 m^2 (2000 ft^2).

The placenta aggressively takes up nutrients from the mother's blood (Pecorari 2002). Glucose, the primary fuel for development and the material from which infant fat is synthesized, diffuses from mother to fetus. Some of the more complex nutrients cross by active transport. At the same time, the mother's body increases the levels of nutrients in her blood. A temporary resistance to insulin permits carbohydrates to circulate for a longer time, while lipids are broken down at a higher rate both to provides her own body with energy and to provide fatty acids to the fetus. The ratio of lipids to proteins circulating in the mother's blood is greater in humans than in other species, consistent with the unique accumulation of fat in human fetus (Père 2003). A special form of hemoglobin is synthesized by the fetus that has a greater affinity for oxygen than adult hemoglobin. Thus, oxygen preferentially crosses the placental barrier from mother to fetus.

In addition to nutrient transfer, the placenta serves other critical functions, including synthesis of macromolecules for the fetus and the secretion of a number of important hormones that act on both the fetus and the mother. These help to sustain the pregnancy and prepare the mother for lactation following delivery. Placental estrogen and progesterone supplement the mother's own hormones. Estrogen stimulates the further growth of the maternal uterus, external genitalia, and breasts. Progesterone maintains the endometrium and suppresses the contractions of labor. There is evidence from nonhuman mammals that the hormones also affect the mother's brain, triggering maternal behavior.

LABOR AND DELIVERY

Delivery for most mammals appears effortless. The mother seeks out a secluded place and bears her offspring quietly and with little apparent effort. For modern humans, delivery is a time of pain and risk. The great muscular effort a woman must put forth earns this event its sobriquet "labor." The difficulty, of course, lies with the large size of the infant brain and its tight fit through the birth canal.

Labor: A Positive Feedback Mechanism

The exact trigger for the onset of labor is not known, but it appears to vary in different mammalian species. The cause of labor in human mothers is best understood at the convergence of several physiological states contributed by both the mother and the fetus (Ellison 2001; Nathanielsz 1996; Smith 1999).

The uterus actively contracts to expel the fetus, but these contractions are regulated through hormonal signals and reflex loops rather than the central nervous system. During gestation, the hormones progesterone and relaxin suppress uterine contractions. Oxytocin and prostaglandins released by the placenta initiate contractions. During the course of a pregnancy, the uterus is kept quiet by a delicate balance of these hormones.

Several changes occurring toward the end of gestation upset that balance. Steroids, prostaglandins, and oxytocin contribute to ripening of the uterine cervix and increase its sensitivity to stretching. Stretching will trigger reflexive contraction. Placental progesterone and estrogen cause an increase in the number of oxytocin receptors in the uterus, thus increasing its responsiveness to the hormone. Estrogens facilitate the formation of gap junctions, or intercellular communications, within the muscular wall of the uterus, promoting a synchronous contraction of the tissue.

The growing fetus causes increasing stretching and pressure on the uterus, stimulating reflexive contractions. Corticotropin-releasing hormone manufactured by the placenta and androgens and cortisol produced by the fetus act to alter the balance between estrogen and progesterone in the mother. Additional oxytocin is released by the fetus and by the uterus itself. All of these factors lead to an onset of uterine contractions, amplified through positive feedback—increased pressure on the cervix leads to an increase in oxytocin secretion. Tentative and isolated uterine contractions (Braxton-Hicks contractions) may occur weeks before delivery. They increase in strength and frequency over time, but true labor is not defined until contractions are sustained and regular. Cortisol, a normal stress hormone, also has a critical function preparing the infant's organs for life outside the womb (Lagerkrantz 1996; Lagerkrantz and Slotkin 1986).

It has been argued that the human fetus is ultimately in control of pregnancy, from the first release of chorionic gonadotropin to delivery. Hormones of the chorion and placenta manipulate the mother's body to maintain the womb, to provide nutrients to the fetus, and to prepare the mother for nursing and maternal care. Further evidence suggests that the placenta manipulates the mother's immune system to prevent immune rejection. At the end of gestation, the fetus probably initiates labor. It may be argued that the human fetus is ultimately in control of pregnancy. Although literature abounds with metaphors and explicit discussions of antagonism between fetus and mother, clearly a successful pregnancy depends on appropriate communication and collaboration. Natural selection has acted on both the fetal and maternal stages of life to bring this about.

The Anatomy of Delivery

Delivery of an infant involves its passage through the pelvis and perineum. The lower part of the pelvis defines the birth canal and presents the greatest constriction on the head of the fetus. The soft tissues of the perineum offer less resistance but must stretch considerably. The entrance of the birth canal is the pelvic brim, a ridge in the coxal bone extending from near the pubic symphysis to the sacrum (Figure 19.2). Above the brim, the iliac blade flares to the side forming the "false pelvis," which presents no obstacle. Below the brim, the pubis and ischium narrow only slightly to create the "true pelvis." The primary constrictions are caused by the ischial spines and the coccyx projecting into the canal.

The birth canal is closed inferiorly by a layer of muscle called the pelvic diaphragm. The diaphragm is composed of two paired muscles, levator ani and

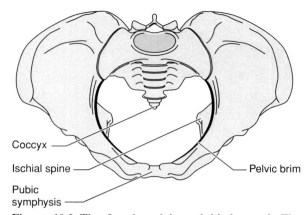

Coccyx — Ischial spine — Pubic symphysis — Pelvic brim

Figure 19.2 The female pelvis and birth canal. The entrance to the birth canal is marked by the pelvic brim. The coccyx and ischial spines and attached ligaments protrude into the canal and limit the size of its passage.

coccygeus, whose fibers arise from the pubis and ischial spine and converge on the midline. Only the urethra, vagina, and rectum pass through the diaphragm. Inferior to the diaphragm is the perineum, a space containing several layers of connective tissues and small muscles. Embedded in the perineum are the external genital organ and their erectile tissues plus the nerves and blood vessels that supply them.

Muscles of the perineum arise mostly from the ischial tuberosities and coccyx. When these landmarks are close together, the soft tissues are shorter and the perineum and diaphragm better supported (Abitbol 1988). Moving the bony processes outward increases the size of the birth canal, but weakens the muscular floor of the pelvis. The design of the female perineum is thus a compromise between these competing demands. Males, who need not be concerned with birth passages, have a much more restricted pelvic outlet and a correspondingly stronger perineum.

The head of a fetus descending the birth canal must negotiate a passage smaller than itself. Several mechanisms make this possible. First, the pelvic joints loosen and open slightly. The hormone relaxin produced by the corpus luteum in the ovary causes the collagen fibers of the sacroiliac joints and pubic symphysis to lengthen slightly. At the risk of some instability to these joints, the maternal pelvis increases its maximum diameter. Second, the head of the fetus deforms from the pressure. Since the membranous bones of the skull are not fully ossified, they are able to move together along the sutures and at the fontanelles. The fontanelles are generally gone in full-term fetuses of other primates. Even with these compromises, passage through the birth canal requires several partial rotations of the fetus to work first the head, then the shoulders past the constrictions (Trevathan 1987; Rosenberg and Trevathan 2001).

Aside from pain, another great consequence of a difficult delivery is the need for assistance. Human birth occurs in a variety of cultural settings, a variety of positions for the mother, and with or without medical intervention. It rarely occurs alone. The modern condition of delivering large-brained infants could only have evolved in a context of social support (Trevathan 1987; Rosenberg and Trevathan 2001).

THE EVOLUTION OF BIRTH

For most primates, the passage of the fetus through the birth canal is relatively easy. Neonatal brain size is closely correlated with the mother's body size, with larger species having relatively smaller heads and easier deliveries (Leutenegger 1982; Lindburg 1982). Two species have difficult deliveries. Squirrel monkeys have a tight fit between the infant head and the mother's pelvis. Probably squirrel monkeys are a "dwarf species," one that has substantially reduced its body since its ancestor. Fetal brain growth has been more conservative and thus is oversized for the birth canal.

Humans are the other exception, with difficulties caused by brain expansion rather than body reduction. The shape of the modern human pelvis clearly represents a compromise between its obstetric obligations and the requirements of support and locomotion. The fossil record presents an interesting glimpse into the process by which this compromise was achieved.

Two reasonably complete, though distorted, female pelves have been recovered and described from australopithecines. The first, from Sterkfontein (Sts 14), is assigned to *Australopithecus africanus*. The second belongs to the famous "Lucy" skeleton of *A. afarensis* (AL 288-1). Although reconstructions of these specimens vary, both define a birth canal that is very wide from side to side but narrow from front to back (Figure 19.3). This pattern has been interpreted as driven by locomotor, rather than obstetric, demands (Abitbol 1987b, 1991, 1995; Berge et al. 1984; Leutenegger 1987; Rak 1991). Bipedalism requires shifting the position of the sacroiliac joint as close to the acetabulum as possible. Widening the pelvis laterally gives more support to the abductors of the hip. It is difficult

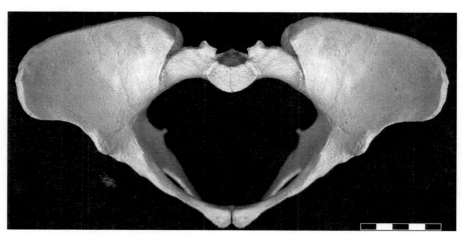

Figure 19.3 Reconstruction of a female australopithecines pelvis (Al-288-1 "Lucy"). The birth canal is relatively wide but short from front to back compared with modern humans. Nonetheless, it could comfortably accommodate an ape-sized fetus. (Photo courtesy of Martin Häusler and Peter Schmidt.)

to appreciate any difficulty this arrangement might have caused in delivery for australopithecines since the neonate head size is unknown.

At what point did the fit between birth canal and fetal size become so tight as to require fetal rotations during delivery? Some scholars argue they were necessary for australopithecines (e.g., Berge et al. 1984); others disagree, according to how they reconstruct the pelvis and the size of early hominin infants (e.g., Abitbol 1996; Leutenegger 1972, 1974, 1987; Rosenberg 1992). Tague and Lovejoy (1986) suggested that australopithecine birth could be accomplished relatively easily, but with the fetus facing sideways the entire time, a unique orientation among mammals. However, the shoulders may have passed through more easily if the fetus faced either forward or backward, suggesting a possible rotation (Rosenberg and Trevathan 1996; Trevathan and Rosenberg 2000).

With the appearance of *Homo*, the brain increased substantially and the pelvis began to assume its modern shape. Certainly by early *H. neanderthalensis* and early *H. sapiens* problematic births would have been encountered and addressed by still more compromise. Gestation is cut short and the baby is born earlier in brain development (Berge et al. 1984; Leutenegger 1982; Montagu 1961; but see Lazar 1986 and Lindburg 1982). The downside is that the infant spends its first years in a quiet helpless and dependent state. If altriciality, the condition in which much of the brain growth is postponed until after birth, appeared early, *H. erectus* and similar medium-brained hominins might have had a relatively easy birthing process (Rosenberg 1992; Ruff 1995). Otherwise, large-brained infants could have caused difficult labors very early.

The abbreviation of gestation has other implications, and the size of the fetal head may not be the only consideration. The growing fetal brain has a tremendous appetite for nutrients that must be satisfied by the mother. At a certain point, the placenta will reach the limits of its ability to transfer nutrients. After that the mother and infant must find a new arrangement if growth is to continue. Ellison (2001) has suggested that birth occurs at this time so that the pair can switch from placental supply to milk and nursing.

LACTATION AND THE BREAST

Lactation and suckling are perhaps the only behaviors found in all mammals and are definitive of the order. (Linnaeus, who created the name Mammalia, was a supporter of women breast-feeding their own children instead of hiring nursemaids; hence his choice of nipples, rather than hair or teeth to define the order.) However, there is a great variability among species in the anatomy of the mammary glands, the composition of milk, and specific feeding behaviors.

The mammary gland is actually a cluster of cutaneous exocrine glands derived from sweat glands. Primitively, these glands probably secreted a liquid into the mother's fur that was then licked up by the infants. Selection favored the inclusion of nutrients in the secretion. By collecting the ducts into a series of discreet nipples, feeding became more efficient. Alveoli deep to the skin store milk temporarily before it is released into the ducts.

The development of the mammary gland is under the influence of many hormones, especially estrogen. Full maturation occurs during the latter part of pregnancy, and the actual production of milk is stimulated by prolactin from the pituitary gland.

THE COMPOSITION OF MILK

For the first three days following birth, a human mother does not produce milk. Instead, she releases small quantities of a thin fluid called colostrum. At one time, medical opinion considered this of no value. Now colostrum is known to be very rich in nutrients, especially antibodies and other proteins and essential fatty acids. During its first days, an infant subsists primarily on its own nutrient reserves, though the weight loss typical in the first week following birth represents shedding of excess fluid from the body. By the time the infant needs nourishment to continue growth, the mother should be able to produce milk on demand.

Human milk is thinner than that of many other mammals. It is relatively rich in free amino acids and long-chain fatty acids, but it is also less dense in calories, proteins, and lipids (Hartmann 1988). The low level of the protein casein, which causes dairy milk to curdle and make cheese or yogurt, allows a human infant stomach to digest quickly. Human milk is adapted for frequent bouts of nursing, day and night. Apparently our species' evolutionary history presumed that the mother and infant would be together continually. The milk is formulated to supply all of the infant's nutritional needs as well as supplementary digestive enzymes, antibodies and other immune factors, hormones, and endorphins.

Human milk contains the highest concentration of carbohydrates of any mammal. Most of this is lactose and is needed to support the large and still-growing brain. Surplus carbohydrate is converted into fat, while fatty acids in milk are often metabolized immediately. Human milk contains among the lowest concentrations of protein and fat of any mammal. Their limited quantities permit the infant to digest milk quickly. Total fat content of human milk is similar to that of ruminants, about 4%, compared with a range from 0% reported in rhinoceros to as much as 50% in pinnipeds and whales (Neville and Picciano 1997). These figures represent species adaptations, including the frequency and intensity of infant suckling.

Human milk composition and value are generally constant, despite the mother's own food intake (Lammi-Keefe and Jensen 1984). The total fat content of human milk varies little. Fasting or severe malnourishment of the mother may cause an increase in fat concentration, presumably to conserve her own glucose stores. Unlike humans, goats and rats showed a marked reduction of milk production with fasting, as much as 60% in 24 hours (Neville and Picciano 1997). Some other species (bears, seals, baleen whales) are well adapted to fasting and do not feed during lactation. Their milk levels are maintained and draw heavily from stored body fat. The human pattern is suited to the continual energy need of the infant to support its growing brain.

BREASTS AND THE FEMALE FIGURE

Human breasts are unusual because of the placement of significant amounts of adipose tissue around the glandular tissue. This fat creates the conspicuous shape of the breast. Although body fat is the source of much of the energy used to produce milk, it is not apparent why humans have any more reason to place the fat next to the glands than do other species. In light of the erotic visual value of breasts to males in many societies, it has generally been argued that their form is the product of sexual selection for the attraction and retention of male attention. Among the wide range of adaptive models that has been offered, there are at least two problems involved: Why are human breasts enlarged and why are breasts erotic? The prevalent model for sexual selection argues that breasts are enlarged because they are erotic. The evidence for this is weak.

BREASTS AS EROTIC ATTRACTORS

One of the first accounts that recognized just how unusual human breasts are was advanced by Morris in *The Naked Ape* (1967). He explained the combination of breasts and pigmented or large fleshy nipples and lips as visual mimics of the sexual skin and labia of our catarrhine ancestors. As such, they attract the same sexual attention from males as did those genuine indicators of ovulation. Although his hypothesis has not been adopted by later scholars, it set a precedent for later arguments in assuming that breasts represent otherwise nonfunctioning sexual attractors.

Why are human breasts commonly (but not universally) a focus of erotic arousal and behavior? There are problems with the adaptationist models and reasons why the breast might be expected to become an incidental focus of sexuality even if it is not the product of sexual selection. The tactile receptors around the areolae are highly sensitive and give pleasurable feedback for reasons best explained by the function of the breast in nursing. Sensory feedback from suckling contributes to several critical functions: the release of prolactin to sustain milk production and oxytocin to release milk, relaxation of the mother to facilitate nursing, and reinforcement of the relationship between the mother and infant (Eibl-Eibesfeldt 1989; Montagu 1971; Trevathan 1987). Given the important roles of play, tactile exploration, and manual stimulation in human foreplay, the female nipple, in particular, is a likely focus of attention during sex. Female response serves as positive feedback to the male. Eibl-Eibesfeldt (1989) suggests the infant bonding mechanism triggered by nipple stimulation and uterine contraction promotes love and sexual bonding when triggered by foreplay and orgasm. This argument should also apply to other species. Indeed, Carpenter (cited by Ford and Beach 1951) observed female macaques in estrus fondling and sucking on their own nipples and interpreted this as autoerotic behavior.

For these reasons, the possibility must be considered that the erotic value of breasts is secondary to other reasons for their prominence. In fact, the evolutionary increase in human breast size may have enhanced its erotic value, the reverse of the usual sexual selection argument. Some authors have noted that not all cultures consider bare breasts to be visually erotic (e.g., Caro 1987; Caro and Sellen 1990; Eibl-Eibesfeldt 1989; Ford and Beach 1951), although this does not exclude a role for the breast in erotic play in those same cultures or in the assessment of physical beauty and attraction. The importance it assumes in Western male sexuality certainly colors attempts by Western anthropologists to understand its evolution.

To argue that sexual selection acts for the enlargement of the breasts, it is necessary to postulate a correlation between breast size and mating or reproductive success. Despite common wisdom concerning a woman's attractiveness in this society, such a correlation has not been documented. Furthermore, the form of the breast is highly variable between and within populations and varies additionally according to age and nutritional and reproductive status. The semispherical shape commonly invoked as most attractive is typical only of some younger women in mostly European populations. Just as a "normal shape" defies definition, it also eludes any single explanation.

Do young adult human females have conspicuous fatty breasts because breasts are sexy? Earlobes are also highly sensitive and receive sexual attention. Although they are not anatomically sexually dimorphic, males and females decorate them differently in many cultures. Morris included earlobes in his list of traits that evolved to encourage sexual intimacy, although it appears that no scholar has pursued his idea seriously. Hair and slender ankles are considered attractive in some cultures and women are required to keep them covered. While the unique anatomical form of the human breasts, earlobes, and long hair may or may not have physiological functions, their specifically erotic roles are probably secondary attributes.

BREASTS AS ADVERTISEMENTS

A second line of argument asserts that breasts are honest advertisements of the fat stores that they contain (Anderson 1987; Barber 1995; Burd 1986; Cant 1981; Gallup 1982). A certain level of energy reserve in the form of body fat is necessary for ovulation and for successful reproduction. Males should select mates according to such an indicator of fertility. Fat, up to a point, therefore is healthy and potentially worth displaying. However, fat is only one of several determining factors of female reproductive potential. Selection for fat by itself is prone to runaway selection that would defeat its purpose. The healthy range of variance of stored fat in which reproduction is not affected is great enough to include nearly all women who are neither starving nor pathologically obese (Huss-Ashmore 1990).

One important role of fat is to emphasize the hourglass figure so celebrated in art and fashion across cultures. In analyzing the attractiveness of this form, Singh (Singh 1993a, 1993b; Singh and Young 1995; also Streeter and McBurney 2003) has proposed that the waist-hip ratio is the most important factor (Figure 19.4). This hypothesis is support by stated preferences among diverse ethnic background in many, but not all cultures (Wetsman and Marlowe 1999). It is both an honest indicator of skeletal (reproductive) maturity and of a healthy amount of fat. It also provides ready gender recognition by exaggerating the already wider pelvis and hip dimensions. Prominent breasts enhance the pattern. An emphasis on attraction to certain ratios of measurements, rather than absolute size, would limited runaway sexual selection and tolerate cultural and individual variation in the erotic interest of specific features. It would also explain the preference for an "ideal" female figure that is more specific than the range allowed by physiological fitness alone.

A related idea suggests breasts are attractive as honest advertisers of reproductive maturity and potential (Marlowe 1997). However, breast size does not correlate significantly with milk production except

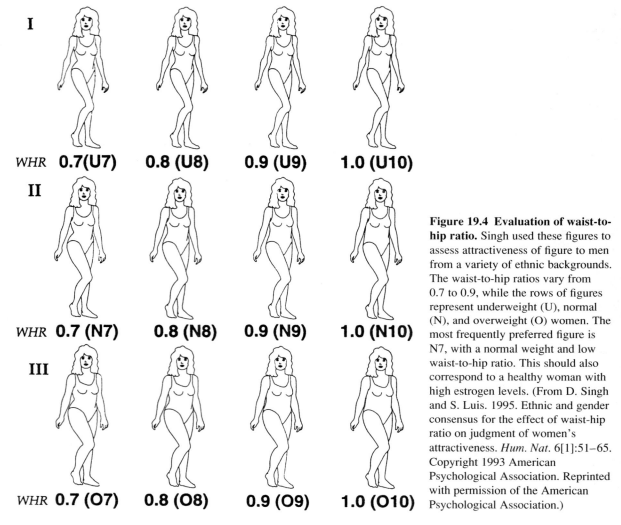

I

| WHR | 0.7(U7) | 0.8 (U8) | 0.9 (U9) | 1.0 (U10) |

II

| WHR | 0.7 (N7) | 0.8 (N8) | 0.9 (N9) | 1.0 (N10) |

III

| WHR | 0.7 (O7) | 0.8 (O8) | 0.9 (O9) | 1.0 (O10) |

Figure 19.4 Evaluation of waist-to-hip ratio. Singh used these figures to assess attractiveness of figure to men from a variety of ethnic backgrounds. The waist-to-hip ratios vary from 0.7 to 0.9, while the rows of figures represent underweight (U), normal (N), and overweight (O) women. The most frequently preferred figure is N7, with a normal weight and low waist-to-hip ratio. This should also correspond to a healthy woman with high estrogen levels. (From D. Singh and S. Luis. 1995. Ethnic and gender consensus for the effect of waist-hip ratio on judgment of women's attractiveness. *Hum. Nat.* 6[1]:51–65. Copyright 1993 American Psychological Association. Reprinted with permission of the American Psychological Association.)

during nursing, a time when a male should be less interested in the woman.

BREASTS AS DECEITFUL ADVERTISEMENTS

Mates may be attracted by the offer of very concrete rewards. Males of many birds and insect species use the presentation of food to win the consent of a female. Individual may also court potential mates with the promise of future favors—food, protection, good genes. As many humans have learned, flirtatious

suitors do not always deliver, but sometimes evolution favors a good liar.

Low et al. (1987) have suggested that the superfluous tissue making up the form of a human breast is an example of such deceitful advertising. It was proposed by Alexander (1971) that breast size might correlate with milk production. If this were true, men who preferred large-breasted women might be favored because of the ability to select good mothers for their children. Low proposed that it would be in the interest of any woman to enlarge her breasts with fat to deceitfully attract the attention of such a male.

By a similar argument, she explains the evolution of fat deposits on the hips. A wide birth canal generally means a lower incidence of complications in labor. A wide pelvis might therefore become a desirable feature in a woman. Once again fat exaggerates the apparent size of the bony pelvis and enables any woman to cheat.

There are many stories of evolutionary deceit, but probably not as many genuine examples as one might imagine. If selection leads one gender to lie about its virtues, selection favors individuals of the other to discern the truth. The lie will not work for long. Many critics have responded to Low's hypothesis (e.g., Anderson 1987; Caro and Sellen 1990). Most important, the assumptions on which it builds have not been confirmed. A successful lie must be founded on some truth, but there is none in this case. The correlation between breast size and milk production has never been confirmed. The external dimension of the pelvis does not correlate with the diameter of the birth canal. If indeed enlarged breasts and hips have evolved because of their attractiveness to men, mates might be seeking the fat itself. Body fat is crucial to a woman's fertility. A male looking to increase his reproductive fitness is better off looking for a female's immediate ability to ovulate than guessing about her future ability to generate more or less milk.

THE PROBLEM OF GENDER RECOGNITION

Sexually dimorphic traits are obvious candidates for hypotheses of sexual selection. Larger antlers can increase the chance that a moose has to mate it they enable him to intimidate his rivals. Explaining colorful plumage or extravagant tails on male birds has presented more of a problem. What interest could a female have in a male whose feathers confer no survival advantage and may even make it more vulnerable to predators?

Enquist and Arak (1993) answered this question in an unconventional way. They constructed a computer simulation of mate preference. How does an individual learn to distinguish potential mates from animals of different species or of the same sex? The artificial brain was programmed to respond to an arbitrary visual pattern. Suppose this "brain" belongs to a female of a species of bird whose males possess the longest tail of any bird of comparable size in the habitat. Merely by correlating its degree of response to the length of a tail, the "brain" is assured of recognizing a mate of the right species. Many generations of such mate selection would lead to sexual selection for longer and longer tails.

This model can be applied to any sensory modality, cuing to a scent, a song, or a movement pattern as easily as to a color or shape. Most importantly, the search pattern need have no relationship to other aspects of fitness. Adaptationists may search in vain for reasons why one trait rather than another happened to be used, but the observable results are completely arbitrary. Arbitrariness is exactly what biologists find in one species after another. Enquist and Arak may have given us a model for a nonexplanation of many puzzling human traits, including breasts, beards, and hair length—if real brains work like their computer program.

Species and gender recognition are nontrivial evolutionary concerns. Proper identification of potential mates is essential for a successful reproductive effort. Ideally, this problem is resolved if individuals have a sexual response only in the presence of certain stimuli or releasers. These stimuli become synonymous with "sexy" traits. Sexy features of potential mates should become overt signals of species, gender, reproductive maturity, health, and other promises of fertility. Singh's hypothesis for the evolution of the female figure makes even more sense when we place it in the context of both gender recognition and the importance of the visual sense for human sexual arousal. As many of the distinctively human traits under discussion are sexually dimorphic, they serve the purpose of gender identification. These include the different figures of men and women resulting from differential developments of shoulders, breasts, waist, and hips; patterns of head and facial hair; voice range; and gait. Gender signals of human anatomy are also commonly enhanced or superseded by cultural fashion and ornamentation.

The species and gender recognition problem may be sufficient to produce such striking examples of sexual dimorphism as a peacock's tail. Some of the less easily explained human traits may have enhanced species recognition in the presence of multiple bipedal hominin species with overlapping habitats. Perhaps early *Homo habilis* used facial hair or breasts to distinguish potential mates from hairless or flat-chested *H. rudolfensis*. This, of course, is complete speculation, and we will never know whether such selection occurred. (On the other hand, we must be ready to accept the fact that the evolution of many traits may never be explainable.)

What is the basis of human gender recognition? What stimuli attract erotic attention in a nonintimate encounter? Among most mammalian species, smells and pheromones are crucial, but the evidence that humans are sensitive to pheromones under natural conditions is poor (see Chapter 12). Under experimental conditions, male scent may cause women to ovulate more regularly. Both sexes appear to respond to the scents by regarding pictures of members of the opposite sex more positively. While such unnatural observations are not necessarily insignificant, human olfactory signals appear to be useless for long-range signalling. No practical value for assessing ovulatory state or even enhancing sexual attraction has been demonstrated under normal nonintimate conditions. While we cannot ignore the dimorphism of the human voice, visual cues appear to predominate in human sexual attraction. The unique patterns of human sexual signaling and arousal reflect the primate shift from olfactory to visual senses. If humans are not always able to identify gender when visual cues are disguised, the female waist-hip ratio and the development of breasts would address the problem.

20

SEX AND HUMAN EVOLUTION

Social behavior is the interaction of an individual with members of the same species outside the parent-offspring bond. For most animals, those interactions, whether friendly or antagonistic, relate to sex and sexual competition. Even the complex behaviors of mammals living in large societies are consistently shadowed by the fact that members of the opposite sex are potential partners while those of the same sex are potential rivals. The same is true of ourselves, and studies of the evolution of sexuality and society are inextricably intertwined.

Of great popular and academic interest is the question to what extent and in what ways has our social and sexual behavior been shaped by selection. Of more immediate interest for this discussion is the extent to which our bodies reflect this selection process.

Humans, as a species, have many distinctive patterns of sexual anatomy and behavior. Males are distinguished by such features as a disproportionately large penis and rapid exhaustion of sperm. Females are characterized by conspicuous deposits of fat in the breasts, hips, and buttocks. Human sexual intercourse is marked by ready arousal from visual stimuli, extended foreplay, longer time to male orgasm, potential for female orgasm, and the mutual interest in intercourse throughout the ovulatory cycle. Both sexes show a "hypersexuality," or exceedingly great sexual drive that applies sex to play, pleasure, status and dominance, manipulation of other individuals, and material gain. Furthermore, this hypersexuality extends sexual behavior well beyond any conceivable reproductive value in such variations as fetishism and long-term homosexual relationships. Mating patterns focus around a long-term pair-bonding that involves romantic love and sexual division of labor.

From observations of nonhuman sexual behavior, we may appreciate how and why the human patterns have diverged. The following discussion builds on three premises: (1) The biological basis of human sexual behavior differs from that of our ape relatives only in degree, although culture may add elaborate trappings. This perspective, based on comparative observations, enables us to highlight the differences for discussion. (2) The pattern of life history strategy described in the next chapter, including the logistical support of large-brained children, has promoted the evolution of a capacity for a long-term pair-bond between reproductive partners. (3) The evolving human brain has increased behavioral flexibility in sexual behavior as much as for any other behavior.

SEXUALITY AND MATING AMONG THE APES

Hypotheses of sexual evolution make certain assumptions about mating strategies and social structure. These assumptions must be identified and considered along with the models of selection. As always, we look to the apes as a baseline for evolutionary change in the human lineage.

The traditional view: Gibbons are famous for their supposed lifelong monogamy, living in nuclear family groups. Although often cited in the past as models for the evolution of the human family, the lack of a larger social framework is a more important difference than any resemblance to human social ideals of family structure. Orangutans are described as solitary animals, associating only for mating. Gorillas form a harem structure in which the stable social group is composed of adult females and their immature offspring dominated by a single adult male. Chimpanzees live in a fluid promiscuous society. Adult males and females associate together but leave and rejoin the troop freely. While short-term consortships may be formed between a single male and female, more often the female mates with many or all of the adult males present during the peak of a single ovulatory cycle (Goodall 1986).

Studies of the lesser known bonobo chimpanzee have documented a much more familiar pattern. Bonobos use sex in a variety of contexts outside of reproduction —for social reassurance or for access to food, for example (de Waal 1995, 1997b; Wrangham 1993). Sexual dimorphism is slight; thus, the bonobo females are less physically dominated by the males and perhaps assume dominant roles themselves (Parish and de Waal 2000). They mate promiscuously by human standards. Bonobos demonstrate that one important human behavior, nonconceptive sex, has its parallels in nonhuman society.

All of these models oversimplify a greater complexity to hominoid behavior. "Monogamous" gibbons are sometimes unfaithful, mating promiscuously or deserting spouses (Gibbons 1998). Instead of living in nuclear human-like families, they are better described as forming small interacting communities of three to five animals (Fuentes 2000). Some orangutans may interact in groups, at least for feeding and travel (Delgado and van Schaik 2000). Chimpanzee and bonobo females may mate in varying degrees of promiscuity (Goodall 1986; Takahata et al. 1996). In the midst of promiscuity, there also is evidence that primates form heterosexual friendships (Fedigan 1982; Smuts 1985). These are not exclusive relationships, but they may exist beyond the needs of reproduction and may be based on individual preferences rather than dominance.

TWO ADAPTATIONIST MODELS

Humans descended from an ancestor that was chimpanzee-like in many dimensions of its biology. Sexually, the common chimpanzee is noted for its promiscuous, male-dominated society, which provides much opportunity for sexual selection. We may assume that in its generalities this was very likely the mating system of the ancestors of the hominin lineage. Two competing behavioral themes dominate interpretations of modern human sexual biology and its evolution from this ancestral state; these are

reinforcement of the pair-bond and the strategies of promiscuous behavior.

The Pair-Bond Model

The pair-bond model contrasts the current behavior of the human species with that of chimpanzees and originally attempted to explain the evolution of monogamy. Its depiction of the evolution of human sexuality is that of a radical transformation in behavior and anatomy. This perspective was presented in detail by Morris in his ground-breaking *The Naked Ape* (1967). In this book, he characterizes modern humanity as hypersexualized in order to create strong internal male-female bonds that provide a suitable environment for raising children. Specific adaptations include sexualization of the skin through enhanced sensitivity and the development of erogenous zones in such places as the lips and ear lobes, visual stimuli of the lips and breasts as buttock mimics, visual displays of scalp and pubic hair, olfactory stimuli through apocrine glands, loss of estrus and continual female receptivity, and female orgasm. These changes in the body are related to the additional time spent in intimate contact, including sexual foreplay, as well as complex communication through facial expression. The solution to cementing the pair-bond, he states, was to "make sex sexier." "[W]e can see that there is much more intense sexual activity in our own species than in any other primate. . . . Clearly the naked ape is the sexiest primate alive" (p. 53).

The model of monogamy so bluntly stated may be criticized for its culture-bound assumptions that monogamy is indeed representative of our species. Morris explicitly accepted the contemporary North American mating system and data on sexual behavior as the species norm and dismissed "whatever obscure, backward tribal units are doing today" (pp. 69–70). His rationale (that the odd customs of remote tribes may have held them back and thus proved themselves unsuccessful [pp. 43–44]) is anathema to modern anthropologists. Although it is currently a truism that monogamy in the strict sense is not a species characteristic and that most cultures accept polygamous relationships, monogamy remains a common and perhaps modal mating structure even within polygamous societies. The pair-bond model today may be more broadly understood to describe the evolution of tendencies for long-term (several years or more) sexual and reproductive relationships. In the context of culture, these relationships, as marriage, assume rules relating to economic interdependence and sexual exclusivity.

Morris' model was updated and elaborated by Fisher (1982), who interprets human sexuality primarily as a female strategy to bargain for male paternal investment. Lovejoy (1981) used a pair-bond model to explain the most fundamental hominin adaptation, bipedalism. In his version, males became bipedal so that they could provision their mates and thus provide adequately for their children. Females evolved sexually to cement this bond. Lovejoy's version of the model has been widely cited in student textbooks and popular writings.

The Promiscuity Model

The alternative approach, the promiscuity model, emphasizes similarities between chimpanzee and human sociosexual behavior and would apply the terms "promiscuous" and "male-dominated society" to both species. In many versions of this view, the ideals of long-term stable and sexually faithful marriage is a social veneer, if not an outright deception, for actual human behavior. The opportunism of males places females on the defensive. Many of the adaptationist explanations in this model reflect the female strategy to maneuver and manipulate relations from a non-dominant position.

Morgan's reply to Morris in her book *The Descent of Woman* (1972) revolted against Morris' idealistic family, reinterpreting many female sexual and reproductive characteristics as adaptations other than for sexual attraction. Frontal sex, rather than an opportunity for intimate bonding, became a dangerous confrontation with a threat of violence. In her sequel, Morgan aptly comments, "It has always been generally recognized that making sex more exciting does not necessarily favor monogamy" (1982, p. 29).

Hrdy (1981, 1997, 1999) brought the alternative perspective of female as counterstrategist into the mainstream of human evolution studies. While accepting some of Morris's argument that sexual attraction reinforces the pair-bond, Hrdy asserts that aggressive female sexuality, not passive receptivity, has been a driving evolutionary force. Humans are not that different from other primates in that they show a peak of sexual activity near ovulation and lower levels during other parts of the cycle. In her model, female orgasm developed to inspire females to higher levels of promiscuous sexual activity to fulfill nonconceptive functions of copulation, including obscuring paternity, discouraging infanticide by males, and facilitating female mobility in the social landscape.

Smith (1984) and other authors returned the emphasis to males by discussing adaptations for sperm competition. According to Smith, not only are human semen quantity and composition reflective of intermale competition but also female sexual attractiveness evolved to further stimulate this competition. Echoes of the theme that promiscuity is a fundamental human behavior continue to appear in both technical and popular literature.

Contrasting the Models

Of course, human mating behavior assumes a wide variety of forms. "Serial monogamy," the practice of having multiple partners through divorce and remarriage, and institutional polygamy are not the same as promiscuity. Both of the former involve pair-bonding for periods of years as offspring pass through the critical period of helpless infancy. They are more consistent with the strategies of the pair-bond model. True promiscuity implies no sexual fidelity and permits males to evade parental responsibility unless and until coerced by females. The promiscuity model tends to interpret human traits as having evolved in the context of strong sexual selection through intermale competition. It emphasizes the differences in male and female reproductive strategies and the conflicts that these differences create. Accordingly, it is heavily dependent on arguments for sexual dimorphism and asymmetric sexual selection.

In many details, this two-fold categorization is artificial. Baker and Bellis (1993) express this well: "Behavioral ecologists view monogamy as a subtle mixture of conflict and cooperation between the sexes," perhaps with promiscuity and bonding as opposing tendencies within a relationship. Yet the two models sometimes find completely different interpretations of modern sexual anatomy and behavior. For example, Morris interprets the loss of estrus and extended female receptivity as an attempt to manipulate the mated male by enticing him into a long-term relationship with frequent sex as a reward. Hrdy observes this as simultaneous manipulation of all the males by confusing paternity to deflect violent domination and infanticide. Morris understands female orgasm as positive feedback to encourage female participation in the sexual bond. Hrdy views orgasm as driving females to promiscuous sex so as to appease and manipulate as many males as possible. Monogamy may result from selection of mates by females based on the anticipated ability and willingness of males to invest in offspring. Promiscuous competition among males predicts sexual selection to act on the ability of males to evade parental responsibility.

Are human relationships fundamentally pair-bonded or promiscuous? It is impossible to identify a single mating structure that characterizes our species. Within and among human societies, polygyny, polyandry, and monogamy may all be found, spiced with different levels of tolerated extramarital sex and reproduction. The flexibility that confounds this question is itself one of the most striking of human strategies. Observations of ape species, cited earlier, show that such behavioral complexity has deep roots.

SEXUAL SELECTION IN THE HUMAN SPECIES

Human males have disproportionately large penises relative to other primates and are commonly described as having disproportionately large testicles and marked dimorphism in body size as well. Females have a unique development of the breasts that, along with a wider pelvis, produces a distinctive figure. Sexual selection has been invoked to explain these patterns and other traits such as facial hair. However, neither human sexual performance nor specific anatomical traits are consistent with a model invoking intense sexual competition. Male and female anatomy may be consistent with stable monogamous or polygynous mating.

SEXUAL DIMORPHISM IN BODY SIZE

Among mammals, sexual dimorphism in body size is generally assumed to be a result of sexual selection. The catarrhine pattern, in which males are typically larger, more aggressive, and better equipped with slashing canines, is understood to be the consequence of male–male competition for status and/or resources, including females, although the relationship is not simple (e.g., Plavcan 2000). Humans display a pattern in which the male has about 20–25% greater body mass than the female. By the standards of the great apes, this is not a huge difference. Chimpanzee dimorphism is comparable to that of humans, but gorilla and orangutan males are about twice the size of the females (Figure 20.1).

What does this information tell us about humans? Martin and May (1981) interpret the relatively low degree of dimorphism in humans as indicative of either unimale polygyny or monogamy, in which there is little direct male–male competition. They explain the similar chimpanzee pattern, occurring in the context of a multimale troop, by a lack of direct competition among related males. Others draw just the opposite conclusion and argue that differences in human body size are large and indicate past (or present) selection in a competitively polygamous society (Hrdy and Bennett 1981; Schröder 1992). There is some evidence that stature correlates with mating success of men and reproductive success of women (Nettle 2002; Pawlowski 2003; Pawlowski et al. 2000). However, as measured by both stated preferences and correlates of reproductive success, contemporary human mate competition most often revolves around wealth as a reflection of status and potential as a provider, as discussed later.

Although this debate leaves us in the dark about the mating pattern of early hominins, it appears that modern humans have a lesser degree of sexual dimorphism than did past hominoids or early hominins. Dental measurements reveal that at least some Miocene hominoids had equal or greater sexual dimorphism than the more dimorphic of modern apes. Recent estimates of sexual dimorphism among Plio-Pleistocene hominins show male body mass exceeded female mass by 20–55% in different species of australopithecines and by 62% in the problematic collection designated *Homo*

SEXUAL DIMORPHISM

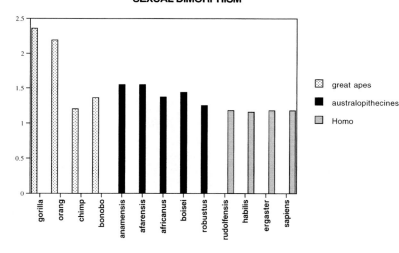

Figure 20.1 Sexual dimorphism in hominin body size. The vertical scale show the ratio of male to female body weight of living great apes, australopithecine hominins, and species of *Homo*. Fossil *Homo* species show a similar degree of sexual dimorphism to that of modern humans, but this is distinctly less than that of the great apes (except chimpanzees) and australopithecines. (Data from McHenry and Coffing 2000.)

habilis. Modern populations of our species thus represent a reduction in sexual selection that was established by the Middle Pleistocene (Arsuaga et al. 1997). This is consistent with the reduced competition indicated by the pattern of sperm production.

It is therefore difficult to read this signal of sexual selection. Depending on one's perspective, dimorphism in human body size either may indicate significant competition for mates or a significant decline in competition for mates or may not refer to sexual selection at all. Present and past natural selection for dimorphism may alternatively be explained as resulting from a division of labor between the sexes (e.g., hunting versus gathering).

EVIDENCE FOR SEXUAL SELECTION OF MEN

Sexual Selection and Genital Size

The human penis is by far longer than that of other primates, including those of the much larger gorilla and orangutan. Species that possess a longer or more ornate penis are more likely those in which the female can exert sexual selection on the males by mating with multiple partners (Dixson 1987). Such species organize in multimale groups or are dispersed so that the males have less control over the females. The

size of the human penis has thus been explained by selection for either sexual or aggressive display by competing males (Diamond 1992b; Lovejoy 1981; Reynolds 1981). That is, it serves as a visual signal to attract women or intimidate other men. Some higher primates do use the erect penis in displays (e.g., it is a standard signal in soliciting copulation with female chimpanzees, although not overtly as competition) but apparently not in dominance interactions. However, "ornateness" in other species may take such forms as elaborate shapes or the presence of spines, papillae, or plates. The human penis, although unusually large, is simple in form. In the absence of modern day or comparative evidence, these hypotheses for human evolution are unsupported.

The size of the male penis has alternatively been interpreted as providing greater pleasure for women and an increased likelihood of inducing orgasm (Fisher 1982, 1992; Morris 1967). The important feature in this context should be the fit and interaction between the male and female genitals, not their absolute size. The length of time to male orgasm has evidently increased in humans relative to that of the apes. The mean duration of coitus has been estimated at 4 minutes in humans versus 1 minute in gorillas, 15 seconds in pygmy chimpanzees, and only 7 seconds in common chimpanzees, where a copulating male risks the intervention of competitors. It may well be that a longer duration of human coitus is more likely to evoke orgasm in the woman. However, female orgasm often has a higher stimulation threshold than male orgasm. An increased sensitivity on the woman's part would have been a more effective adaptation if female orgasm were an important evolutionary objective. This explanation for penis size is therefore weak, although one should not expect all evolutionary adaptations to take the obvious path.

The human penis is proportionately far larger with respect to body size than that of any other primate species. The significance of this fact is a matter of debate, but probably informs us more of female anatomy than of male sexuality. The human penis also differs from most other primates in the lack of a baculum. The baculum, or penis bone, is commonly found in mammals, including nonhuman primates, although it is rudimentary among the great apes. The bone functions to increase the stiffness of the penis. Dixson (1987) found that the baculum is more prominent in species in which the male organ stays in the female for a period after ejaculation. Its absence suggests the male organ has been selected only to reach ejaculation, not to persist in coitus for the sake of pleasing the woman or monopolizing her.

The size of the penis is uncorrelated among human males with body size or frequency of erection or frequency of intercourse; thus, evidence of a role for sexual selection is lacking. Several authors have sensibly concluded that it is best to regard adaptations of the penis as relating to its primary function of inseminating a female. A longer penis will discharge sperm closer to the cervix. Although this may be seen as an adaptation to outcompete rival sperm, anything that increases the chance of fertilization has an advantage in even the most monogamous of species. One need not invoke sexual selection. In seeking to explain the form of the human penis, we could do worse than to contemplate the wisdom of Abraham Lincoln when asked the length of his legs: "Long enough to reach the ground."

The question of penis length may more profitably focus on why the human vagina is deeper than that of other primates (Brown et al. 1995), but few have recognized this as an issue. Among primate species, vagina size does not correlate with body size but might represent a female stategy for selection of sperm. Gallup and Suarez (1983) suggested that with the assumption of upright posture, withdrawing the cervix deeper into the body might impede sperm from draining out of the woman's reproductive tract.

Sperm Competition

Sperm competition is a variation of sexual selection in which the sperm of different males are in competition after ejaculation. Such competition would favor repeated copulation, releasing large numbers of healthy sperm able to find and fertilize an ovum before the competition arrives, and it would favor any adaptation for interference with rival sperm. If fact, the challenges facing any sperm from successful fertilization of an egg are daunting. In order for successful reproduction to occur, the sperm have to swim into and through the uterus, ascend a uterine tube, and find and fertilize the egg. It must struggle in a hostile environment against gravity and against the clock as its nutrient supply runs out.

The upright human posture tilts the vagina so that gravity tends to work against semen—a concern that quadrupedal monkeys, for example, do not have. This danger is reduced by several possible factors. The horizontal position of most human intercourse is unusual among mammals but results in a similar horizontal orientation of the female tract as long as she is lying down. Commonly a strong sensation of sleepiness follows intercourse to encourage this, an effect of opioids released in the brain. These opioids contribute to the pleasurable sensations and strong positive feedback from intercourse.

Approximately 60 seconds after ejaculation, the semen thickens due to the activation of clotting proteins. This state lasts 5 to 15 minutes, after which the semen again becomes less viscous. The coagulation may function to resist drainage. Yet another mechanism that may work against the pull of gravity is the uterine contractions of orgasm. The purpose of orgasm in the female, if any, has been the subject of much debate. Orgasm is accompanied by rhythmic muscular contractions of the wall of the uterus. One interpretation considers this movement to create a suction that draws semen up into the body of the uterus.

The hostile environment of the vagina is adaptive in preventing microbes from establishing colonies in its warm moist environment. The combination of acidity and macrophages is equally destructive of unprotected sperm. Semen must neutralize the acidity with alkaline buffers that are included in the secretions of several male glands, including the seminal vesicles, prostate, and bulbourethral glands. As semen spreads out, the foremost sperm are likely to die, but these pave the way for others following behind them.

Sperm that have overcome gravity and the defenses of the vagina still face the problems of time, distance, and direction. A sperm maintains direction within the uterine tube by following a slight thermal gradient toward the egg (Bahat et al. 2003). Once a sperm is very close to its target, the ovum itself may assist it by releasing a chemical signal (Spehr et al. 2003). Sperm picking up this chemical trail may be able to home in on the egg. From this point, the ovum is not a passive target but rather an active player in binding to the sperm and drawing it to itself (Freedman 1992; Martin 1991). Sperm are viable in the female tract for up to 3 days. Beyond this period, they will die from lack of nutrients and a supporting environment. Those sperm fortunate enough to enter the correct oviduct and to approach an egg will have left behind a trail of dead and dying companions. Of the estimated 500 million sperm released in a single ejaculation, 1000 to 3000 may reach the egg. The others are lost and dying or perhaps sacrificed themselves to create a usable path.

Sperm competition can be expected to occur when females voluntarily or involuntarily mate with more than one male during the period when fertilization is possible (Birkhead 2000; Ginsberg and Huck 1989; Møller and Birkhead 1989). It commonly takes the form of a vaginal plug of some kind to block the progress of rival sperm. The coagulation of human semen might serve such a function, were it not so short lived. Another proposed tactic of human semen is a large percentage of defective sperm that entangle their opponents.

Sperm competition has been hypothesized to be significant among humans (Bellis and Baker 1990; Baker and Bellis 1993a, 1993b; Small 1989; Smith 1984). The argument rests on the observations that there are opportunities for sperm competition to occur in human society (e.g., communal sex, rape, prostitution, promiscuity before or within marriage) and that adaptationist explanations for anatomical characteristics can be found that are consistent with the predictions of sperm competition and sexual selection. Caution is required for such interpretations. Once again, any variation that would increase the probability of conception from a single or multiple matings would be favored by natural selection regardless of whether sperm competition was significant. Furthermore, the mere identification of modern cultural opportunities for sperm competition tells us little about actual selection in the past or present.

Human testes are of moderate size and are rapidly depleted of sperm (Harcourt et al. 1981; Martin and May 1981). Although ejaculate volume is large, the density of sperm in human semen is low (Table 20.1). Chimpanzees, the model for a promiscuous primate, have proportionately the largest testes and largest sperm output among the hominoids. Similarly, seminal vesicle size is medium in humans but large in chimpanzees and other promiscuous species (Dixson 1998). The relatively modest size and output of the human testes contrast markedly with those of the chimpanzee and argue for a very different mating pattern, one in which males are not copulating with great frequency. Furthermore, unlike those of chimpanzees, human stores of sperm can be rapidly depleted. It has been estimated that production of sperm can sustain fertile intercourse at a rate of once every 2 or 3 days in our species (Short 1979). Male fertility peaks in very early adulthood, before the peak of female fertility, and declines gradually thereafter. These observations are not consistent with a high degree of sperm competition, nor with a promiscuous lifestyle. They are consistent with many possibilities, including a mating pattern in which exclusive sexual access to a partner is reasonably assured.

Male mammals are often characterized by their copious production of sperm and by a high number of apparently deformed sperm. Among primates, the percentage of healthy motile sperm corresponds very loosely to mating system and thus to predictions concerning sperm competition. On average, higher percentages of sperm are motile in species with multimale troops than in monogamous species, with values

Table 20.1 Potential Indicators of Male Sexual Competition

Indicator	Human	Chimpanzee	Bonobo	Gorilla	Gibbon	Orangutan	Baboon
Erect penis length (cm)	Large: 13	Moderate: 8		Small: 3		Small: 4	
Male body size (kg)	66	44, 64[4]	45	169	5.5	75	24
Sexual dimorphism:							
M/F body weight × 100	120–125	110		200	100	200	200
Testicle size as % of male body weight[1]	0.079, 0.046–0.055[2]	0.300, 0.018,[3] 0.106–0.189[4]	0.013[2]	0.10, 0.088[2]	0.063, 0.048[3]	0.214	0.502
Testicle size as % of female body weight	0.065	0.330		0.026	0.088	0.069	
Seminal vesicle size	Medium	Large		Small	Small	Large	Large
Ejaculate volume (ml)	4.3 ml	2.1, 3.4[4]	1.8	0.52	1.3	1.1	3.6
Sperm motility (%)	50–60	44.3, 76[4]	45	50	5–7	47	45
Sperm number (× 10^6)	255	657, 214–6300[4]	821	84	144	67.1	143
Mating system	Variable	Multimale	Multimale	Single male	Monogamous	Single male	Multimale

References: Dahl et al. 1993; Dixson 1998; Gould et al. 1993; Harcourt et al. 1981; Harvey and May 1989; Møller 1988; Napier and Napier 1967; Short 1979; Smith 1984; Symons 1979.
[1]Møller 1988, except as noted.
[2]Dahl, et al. 1993.
[3]Short 1979.
[4]Gould et al. 1993.

ranging from 90% motility in some macaques to 5% motility in gibbons (Table 20.1). Approximately 50–60% of human sperm are motile. These fall into a middle range among primates, overlapping those of both multimale and unimale breeding systems.

Although some authors have suggested an adaptive role for defective sperm as passive barriers to rival sperm (Baker and Bellis 1988, 1989; Small 1991), this would be effective in mammals only in conditions of repeated mating with rapid change in partners, which is quite unusual in known human societies. The brief time of viability of sperm in the female reproductive tract (approximately 48 to 72 hours for humans), along with the even briefer life of an unfertilized ovum (approximately 24 hours), severely reduces the opportunities for sperm competition. Furthermore, direct observation of sperm interacting does not reveal any interference behavior (Moore et al. 1999). Consequently, it is difficult to reconcile the overall low productivity of the human testes with any model of sperm competition.

These arguments undermine the notion that sperm competition has been a significant factor in recent human evolution, but more information is needed to put the human pattern into perspective. For instance,

in characterizing human sperm, are we accurately describing the species or observing one ethnic group or merely one atypical population? Given the fact that sperm are highly sensitive indicators of damage from environmental hazards and that some reports suggest the average quality of human semen may be declining (Bribiescas 2001; Carlsen 1992; Swan and Elkin 1999), is the published rate of viability typical of sperm of the human species or only of males in certain populations? Sperm count is known to vary significantly in different parts of the United States and Europe. Until broader and systematic sampling is available, we cannot be certain that we have correctly characterized our species, much less identified an evolutionary pattern. Sperm competition in human ancestry cannot be ruled out, but there appears to be no evidence that it has done anything but diminish since the last common ancestor with other hominoids.

EVIDENCE FOR SEXUAL SELECTION AMONG WOMEN

Anthropological literature has extensive discussions of the evolution of breasts and figures (see Chapter 19), concealed ovulation (see later), orgasm, and other aspects of female behavior that may be explained by sexual selection. The pattern continues in the widespread cultural emphasis on female displays through costume, cosmetics, and body mutilation. If these do indeed represent sexual competition among females, then the appearance of ornamentation on the females rather than on the males of the human species is most unusual among higher vertebrates. Strong sexual selection on human females makes sense only if some males are much more desirable than others and if those males are in relatively short supply. The implication may be carried further that such males are investing significantly in their offspring and/or partner and thus limit their own mating opportunities.

SEXUAL SELECTION AND CONTEMPORARY MATE SELECTION

What is the evidence for sexual selection from mating behavior? Numerous cross-cultural studies in mate selection have been published (e.g., Bereczkei and Csanaky 1996; Betzig et al. 1988; Buss 1989, 1993, 1994a, 1994b, 2000; Jones and Hill 1993; Pérusse 1994; Symons 1979). Their conclusions are reasonably consistent but do not support the hypothesis that sexual selection is a measurable agent directing the evolution of the species at the present. The apparent selection for status by women and fertility by men admit to other interpretations.

Women: Choosing Status

Along with health, the trait that women admire most in prospective husbands is status (Cashdan 1996), which may be measured on the basis of cash, land, livestock, political power, or title. Although other traits, such as ambition (representing the potential to achieve status?), were ranked by women as very important, status is the trait that has been documented to correlate with marriageability and polygamy and presumably with reproductive success.

Why do women prefer mates with higher status? Among nonhuman primates, there is a rough but imperfect correlation between male status and reproductive success. The preference is not unique to our species. Higher rank has material benefits for both sexes in terms of access to resources. Lower status correlates with stress, which can have a direct effect on fertility (de Catanzaro and MacNiven 1992; Sapolsky 1990). Both of these benefits may translate to enhanced reproductive success in higher status individuals (Wade and Schneider 1992; Wasser 1990; Wasser and Barash 1983).

Male wealth and status will not influence male evolution unless they correlate with genes. Although social rank is often hereditary, the historical longevity of a dynasty cannot be equated with evolutionary time. One might inquire what genetic traits contribute to high status. The obvious traits—ambition, intelligence, political skills—are not limited to males and have potential selective value for any individual in the species. Even when intermale competition is direct and physical, the outcome in historic times has been more likely to be decided by the ability to manipulate cultural artifacts (weapons) than by somatic characteristics.

Men: Choosing Reproductive Potential

The most consistently reported preferences of men for wives relate to youth and health. Facial "beauty" is manifest as symmetry and averageness (Jones and Hill 1993; Langlois and Roggman 1990; Thornhill and Gangestad 1993; Watson and Thornhill 1994; see Perrett et al. 1994 and Swaddle and Cuthill 1995 for contrary views). These traits may reflect sound genetic development and a history of good health, because malnutrition and disease may distort facial and body development (Rhodes et al. 2001). Youth and beauty therefore relate to potential fertility (Grammar et al. 2003). If beauty correlates in females with mating or reproductive success (neither has been demonstrated), sexual selection may be occurring.

Attractive females are not the only ones to reproduce, but physical beauty may be one factor determining high-status marriages. It may be that only the extreme physical variants in the society are excluded from marriage and reproduction, but the role of status confounds any simple attempt to measure success by counting babies.

Youth, of course, cannot be selected, but youthful-looking features can be. Several secondary female characteristics—skin texture, facial shape and hairlessness, pitch of voice—may be understood as both attractive and neotenous (Jones 1995). The importance of youthfulness among men's preferences is revealing. If humans evolved a tendency to pair-bond for only short periods or if males are pursuing a strategy of minimal investment, a man should prefer a woman at the peak of her fertility who is best able to raise an infant. This implies a preference not only for health and fertility but specifically for full reproductive, social, and skeletal maturity; for experience; and for self-sufficiency. Teenagers need not apply, because peak ovarian function normally occurs in the late 20s. Virgins are poor candidates also, because their fertility is unproved and because first-time mothers (among nonhuman primates) are less successful at raising infants.

However, this prediction is not what has been observed. Preferred ages for wives are either in early maturity or slightly younger than the husband. A preference for virginity appears to be widespread. Pursuit of virginal youth makes reproductive sense if men are intending to make a long-term commitment; only then might the earlier years of irregular ovulation and inexperience pay off in terms of total fertility.

SUPPORT FOR THE PAIR-BOND MODEL

The scenarios of male–male competition observed in nonhuman primates and predicted by many models of human sexual evolution are not supported by the evidence of contemporary male anatomy. The argument for modern day sexual selection is weakened by the lack of evidence correlating mating behaviors with genetic traits. However, the human pattern of derived female anatomy suggests that if these traits are the product of sexual selection, then female reproductive success should exhibit a high variance. The limiting factor for female reproductive success is less likely to be access to males than obtaining a commitment of paternal investment by males of high status. If the role of the males in raising offspring made a significant difference in their survival, females should compete among themselves for males willing to commit themselves as husbands and fathers.

This observation plays into the elaborate scenarios for the evolution of human monogamy proposed by Morris, Fisher, and Lovejoy. Although these authors present different versions of a similar story, all agree that the narrative of human evolution involved the formation of the nuclear family when husband and wife began to cooperate economically. Cooperation dramatically enhanced reproduction, perhaps by enabling the woman to care for a slowly developing large-brained infant. The economic support provided by the male enables a woman to increase her fecundity. The pair-bonds were reinforced by sexuality with the female repaying the male by satisfying his desires.

This idyllic picture is marred by the fact that it does not describe modern human society. Promiscuity, polygamy, and divorce are common in diverse cultures. Men do not consistently seek sexual satisfaction from their wives. Furthermore, mating competition does not necessarily correlate with reproduction. There is a disparity between status and reproduction in some cultures, especially among females (Elia 1986; Pérusse 1993; Vining 1986).

HUMAN SEXUAL BEHAVIOR

THE PROBLEM OF ESTRUS

When a female chimpanzee ovulates, every member of the troop knows it. For several days before and after ovulation, her perineum becomes visibly swollen and pink. To a human observer, it appears painfully inflamed. During this period of estrus, she attracts disproportionate attention from the males and is generally eager for sex. Estrus has been defined by three changes in behavior: the female is more attractive to males; she is more likely to make advances to them; and she is more receptive to copulation (Hrdy and Whitten 1986).

Estrus is widely distributed among Old World monkeys and apes. It may be signaled by visible changes in the female, by the release of hormones, or by changes in her behavior. From an adaptationist perspective, estrus makes sense. It is to the female's advantage not to bother with male attention when she is not fertile. The males should not waste their time, energy, and threat of violent competition if conception is not possible. The evolutionary puzzle is why humans do not display estrus. A woman's ovulatory cycle is most conspicuously indicated by menses, a potential signal of nonfertility. Equally distinctive is the fact that both males and females may be willing to have sex at any point of the cycle.

Anthropologists have traditionally interpreted the human condition as a loss of estrus and a concealment of ovulation. These are widely held to represent an evolved functional complex to manipulate the behavior of the male. Numerous explanations on this theme have been published (see reviews by Daniels 1983; Gray and Wolfe 1983; Sillén-Tullberg and Møller 1993). Perhaps the most widely cited view has been that suppressing estrus to conceal ovulation is a strategy of females to reinforce the pair-bond by maintaining the sexual interest of males (Burd 1986; Elia 1986; Fisher 1982; Gallup 1982; Lovejoy 1981; Morris 1967; Szalay and Costello 1991; Wilson 1978). However, long-term pair-bonds in other species are characterized by reduced, not increased, sexuality. Alternatively, disguising ovulation may be a ploy to discourage attention from undesirable males (Strassman 1981).

Another approach suggests that concealed ovulation is a mechanism to defuse male competition and facilitate cooperation. Proponents argue that the loss of estrus would cause males to express less interest in neighboring females—just the opposite expectations of those expressed by supporters of the pair-bond model. Other authors have further pursued the theme of selection against violent, competitive, or selfish male behavior. Males uncertain of paternity of the offspring would be forced to treat them as possible sons (Heistermann et al. 2001; Hrdy 1981). This could be important in species where infanticide by males is common. Males might further be forced to maintain a long-term association with the female to preclude cuckoldry (Alexander and Noonan 1979; Smith 1984; Turke 1984, 1988; Soltis 2002), or they might be discouraged enough to leave the beleaguered female and her offspring alone (Cann and Wilson 1982).

Yet others have inverted this idea of deception to argue that concealed ovulation functions to manipulate the woman herself and thwart her efforts to limit her own fertility (Burley 1979; Daniels 1983; Freeman and Wong 1995).

Is Ovulation Concealed?

From whom is ovulation being concealed? Women apparently signal ovulatory cycling to one another subconsciously by pheromones (Stern and McClintock 1998). If so, then ovulatory state could not be concealed from males if males had a biological interest in knowing about it. Studies suggests men can detect the periovulatory period by the woman's odor (Pawlowski 1999; Singh and Bronstad 2001). Furthermore, several studies demonstrate women have sex slightly more frequently around ovulation (Hedricks et al. 1994a; Udry and Morris 1968) or are more interested in it then (Gangestad et al. 2002), so it is not entirely concealed from themselves. Occasionally a woman may feel a twinge of ovarian pain at the moment of ovulation, a sensation called *mittelsmertz*, or detect a change in cervical mucus. A survey of college students showed that about half of women and men were always or usually aware of when ovulation occurred in themselves or their girlfriends (Small 1996). Menstruation, of course, is another blatant signal of timing in the ovulatory cycle. Clearly, there are indicators of ovulation present. If evolutionary advantage were to be gained by either partner by monitoring ovulation, sensitivity to it would have been maintained in a more obvious way. This conclusion rules out several of the above hypotheses.

A second issue is whether estrus has been suppressed or extended throughout the menstrual cycle. If the human pattern functions to reduce male competition and/or interest, it is to be expected that estrus signals, which are powerful arousal stimuli for male primates, have been suppressed. If the continuous attractiveness, affiliative behavior, and receptivity are supposed to maintain the interest of one or more males beyond the period of fertility, then it should be assumed that estrus has been extended. These two evolutionary models, while seemingly converging on the concept of a reinforced pair-bonded relationship, are mutually incompatible. The question can be stated more bluntly: Is human sexuality permanently turned down or turned up? Contemporary human behavior supports the latter idea.

The argument for an extended (versus suppressed) estrus is strengthened by the observation of a parallel development among pygmy chimpanzees. Bonobos have prolonged estrus display and period of sexual activity relative to those of the common chimpanzee (Blount 1990; Dahl 1986). The common chimpanzee itself shows some extension of estrus beyond ovulation, and extension of estrus may simply be a phylogenetic trend in anthropoid lineages.

Although most theories of "concealed ovulation" focus on long-term strategies (e.g., Soltis 2002), such as the pair-bond and the need to commit male investment, observations of nonhuman primates suggest short-term gains of the female that accompany estrus or nonconceptive sex, including temporary rise in status and obtaining meat (Blount 1990; Goodall 1986; Savage-Rumbaugh and Wilkerson 1978; Smuts 1985; Thompson-Handler et al. 1984; Wrangham 1993). In addition, nonconceptive sex may occur in the context of social tension as a gesture of reassurance or appeasement and as a means for constructing alliances with the opposite or same sex. There are reproductive consequences. If sexual strategies such as a prolonged period of attractiveness can enhance success in political manipulation and gain access to material resources, they can enhance reproductive success. However, we would now understand sexuality as a general political tool to be used as the occasion merits. Highly specific objectives such as avoiding incest are unnecessary.

Why do males respond to nonconceptive sex? A model that considers males to be fundamentally promiscuous and opportunistic views nonconceptive sex as deceptive and regards the males as enticed into relationships by the hope of fathering offspring. Models emphasizing the human pair-bond model understand males to be responding to sexual gratification within the bond, thus reinforcing and cementing the relationship. To the extent that sexual pleasure may be enjoyed independently of reproduction, males may respond simply for the sake of gratification, whether pair-bonded or not. Because nonconceptive sex and copulation outside of estrus may be observed in our anthropoid relatives, it is reasonable to speculate that our own pattern predates the human pair-bonding strategy.

THE PLEASURE OF SEX

Humans are "hypersexed." Many, if not most, encounters between adults of opposite sex involve flirtation and/or a sensitivity to gender. Males and females copulate without regard to ovulation and fertility. They devote a great deal of time, energy, and money in the pursuit of sexual pleasure. These characteristics of the species raise several important concerns for evolution, including nonconceptive sex and a pervasive concern with being sexually attractive.

Pleasure is a significant motivating force in the mind, and its evolution with regard to sex is no mystery. Sexual pleasure is the product of a basic reproductive drive as it interacts with the complex mammalian forebrain to give positive feedback to a biologically significant act. Sexual pleasure is therefore a proximate mechanism that has evolved in support of the ultimate objective of reproduction. Does it matter whether we describe sexual behavior in terms of gratification or reproduction? Absolutely. Specific behavior decisions are made at the proximate level. Gratification explains a wider range of behaviors than does ultimate reproduction, including some behaviors, such as contraception, that run counter to reproductive interests. We cannot and should not interpret each of the acts by which humans receive sexual pleasure in terms of reproductive competition, yet that has been done.

Male masturbation has been explained as an adaptive behavior that clears the reproductive tract of old and less viable sperm (Smith 1984). Sexual coercion and rape have been interpreted as a desperate reproductive strategy of men who are unable to attract the voluntary consent of women (Starks and Blackie 2000; Thornhill and Thornhill 1983). Extramarital promiscuity and infidelity are described as male strategies to father additional offspring and a female strategy to find better genes for her children. Prostitution is described as an arena for sperm competition and thus an extension of male competition. Homosexual attraction, the most sterile common expression of human sexuality, has been interpreted as enhancing inclusive fitness. In the absence of data supporting these as viable reproductive strategies, all of these behaviors are better understood as outlets for sexual energy rewarded by the positive feedback of sexual pleasure. Common sociobiological explanations assume that such behaviors evolved and are adaptive. It is more likely that they did not evolve, that they are not discrete selectable traits, that they have no identifiable genetic basis, and that they serve no purpose beyond satisfying desire and other psychological needs.

Men and women pursue sex differently and express it culturally in different ways, but it is clear that both sexes seek and derive pleasure from it. From the auto-erotic or homosexual behavior of captive and laboratory animals (Chevalier-Skolnikoff 1976; Ford and Beach 1951), we can infer that humans are not unique in enjoying sex. Many mammals do not appear to pursue nonconceptive sexual gratification and are only seasonally or cyclically active. Humans may exhibit a greater elevation of hormone activity and sexual interest than most other species, but this tendency is conspicuously shared with bonobo chimpanzees and possibly some nonprimate species.

The human drive for sexual pleasure extends beyond the pair-bond. It continues beyond reproduction and even beyond copulation itself. Our sexual interest pervades literature, visual art, advertising, eating, and anthropology and psychology journals. It forms a major part of our leisure entertainment and influences nearly all adult interpersonal relationships. Certainly some aspects of human sexuality driven by pleasure are nonadaptive and are even self-destructive. Human sexuality is not an ingenious strategy to retain mates. It is an important preadaptation of pair-bonding, at best. Although sexual play and intimacy may continue to be important contributors to the success of a pair-bond, it is unnecessary to suggest that they evolved for that role.

A rat in a cage able to press a lever to stimulate a "pleasure center" in the medial forebrain bundle of its brain will do so up to 100 times a minute. Should we expect humans to behave otherwise, given a similar opportunity? Programmed to be attracted to signs associated with mature individuals of a given sex, we find suggestions of sexuality spread throughout our environment. Human culture gives us an unprecedented amount of discretionary energy and time to expend in the pursuit of sexual gratification. Individuals in Western culture have taken this to an extreme level of expression.

THE OBSCURING OF ESTRUS: SUPPORT FOR THE PROMISCUITY MODEL?

The story that emerges from these ideas is that the estrus period of enhanced receptivity was expanded across the ovulatory cycle because of its benefit for female reproductive well-being. In the process, ovulation was not *concealed*—that term suggests purposeful strategy—but merely *forgotten*. With emphasis on nonconceptive sex, signals of ovulation become irrelevant. As human sexuality became more complex, accompanying the enlargement of the brain, the value of sex for sensual gratification increased. Males responded to the increased attractiveness of females, and the strategy of offering sex for social and material benefits became even more effective.

Although the strategy of copulation for material gain is currently viewed as the last resort of society's failures, the emphasis on female attractiveness and ornamentation continues at all levels of society in most cultures. Sexual attractiveness appears to have replaced sexual receptivity as a social lubricant that enables women to maneuver in a species whose culture is dominated by males. It is difficult to argue that the vast majority of flirtation has much direct reproductive significance. For both sexes, operating within the confines of culture, sexual attractiveness and attraction have become tools that are independent of their adaptive origins.

HUMAN MATING SYSTEMS ARE FLEXIBLE

The evolutionary expansion of the brain in primates is accompanied by increasing flexibility of behavior. This applies no less to sexual strategies than to other behaviors. One consequence is increasing variability in the patterns we can observe. Female chimpanzees, for example, are able to engage in either consortships or promiscuous behavior as alternate mating strategies within a multimale band. It is similarly impossible to identify a single mating structure that characterizes our species.

In the past two centuries, Western nations have undergone a "demographic transition," in which parents have chosen to have fewer children. This phenomenon appears to reflect rational choices by parents to pursue gratification from a higher standard of living, regardless of the sacrifice in Darwinian success. In societies around the world, individuals may pursue abortion, infanticide, prostitution, homosexuality, celibacy, or divorce. These behaviors may be given adaptive explanations. However, when predictions of sociobiological strategy and predictions of rational economic choice coincide, is there any need to invoke biological selection?

CONCLUSIONS

The theory of sexual selection is well developed and readily cited to describe human sexual anatomy and behavior. However, observations of human behavior do not easily fit with the theory. The tremendous individual and intercultural variability in mating patterns and sexual behavior seemingly provides support for any hypothesis, yet negates any single pattern for the species.

The importance of cooperative effort and extended parental investment in raising dependent children appears to have shaped human pair-bonding. Women commonly seek mates able and willing to invest and men prefer wives with long-term reproductive potential. However, pair-bonding is not equivalent to sexual behavior.

Human sexuality appear to be an assortment of proximate behaviors removed from their ultimate objective of reproductive success. Central to human sexual behavior is the fact that sexual pleasure has become independent of any drive for increasing reproductive success. The motivation of gratification alone explains a large number of human sexual behaviors. Ovulation became irrelevant, not concealed, as female sexual receptivity was extended over the ovulatory cycle to exploit the advantages of nonconceptive sexual pleasure. In modern society, these economic and political advantages for women may be pursued by an emphasis on sexual attractiveness independently of copulation.

Finally, it is essential to recognize that sexual strategies, as well as contemporary human mating and reproductive patterns, are characterized by the same diversity and flexibility seen in other human behavior systems. Individual behaviors do not need an adaptive explanation. From an evolutionary perspective, it is no more possible to explain divorce patterns or homosexuality in the species than it is to explain why one person seeks a divorce and another does not.

In the past, two lines of discussion have debated monogamy and promiscuity as the evolved human reproductive pattern. The data appear to be inconsistent with a competitive highly promiscuous strategy. They accord better with a mating pattern involving the long-term pair-bonds needed to raise altricial human children, but this is not necessarily support for the notion of rigid monogamy. Clearly each of these behaviors is represented in our species, but is either one a useful characterization of our species? Can we objectively describe ourselves in ways that do not automatically fulfill our expectations? Can we make generalizations that are not merely reflections of our ideals or frustrations?

The two extremes differ fundamentally in how they interpret contemporary human society. Is the male–female relationship better described as competitive or cooperative? Are modern human relationships better described as life-long bonds with occasional lapses or as a coercion of opportunistic individuals into a restrictive pattern? These are political rather than scientific questions, but they have much to do with how we construct and receive models for our evolution. Evolutionary narratives must have a beginning point and an end point, yet if we cannot agree on where we have arrived, we certainly will not agree on how we got here.

Humans apply and manipulate the facts of sexual anatomy and physiology to a variety of cultural mating systems and to individual rational strategies of social and sexual behavior. "Where we have arrived" is not a single mating structure but rather a variety of cultural systems and a flexibility of complex rational behavior that is the hallmark of the human species.

21

THE HUMAN STRATEGY

Life History

Twenty-first century Americans face an overwhelming array of choices, responsibilities, and pressures. We are told to budget—budget our money, budget our time, budget our effort—to balance out the many simultaneous tasks we undertake. Imagine planning your financial budget for a lifetime. How long should your education last and how much can you afford to pay for schooling? How much money should you spend on yourself for everyday needs and comforts? How much should you spend on food? Should you buy expensive, higher-quality food or cheap, less-nutritious food? When should you start to have children? How many offspring should you try to have? How much should you invest in each offspring? What income can you count on? How much money should you set aside for the future? How long should you count on living? This budgeting process consists of a long series of compromises.

These questions are not new or restricted to humans. All organisms have to budget their resources to allow for growth, maintenance, and reproduction. Each species commits to a specific growth trajectory, diet, metabolism, reproductive pattern, and brain size. Each species must answer the questions given earlier in terms of nutrients, not dollars. How each species allocates its available time and energy is its life history strategy. Thus far, this text has largely concentrated on understanding the evolution of individual traits and organ systems. In this final discussion, we will attempt to place them in a broader perspective as part of a strategy for living.

THE r AND K STRATEGIES

Ecologists have long recognized that there is a basic correlation across species between the proportion of energy invested in growth and development and the number of offspring produced. If one were to (over)simplify the problem, one could attempt to describe any given species along a continuum between two extremes—minimal investment in individual growth and very high reproductive output versus maximal investment in individual growth and very low reproductive output. These two tendencies appear to relate to two variables in a standard equation for population growth (Figure 21.1). In this equation, "r" refers to the rate of increase in a population and "K" refers to the carrying capacity, a theoretical maximum sutainable population size. A species with a high reproductive rate—an "r strategist"—will prosper when the initial population density is low and the environment is unstable. A species with high investment in individual growth—a "K strategist"—will be competitive when the population density is high, approaching K, and the environment is stable and predictable.

This is a very simplistic dichotomy with minimal value when applied literally, because it reduces a great many factors to a single dimension (Table 21.1). However, the concept can be very useful as a first approximation, because so many variables correlate with these strategies (Harvey and Clutton-Brock 1985; Harvey et al. 1989; Lee 1989; Martin and MacLarnon 1990; Pianka 1970; Stearns 1976). The concept is most useful when comparing two possible organisms with each other. For example, the bottlenosed dolphin has some superficial similarities to yellowfin tuna in such aspects as habitat and body form but follows a very different reproductive strategy. A dolphin gives birth to a single very mature offspring every 2 to 3 years, following a 12-month gestation period and nursing for up to 18 months. That calf, if a female, will not reach sexual maturity until it is about 6 years old. A tuna may produce as many as 10 million eggs at a time, spawning almost daily during the appropriate

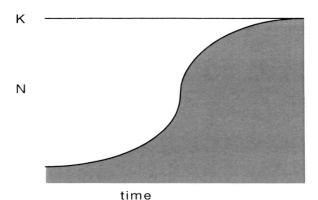

time

Figure 21.1 The Lotka-Volterra equation for logistical population growth. The equation that predicts the rate of population growth is

$$r_N = r_0 N \frac{(K - N)}{K}$$

where r_N is the rate of population increase (as a percentage of current population size); r_0, "natural" or maximum rate of population growth; N, current population size; and K, carrying capacity of the environment, based on finite resources.

Visual inspection of this equation reveals that when N is very small, the effect of limited resources is negligible. As N approaches K, r_N declines. When N = K, there is no growth ($r_N = 0$); when N > K, the population declines because the demand exceeds available resources.

The logistic growth equation may also be expressed graphically (below). Growth is maximum in the middle range when N is much less than K.

season but producing very small young (eggs are about 1.0 mm in diameter). The dolphin is relatively a K strategist, and the tuna, an r strategist.

There are many exceptions to this pattern because species do not easily arrange themselves on a single linear scale. An oak tree, for example, exhibits slow maturation and large body size in the competitive environment of a K strategist, yet the reproductive output of a mature oak tree is immense if measured in the shear numbers of acorns it can produce. Typical of r strategists, individual acorns represent minuscule investments by the parent trees and experience very high morality rates before the trees can mature. Reproductive strategies and many life history traits correlate most closely to juvenile mortality levels (Promislow and Harvey 1990). If offspring mortality

Table 21.1 Correlates of r and K Strategies

Correlates of "r strategy"
  Highly variable or unpredictable environments
  Density independent mortality
  High maximum rate of population increase
  Population size variable, fluctuating; usually below carrying capacity
  Competition centers on rate of reproductive output and colonization of new areas
  Species success depends largely on habitat fluctuations and chance
  High resource allocation to reproduction
  Large brood or litter size
  Small parental investment in individual offspring
  Offspring small
  High offspring mortality
  Small body size
  Rapid development and maturation
  Short life span
For mammals, these features may specifically include
  Shorter gestation
  Smaller birth weight
  Early weaning
Correlates of "K strategy"
  Stable and predictable environments
  Density dependent mortality
  Low maximum rate of population increase
  Population size stable, near carrying capacity
  Competition centers on resources
  Species success depends on quality of offspring
  High resource allocation to growth and maintenance
  Small brood or litter size
  Large parental investment in individual offspring
  Offspring large
  Low offspring mortality
  Larger body size
  Slow development and maturation
  Long life span
For mammals, these features may specifically include
  Longer gestation
  Larger birth weight
  Later weaning

rate is predictably high, it is adaptive for a parent to produce many young and to invest less in each one.

TRENDS

If we compare other organisms with ourselves, the history of life appears to be one of steady increases in animal body size and complexity. (A more objective and less anthropocentric viewer observes an increasing *range* of body sizes and designs over time.) Larger size and greater complexity are indicators of K selection. Indeed, if we follow the traditional *scala naturae* that arranges animals from simple to complex, life history strategy is a one of the many trends we observe. Most marine invertebrates, for example, release huge numbers of very small and simple eggs or larvae into the environment. The lower vertebrates show greater parental investment—their young develop brains and other organ systems from nutrients provided by the parents before they must fend for themselves.

Tetrapods encountered a more variable and hostile environment than their aquatic ancestors. Any offspring that was independent of its parents had to be able to walk and find food on its own. New minimum standards of neuromuscular complexity were demanded, and the result was a slowing of maturation rate and further reduction in the number of offspring produced by any given female.

The mammalian strategy increases metabolism and tolerance of environmental variability. With this comes a larger brain, more complex behavior, and the need for a greater calorie flow. Mammals invest more greatly in their offspring, grow more slowly, and produce fewer young. Among the mammals, these trends are most pronounced among very large species and primates.

HUMAN LIFE HISTORY

STAGES OF LIFE HISTORY

The life of a mammal may be divided into six stages, for the purposes of discussion and comparison: fetus, infant, juvenile, subadult, adult, and postreproductive. The different ways in which species allocate resources to each phase define a life history. In general, the apes have extended each phase of life, and humans have carried that process even further (Figure 21.2).

Fetal Life

The period of fetal life is the gestation period from the mother's perspective. Obviously the offspring is totally dependent on its mother, receiving nourishment necessary for growth and development. The more

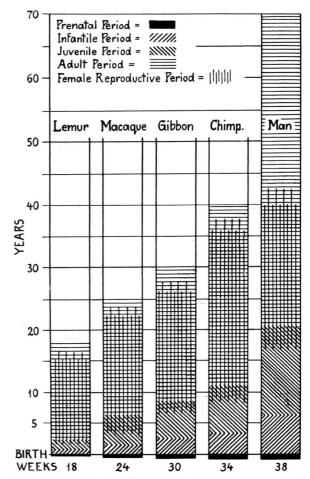

Figure 21.2 Periods in life histories of primates compared. This chart shows a rough relationship between life span, length of parts of life, and body and brain size. Humans are particularly notable for the extended times of immaturity and postreproductive life. (From A. Schultz. *The Life of Primates.* Universe Books. Copyright 1969 by Adolph H. Schultz.)

energy and time allocated to the fetus, the better are its chances of survival after birth. However, gestation places a considerable burden on the mother in terms of energy expenditure, encumbrances that may make her more vulnerable to predators, and opportunity costs of delaying the next offspring.

How long should gestation last? The mother and fetus have different perspectives on this question. Because the maturing fetus represents an increasing cost and encumbrance on the mother, there should be a point at which the cost to her should outweigh the advantages to the fetus of remaining in the womb (Trivers 1974). However, the fetus probably faces better chances of survival where it is and should put off birth as long as possible. The timing of birth is therefore a compromise. Birth is triggered by signals and physiological changes involving both the mother and the fetus.

Is human gestation length exceptional? Among mammals, gestation period can be estimated by adult body size and longevity. Human gestation is not extreme in that regard. However, it is unusually short in proportion to our brain size (which is extreme for our body size) and pattern of brain growth (Figure 21.3). In other primates, the growth of the fetal brain occurs at a high rate until the time of birth; at that point, the rate of increase slows dramatically. The infant human brain, however, continues to grow at the fetal rate through the first 12 months after birth (Gould 1977; Martin 1983b; Vrba 1998). At birth, it is only about 25% of adult size, compared with 47% for the

chimpanzee brain (Foley and Lee 1991). As observed in Chapter 19, this shift from prenatal to postnatal growth is probably related to the ability of the placenta to transfer nutrients from the mother to the offspring. The placenta has a finite capacity for delivering energy to the fetus, but the growing brain requires a constantly increasing rate of supply. At a certain point, it is necessary to switch from placental supply to breastfeeding, in which the quantity of nourishment can be significantly increased. This crossover point apparently occurs at about 9 months for the human fetus, and birth occurs accordingly. (Of course, the rest of the organ systems have to be ready as well.) Based on the pattern of brain growth, the predicted human gestation period has been calculated to be about 21 months, and the immediate postnatal period might be considered a continuation of gestation (Portmann, cited in Gould 1977).

Infancy and Nursing

Infancy is a period of continued dependency on the mother for nourishment through milk. The mother may be freed of carrying her offspring around (in some species), but she continues to face a considerable energetic obligation to produce milk on demand. As the infant grows, it is able to develop muscular and locomotor coordination in the real world and to explore environmental and social challenges without having to face them entirely on its own.

How long should an infant continue to nurse? Again, there is a conflict between the interests of the infant and those of the mother. The mother's breast represents a free, safe meal for the infant, but a nursing infant is a significant cost to the mother's body. She must build up energy reserves and reset her hormonal balance before she can reproduce again. It is in the mother's interest to stop nursing as soon as her infant can reasonably survive on its own.

Several attempts are reported to find patterns among mammals that would predict age at weaning. In a review of these predictors—including age at quadrupling of birth weight, achieving one third of adult weight, six times gestation length, proportion of adult female body weight, eruption of first permanent molar— Dettwyler (1995) concludes that "natural" human weaning should occur between 2.5 and 6 years of age.

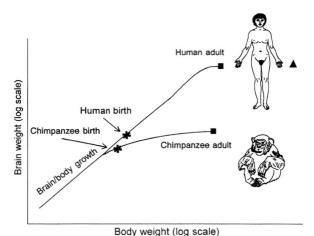

Figure 21.3 Growth trajectories of humans and chimpanzees. The growth trajectories of humans and other primates (such as the chimpanzee) are similar in shape but differ in timing relative to birth and weaning. Humans are unique in that the brain continues to grow significantly after birth. (From B. Bogin. 2001. *The Growth of Humanity*. Copyright Wiley-Liss. Reprinted with permission of Wiley-Liss, Inc.)

Her survey of average age at weaning in 64 traditional societies shows a pattern of weaning at just under 3 years. This finding, that the period of human nursing is relatively short, comes as a surprise in light of the argument given earlier that part of fetal development has been pushed into infancy, yet the human period of nursing is shorter than that of any of the apes (Key 2000). It is appropriate, therefore, that the next phase of life history, childhood, is redefined for humans.

Juvenile Life and Childhood

Infancy ends with weaning for nonhuman species. From that point on, an individual must forage independently in order to survive, although it may still live in a social group. As long as it is reproductively immature, it is considered a juvenile. The juvenile period of life tends to be emphasized in highly social animals, such as canids, elephants, lions, and primates. These species also tend to have larger brains and to use this period for learning social and technical skills. For some species, especially primates, the individual may maintain a closer relationship with its family while it continues to the process of socialization. Nonetheless, a juvenile should be able to survive without direct support from adults.

Human childhood is a unique transition period between the total dependency of infancy and the relative independence of juvenile life (Bogin 1988, 1990, 1997). Only human children are usually unable to obtain their own food. They require an unusually high-energy diet because of their immature digestive tract and because of continued brain growth, which is completed toward the end of childhood, when individuals are capable of independent foraging. During this time, the responsibilities of caring for a child are usually shared widely by other adults in the community, called *alloparents*, or "other parents." Early childhood may be understood as compensating for the gap between the predicted and actual times of weaning, when the child can continue to rely on the mother and alloparents, but her body is relieved of the increasing energetic and nutritional burden of providing all of its food.

Subadults and Maturity

A subadult (cf. adolescent) is in the process of becoming sexually mature but not yet reproducing. It takes time for the reproductive system to mature to a functioning stage. For example, females may have irregular menstrual cycles before they begin to ovulate and become fertile. The period between first menstruation (menarche) and first conception, or possible conception, is sometimes referred to as adolescent infertility. During the subadult period, mating skills are learned and added to the growing body of social skills. Once an individual begins to reproduce or is fully capable of it, he or she is considered an adult.

At what age should an individual become sexually mature? Once reproduction begins, mating, gestation, and nursing place such an energy burden on both males and females that body growth is no longer practical. Thus, sexual maturation usually coincides with achieving final body size. It is logical that final development of both the skeleton and the gonads is regulated by the same factors—sex hormones. The advantages of putting off adulthood are continued growth and storage of nutrients, to increase both future reproductive potential and survival. However, two factors favor an earlier start to reproducing: a limited life span and the possibility of sudden death. If life expectancy does not increase in proportion to age of maturity, then delayed maturation occurs at the cost of reproduction. Then there is the chance that a predator will erase all of the investment in a single moment.

Postreproductive Life

How long should an individual continue to reproduce? By the Darwinian scorecard, the production of new offspring should continue as long as possible—that is, until death. For most mammals, that is what we observe, although reproductive decline is a common aspect of aging in both sexes and in different species (Caro et al. 1995; Packer et al. 1998; Paul et al. 1993). Males exhibit a gradual long-term decline in fertility. In most species, if male reproduction terminates before death, it is because the older males can no longer compete with younger ones for mating opportunities. The decline manifests itself among females as lower age-specific fertility and greater interbirth intervals—indicators possibly of irregular ovulation or failed conception. However, humans are unusual in experiencing menopause, a total and permanent cessation of ovulations, while a significant portion of the life span still lies ahead. Because the human female postreproductive span is an unusual pattern, it has attracted considerable attention (e.g., Pavelka and Fedigan 1991).

Is human menopause normal for the species? As experienced by women in Western society, menopause is a specific life event, occurring about age 50 and accompanied by predictable symptoms that include hot flashes, dizziness, and mood swings. Women in nonindustrial societies who reproduce more or less continually may not experience such symptoms. Instead, they may simply not resume cycling following their final child. Thus, reproductive cessation is normal for women who live beyond ages 45 to 50. It becomes evolutionarily relevant if a significant number of women survive to that age. Such is true of most societies today, but the past is less well known.

Aging is understood to be the consequence of accumulated damage to the body, beginning at the molecular level (Kirkwood and Austad 2000). As cell lines specialize, proteins alter, and chromosomes mutate, repair mechanisms are unable to keep up with demands and functions deteriorate. This process determines maximum longevity of an individual or a species. The aging process that appears to lead to menopause is the simple fact that the ovaries run out of eggs. At birth, the ovary already contains all of the eggs a woman is ever going to have. Before puberty, the eggs are contained in follicles in a state of suspended activity, part way through the first meiotic cell division. Each month, when the ovulatory cycle begins, surges of gonadotropins from the pituitary "awaken" a number of these follicles, and the eggs within them resume cell division. Normally only one of these is permitted to mature fully, and the others are destroyed. Thus, although a woman is born with potentially millions of eggs, follicles are continually used or lost throughout her reproductive life until, at about age 50, she runs out. As the essential interaction between ovary and endocrine system is upset, reproductive functions cease and menopause occurs (Ellison 2000; Leidy 1994; O'Connor et al. 2001; Wise et al. 1996).

Menopause and fertility decline may also be understood as a secondary outcome of reproductive strategy. The lifetime reproductive potential of animals is finite. In both males and females, earlier or more intense reproductive investment means a sacrifice of longevity and/or reproduction in later life. The observed age of menopause may represent the result of that compromise (Peccei 1995; Packer et al. 1998).

If aging is to be a sufficient evolutionary explanation for menopause, we must assume that an alternative pattern of continued reproduction has not been available on which natural selection might have acted. This may well be true, but it is also possible that menopause is a superior strategy—but only for humans and the few other species that experience it. As a woman's body ages, the chances of a new baby being healthy and born (and nursed) successfully decline. When would it benefit her reproductive output to stop having more babies altogether?

One possible answer is when she can spend her resources more effectively on her children or grandchildren that have already been born. In the so-called "grandmother" hypothesis, a postmenopausal woman can share knowledge and wisdom, help care for grandchildren, and help forage for food. Her inclusive fitness might be higher in such a role than if her energies were diverted into increasingly unsuccessful or risky pregnancies (Diamond 1996; Hawkes 2003; Hawkes et al. 1997, 1998; Hill and Hurtado 1991; O'Connell et al. 1999). Such hypotheses depend on the sharing economy that is unique to human culture and would not apply to other species (Packer et al. 1998).

Studies among hunter-gatherers show that many women to survive to menopausal age (Blurton-Jones et al. 2002) but that they have not demonstrated that grandmothers can gain more by helping than by having more offspring (Hill and Hurtado 1991, 1999; Peccei 2001a, 2001b; Shanley and Kirkwood 2001). Any adult or group sharing could make an equally significant contribution to child support. Observations of other species that have menopause (Alvarez 2000; Packer et al. 1998) but not a corresponding investment in offspring likewise undermine the model. In the absence of good information about past mortality rates (did many early women survive to menopause?) and social structure (were other adults around to help?), a definitive test of this model may not be possible. The contributions that grandmothers and other adults make in helping to raise children may well be secondary to an aging process that will shut down reproduction (Alvarez 2000; Peccei 1995, 2001b; Shanley and Kirkwood 2001).

It is therefore likely that menopause does represent the natural aging process, but it may also be true that the economic dependence characteristic of human society developed more easily because post-reproductive individuals were available.

LONGEVITY AND EVOLUTION OF THE MODERN LIFE SPAN

How long should a person live? The life expectancy in a given society can change rapidly as social stability, standard of living, and health care evolve. Life expectancy in Western countries has shown almost a linear increase over the past two centuries (Oeppen and Vapel 2002; Tujipurkar et al. 2000), whereas in the author's lifetime, life expectancy in the United States has increased from the upper 60s to more than 80 years of age. Of course, that varies considerably with sex, income, and race. Life expectancy is not the same as longevity, the maximum age observed in a species. It has been argued that a human population has a potential life expectancy of 100, under ideal conditions and that individuals can reach a maximum of 120 years. These observed changes are due to environmental factors, not selection for genes. However, there is accumulating evidence of genetic variations that could increase life span (Perls and Fretts 2001; Perls et al. 2002). Thus, we should assume our current longevity has evolved and has the possibility of further change in the future.

Old age is not necessarily a worthwhile goal of a species life history strategy. Recall that life history is the strategic allocation of energy among growth, maintenance, and reproduction. We would expect that higher investment in growth or reproduction would reduce life span, and that is what has been observed in many species. Limiting the number of calories in the diet to a level sufficient for health but less than an individual would freely consume may limit growth but extends the life span of many animals, including insects and rodents (e.g., Kirkwood et al. 2000) and probably humans (Lummaa and Clutton-Brock 2002; Roth et al. 2002). Similarly, early or greater investment

in reproduction correlates with reduced life span in flies and mammals (Promislow and Harvey 1990). In limited studies of human societies, negative correlations between age of reproduction and life expectancy and between number of sons and life expectancy have been observed (Doblhammer and Oeppen 2003; Helle et al. 2002a, 2002b; Westendorp and Kirkwood 1998).

Other factors relevant to humans also correlate with longevity, particularly brain size. A relatively larger brain is accompanied by slower development and maturation and a longer life span (Sacher 1975). Not surprisingly, therefore, humans have the greatest known life span among mammals, and brain expansion must be considered an important factor in human life history evolution (Holliday 1996; Kaplan and Robson 2002). This correlation has been used to predict longevity among fossil hominins (Helmuth 1999). Australopithecines would resemble apes in this regard, and only with *Homo erectus*–grade hominins would there be a significant increase in life span. Interpretations of skeletal growth has pushed the human pattern later, to after *H. erectus* (Bogin and Smith 1996; Smith and Tomkins 1995).

Dental Maturation in Early Hominins

What is needed to answer the question of life histories among extinct hominins is a patterned sequence of developmental events that is identifiable in the fossils and that also clearly distinguishes humans from apes. Dental development may represent such a sequence (Smith 1989, 1991, 1992). The sequence of human tooth eruption is different from that of closely related primates and is reasonably predictable. A dental anthropologist can estimate the age of a human child on the basis of which teeth have erupted and which have not. Significantly, the teeth of humans and apes erupt in a slightly different sequence, due primarily to the longer time it takes for the larger canines and incisors of apes to develop (Simpson et al. 1990). Several anthropologists have attempted to estimate the ages of immature australopithecines from this pattern and to make inferences about maturation rates and patterns. Additional lines of research now use data on crown and root formation and the histology of enamel.

One of the first systematic examinations of early hominin dental maturation was that of Mann (1975), who concluded that the sequence and rate of tooth formation and eruption were essentially like those of modern humans. The inference that a prolonged infancy and childhood supported a larger australopithecine brain fit in with prevailing views that early hominins already displayed to some degree the characters which define modern humans. However, later studies that included additional fossils found an apelike sequence of tooth eruption (e.g., Anemone 2002; Bromage 1987; Conroy and Kuykendall 1995; Conroy and Vannier 1987; Smith 1986, 1991, 1992, 1994; Smith et al. 1995). These conclusions fit into the current general perception of australopithecines as fundamentally apes that had acquired a form of bipedalism. Furthermore, observations of heavy wear on the deciduous teeth has been cited to argue for short infancy and interbirth periods (Aiello et al. 1991). However, the use of tooth eruption sequence alone is of questionable significance, because the early appearance of anterior dentition reflects the reduction in hominin anterior dentition and not necessarily a change in maturation rate (Simpson et al. 1990, 1991, 1992). Furthermore, eruption is influenced by such factors as crowding in the jaw that also are not related to life history patterns.

Healthy debate in a field often spurs new lines of research, and the discussions around this problem are a good example. Researchers have moved to the histological level for indicators of rates of growth (Beynon et al. 1998; Dean 2000). Both crown and root formation is slowed in human children (Anemone et al. 1991, 1996; Dean and Beynon 1991; Reid et al. 1998), so, for example, there are clearer intervals between the development of individual molars. In contrast, the teeth of chimpanzees develop more quickly and there is greater overlap in crown formation of different molars. Microstudies of enamel patterns found that markings in the tooth, called perikymata, represent daily and perhaps weekly indicators of growth, much like the rings of a tree trunk (Dean 1987; FitzGerald 1998; Risnes 1998). Absolute counts of these suggest short development periods in australopithecine fossils and differences from humans (Beynon and Dean 1988; Beynon and Wood 1987; Dean 2000; Dean and Reid 2001; Dean et al. 1993, 2001).

These differences from modern humans persist in species early *Homo*, including *Homo habilis*, *H. ergaster*, and *H. erectus* (Dean and Reid 2001; Dean et al. 2001). Chronologically, the next good sample of juvenile dentition comes from the Middle Pleistocene site at Atapuerca about 300,000 years ago. This population of *H. heidelbergensis* shows an eruption pattern similar to that of modern humans, although canine formation is slightly delayed (Bermudez de Castro and Rosas 2001). Neanderthals are also reported to match the modern pattern (Smith 1994).

Tentatively, therefore, we have evidence that important features of modern life history strategy appear only relatively late in the human lineage, sometime after 1 million years and after brain size grew to about 1200 cm^3. Because the studies of enamel histology appear to be more precise than those of eruption sequence and gross tooth formation, future studies may push that landmark of our evolution closer to the present.

FORAGING, ENERGY, AND SOCIETY

In our society, very young and very old people are not expected to provide for themselves. This is no less true in hunter-gatherer societies, where neither children nor the aged accumulate enough calories from daily hunting and gathering to provide for their own needs. The pattern is in marked contrast to what is observed in other mammals and tells us something very important about our species: food sharing is both unique and absolutely essential.

Kaplan and colleagues (2000) graphically illustrated the ability of chimpanzees and Ache foragers to provide food (Figure 21.4). Chimpanzees consume what they collect; thus, their productivity and consumption are roughly equivalent from the time of weaning, about age 5, until death. Ache males, not surprisingly, begin to hunt effectively when they are between 15 and 20 years old. From that time, they can obtain more than twice as many calories as they consume. This level of productivity may be sustained until about age 50, at which time production decreases, after age 60, they are dependent on younger members of the society, although senior citizens may still contribute to the band through their experience and specific skills.

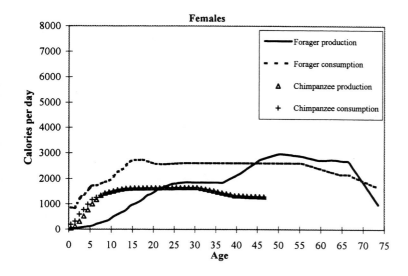

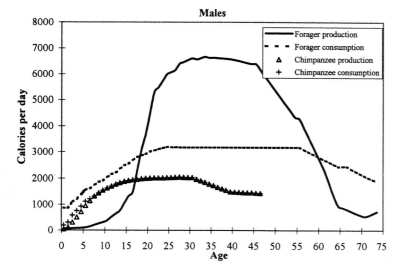

Figure 21.4 Comparison of production and consumption in human hunter-gatherers and chimpanzees at different ages. These graphs show that young and old humans are dependent on others to supplement the food that they can obtain for themselves. Significantly, women produce less than they consume through their childbearing years. (From H. Kaplan et al. 2000. A theory of human life history evolution: Diet, intelligence, and longevity. *Evol. Anthropol.* 9[4]:156–185. Reprinted with permission of Hilliard Kaplan and Wiley Liss, Inc.)

Ache women were not self-sufficient. Average calorie consumption is comparable to that of larger men because of costs of child-bearing and nursing. However, reproducing and child care also interfere with foraging and limit the amount of food a woman brings home. Between ages 20 and 40, an average woman produces about two thirds of what she consumes. Only after she stops having children, from age

45, does her productivity exceed her consumption. This postreproductive period, from 45 to 65, is the time when a "grandmother" could provide needed assistance to her own daughters and grandchildren.

Of course, postmenopausal women are not the only source of food for a busy mother. According to these figures, her husband is overproducing enough to support her and her children. It is impossible to reconstruct the social organization of prehistoric hominins, nor are there equivalent data available for the rest of modern humans. However, it is clear that economic networks have long been an essential part of human society and that food sharing is universal. Lovejoy (1981) proposed that the ability of a male to provide supplementary calories to a female not only accounted for bipedalism and family structure but also represented a demographic breakthrough that established the evolutionary success of the hominin line. Aiello and Key (2002) pointed out the comparable advantage that provisioning of children would have had by replacing nursing and shortening the birth interval. It would not matter who provided the additional calories. For various reasons, specifics of Lovejoy's hypothesis concerning bipedalism have

not held up, but his recognition of the economic solution to the demographic hurdle that hominins crossed has been supported by more recent reviews of life histories (Hill 1993; Kaplan et al. 2000; Key 2000).

Perhaps more critical is the vulnerability of the children to the reasonably high possibility that one or both parents will die (Kennedy 2003). Orphans always have been a common occurrence and extended family and adoptive parents would become important as fathers and grandmothers in raising them. What is unusual is that humans are willing to adopt the children of others. The adage "it takes a village to raise a child" becomes particularly apt. Once again, we see that the ultimate human strategy is behavioral and adaptive flexibility.

GLOSSARY

abductor hallucis muscle   Small muscle on the medial side of the foot inserting into the first toe.

acetabular joint   Hip joint, between the coxal bone and the femur.

acetabulum   Deep socket in the coxal bone for the head of the femur.

accessory motion   Motion that may occur at a joint when the two bones have been separated by external forces.

acrosome   Coating of digestive enzymes on the head of the sperm that allows it to penetrate the protective materials surrounding an egg.

adaptation   Trait that increases the ability of an organism to survive and reproduce.

adenohypophysis   Anterior endocrine portion of the pituitary gland.

adipocyte   "Fat cell", primarily a vacuole containing stored lipids.

adipose tissue   Connective tissue in which fat cells, or adipocytes, predominate.

adrenalin   Hormone epinephrine.

adrenocorticotropic hormone (ACTH)   Hormone of the adenohypophysis that stimulates activity by the cortex of the suprarenal gland.

Allen's rule   Allen's rule asserts that species or populations living in tropical and warmer climates have longer extremities, increasing surface area and exposure.

alveolus   Minute air chamber in the mammalian lung in which gas exchange occurs.

ameloblast   Specialized embryonic cell that lays down enamel.

amnion (amniotic membrane)   Membrane surrounding and protecting an embryo from dehydration and other hazards. The amnion made it possible for vertebrates to reproduce on land.

amorphous bone   Developmentally the first bone laid down, without specific orientation of its tissue.

amygdala   Basal ganglion in the temporal lobe functioning most conspicuously in emotion and memory.

anadromous   Able to thrive in aquatic habitats of a wide range of salinity. It has been proposed that early vertebrates were anadromous.

angular (rotational) acceleration   Acceleration in a rotational manner about an axis.

angular bone   Small bone at the back of the lower jaw of early tetrapods. In mammals, the angular becomes the tympanic bone (not distinct in humans).

annulus fibrosus   Ring of fibrocartilage that makes up the outer portion of an intervertebral disc.

antebrachium   Forearm, from wrist to elbow.

aorta   Primary artery conducting blood from the heart to the body tissues.

aphasia   Any disorder of the brain interfering with the processing of language.

apocrine gland   Type of gland in which the secretions are released by pinching off a part of the cell contents within a portion of membrane.

aponeurosis   Flat sheet of fibrous tissue functioning as a tendon for a muscle.

arachidonic acid (AA)   Long-chain polyunsaturated fatty acid of the omega-6 family that is used in cell membranes.

arcade   Shape of the tooth row as it lies on the jaw.

archicortex   Primitive and ancient portion of cerebral cortex involved in limbic functions.

archinephric duct   Duct draining urine from a holonephros.

arrector pili muscle   Smooth muscle attached to each hair follicle responsible for raising the hair to an upright position.

articular bone   Small bone at the back of the mandible in early tetrapods, articulating with the quadrate on the cranium. In mammals, the articular becomes the malleus in the middle ear.

articular cartilage   Layer of hyaline cartilage covering the joint surface at a synovial joint.

association area   One of several regions of the cerebral cortex that receive diverse sources of information and appear to be responsible for a higher-level processing rather than sensation or output.

atlantooccipital joint   Joint between the occipital bone of the skull and the first vertebra (atlas).

atlas   First vertebra, on which the cranium rests.

atrium   First chamber(s) encountered by blood entering the heart (in tetrapods) or the homologous chamber in other vertebrates.

autonomic nervous system   Functional division of the nervous system that regulates the performance of visceral organs, glands, and vessels.

axis   Second vertebra, C2.

axis of rotation   Imaginary line about which a bone rotates during normal joint movement.

axon   Long neuron fiber conducting signals away from the cell body.

baculum   Bone within the penis in many nonhuman species.

basal ganglion (striatum)   One of several pairs of nuclei deep within the cerebrum acting as coprocessors of diverse cortical functions.

basicranium   Bottom of the braincase, primarily consisting of the occipital bone and parts of the temporal and sphenoid bones.

basilar membrane   Membrane to which hair cells of the cochlea are anchored.

Bergman's rule   Bergman's rule states that species or populations living at high latitudes tend to increase their body mass.

biceps muscle   Flexor muscle of the forearm arising from the scapula, passing in front of the elbow and inserting on the radius.

bicuspid   Having two cusps. Human premolars are bicuspid, but the more anterior lower premolar of apes has only one cusp.

bile   Digestive secretion of the liver containing waste products and emulsifiers to break apart lipids in food.

brachialis muscle   Flexor muscle of the forearm passing in front of the elbow and inserting on the ulna.

brachium   Upper arm, from elbow to shoulder.

brainstem   Informal caudal portion of the brain that includes "primitive" functions of homeostatic regulation.

branchial arch   Tissues surrounding and defining the slits in the pharynx of primitive chordates or the gill slits of fish. Most vertebrates have six pairs of arches at an early stage of development, each of which is supported by skeletal elements and other tissues.

Broca's area   Region of the frontal lobe associated with complex motor patterns, including speech.

browridge (supraorbital torus)   Prominent bony crest above the orbit in the great apes and many early hominins.

calcaneal tuberosity   Bone of the heel, a weight-bearing process of the calcaneus.

calcaneus   Heel bone, a tarsal at the back of the foot.

callus   Thickening of the epidermis induced by increased wear and mechanical stimulation.

canine (tooth)   Tooth just lateral to the incisors. In many mammals, the canine is long and has a single point.

capillary   Microscopic vessel connecting arteries with veins and completing the vascular circuit. Nutrients and oxygen leave the bloodstream from the capillaries to supply the body tissues.

capitate   One of the carpal bones, at the base of the fourth metacarpal.

capitular facet   Articular surface on the humerus for the head of the radius.

carnivory   Referring to a diet that consists primarily of the flesh of other mammals.

carpal   One of the small bones of the wrist and proximal hand. Humans have eight carpals in each hand.

carrying angle   Angle formed by the shaft of the femur in its normal position and a perpendicular line. Humans have a significant carrying angle to better position the knees under the center of body mass.

cartilaginous joint   Joint at which bones are connected by a form of cartilage.

caudal   Pertaining to the tail.

caudate   One of the basal ganglia functioning most conspicuously in movement.

cavernous sinus   Dural sinus lying inferior to the hypothalamus, on either side of the hypophysis.

cecum   Saclike proximal end of the colon.

central canal   Hollow cavity of the developing spinal cord. In the adult, the central canal is a closed "potential space."

central nervous system   Brain and spinal cord considered together.

cephalization   Formation of a head, defined by a concentration of sense organs and the mouth at the anterior end of an animal.

cerebellar peduncle   Large bundle of fibers communicating between the cerebellum and the brainstem.

cerebellum   Structure of the brain stem responsible for coordinating sensory and motor information to smooth motor performance.

cerebral aqueduct   Communicating pathway through the midbrain between the third and fourth ventricles.

cerebrum   Most rostral part of the brain; the part of the mammalian brain responsible for sensory integration and consciousness.

cervical   Pertaining to the neck.

cholecystokinin   Hormone of the digestive tract that stimulates release of pancreatic enzymes and bile and communicates with the brain.

chondrocranium   Portion of the cranium derived from cartilage that surrounded and protected the brain and sensory organs.

chondrocyte   Type of cell specific to cartilage that secretes chondromucoprotein.

chorion   Outermost layer of embryonic membranes, in contact with the maternal tissues of the uterus.

circadian rhythm   Pattern of behavior and body physiology that cycles on a daily basis.

clavicle   Collar bone; a slender bone connecting the rest of the upper limb to the sternum.

close-packed position   Position of a joint in which congruence, ligament tension, and stability are all maximized and movement is inhibited.

coccyx   Rudimentary caudal spine in apes and humans. Usually three to five caudal vertebrae fuse to make up the human coccyx.

cochlea   Portion of the inner ear that registers and reports sound to the brain.

coelom   Body cavity, of which the peritoneal cavity is a division.

collagen   Protein found in connective tissues of all animals that anchors cells in place and provides tensile strength to tissues.

collecting duct   Last part of the nephron tubule carrying urine from numerous nephrons through the pyramid to the renal pelvis. Water is drawn through the walls of the collecting duct to concentrate the urine.

colliculus   One of two paired prominent nuclei on the dorsum of the mammalian midbrain. The superior and inferior colliculi are homologous to the optic lobe and auditory lobe of other vertebrates.

colon   Large intestine, responsible for storage and compaction of feces and the recovery of water and a few nutrients.

colostrum   Protein-rich milk produced during the first few days after birth.

concentric contraction   Contraction of a muscle in which the force of contraction exceeds resisting forces and the length of its fibers is allowed to shorten.

condyles of the femur (medial and lateral)   Processes at the bottom of the femur that articulate with the tibia.

cone   Type of primate photoreceptor that is sensitive to a narrower range of wavelengths and so is identified with color vision.

congruence   Measure of the extent of contact between two bones at a diarthrosis.

conjunct motion   Motion at a joint that occurs in addition to its primary rotation. Conjunct motion is unavoidable and is created by the shapes of the surfaces of the joint.

connective tissue   One of the fundamental classes of animal body tissues. Connective tissues consist of cells suspended within a noncellular matrix. The nature of the matrix and types of cells vary in different types of connective tissue.

contrahentes muscle   Muscle within the hand and foot of many primates but not humans. Slips of the muscle insert on metapodials and draw them together.

contralateral   Pertaining to the opposite side of the body. For example, the cerebral cortex receives sensory information from the contralateral side of the body.

conus arteriosus   Fourth and final chamber of the heart in lower vertebrates and vertebrate embryos. The conus arteriosus divides in mammals to form the first part of the aorta and pulmonary trunk.

convergent evolution   Similar evolution of traits occurring in distantly related lineages so that non-homologous derived traits possess comparable form and functions.

conversion mechanism   Ability of the midtarsal and subtalar joints of the foot to assume either of two configurations, one mobile and the other rigid.

coracoid process   Process of the scapula medial to the shoulder joint and deep to the clavicle that receives the attachment of pectoralis minor.

coronal plane   Anatomical plane that would divide the body into ventral and dorsal portions.

coronoid process   Crest of bone just in front of the trochlear notch of the ulna.

corpus callosum   Bundle of nerve fibers communicating between the two cerebral hemispheres.

corpus luteum   Structure that develops in the ovary from follicular cells following ovulation. The corpus luteum secretes progesterone to sustain the endometrium of the uterus for pregnancy.

corpuscle (of a nephron)   Proximal part of a nephron, consisting of a capillary bed, or glomerulus, in which filtration occurs and a surrounding capsule in which the filtrate is contained.

cortex   Outer layer of tissue of an organ, especially of the brain.

cortical bone   Dense bone on the surface of skeletal elements.

corticobulbar fibers   Descending pathway from the cerebral cortex to the various nuclei and parts of the brainstem and cerebellum.

corticoids   Hormones of the suprarenal cortex, of which cortisol is an example.

corticospinal tract   Descending pathway from the cerebral cortex to the spinal cord carrying commands for voluntary movements.

costovertebral articulation   Joint between a rib and a vertebra.

countercurrent exchange   Vascular network within the body involving two pathways for fluids traveling in opposite direction in close contact with one another. A countercurrent exchange can efficiently exchange heat, oxygen, or other substances to maintain high gradients between tissues.

coxal bone (innominate bone)   Formed by the fusion of the ilium, ischium, and pubis, the coxal bone is the larger bone of the pelvis.

cranial nerve   One of 12 pairs of nerves emerging from the brain.

cranium   Skeleton of the head minus the lower jaw.

cribriform plate   Region of the ethmoid bone with many small passages through which the olfactory nerve communicates between the nasal cavity and the anterior cerebral cavity.

cruciate ligaments   Anterior and posterior cruciate ligaments are two important ligaments between the condyles of the knee joint.

crus   "Leg" or lower leg, from ankle to knee.

cuboid   Tarsal bone on the lateral side of the foot.

cuneiform   One of three tarsals, arrayed at the bases of the first three metatarsals.

cupula   Part of the macula of the inner ear. The cupula is a gelatinous structure in which the otoliths are embedded. It is affixed to the tips of the cilia of the hair cells and moves independently of the cells, thus bending the cilia.

Darwinism   Theory that organisms evolve gradually through natural selection.

decussation   Symmetrical crossing of the midline by fibers of the brain or spinal cord.

degrees of freedom   Number of different and independent axes about which movement at a joint may occur. The maximum degrees of freedom for a joint is three.

deltoid ligament   Largest ligament of the ankle, on the medial side.

deltoid muscle   Large muscle of the side of the shoulder. The deltoid is the primary abductor of the shoulder.

dentary   Tooth-bearing bone of the lower jaw. In mammals, the dentary is the only bone of the jaw and is equivalent to the mandible.

denticle   Toothlike structures in the skin of various lower vertebrates. Teeth probably evolved from denticles.

dentin   Underlying tissue of a tooth, composed of mineral interlaced with collagen fibers.

derived trait   Trait that has changed from the homologous version of the trait present in an ancestral species. The opposite of a derived trait is a primitive trait.

dermatocranium   Portion of the cranium derived from dermal or membranous bone, consisting of parts of the braincase, face, and jaws.

dermis   Connective tissue of the skin lying immediately deep to the epidermis.

diaphysis   Shaft of a long bone.

diarthrosis   Synovial joint, in which two bones are separated by a synovial cavity and connected by synovium and ligaments.

diastema   Gap, particularly the space between the canine and adjacent premolar in many nonhuman anthropoids.

diencephalon   Region of the brain that includes the thalamus, hypothalamus, and smaller structures. In the human brain, the diencephalon is almost entirely hidden by the cerebrum.

diffusion   Passive flow of molecules in solution tending to create a uniform distribution or density.

digitigrade   Term describing an animal that walks placing the digits flat on the ground while the palm or sole is held above the ground.

diphyodont   Mammalian condition of possessing two generations of teeth (deciduous and permanent).

docosahexaenoic acid (DHA)   Long-chain polyunsaturated fatty acid of the omega-3 family that is preferentially incorporated into the membranes of neurons.

dopamine   Neurotransmitter of the brain whose imbalance relates to a variety of mood disorders.

dorsal column   Ascending pathway of the spinal cord carrying messages about touch and pressure senses.

dorsal spine   Process for attachment of muscle and ligament projecting from the vertebral arch dorsally and on the midline.

dorsum sellae   Process of bone on the floor of the cranial fossa that rises anteriorly from the foramen magnum.

duodenum   Proximal portion of the small intestine where most of the digestive enzymes are mixed with the food.

dural sinus   One of many blood-filled passages within the dura mater. Blood from the brain drains through the dural sinuses and then leaves the skull via the internal jugular vein.

eccentric contraction   Action of a muscle in which the force of contraction is less than opposing forces and the muscle length increases in length.

eccrine gland   Type of gland that secretes directly across its cell membrane through a process of filtration. The sweat glands in the skin are eccrine glands.

elastic cartilage   Type of cartilage containing significant quantities of the protein elastin. Elastic cartilage is found in the ear and nose, where it can tolerate a great deal of temporary deformation.

elasticity   Ability of an object to resume its original shape after it has been deformed by a stress.

elastin   Protein found in many connective tissues that provides elasticity to those tissues.

electrolyte   Dissolved ion or charged particle. Many molecules, such as salt, dissolve in water by dissociating into charged particles.

embryo   Early stage of animal development in which most organs of the body are not yet defined or functional.

endometrium   Internal lining of the uterus in which an embryo implants and which maintains interaction with the placenta throughout pregnancy.

emissary foramen   Any of a number of mostly variable passages conducting veins through the bones of the braincase.

emissary vein   One of many, mostly variable, veins draining from the scalp through the skull into the dural sinuses.

enamel   Hard outer layer of a tooth, composed almost entirely of crystalline hydroxyapatite.

encephalization   Relative expansion of brain size.

encephalization quotient   Measure of the size of the brain relative to body mass.

endocast   Cast of the inner surface of the cranial fossa.

endocrine system   Traditional body system consisting of miscellaneous hormone-secreting glands.

endolymph   Fluid contained within the membranous labyrinth.

endorphin   Neurotransmitter released in response to stress or pain that has a calming effect on the brain.

endothermy   Ability of an organism to regulate its body temperature by the internal generation of heat.

epaxial muscles   Muscles of the back, supplied by the dorsal rami of the spinal nerves.

epidermis   Epithelium that comprises the outermost layer of skin.

epiglottis   Cartilage at the top of the respiratory tract that deflects liquid and solid particles from the larynx and into the esophagus.

epinephrine (adrenalin)   Hormone of the suprarenal gland that participates in body arousal during the stress response.

epiphysis   Articular end of a long bone, which ossifies independently of the diaphysis.

epithalamus   Diverse cluster of miscellaneous nuclei and tracts in the diencephalon, including the pineal gland and habenula.

epithelium   Class of tissue that forms boundary layers of the body. Epithelia consist of one or more layers of cells anchored to one side of a noncellular basement membrane.

erector spinae   Set of longitudinal muscles on either side of the spine on the back.

erythrocytes   Red blood cells.

esophagus   Portion of the digestive tract between the pharynx and stomach responsible primarily for the transport of food to the abdomen.

estrogen   One of a class of female hormones secreted by the ovary.

estrus   State of females of some species near the time of ovulation in which they signal fertility and heightened receptivity to males.

ethmoid bone   Bone of the center part of the face, including the walls of the nasal cavity and parts of the orbits.

eumelanin   Dark variant of the pigment melanin commonly found in dark-skinned people.

eustachian tube   Air-filled passage between the middle ear and the pharynx.

evolution, biological   Change in gene frequencies or in the expression of heritable traits in a population through time.

expressive aphasia   Class of language disorders of the brain interfering with the expression of language.

external auditory meatus   Passage of the ear within the temporal bone from the pinna to the eardrum.

fascia lata   Tough sheet of fascia and aponeurosis deep to the skin of the thigh, especially on the lateral part.

faunivory   Referring to a diet that consists primarily of other animals, including invertebrates.

femur   Bone of the thigh, or the thigh itself.

fenestra   "Window" or opening in a flat bone.

fetus   Later stage of vertebrate development, especially mammalian, before birth in which all of the major organs of the body are functioning.

fibrocartilage   Type of cartilage containing significant quantities of the protein collagen. Fibrocartilage has a great tensile strength and is commonly found in fibrous joints.

fibrocyte   Type of cell that manufactures collagen fibers.

fibrous joint   Joint at which bones are connected primarily by collagen fibers.

fibula   Smaller and more lateral bone of the lower leg.

fibular collateral ligament   Ligament of the knee on the fibular side of the knee joint.

filtration   Filtering of a fluid under pressure.

flexor digitorum brevis muscle   Small muscle on the sole of the foot inserting into and flexing the toes.

flexor digitorum longus muscle   Deep muscle of the calf inserting into and flexing the toes.

flexor fibularis muscle   Deep muscle of the calf inserting into and flexing the toes. This is called the flexor hallucis longus in humans.

flexor hallucis longus muscle   Deep muscle of the calf inserting into and flexing the toes, especially the first one.

flexor pollicus brevis   Short flexor muscle of the thumb.

flexor pollicus longus   Long muscle of the forearm inserting into the thumb.

flexor tibialis muscle   Deep muscle of the calf inserting into and flexing the toes. This is called the flexor digitorum longus in humans.

folivory   Referring to a diet that consists primarily of leaves.

fontanelle   Unossified region of a neonatal skull at the intersection of the bones of the braincase. Normally human infants have two fontanelles, anteriorly and posteriorly over the brain.

foramen magnum   Large passage from the underside of the skull through which the spinal cord connects to the brain.

forebrain   Prosencephalon, the most rostral vesicle of the brain.

fourth ventricle   Chamber between the cerebellum and pons filled with cerebrospinal fluid.

frontal bone   Bone of the forehead, including the top of the orbit and front of the braincase.

frontal lobe   Division of the cerebral cortex next to the frontal bone.

frugivory   Referring to a diet that consists primarily of fruit and selected plant parts.

gastrocnemius muscle   Most superficial muscle of the calf and part of the triceps surae.

glenoid fossa   Shallow socket on the scapula that receives the head of the humerus to form the shoulder joint.

glial cell   Support cell of the brain. There are many types of glial cells that help to maintain the health and function of the neurons and may participate in information processing.

glomerulus (in olfactory bulb)   Cluster of neurons where fibers reporting a specific class of smells converge. (**in nephron**) Capillary bed within the corpuscle of a nephron in which filtration takes place.

glutamic acid   Amino acid to which human taste is sensitive. Glutamic acid is present in the seasoning monosodium glutamate (MSG) and in meats.

gluteus maximus muscle   Major muscle of the human buttock. It is much smaller and more lateral on other mammals.

gluteus medius muscle   Muscle of the lateral iliac blade and abductor of the hip.

gluteus minimus muscle   Deep muscle of the lateral iliac blade and abductor of the hip.

glycolysis   Initial step in the metabolism of glucose by the cell, performed without requiring oxygen.

gray matter   Any part of the central nervous system in which cell bodies predominate. The cerebral cortex and nuclei of the brainstem and spinal cord are gray matter.

greater trochanter   Process of the femur that is prominent lateral to the hip joint.

guidepost cells   Embryonic cells in the nervous system whose function is to cue developing axons on the appropriate direction to their targets.

gyrus   Ridge of gray matter of the cerebral cortex.

habenula   Small nucleus of the epithalamus involved in reproductive behavior (as observed in nonhuman mammals).

hair cell   Mechanoreceptor of the inner ear that is stimulated or inhibited by the bending of hair-like processes called cilia.

hallux   Great toe.

hamstring muscles   Muscles of the posterior thigh and primary extensors of the thigh: semitendinosus, semimembranosus, and biceps femoris.

Haversian canal   Hollow passage at the center of an osteon containing blood vessels to nourish the osteocytes.

helicoidal plane of occlusion   Unique pattern of wear on molar teeth in which planes of wear change from the back of the arcade toward the front.

hepatic duct   Duct system of the liver carrying bile from the liver tissue to the gallbladder and to the duodenum.

hepatic portal vein   Vein draining the abdominal portion of the digestive tract. The portal vein carries newly absorbed nutrients directly to the liver for processing.

hepatocyte   Cell of the liver.

heritability   Possibility that a trait may be genetically transmitted from one generation to the next. Heritability is a necessary condition for natural selection.

heterochrony   Changes in the rate or degree of development of a trait in a species over time.

heterodonty   Condition of having teeth of different form and function in different parts of the mouth.

heterotroph   Organism that must obtain its chemical energy from organic molecules synthesized by other organisms. All animals are heterotrophs.

hippocampus   Area of cortex of the medial portion of the temporal lobe associated with memory recording and recall.

holocrine gland   Type of gland in which entire cells containing the secretory product are released. The sebaceous glands of skin are holocrine glands.

holonephros   Primitive vertebrate kidney composed of nephrons from all segments of the trunk.

homeostasis   Physiological *status quo* in which an organism balances and resists changes in temperature, water and ion balance, energy supply, and other vital properties.

homeothermy   Ability of an organism to maintain a constant body temperature.

hominid   Member of the family Hominidae, including chimpanzees, gorillas, humans, and fossil hominins.

hominin   Member of the taxon that includes humans and our fossil relatives, distinct from living apes.

hominoid   Member of the taxon that includes living apes and humans and a diverse radiation of their relatives during the Miocene Epoch.

homology   Relationship of two traits that are derived from a common ancestral trait. The identification of homologies between two species is good evidence that the species shared a common ancestor.

hormone   One of many chemical signals released by cells to communicate information or instructions to distant cells in the body.

Hox genes   Genes that regulate the differentiation of body segments during development of body form. Their products control the transcription of other genes.

humerus   Bone of the upper arm.

humoral   Pertaining to the body fluids or humors. In particular, "humoral signals" refer to hormones circulating in the bloodstream.

hyaline cartilage   Embryonic form of cartilage found in bones that have not fully ossified, on joint surfaces, and other places in the body.

hyoid bone   Free-floating bone suspended by muscles at the top of the neck, between the mandible and the larynx.

hypaxial muscles   Muscles of the ventral trunk and limbs, supplied by the ventral rami of the spinal nerves.

hypocone   Fourth cusp commonly added to the three primitive cusps on the tribosphenic molar. The hypocone may change the shape of the molar from a triangle to a rectangle.

hypodermis (superficial fascia)   Layer of loose connective and adipose tissue lying immediately deep to the dermis. The hypodermis is the layer that permits our skin to glide over underlying structures.

hypoglossal canal   Passage for the hypoglossal nerve through the occipital bone.

hypoglossal nerve   Twelfth cranial verve, providing motor control over the tongue.

hypophyseal fossa   Small hollow area on the midline of the center of floor of the cranial fossa anterior to the dorsum sellae. The hypophyseal fossa holds the pituitary gland.

hypophysis   Pituitary gland.

hypothalamus   Cluster of nuclei in the diencephalon regulating autonomic functions of the body.

hypothenar muscles   Collective mass of muscles at the base of the little finger and mostly inserting upon it.

ileum   Distal part of the small intestine where much of the absorption of food occurs.

iliac blade   Broad flat area of the ilium.

iliac fossa   Hollow area on the internal surface of the iliac blade of the pelvis. Iliacus muscle occupies this fossa.

iliac pillar   Thickened column of bone in the human ilium from the top of the lateral ilium to the acetabulum.

ilium   Broad upper bone of the (human) coxal bone, or more anterior part of a quadruped pelvis.

incisor   One of the anterior mammalian teeth. Humans have two incisors on each side of each jaw.

incus   "Anvil"; the middle of the three ossicles of the middle ear.

infraspinous fossa   Hollow area on the dorsal surface of the scapula, below the spine. The infraspinatus muscle occupies this fossa.

infundibulum   Stalk of nerve fibers connecting the pituitary gland to the hypothalamus. Its fibers conduct hormones of the neurohypophysis to the gland for release.

inner ear   Neural structures of hearing and equilibrium contained with a cavity of the petrosal bone.

insectivory   Referring to a diet that consists primarily of insects and small invertebrates.

intermuscular septum   Connective tissue defining and joining segmented bands of muscle in the body wall of lower vertebrates. Muscle fibers arise and insert on the septa so that they may contract as one unit as well as separately.

internal carotid artery   Major paired artery supplying the brain.

interneuron   Neuron that receives input and sends output to other neurons; it has no direct communications outside the central nervous system.

interosseous membrane (of the leg)   Membrane connecting the tibia and the fibula for most of their lengths.

interphalangeal joint   Articulation between two phalanges within a finger.

intervertebral disc   Fibrous structure separating and spacing the bodies of adjacent vertebrae. The disc is composed of a ring of fibrocartilage and a central nucleus of fluid.

intervertebral foramen   Passage between adjacent vertebral arches through which the spinal nerve emerges.

intestinal crypts   Pits in the walls of the small intestine among the villi, from which digestive enzymes are secreted.

ischial tuberosity   Process of the ischium from which hamstrings arise. The ischial tuberosity bears weight in a sitting posture.

ischium   Lower or posterior bone of the coxal bone.

isometric contraction   Action of a muscle in which the force of contraction balances resisting forces and its length does not change.

jejunum   Middle part of the small intestine where much of the breakdown of food occurs.

K strategy   Life history strategy to invest in fewer, but more competitive, offspring.

keratin   Waxy protein produced by cells of the epidermis that provides strength and water—resistant properties to the outer layer of skin.

knuckle-walking   Mode of quadrupedalism displayed by chimpanzees and gorillas in which fingers are bent onto the palm and weight is borne on the heads of the proximal phalanges.

kyphosis   Posterior curvature of the spine. The thorax has a natural kyphosis.

lamellar bone   Secondary bone that has been reshaped to align its internal structure with stresses.

larynx   "Voice box"; a chamber at the top of the trachea and behind the thyroid cartilage on the front of the neck.

lateralization   Functional asymmetry of the nervous system.

ligamentum nuchae   Elastic ligament dorsal to the neural spines of the cervical vertebrae that helps to lift and support the weight of the head.

limbic system   Conceptual division of the brain that handles olfaction, emotion, and memory.

linear acceleration   Acceleration in a straight line.

linoleic acid   Essential dietary fatty acid of the omega-6 family from which longer-chain lipids are synthesized.

linolenic acid   Essential dietary fatty acid of the omega-3 family from which longer-chain lipids are synthesized.

locus ceruleus   Nucleus of the midbrain that triggers periods of rapid-eye-movement sleep.

longitudinal arch of the foot   Arch on the medial side of the sole of the foot, from the heel to the head of the first metatarsal.

loop of Henle   Portion of the nephron tubule in mammals from which electrolytes are actively removed to concentrate them in the renal pyramid.

lordosis   Ventral curvature of the spine. The lumbar lordosis of the lower back is a unique characteristic of humans.

lumbar   Pertaining to the small of the back, and especially to the vertebrae there.

macrophage   Type of cell of the immune system that engulfs and destroys other cells and particles.

macula   Dense concentration of sensory receptors. There is a macula in the retina corresponding to the center of the visual field, and several in the inner ear reporting changes in equilibrium.

malleolus   Prominence of bone on either side of the ankle. The medial malleolus belongs to the tibia. The lateral melleolus is a part of the fibula.

malleus   "Hammer"; the most lateral of the three ossicles of the middle ear.

masseter muscle   Powerful muscle of mastication on the cheek below the zygomatic arch.

mastoid process   Process of bone on the underside of the skull just behind the ear.

maxilla bone   Bone of the face, from the bottom of the orbit to the mouth. In humans, this comprises part of the palate and contains the alveoli for the upper teeth.

mechanoreceptor   One of many types of sensory receptors that is sensitive to touch, pressure, or stretching.

median forebrain bundle   Neural tract participating in the reward system of the brain.

medulla oblongata   Most caudal part of the brainstem.

Meissner's corpuscle   Sensory ending in the skin providing a fine perception of touch.

melanin   Primary brown pigment of the skin, produced by melanocytes. Melanin is also present in hair, in the iris, and in the substantia nigra of the brain.

melanocyte   Pigment-producing cell of the epidermis.

melanosome   Organelle of melanocytes in which molecules of pigment are concentrated.

melatonin   Hormone secreted by the pineal gland that helps to regulate circadian behaviors.

membranous labyrinth   Membrane lining the inner ear cavity of the petrosal bone.

meninges   Three layers of connective tissue surrounding, isolating, and protecting the central nervous system.

mesencephalic locomotor region   Region of the midbrain of the cat and other nonprimate mammals that generates patterns of muscle activity appropriate for locomotion.

mesencephalon (midbrain)   Middle vesicle of the brain.

mesentery   Double layer of peritoneum connecting the small intestine to the posterior abdominal wall and containing blood vessels, nerves, and other support structures. The term is commonly used for any such fold supporting the digestive organs.

mesoderm   One of the original three embryonic tissue layers. The mesoderm gives rise to the connective tissues of the body, including muscle and bone.

mesonephros   Embryonic amniote kidney formed from nephrons of middle segments of the body.

metacarpal   One of five long bones of the hand, at the base of each finger.

metacarpophalangeal joint   Articulation at the base of a finger between the metacarpal and the proximal phalanx.

metanephros   Mature amniote kidney formed from nephrons of posterior segments of the body.

metapodial   Bone within the hand (metacarpal) or foot (metatarsal) at the base of each digit.

metatarsal   One of five long bones within the foot, at the base of each of the toes.

metatarsophalangeal joint   Articulation between the metatarsals and the proximal phalanges of the toes.

metencephalon   Middle portion of the developing brainstem including the pons and cerebellum.

microvillus   Finger-like projection of the cell membrane of cells of the mucosa. Microvilli increase the absorptive surface area of the cells.

molar   One of the posterior teeth and the location of most chewing for most mammals. Humans have three molars on each side of each jaw.

midbrain   Middle division of the brain and rostral part of the brainstem containing nuclei for visual and auditory processing in lower vertebrates.

middle ear   Air-filled middle chamber of the ear in which vibrations are amplified.

mucosa   Epithelial lining of the gut containing digestive glands and absorptive cells.

muscularis   Layer of smooth muscle of the gut that mechanically churns food and pushes it along the tract.

myelencephalon (medulla oblongata)   Most caudal division of the brain.

myelin   Whitish insulation on nerve axons that increases speed and efficiency of transmission.

myomere   Muscle tissue belonging to a body segment, especially in its embryonic state.

natural selection   Mechanism by which species may evolve to become better adapted to its environment. Natural selection occurs when heritable variations in the species correlate with differential reproductive success.

navicular   Tarsal bone on the medial side of the foot.

neocortex   Evolutionarily recent portion of the cerebral cortex, especially developed in mammals where it contributes the nonlimbic functional areas.

neostriatum   Part of the basal ganglia including the candate and putamen.

neoteny   Retention of juvenile characteristics into adult form. This is one form of paedomorphosis.

neural tube   First embryonic formation of the nervous system. A portion of the ectoderm forms a hollow neural tube, which will then mature into the central nervous system.

neuroendocrine system   Intercommunicating brain structures, neural pathways, and endocrine glands responsible for autonomic and homeostatic functions of the body.

neuron   Nerve cell participating in the information network of the brain. Neurons are capable of responding to and sending messages using neurotransmitters.

neurotransmitter   Chemical signal released by one neuron to communicate with another, usually at a synapse.

norepinephrine   Hormone of the suprarenal gland that participates in body arousal during the stress response.

notochord   Stiff rod of connective tissue underling the spinal cord of chordates and providing anchorage for segmented muscles. The notochord is the evolutionary and developmental precursor of the vertebral column.

nucleus (of the brain)   Any spatially and functionally defined cluster of neurons within the brain.

nucleus pulposus   Jelly-like remnant of the notochord that forms the center of an intervertebral disc. The nucleus becomes more fibrous with age.

obturator foramen   Large opening on the side of the bony pelvis. All except a small passage is normally covered by membrane in life.

occipital bone   Bone of the underside and posterior of the braincase, surrounding foramen magnum.

occipital condyles   Articular processes of the occipital bone by which the cranium contacts the first vertebra of the spinal column.

occipital lobe   Division of the cerebral cortex next to the occipital bone.

odontoblast   Specialized cell that lays down dentin.

odontoid process   Tooth-like process of the body of the axis. The odontoid process passes inside the ring of the atlas and helps to guide the motion of nodding "no."

olecranon process   Proximal end of the ulna that projects posterior to the humerus.

olfaction   Sense of smell.

olfactory bulb   Nucleus of the forebrain that receives and processes olfactory information.

olfactory epithelium   Epithelium of the upper nasal cavity where smell receptors neurons are located.

olfactory lobe   Portion of the cerebrum responsible for processing smell, particularly in lower vertebrates.

omnivory   Referring to a diet that includes significant amounts of both plant and animal foods.

ophthalmic vein   Vein draining blood from the orbit and the skin of the face around it posteriorly into the cavernous sinus.

opistonephros   Mature vertebrate kidney in most fish and amphibians formed from nephrons of middle and posterior segments of the body.

opposition   Ability of the thumb or first toe to rotate and adduct against the other digits. All primates have opposition, broadly defined, but only humans have a dexterous pad-to-pad contact between the thumb and fingers.

opsin   Any of several of types photosensitive pigment molecules.

optic chiasm   Place at which the optic nerves intersect and cross the midline. About half of their fibers are exchanged at the chiasm.

optic lobe   Part of the midbrain that processes vision in nonmammalian vertebrates; homologous to the superior colliculus.

optic nerve   Second cranial nerve, conveying visual information from the eye to the brain. Technically the optic nerves end at the chiasm where the fibers are continued in the optic tracts.

optic tract   Continuation of the visual pathway toward the brain, from the optic chiasm to the thalamus.

organ of Corti   Apparatus within the cochlea that converts vibrations to neural impulses. The organ of Corti contains the basilar and tectorial membranes and the hair cells.

os centrale   Small carpal bone in the wrists of many anthropoid primates. The os centrale has been fused into the capitate in humans and African apes.

osmotic pressure   Measure of disequilibrium in the concentration of dissolved substances in two fluids separated by a membrane. Water will tend to flow across such a gradient.

ossicles of the middle ear   Three small bones that transmit sound in the middle ear, the malleus, incus, and stapes.

ossification   Process by which osteocytes displace another tissue and create bone.

osteoblast   Specialized osteocyte capable of secreting hydroxyapatite and creating additional bone.

osteoclast   Specialized osteocyte capable of resorbing bone and breaking down hydroxyapatite.

osteocyte   Type of cell specific to bone that secretes or breaks down hydroxyapatite.

osteon   Structural unit of lamellar bone consisting of a Haversian canal containing a nutrient blood vessel and surrounded by concentric layers of bone.

otolith   Calcium carbonate crystal within the inner ear that help to register changes in linear movement.

outer ear   Portion of the ear lateral to the eardrum, including the pinna and ear canal.

oval window   Foramen in the wall of the middle ear connecting to the inner ear. The incus vibrates against a membrane over the oval window to transmit sound vibration to the perilymph on the other side.

ovulation Process in which an egg ripens and is expelled from the ovary to be fertilized.

oxytocin Pituitary hormone that stimulates contraction of the walls of the uterus during labor.

paedomorphosis Change in a developmental pathway that results in an adult expression of a trait to more closely resemble the juvenile form of that trait.

palate Roof of bone that separates the oral cavity from overlying structures. The **primary palate** in more primitive vertebrates separates the mouth from the brain cavity. The **secondary palate** in mammals separates the mouth from the nasal cavity.

paleocortex Primitive and ancient portion of cerebral cortex involved in limbic functions.

paleostriatum (globus pallidus, in mammals) Part of the basal ganglia.

palmigrade Term describing an animal with a mode of walking in which the palm of the hand is placed flat on the ground with each step.

pancreas Accessory gland developing out of the duodenum and producing powerful digestive enzymes.

parallel evolution Evolution of two related lineages along similar paths so that they possess similar derived traits.

paranasal sinuses Cavities in various bones of the facial skeleton lined by mucous membrane and normally filled with air.

parasympathetic Pertaining to the division of the autonomic nervous system responsible for body maintenance.

parietal bone Bone of the upper lateral wall of the braincase.

parietal foramen Small foramen for an emissary vein through the parietal bone on top of the skull near the midline.

parietal lobe Division of the cerebral cortex next to the parietal bone.

parsimony Desirable simplicity of a scientific explanation. A more parsimonious hypothesis requires fewer unproved assumption.

patella Knee cap.

pectoral muscles Pectoralis major and minor are muscles of the chest that connect the upper limb and ribcage.

pectoralis minor muscle Deep muscle of the pectoral group on the chest.

pelvic diaphragm Thin layer of muscle tissue that closes off the bottom of the pelvic cavity.

pelvic inlet Entrance to the birth canal defined by the internal circumference of the pelvis.

pennate muscle Muscle form in which fibers attach obliquely to a tendon and the direction of pull. A pennate muscle has the potential of greater contractile strength with reduced distance of contraction.

peramorphosis Change in a developmental pathway that results in an adult expression of a trait to diverge to a greater degree from the juvenile form of that trait.

perikyma Daily "growth line" found in tooth enamel that can be used to estimate the length of time the tooth developed.

perilymph Lymph that lies between the membranous labyrinth of the inner ear and the surrounding bone.

perineum   Region of the body inferior to the pelvic diaphragm, containing external genitalia, among other structures.

periodontal ligament   Connective tissue anchoring a tooth with a socket on the jaw.

peristalsis   Pattern of contractions of the muscles of the gut to push food along the digestive tract.

peritoneal cavity   Body cavity defined by the peritoneum and surrounding the visceral organs permitting the body wall to move independently over them.

peritoneum   Serous membrane lining the abdominal cavity and visceral organs.

peritricial nerve ending   Sensory nerve ending at the base of a hair follicle reporting movement of the hair.

peroneal tubercle   Protuberance on the lateral side of the calcaneus. The tendon of peroneus longus is anchored to the tubercle.

peroneus longus muscle   Muscle of the lateral ankle that plantar flexes and everts the foot.

petrous portion of the temporal bone   Dense portion of the temporal bone housing the inner and middle ear cavities.

phalanx   Bone of the finger or toe. There are 14 phalanges in each hand and foot.

pharynx   Portion of the digestive tract that transitions between the mouth and the esophagus. The human pharynx lies behind the mouth and nasal cavity.

pheomelanin   Yellowish variant of the pigment melanin commonly found in lighter-skinned people.

pheromone   Chemical signal communicating between two individuals of the same species.

photon   Smallest unit of light.

photoreceptor   Sensory receptor for radiant energy in the wavelengths of visible light. Photoreception is the basis for vision.

pineal gland   Small gland of the epithalamus that secretes the hormone melatonin.

pinna   Fleshy part of the ear on the outside of the skull.

pisiform   Carpal bone prominent on the ulnar side of the wrist.

pituitary gland   Endocrine gland on the underside of the hypothalamus releasing a variety of hormones that regulate growth, metabolism, and reproduction through their control of other endocrine glands in the body.

plantar aponeurosis   Strong deep layer of fascia on the sole of the foot.

plantar flexor muscles   Long muscles of the calf that assist in plantar flexion of the ankle. (Plantar flexion points the toes down.)

plantaris muscle   Plantar flexor of the calf and ankle. This muscle is very small in humans.

plantigrade   Term describing an animal with a mode of walking in which the sole of the foot is placed flat on the ground with each step.

plasma   Fluid portion of blood that excludes the blood cells.

plasticity   Tendency of an object to respond to a strain by permanently changing its shape.

plicae circulares   "Circular folds" of mucosa and submucosa on the inner lining of the small intestine.

pluripotency   Capacity of an embryonic or stem cell to differentiate into a variety of cell types.

podial   One of the small bones of the wrist (carpal) or foot (tarsal).

poikilothermy   Tendency of an organism to permit its body temperature to vary with that of the environment.

pons   Part of the brainstem to which the cerebellum attaches.

popliteus muscle   Small deep muscle on the back of the knee.

posterior cranial fossa   Lower and posterior portion of the cranial cavity. The posterior cranial fossa lies on the occipital bone and is occupied by the cerebellum and brainstem.

postorbital bar   Process of bone contributing to the lateral side of the primate orbit, derived from the zygomatic and frontal bones.

postorbital bone   Small bone posterior to the orbit in the skull of early tetrapods.

postorbital constriction   Absolute narrowing of the cranium between the facial skeleton and the braincase. A postorbital constriction is typical of most hominoids, including fossil hominins, and was eliminated only with the expansion of the brain late in human evolution.

prehensile tail   Tail of many New World monkeys that is able to curl and grip supports.

premolar   Type of tooth between the canine and the molars. Humans have two premolars on each side of each jaw.

primary vesicle   One of three initial divisions (prosencephalon, mesencephalon, rhombencephalon) of the embryonic brain.

primitive trait   Trait that resembles the homologous trait present in an ancestral species. The opposite of a primitive trait is a derived trait.

progesterone   Hormone secreted by the corpus luteum of the ovary to sustain the endometrium of the uterus for implantation and pregnancy.

prolactin   Pituitary hormone that stimulates continued production of milk from the breast.

pronephros   Embryonic vertebrate kidney formed from nephrons of anterior segments of the body.

proprioception   Senses of muscle tone, body position, and motion, as monitored from muscles, joints, and connective tissues.

prosencephalon (forebrain)   Anterior or rostral division of the embryonic brain that will develop into the cerebrum (telencephalon) and diencephalon.

pseudogene   Gene that is recognizable in relation to homologues elsewhere in the genome or in other species, but which has been altered so that it no longer functions.

pterygoid muscles   Medial and lateral pterygoid muscles are muscles of mastication lying medial to the mandible.

pterygoid plates   Processes of the sphenoid bone in the skull from which the pterygoid muscles arise.

pubis   Ventral bone of the coxal bone that articulates across the midline at the pubic symphysis.

pulmonary trunk   Major vessel leaving the right ventricle to carry deoxygenated blood to the lungs.

pulmonary veins   Veins returning from the lungs carrying oxygenated blood into the left atrium.

punctuated equilibrium   Model of evolutionary change in which change is uneven in time: long periods of stasis (equilibrium) are interrupted (punctuated) by short bursts of rapid change.

putamen   One of the basal ganglia, functioning most conspicuously in movement.

pyloric sphincter   Sphincter muscle closing the digestive tract between the stomach and duodenum to regulate the rate of movement of food through the tract.

pyramid   Prominent bundle of corticospinal fibers on the ventral surface of the medulla.

quadrate bone   Small bone of the early tetrapod cranium articulating with the lower jaw. In mammals, the quadrate becomes the incus in the middle ear.

quadriceps femoris muscles   Group of four muscles (vastus medialis, vastus intermedius, vastus lateralis, and rectus femoris) on the anterior thigh, inserting by a common tendon through the patella onto the tibia.

r strategy   Life history strategy to invest small amounts of energy in individual offspring, but to produce large numbers of them.

radiocarpal joint   Wrist joint proper. (A second articulation between two rows of carpal bones is usually also involved in "wrist" movements, as commonly understood.)

radius   Bone on the lateral side (thumb side) of the forearm.

receptive aphasia   Class of language disorders of the brain interfering with the comprehension of language.

receptor   Molecule on the surface of a cell that is designed to bind with a certain class of other molecules and signal their presence to the cell.

rectus abdominis muscle   Longitudinal muscle of the abdominal wall, on either side of the midline.

red nucleus   Nucleus of the midbrain participating in motor control.

REM (rapid eye movement) sleep   Stage of sleep characterized by stimulation of the cerebral cortex, eye movements, and (often) dreaming.

renal pelvis   Membranous chamber within the kidney in which urine collects.

renal pyramid   Concentration of nephron tubules within the kidney where the urine is concentrated.

reproductive success   Number of viable offspring an individual produces.

respiration   Exchange of gases, whether at the level of the organism or of the cell.

rete mirabile   Type of vascular countercurrent exchange in which a larger vessels divide into numerous smaller ones for more efficient exchange.

reticular formation   Primitive nuclei of the brainstem that oversee diverse functions including homeostasis, muscle tone, and states of consciousness.

reticulospinal tract   Descending pathway from the reticular formation to the spinal cord relating to autonomic functions and muscle tone.

retina   Layer of light-sensitive receptors in the eye.

retinotopic map   Pattern of information arrival and processing in the visual cortex in which the spatial relations are the same as those within the retina and within the visual fields.

rhodopsin Photosensitive pigment of rods.

rhombencephalon Posterior or caudal division of the embryonic brain that will develop into the pons and cerebellum (metencephalon) and medulla (myelencephalon).

rod Type of primate photoreceptor extremely sensitive to light and responsive to a broad spectrum of wavelengths.

rotator cuff Four muscles of the shoulder joint (supraspinatus, infraspinatus, teres minor, and subscapularis) that rotate the humerus and, collectively, maintain the integrity of the joint.

saccule Chamber of the inner ear that participates in hearing and equilibrium.

sacral angle Angle formed by the body of the sacrum relative to the rest of the spine. A substantial angle corresponds to erect human posture.

sacrum Fused series of vertebrae to which the coxal bone attaches.

sagittal plane Anatomical plane that divides the body into right and left halves, or any plane parallel to that one.

scapula Shoulder blade.

scaphoid Carpal bone in the wrist that articulates with the radius.

scoliosis Lateral curvature of the spine. A small degree of scoliosis is common; a large scoliosis is a medical condition.

sebaceous gland Glands of the nonhairy skin, particularly in the face and neck, that produce an oily secretion.

sebum Oily secretion of sebaceous glands in the skin.

secondary compound Plants manufacture secondary compounds by modifying metabolic by-products. Such compounds may be pharmacologically active or bad tasting to discourage herbivores from eating their leaves.

secondary sexual characteristic Trait that differentiates the sexes but is not directly involved in copulation or reproduction.

secondary vesicle One of five secondary divisions of the developing brain (telencephalon, diencephalon, mesencephalon, metencephalon, myelencephalon).

sectorial Ability to cut, referring especially to the form of the elongated anterior lower premolar of most Old World monkeys and apes. That tooth hones and sharpens the upper canine. Human premolars are not sectorial.

semicircular canal Circular portion of membranous labyrinth containing receptors sensitive to angular acceleration.

seminal vesicle Gland of the male reproductive tract that contributes much of the fluid volume of semen.

serosa Type of membrane lining the body wall and visceral organs and secreting serous fluid. Serous membranes, including the pericardium, pleurae, and peritoneum, define the body cavities.

serous fluid Watery body fluid of intercellular spaces containing proteins, but few cells.

serratus anterior muscle Muscle of the lateral side of the ribcage that passes deep to the scapula and inserts upon it.

sexual dimorphism   Pattern of anatomical characteristics or of body size by which males and females of a species may be differentiated.

sexual selection   Selection acting on individuals of a species according to their ability to attract and mate with other members of the opposite sex.

shear   Type of stress in which forces are applied from opposite direction but are not aligned.

simian shelf   Horizontal buttress of bone posterior to the mandibular symphysis expressed to varying degrees among the apes.

sinus venosus   First chamber of the heart in lower vertebrates and vertebrate embryos. The sinus venosus is formed by the confluence of the venae cavae and later becomes part of the atrium (in tetrapods).

sinusoids   Capillary-sized divisions of the portal hepatic vein within the liver bringing nutrient-laden venous blood in close proximity to the liver cells. The sinusoids are drained by the hepatic vein.

soleus muscle   Deeper portion of the triceps surae of the calf.

somatic senses   Senses of the body wall monitoring the external environment; principally, touch, pressure, pain, temperature, and proprioception.

somatomedin   Hormone intermediate released by the liver in response to circulating growth hormone. Somatomedins stimulate growth and cell division in body tissues.

somatosensory cortex   Cortex of the parietal lobe that receives ascending input relating to the somatic senses and where conscious perception appears to occur.

somatotopic organization   Pattern of organization of motor and sensory pathways and nuclei by which the body is mapped onto the spatial arrangement of neurons and fibers.

somite   Unit of embryonic mesodermal tissue that will develop into a segment of the body wall.

species   Population of organisms that can reproduce among themselves but not with other organisms.

sperm competition   Competition among sperm of different males that occurs within the female reproductive tract after copulation.

spermatogenesis   Process by which sperm are produced in the testis.

sphenoid bone   Bone in the center of the cranium that contributes both to the braincase and to part of the orbits.

spinal nerve   One of the paired segmental nerves emerging from the spinal cord.

spinothalamic tract   Ascending pathway of the spinal cord carrying pain and temperature senses.

splanchnocranium   Portion of the skeleton derived from the branchial arches, including the hyoid bone, styloid process, and cartilages of the neck.

squamosal portion of the temporal bone   Flat region of the temporal bone on the side of the braincase.

squamous epithelium   Epithelium consisting primarily of flattened cells. The epidermis, alveoli in the lungs, and the esophagus are all lined with squamous epithelium.

stapedius muscle   Muscle attaching to the incus to dampen excessive vibration of the membrane of the oval window.

Starling's Law   Starling's Law states that muscle contracts more forcefully if it is first stretched. The law is applied to the heart to explain how each successive chamber can contract more forcefully than the preceding chamber.

stereocilia Cilia on hair cells that are linked to ion channels. Bending the cilia may open or close the channels.

stereoscopic vision Ability to perceive objects in depth due integrated processing of information from overlapping visual fields of both eyes.

sternocleidomastoid muscle Large straplike muscle on the side of the neck, from behind the ear to the top of the sternum.

sternum Bone of the anterior chest to which the ribs attach.

strain Amount of deformation, or change in shape, in an object in response to a stress.

stress Measure of the intensity of a force applied to a structure or object.

styloid process Slender downward-pointing process of the temporal bone from which muscles and a ligament of the neck arise.

submucosa Layer of the gut underlying the mucosa and containing blood vessels, nerves, and muscle supporting the mucosa.

sulcus Groove between gyri of the cerebral cortex.

suprarenal gland Endocrine gland residing near the kidney producing a large number of hormones involved in the regulation of minerals and circulating metabolites. The cortex and medulla of the suprarenal gland have different hormones and function independently.

sustentaculum tali Shelf of bone on the medial side of the calcaneus, supporting the talus.

swim bladder Gas-filled sac in many bony fish, homologous to the lungs. The amount of gas in the bladder is regulated to adjust the buoyancy of the fish.

sympathetic Pertaining to the division of the autonomic nervous system responsible for body arousal and the mobilization and expenditure of energy.

symphysis Midline articulation of paired bones. The **mandibular symphysis** is the point of fusion of right and left halves of the lower jaw in anthropoids and other mammalian groups. The **pubic symphysis** is a fibrous joint between right and left halves of the pelvis.

synapse Site of chemical or electrical communication between neurons.

synapsid Member of a group of early tetrapods at the base of the lineage leading to therapsids and later to mammals.

synovial fluid Lubricating fluid found within and secreted by synovial membranes.

synovial joint Diarthrosis.

synovium Synovial membrane surrounding a cavity filled with synovial fluid. Synovia are used to reduce friction in diarthroses, around ligaments, and under other moving tissues.

talus Ankle bone, a tarsal connecting the leg with the rest of the foot.

tarsal One of the small bones of the proximal foot. Humans have seven tarsals in each foot.

taxon Formal grouping of organisms at any taxonomic level. *Homo sapiens*, primates, mammals, and animals are examples of taxa.

tectorial membrane Membrane within the cochlea to which cilia of hair cells are attached.

telencephalon   Cerebrum, the most rostral vesicle of the brain.

temporal bone   Part of the braincase that includes the ear region, mastoid process, and the upper part of the temporomandibular joint.

temporal lobe   Division of the cerebral cortex next to the temporal bone.

temporalis fascia   Tough layer of membrane covering the temporalis muscle and from which some of the muscle fibers arise.

temporalis muscle   Chewing muscle on the side of the skull that can be felt above the zygomatic arch.

tendo calcaneus   Achilles tendon, the tendon of the triceps surae at the back of the ankle.

tensor fasciae latae muscle   Muscle lateral to the hip inserting into the fascia lata.

tensor tympani muscle   Muscle attaching to the malleus to dampen excessive vibration of the eardrum.

teres major muscle   Shoulder adductor lying mostly inferior to the scapula.

testosterone   Primary male hormone secreted by the testis.

tetrapod   Member of the division of terrestrial vertebrates with four limbs, and their descendants.

thalamus   Cluster of nuclei in the diencephalon monitoring and processing information entering the cerebrum.

thecodonty   Method of anchoring teeth on the jaws by placing them in sockets, or alveoli, in the bone.

thenar muscles   Collective mass of muscles at the base of the thumb ("ball" of the hand) and inserting upon it.

Theory of Evolution   Argument that life evolves and that modern species are descended from a common ancestor.

therapsid   "Mammal-like reptile" belonging to a taxon that flourished in the Permian Period and was ancestral to later mammals.

thermoregulation   Control of body temperature.

thyroid gland   Endocrine gland of the neck whose hormones regulate metabolic activity.

tibia   Larger and more prominent bone of the lower leg.

tibial collateral ligament   Ligament of the knee on the medial side of the knee joint.

tibial tuberosity   Prominence of bone on the front of the head of the tibia, just below the knee.

tibialis anterior muscle   Major muscle of the anterior leg and dorsiflexor of the ankle.

tibialis posterior muscle   Deep plantar flexor muscle of the calf.

tone   Normal state of partial contraction of a muscle at rest.

trabecular bone   Light porous network of bone inside skeletal elements.

trait   Any observable characteristic of an individual, whether anatomical, physiological, molecular, or behavioral.

transverse foramen   Foramen at the base of the transverse process of cervical vertebrae. The vertebral artery ascends through the transverse foramen.

transverse process (transverse spine)   Process for attachment of muscle and rib projecting laterally from the vertebral arch.

trapezius muscle   Muscle of the back and shoulder, drawing the scapula toward the spine.

tribosphenic molar   Primitive mammalian molar form with three high cusps arranged in a triangle. Modern mammalian teeth are derived from this pattern by the addition or elimination of cusps and ridges.

triceps surae muscle   Collective term for the superficial calf muscles, the gastrocnemius, and the soleus.

trochlear facet   Articular surface on the humerus for the ulna.

trochlear notch   Socket in the ulna for the humerus.

tympanic bone   Part of the housing of the middle ear in some mammals.

tympanic membrane   Eardrum.

ulna   Bone on the medial side (little finger side) of the forearm.

umami   Distinct taste modality sensitive to the amino acid glutamic acid.

umbilical artery   Paired artery carrying blood from the fetus to the placenta.

umbilical vein   Unpaired vein carrying oxygen- and nutrient-rich blood from the placenta to the fetus.

unguligrade   Term describing some mammals (e.g., horses, deer) with a mode of walking in which weight is borne on the tips of the toes.

urea   Waste product containing nitrogen, resulting from the metabolism of protein.

ureter   Muscular tube draining urine from the kidney to the bladder.

utricle   Chamber of the inner ear that registers linear acceleration.

variation   Existence of multiple expressions of traits in a population, including anatomical, physiological, molecular, and behavioral traits. Variation is a necessary condition for natural selection and evolution.

vasae rectae   Network of small blood vessels in the renal pyramid that use a countercurrent exchange to concentrate salt within the pyramid.

venae cavae   Major veins of the body converging on the heart. Humans have two venae cavae, superior and inferior.

ventilation   Flow of air in and out of the lungs.

ventricle   Second chamber(s) encountered by blood entering the heart (in tetrapods) or the homologous chamber in other vertebrates. The mammalian ventricle pumps blood out of the heart.

vermiform appendix   "Worm-like" extension of the cecum containing immune tissues.

vertebral arch   Partial ring of a vertebra that surrounds the spinal cord, attaching on either side to the vertebral body.

vestibular organ   Sensory apparatus of the inner ear that detects changes in equilibrium.

vestibular system   Nuclei and pathways associated with the perception of body equilibrium and responses to it.

vestibulospinal tract   Descending pathway from the vestibular nuclei to the spinal cord contributing to posture and reflexive movements to maintain equilibrium.

vibrissa   Long sensory hair ("whisker") around the face of most mammals.

villus   Finger-like projection, especially on the lining of the small intestine.

vocal fold ("vocal cord")   Folds of tissue within the larynx capable of closing of the air passage or regulating the flow of air through it.

vocal ligament   Ligament within the vocal fold that maintains tension to determine the frequency of vibration of the voice.

vocalis muscle   Muscle within the vocal fold that can adjust tension of the vocal ligament.

vomeronasal organ (VNO)   Secondary organ of smell located near the bottom of the nasal septum. The functional significance of the human VNO is debated, but it may monitor pheromones and influence behavior.

Wernicke's area   Region of the cortex in the posterior temporal lobe that relates to the analysis of sound, including speech.

white matter   Any part of the central nervous system in which myelinated fiber tracts predominate. The internal part of the cerebrum and outer part of the spinal cord are white matter.

windlass effect   Tendency of a ligament to increase its force and distance of pull when wrapped about a process of bone. The windlass effect is seen in plantar flexor muscles crossing the ankle and in the long head of biceps crossing the head of the humerus.

zygapophysis   Paired articular processes by which one vertebral arch contacts the arch above or below it.

zygomatic arch   "Cheek bone" on the lateral side of the face. Masseter muscle arises from this arch.

zygomatic bone   Bone of the lateral face and orbit that comprises most of the zygomatic arch.

BIBLIOGRAPHY

Abitbol, M.M. 1987a. Evolution of the sacrum in hominoids. *Am. J. Phys. Anthropol.* 74:65–81.

Abitbol, M.M. 1987b. Obstetrics and posture in pelvic anatomy. *J. Hum. Evol.* 16:243–255.

Abitbol, M.M. 1988. Evolution of the ischial spine and of the pelvic floor in the hominoidea. *Am. J. Phys. Anthropol.* 75:53–67.

Abitbol, M.M. 1991. Ontogeny and evolution of pelvic diameters in anthropoid primates and in *Australopithecus afarensis*. *Am. J. Phys. Anthropol.* 85:135–148.

Abitbol, M.M. 1995. Reconstruction of the STS 14 (*Australopithecus africanus*) pelvis. *Am. J. Phys. Anthropol.* 96:143–158.

Abitbol, M.M. 1996. *Birth and Human Evolution*. Westport, CT: Bergin and Garvey.

Aboitiz, F. 1996. Does bigger mean better? Evolutionary determinants of brain size and structure. *Brain Behav. Evol.* 47:225–245.

Ackshoomoff, N.A., and E. Courchesne. 1992. A new role for the cerebellum in cognitive operations. *Behav. Neurosci.* 106(5):731–738.

Aerts, P., R. van Damme, L. van Elsacker, and V. Duchêne. 2000. Spatio-temporal gait characteristics of the hind-limb cycles during voluntary bipedal and quadrupedal walking in bonobos (*Pan paniscus*). *Am. J. Phys. Anthropol.* 111:503–517.

Agrawal, A.F. 2001. Sexual selection and the maintenance of sexual reproduction. *Nature* 411:692–695.

Ahlberg, P.E., and A.R. Milner. 1994. The origin and early diversification of tetrapods. *Nature* 368:507–514.

Ahrensburg, B., A.M. Tiller, B. Vandermeersch, H. Duday, L.A. Schepartz, and Y. Rak. 1989. A Middle Paleolithic human hyoid bone. *Nature* 338:758–760.

Ahrensburg, B., L.A. Schepartz, A.M. Tiller, B. Vandermeersch, and Y. Rak. 1990. A reappraisal of the anatomical basis for speech in Middle Paleolithic hominids. *Am. J. Phys. Anthropol.* 83:137–146.

Aiello, L., and C. Dean. 1990. *An Introduction to Human Evolutionary Anatomy*. London: Academic Press.

Aiello, L.C., and C. Key. 2002. Energetic consequences of being a *Homo erectus* female. *Am. J. Hum. Biol.* 14:551–565.

Aiello, L.C., and P. Wheeler. 1995. The expensive-tissue hypothesis. *Curr. Anthropol.* 36(2):199–221.

Aiello, L.C., C. Montgomery, and C. Dean. 1991. The natural history of deciduous tooth attrition in hominoids. *J. Hum. Evol.* 21:397–412.

Aiello, L.C., N. Bates, and T. Joffe. 2001. In defense of the expensive tissue hypothesis. In D. Falk and K.R. Gibson, eds. *Evolutionary Anatomy of the Primate Cerebral Cortex*. Cambridge: Cambridge University Press. Pp. 57–78.

Alba, D.M., S. Moyà-Solà, and M. Köhler. 2003. Morphological affinities of the *Australopithecus afarensis* hand on the basis of manual proportions and relative thumb length. *J. Hum. Evol.* 44:225–254.

Alexander, R.D. 1971. The search for an evolutionary philosophy of man. *Proc. R. Soc. Victoria* 84(1):99–120.

Alexander, R.D., and K.M. Noonan. 1979. Concealment of ovulation, parental care, and human social evolution. In N.A. Chagnon and W. Irons, eds., *Evolutionary Biology and Human Social Behavior: An Anthropological Perspective*. North Scituate, MA: Duxbury Press. Pp. 436–453.

Alexander, R.M. 1991. Characteristics and advantages of human bipedalism. In J.M.V. Rayner and R.J. Wootton, eds., *Biomechanics in Evolution*. Cambridge: Cambridge University Press. Pp. 255–266.

Alexander, R.M., and R.F. Ker. 1990. Running is priced by the step. *Nature* 346:220–221.

Allen, L.L., P.S. Bridges, D.I. Evon, K.R. Rosenberg, M.D. Russell, L.A. Schepartz, V.J. Vitzthum, and M.H. Wolpoff. 1982. Demography and human origins. *Am. Anthropol.* 84:888–896.

Allin, E.F. 1975. Evolution of the mammalian middle ear. *J. Morphol.* 147:403–438.

Allman, J. 1977. Evolution of the visual system in the early primates. *Progr. Psychobiol. Physiol. Psychol.* 7:1–53.

Allman, J.M. 1999. *Evolving Brains*. New York: Scientific American Library.

Alvarez, H.P. 2000. Grandmother hypothesis and primate life histories. *Am. J. Phys. Anthropol.* 113:435–450.

Anderson, J.L. 1987. Breasts, hips and buttocks revisited: honest fat for honest fitness. *Ethol. Sociobiol.* 9:319–324.

Andrews, P.W., S.W. Gangestad, and D. Matthews. 2002. Adaptationism—how to carry out an exaptationist program. *Behav. Brain Sci.* 25:489–553.

Anemone, R.L., E.S. Watts, and D.R. Swindler. 1991. Dental development of known-age chimpanzees, *Pan troglodytes* (Primates, Pongidae). *Am. J. Phys. Anthropol.* 86:229–241.

Anemone, R.L., M.P. Mooney, and M.I. Siegel. 1996. Longitudinal study of dental development in chimpanzees of known chronological age: Implications for understanding the age at death of Plio-Pleistocene hominids. *Am. J. Phys. Anthropol.* 99:119–133.

Anemone, R.L. 2002. Dental development and life history in hominid evolution. In N. Minugh-Purvis and K.J. McNamara, eds., *Human Evolution through Developmental Change*. Baltimore, MD: Johns Hopkins University Press. Pp. 249–280.

Ankel-Simons, F. 2000. *An Introduction to Primate Anatomy*. New York: Academic Press.

Aoki, C., and P. Siekevitz. 1988. Plasticity in brain development. *Sci. Am.* 259(Dec):56–64.

Arcadi, A.C. 2000. Vocal responsiveness in male wild chimpanzees: Implications for the evolution of language. *J. Hum. Evol.* 39:205–223.

Ardrey, R. 1961. *African Genesis*. New York: Dell Publishing Co.

Armstrong, E. 1985. Relative brain size in monkeys and prosimians. *Am. J. Phys. Anthropol.* 66:263–273.

Armstrong, E. 1990. Brain, bodies and metabolism. *Brain Behav. Evol.* 36:166–176.

Arsuaga, J.L., J.M. Carretero, C. Lorenzo, A. Gracia, I. Martinez, J.M. Bermudez de Castro, and E. Carbonell. 1997. Size variation in Middle Pleistocene hominids. *Science* 277:1086–1088.

Arthur, W. 2002. The emerging conceptual framework of evolutionary developmental biology. *Nature* 415:757–764.

Ashton, E.H., and C.E. Oxnard. 1963. The musculature of the primate shoulder. *Trans. Zool. Soc. Lond.* 29(7):553–650.

Ashton, E.H., and C.E. Oxnard. 1964. Functional adaptations in the primate shoulder girdle. *Proc. Zool. Soc. Lond.* 142:49–66.

Aspden, R.M. 1987. Intra-abdominal pressure and its role in spinal mechanics. *Clin. Biomech.* 2:168–174.

Aspden, R.M. 1992. Review of the functional anatomy of the spinal ligaments and the lumbar erector spinae muscles. *Clin. Anat.* 5:372–387.

Axel, R. 1995. The molecular logic of smell. *Sci. Am.* 273(Oct):154–159.

Backwell, L.R., and F. d'Errico. 2001. Evidence of termite foraging by Swartkrans early hominids. *Proc. Natl. Acad. Sci. U.S.A.* 98(4):1358–1363.

Bahat, A., I. Tur-Kaspa, A. Gakamsky, L.C. Giojalas, H. Breitbart, and M. Eisenbach. 2003. Thermotaxis of mammalian sperm cells: A potential navigation mechanism in the female genital tract. *Nat. Med.* 9(2):149–150.

Bahn, P.G. 1996. New developments in Pleistocene art. *Evol. Anthropol.* 4(6):204–215.

Baker, M.A. 1982. Brain cooling in endotherms in heat and exercise. *Annu. Rev. Physiol.* 44:85–96.

Baker, R.R., and M.A. Bellis. 1988. "Kamikaze sperm" in mammals. *Animal Behav.* 36(3):936–939.

Baker, R.R., and M.A. Bellis. 1989. Elaboration of the kamikaze sperm hypothesis: a reply to Harcourt. *Animal Behav.* 37(5):865–867.

Baker, R.R., and M.A. Bellis. 1993a. Human sperm competition: Ejaculate adjustment by males and the function of masturbation. *Animal Behav.* 46:861–885.

Baker, R.R., and M.A. Bellis. 1993b. Human sperm competition: Ejaculate manipulation by females and a function for the female orgasm. *Animal Behav.* 46:887–909.

Balter, V., A. Person, N. Labourdette, D. Drucker, M. Fox, and B. Vandermeersch. 2001. Les Néandertaliens étaient-ils essentiellement carnivores? Résultats préliminaires sur les teneurs en Sr et en Ba de la paléobiocénose mammalienne de Saint-Césaire. *C. R. Acad. Sci. IIa.* 332(1):59–65.

Bao, S., V.T. Chan, and M.M. Merzenich. 2001. Cortical remodelling induced by activity of ventral tegmental dopamine neurons. *Science* 412:79–83.

Barber, N. 1995. The evolutionary psychology of physical attractiveness: Sexual selection and human morphology. *Ethol. Sociobiol.* 16:395–424.

Barham, L.S. 2002. Systematic pigment use in the Middle Pleistocene of south-central Africa. *Curr. Anthropol.* 43(1):181–190.

Bartlett, D. 1989. Respiratory functions of the larynx. *Physiol. Rev.* 69(1):33–57.

Barton, N.H., and B. Charlesworth. 1998. Why sex and recombination? *Science* 281:1986–1990.

Barton, R.A. 1996. Neocortex size and behavioural ecology in primates. *Proc. R. Soc. Lond. B Biol. Sci.* 263:173–177.

Barton, R.A., and P.H. Harvey. 2000. Mosaic evolution of brain structure in mammals. *Nature* 405:1055–1058.

Bateman, A., A. Singh, T. Krahl, and S. Solomon. 1989. The immune-hypothalamic-pituitary-adrenal axis. *Endocr. Rev.* 10:92–112.

Bednarik, R.G. 1995. Concept-mediated marking in the Lower Paleolithic. *Curr. Anthropol.* 36(4):605–670.

Bednarik, R.G. 2003. A figurine from the African Acheulian. *Curr. Anthropol.* 44(3):405–413.

Begun, D.R. 2003. Planet of the apes. *Sci. Amer.* 289(Aug):74–83.

Bekoff, M. 2000. Animal emotions: Exploring passionate natures. *BioScience* 50(10):861–870.

Bellis, M.A., and R.R. Baker. 1990. Do females promote sperm competition? Data for humans. *Animal Behav.* 5:997–999.

Bereczkei, T., and A. Csanaky. 1996. Mate choice, marital success, and reproduction in a modern society. *Ethol. Sociobiol.* 17:17–35.

Berge, C. 1994. How did the australopithecines walk? A biomechanical study of the hip and thigh of *Australopithecus afarensis*. *J. Hum. Evol.* 26:259–273.

Berge, C., R. Orban-Segebarth, and P. Schmid. 1984. Obstetrical interpretation of the australopithecine pelvic cavity. *J. Hum. Evol.* 13:573–587.

Berger, L.R. 2002. Early hominid body proportions and emerging complexities in human evolution. *Evol. Anthropol.* 11(Suppl. 1):42–44.

Berger, L.R., and P.V. Tobias. 1996. A chimpanzee-like tibia from Sterkfontein, South Africa, and its implications for the interpretation of bipedalism in *Australopithecus africanus*. *J. Hum. Evol.* 30:343–348.

Bermudez de Castro, J.M., and A. Rosas. 2001. Pattern of dental development in Hominid XVII from the Middle Pleistocene Atapuerca-Sima de los Huesos site (Spain). *Am. J. Phys. Anthropol* 114:325–330.

Bernstein, H., H.C. Byerly, F.A. Hopf, and R.E. Michod. 1985. Genetic damage, mutation, and the evolution of sex. *Science* 229:1277–1281.

Betzig, L., M. Borgerhoff Mulder, and P. Turke, eds. 1988. *Human Reproductive Behavior: A Darwinian Perspective*. Cambridge: Cambridge University Press.

Beuchat, C.A. 1993. The scaling of concentrating ability in mammals. In J.A. Brown, R.J. Balment, and J.C. Rankin, eds., *New Insights into Vertebrate Kidney Function*. Cambridge: Cambridge University Press.

Beynon, A.D., and M.C. Dean. 1988. Distinct dental development patterns in early fossil hominids. *Nature* 335:509–514.

Beynon, A.D., and B.A. Wood. 1987. Patterns and rates of enamel growth in the molar teeth of early hominids. *Nature* 326:493–496.

Beynon, A.D., C.B. Clayton, F.V.R. Rozzi, and D.J. Reid. 1998. Radiological and histological methodologies in estimating the chronology of crown development in modern humans and great apes: A review, with some applications for studies on juvenile hominids. *J. Hum. Evol.* 35:351–370.

Bhatnagar, K.P., and T.D. Smith. 2001. The human vomeronasal organ. III. Postnatal development from infancy to the ninth decade. *J. Anat.* 199:289–302.

Biggerstaff, R.H. 1977. The biology of the human chin. In A.A. Dahlberg and T.M. Graber, eds., *Orofacial Growth and Development*. The Hague: Mouton. Pp. 71–87.

Birkhead, T. 2000. Hidden choices of females. *Nat. Hist.* 109(Nov):66–71.

Blalock, J.E. 1989. A molecular basis for bidirectional communication between the immune and neuroendocrine systems. *Physiol. Rev.* 69:1–32.

Blanton, P.L., and N.L. Biggs. 1968. Eighteen hundred years of controversy: The paranasal sinuses. *Am. J. Anat.* 124:135–148.

Bloch, J.I., and D.M. Boyer. 2002. Grasping primate origins. *Science* 298:1606–1610.

Bloch, K. 1995. Why Tabby takes to tuna. *Harvard Mag.* 98(Sept–Oct):50–53.

Blount, B.G. 1990. Issues in bonobo (*Pan paniscus*) sexual behavior. *Am. Anthropol.* 92:702–714.

Blum, H.F. 1945. Physiological effects of sunlight on man. *Physiol. Rev.* 25:483–530.

Blum, H.F. 1961. Does the melanin pigment of human skin have adaptive value? *Q. Rev. Biol.* 36:50–63.

Blum, H.F. 1968. Vitamin D, sunlight, and natural selection. *Science* 159:652–653.

Blurton-Jones, N.G., K. Hawkes, and J.F. O'Connell. 2002. Antiquity of postreproductive life: Are there modern impacts on hunter-gatherer postreproductive life spans? *Am. J. Hum. Biol.* 14:184–205.

Bocherens, H., M. Fizet, A. Mariotti, B. Lange-Badre, B. Vandermeersch, J.P. Borel and G. Bellon. 1991. Isotopic biogeochemistry (^{13}C, ^{15}N) of fossil vertebrate collagen: application to the study of a past food web including Neandertal man. *J. Hum. Evol.* 20:481–492.

Bocherens, H., D. Billiou, A. Mariotti, M. Toussaint, M. Patou-Mathis, D. Bonjean, and M. Otte. 2001. New isotopic evidence for dietary habits of Neanderthals from Belgium. *J. Hum. Evol.* 40:497–505.

Boesch, C. 1991. Handedness in wild chimpanzees. *Int. J. Primatol.* 12(6):541–558.

Boesch, C. 2002. Cooperative hunting roles among Taï chimpanzees. *Hum. Nat.* 13(1):27–46.

Boesch, C. 2003. Is culture a golden barrier between human and chimpanzee? *Evol. Anthropol.* 12(2):82–91.

Boesch, C., and H. Boesch. 1989. Hunting behavior of wild chimpanzees in the Taï National Park. *Am. J. Phys. Anthropol.* 78:547–573.

Boesch, C., and M. Tomasello. 1998. Chimpanzee and human culture. *Curr. Anthropol.* 39(5):591–614.

Bogin, B. 1988. *Patterns of Human Growth*. New York: Cambridge University Press.

Bogin, B. 1990. The evolution of human childhood. *BioScience* 40(1):16–25.

Bogin, B. 1997. Evolutionary hypotheses for human childhood. *Yrbk. Phys. Anthropol.* 40:63–89.

Bogin, B., and B.H. Smith. 1996. Evolution of the human life cycle. *Am. J. Hum. Biol.* 8:703–716.

Bookheimer, S. 2002. Functional MRI of language: New approaches to understanding the cortical organization of semantic processing. *Annu. Rev. Neurosci.* 25:151–188.

Bower, B. 1994. Language without rules. *Sci. News* 145:346–347.

Bower, B. 1988. Chaotic connections. *Sci. News* 133: 58–59.

Bower, J.M., and L.M. Parsons. 2003. Rethinking the "lesser brain." *Sci. Am.* 289(Aug):50–57.

Brace, C.L. 1964. Structural reduction in evolution. *Am. Nat.* 97:39–49.

Brace, C.L. 1967. Environment, tooth form, and size in the Pleistocene. *J. Dent. Res.* 46:809–816.

Brace, C.L. 1971. Post-Pleistocene changes in the human dentition. *Am. J. Phys. Anthropol.* 34:191–204.

Brace, C.L. 1986. Egg on the face, *f* in the mouth, and the overbite. *Am. Anthropol.* 88:695–697.

Brace, C.L. 1992. Modern human origins: narrow focus or broad spectrum? David Skomp Distinguished Lectures in Anthropology, Indiana University.

Brace, C.L., S.L. Smith, and K.D. Hunt. 1991. What big teeth you had Grandma! Human tooth size, past and present. In M.A. Kelley and C.S. Larsen, eds., *Advances in Dental Anthropology*, New York: Wiley-Liss. Pp. 33–57.

Bramble, D.R. 1989. Axial-appendicular dynamics and the integration of breathing and gait in mammals. *Am. Zool.* 29:171–186.

Bramble, D.R., and D.R. Carrier. 1983. Running and breathing in mammals. *Science* 219:251–256.

Bramble, D.R., and F.A. Jenkins. 1993. Mammalian locomotor-respiratory integration: Implications for diaphragmatic and pulmonary design. *Science* 262:235–240.

Branda, R.F., and J.W. Eaton. 1978. Skin color and nutrient photolysis: An evolutionary hypothesis. *Science* 201:625–626.

Bribiescas, R.G. 2001. Reproductive physiology of the human male. In P.T. Ellison, ed., *Reproductive Ecology and Human Evolution*. New York: Aldine de Gruyter. Pp. 107–133.

Bromage, T.G. 1987. The biological and chronological maturation of early hominids. *J. Hum. Evol.* 16:257–272.

Broadhurst, C.L., S.C. Cunnane, and M.A. Crawford. 1998. Rift Valley lake fish and shellfish provided brain-specific nutrition for early *Homo*. *Br. J. Nutr.* 79:3–21.

Broadhurst, C.L., Y. Wang, M.A. Crawford, S.C. Cunnane, J.E. Parkington, and W.F. Schmidt. 2002. Brain-specific lipids from marine, lacustrine, or terrestrial food resources: Potential impacts on early African *Homo sapiens*. *Comp. Biochem. Physiol. B Biochem. Mol. Biol.* 131:653–673.

Brown, L., R.W. Shumaker, and J.F. Downhower. 1995. *Am. Nat.* 146(2):302–306.

Bruckner, J.S. 1992. *The Human Subtalar Joint: A Theme on Variations*. Ph.D. Dissertation, Indiana University.

Brues, A.M. 1966. "Probable mutation effect" and the evolution of hominid teeth and jaws. *Am. J. Phys. Anthropol.* 25:169–170.

Brues, A.M. 1977. *People and Races*. New York: Macmillan.

Bruner, E., G. Manzi, and J.L. Arsuaga. 2003. Encephalization and allometric trajectories in the genus *Homo*: Evidence from the Neandertal and modern lineages. *Proc. Natl. Acad. Sci. U.S.A.* 100:15335–15340.

Bunn, H.T. 2001. Hunting, power scavenging, and butchering by Hadza foragers and by Plio-Pleistocene *Homo*. In C.B. Stanford and H.T. Bunn, eds., *Meat-Eating and Human Evolution*. Oxford: Oxford University Press. Pp. 199–218.

Burd, M. 1986. Sexual selection and human evolution: All or none adaptation? *Am. Anthropol.* 88:167–172.

Burley, N. 1979. The evolution of concealed ovulation. *Am. Nat.* 114(6):835–858.

Burling, R. 1993. Primate calls, human language, and nonverbal communication. *Curr. Anthropol.* 34(1):25–53.

Burton, J.H., and L.E. Wright. 1995. Nonlinearity in the relationship between bone Sr/Ca and diet: paleodietary implications. *Am. J. Phys. Anthropol.* 96:273–282.

Buss, D.M. 1989. Sex differences in human mate preferences: Evolutionary hypotheses tested in 37 cultures. *Behav. Brain Sci.* 12:1–49.

Buss, D.M. 1993. Sexual strategies theory: An evolutionary perspective on human mating. *Psychol. Rev.* 100(2):204–232.

Buss, D.M. 1994a. The strategies of human mating. *Am. Sci.* 82(May–June):238–249.

Buss, D.M. 1994b. *The Evolution of Desire: Strategies of Human Mating*. New York: Basic Books.

Buss, D.M. 2000. Desires in human mating. *Ann. New York Acad. Sci.* 907:39–49.

Butynski, T.M. 1982. Vertebrate predation by primates: A review of hunting patterns and prey. *J. Hum. Evol.* 11:421–430.

Cabanac, M. 1986. Keeping a cool head. *News Physiol. Sci.* 1(4):41–44.

Cabanac, M. 1992. Pleasure: The common currency. *J. Theor. Biol.* 155:173–200.

Cadien, J.D. 1972. Dental variation in man. In S.L. Washburn and P. Dolhinow, eds., *Perspectives on Human Evolution*, Vol. 2. New York: Holt, Rinehart, and Winston. Pp. 199–222.

Cahill, L., and J.L. McGaugh. 1998. Mechanisms of emotional arousal and lasting declarative memory. *Trends Neurosci.* 21:294–299.

Caicedo, A., and S.D. Roper. 2001. Taste receptor cells that discriminate between bitter stimuli. *Science* 291:1557–1560.

Calcagno, J.M. 1989. *Mechanisms of Human Dental Reduction: A Case Study from Post-Pleistocene Nubia.* University of Kansas Publications in Anthropology 18. Lawrence, KA: University of Kansas.

Calder, P.C., D.J. Harvey, C.M. Pond, and E.A. Newsholme. 1992. Site-specific differences in the fatty acid composition of human adipose tissue. *Lipids* 27(9):716–720.

Caldwell, M. 2000. Polly wanna PhD? *Discover* 21(Jan):70–75.

Calvin, W.H. 1991. *The Descent of Mind: Ice Age Climates and the Evolution of Intelligence.* New York: Doubleday.

Cann, R.L., and A.C. Wilson. 1982. Models of human evolution. *Science* 217:303–304.

Cant, J.G.H. 1981. Hypothesis for the evolution of human breasts and buttocks. *Am. Nat.* 117:199–204.

Cantalupo, C., and W.D. Hopkins. 2001. Asymmetric Broca's area in great apes. *Nature* 414:505.

Capaday, C. 2002. The special nature of human walking and its neural control. *Trends Neurosci.* 25(7):370–376.

Caramazza, A., and A.E. Hillis. 1991. Lexical organization of nouns and verbs in the brain. *Nature* 349:788–790.

Carey, J.W., and A.T. Steegman. 1981. Human nasal protrusion, latitude, and climate. *Am. J. Phys. Anthropol.* 56:313–319.

Carlsen, E., A. Giwercman, N. Keirding, and N.E. Skakkebæk. 1992. Evidence for decreasing quality of semen during the past 50 years. *Br. Med. J.* 305:609–613.

Caro, T.M. 1987. Human breasts: Unsupported hypotheses reviewed. *Hum. Evol.* 2(3):271–282.

Caro, T.M., and D.W. Sellen. 1990. The reproductive advantages of fat in women. *Ethol. Sociobiol.* 11:51–66.

Caro, T.M., D.W. Sellen, A. Parish, R. Frank, D.M. Brown, E. Voland, and M. Borgerhoff Mulder. 1995. Termination of reproduction in nonhuman and human female primates. *Int. J. Primatol.* 16(2):205–220.

Carrier, D.R. 1984. The energetic paradox of human running and hominid evolution. *Curr. Anthropol.* 25(4):483–495.

Carrier, D.R. 1987. The evolution of locomotor stamina in tetrapods: Circumventing a mechanical constraint. *Paleobiology* 13(3):326–341.

Carrier, D.R., N.C. Heglund, and K.D. Earls. 1994. Variable gearing during locomotion in the human musculoskeletal system. *Science* 265:651–653.

Cartmill, M. 1974a. Rethinking primate origins. *Science* 184:436–443.

Cartmill, M. 1974b. Pads and claws in arboreal locomotion. In F.A. Jenkins, ed., *Primate Locomotion.* New York: Academic Press. Pp. 45–83.

Cartmill, M. 1979. The volar skin of primates: Its frictional characteristics and their functional significance. *Am. J. Phys. Anthropol.* 50:497–510.

Cartmill, M. 1993. *A View to a Death in the Morning: Hunting and Nature through History.* Cambridge, MA: Harvard University Press.

Cartmill, M. 1998. The gift of gab. *Discover* 19(Nov):56–64.

Cashdan, E. 1996. Women's mating strategies. *Evol. Anthropol.* 5(4):134–143.

Cauna, N. 1954. Nature and functions of the papillary ridges of the digital skin. *Anat. Rec.* 119:449–468.

Cela-Conde, C.J., and F.J. Ayala. 2003. Genera of the human lineage. *Proc. Natl. Acad. Sci. U.S.A.* 100:7684–7689.

Chamberlain, J.G. 1996. The possible role of long-chain, omega-3 fatty acids in human brain phylogeny. *Perspect. Biol. Med.* 39(3):436–445.

Chaplin, G., and N.G. Jablonski. 1998. Hemispheric difference in skin color. *Am. J. Phys. Anthropol.* 107:221–224.

Chase, P.G., and H.L. Dibble. 1987. Middle Paleolithic symbolism: A review of current evidence and interpretations. *J. Anthropol. Archaeol.* 6:263–296.

Chen, J.-Y., D.J. Bottjer, P. Oliveri, S.Q. Dornbos, F. Gao, S. Ruffins, H. Chi, C.-W. Li, and E.H. Davidson. 2004. Small bilaterian fossils from 40 to 55 million years before the Cambrian. *Science* 305:218–222.

Cheney, D., R. Seyfarth, and B. Smuts. 1986. Social relationships and social cognition in nonhuman primates. *Science* 234:1361–1366.

Cheour, M., O. Martynova, R. Näätänen, R. Erkkola, M. Sillanpää, P. Kero, A. Raz, M.-L. Kaipio, J. Hiltunen, O. Aaltonen, J. Savela, and H. Hämäläinen. 2002. Speech sounds learned by sleeping newborns. *Nature* 415:599–600.

Chevalier-Skolnikoff, S. 1976. Homosexual behavior in a laboratory group of stumptail macaques (*Macaca arctoides*): Forms, contexts, and possible social functions. *Arch. Sex. Behav.* 5:511–527.

Chiarelli, B. 2001. Pheromonal communication and socialization. In D. Falk and K.R. Gibson, eds. *Evolutionary Anatomy of the Primate Cerebral Cortex.* Cambridge: Cambridge University Press. Pp. 165–176.

Chivers, D.J., and C.M. Hladik. 1980. Morphology of the gastrointestinal tract in primates: Comparisons with other mammals in relation to diet. *J. Morphol.* 166:337–386.

Churchill, S.E. 1996. Particulate versus integrated evolution of the upper body in Late Pleistocene humans: A test of two models. *Am. J. Phys. Anthropol.* 100:559–583.

Clack, J.A. 1989. Discovery of the earliest-known tetrapod stapes. *Nature* 342:425–427.

Clack, J.A. 2002a. *Gaining Ground: The Origin and Evolution of Tetrapods.* Bloomington, IN: Indiana University Press.

Clack, J.A. 2002b. An early tetrapod from "Romer's Gap." *Nature* 418:72–76.

Clandinin, M.T., J.E. Chappell, T. Heim, P.R. Swyer, and G.W. Chance. 1981. Fatty acid utilization in perinatal de novo synthesis of tissues. *Early Hum. Dev.* 5:355–366.

Clarke, R.J., and P.V. Tobias. 1995. Sterkfontein Member 2 foot bones of the oldest South African hominid. *Science* 269:521–524.

Clemens, T.L., S.L. Henderson, J.S. Adams, and M.F. Holick. 1982. Increased skin pigmentation reduces the capacity of the skin to synthesize vitamin D. *Lancet* 1:74–76.

Clutton-Brock, T.H., and P.H. Harvey. 1980. Primates, brains, and ecology. *J. Zool.* 190:309–323.

Coates, M.I., J.E. Jeffery, and M. Ruta. 2002. Fins to limbs: What the fossils say. *Evol. Dev.* 4(5):390–401.

Colegrave, N. 2002. Sex releases the speed limit on evolution. *Nature* 420:664–666.

Conklin-Brittain, N.L., R.W. Wrangham, and C.C. Smith. 2002. A two-stage model of increased dietary quality in early hominind evolution: The role of fiber. In P.S. Ungar and M.F. Teaford, eds., *Human Diet: Its Origin and Evolution.* Westport, CT: Bergin and Garvey. Pp. 61–76.

Conroy, G.C. 1997. *Reconstructing Human Origins: A Modern Synthesis.* New York: W.W. Norton.

Conroy, G.C., and K. Kuykendall. 1995. Paleopediatrics: Or when did human infants really become human? *Am. J. Phys. Anthropol.* 98:121–131.

Conroy, G.C., and M.W. Vannier. 1987. Dental development of the Taung skull from computerized tomography. *Nature* 329:625–627.

Coon, C.S. 1982. *Racial Adaptations.* Chicago: Nelson Hall.

Coppens, Y. 1994. East side story: The origin of humankind. *Sci. Am.* 270(May):88–95.

Cordain, L., J.B. Miller, S.B. Eaton, N. Mann, S.H.A. Holt, and J.D. Speth. 2000. Plant-animal subsistence ratios and macronutrient energy estimations in worldwide hunter-gatherer diets. *Am. J. Clin. Nutr.* 71:682–692.

Cordain, L., B.A. Watkins, and N.J. Mann. 2001. Fatty acid composition and energy density of foods available to African hominids: Evolutionary implications for human brain development. In A.P. Simopoulos and K.N. Pavlou, eds., *Nutrition and Fitness: Metabolic Studies in Health and Disease. World Rev. Nutr. Diet* 90:144–161.

Cordain, L., S.B. Eaton, J.B. Miller, N. Mann, and K. Hill. 2002. The paradoxical nature of hunter-gatherer diets: Meat-based, yet atherogenic. *Eur. J. Clin. Nutr.* 56(Suppl 1):S42–S52.

Correia, H.R., S.C. Balseiro, E.R. Correia, P.G. Mota, and M.L. de Areia. 2004. Why are human newborns so fat? Relationship between fatness and brain size at birth. *Am. J. Hum. Biol.* 16:24–30.

Corruccini, R.S., and H.M. McHenry. 1980. Hominid femoral neck length. *Am. J. Phys. Anthropol.* 52:397–398.

Corruccini, R.S., and H.M. McHenry. 2001. Knuckle-walking hominid ancestors. *J. Hum. Evol.* 40:507–511.

Cotman, C.W., ed. 1987. *The Neuro-Immune Endocrine Connection*. New York: Raven Press.

Courtiss, E.H., T.J. Gargan, and G.B. Courtiss. 1984. Nasal physiology. *Ann. Plast. Surg.* 13:214–223.

Cowles, R.B. 1959. Some ecological factors bearing on the origin and evolution of pigment in the human skin. *Am. Nat.* 93:283–293.

Crawford, M.A. 1992. The role of dietary fatty acids in biology: Their place in the evolution of the human brain. *Nutr. Rev.* 50(4):3–11.

Crelin, E.S. 1987. *The Human Vocal Tract: Anatomy, Function, Development and Evolution*. New York: Vantage Press.

Crompton, A.W., and K. Hiiemäe. 1969. How mammalian molar teeth work. *Discovery* (Yale Peabody Museum) 5(1):23–34.

Crompton, R.H., Y. Li, W.J. Wang, M.M. Guther, and R. Savage. 1998. The mechanical effectiveness of erect and "bent-hip, bent-knee" bipedal walking in *Australopithecus afarensis*. *J. Hum. Evol.* 35:55–74.

Cronin, J.E., N.T. Boaz, C.B. Stringer, and Y. Rak. 1981. Tempo and mode in hominid evolution. *Nature* 292:113–122.

Cubelli, R. 1991. A selective deficit for writing vowels in acquired dysgraphia. *Nature* 353:258–260.

Cunnane, S.C. 1980. The aquatic ape theory reconsidered. *Med. Hypotheses* 6:49–58.

Cunnane, S.C., L.S. Harbrige, and M.A. Crawford. 1993. The importance of energy and nutrient supply in human brain evolution. *Nutr. Health* 9:219–235.

Curnoe, D., and A. Thorne. 2003. Number of ancestral human species: A molecular perspective. *Homo* 53(3):201–224.

Currey, J. 1984. *The Mechanical Adaptations of Bones*. Princeton, NJ: Princeton University Press.

Daegling, D.J. 1993. Functional morphology of the human chin. *Evol. Anthropol.* 1(5):170–177.

Dahl, J.F. 1986. Cyclic perineal swelling during the intermenstrual intervals of captive female pygmy chimpanzees (*Pan paniscus*). *J. Hum. Evol.* 15:369–385.

Dahl, J.F., K.G. Gould, and R.D. Nadler. 1993. Testicle size of orang-utans in relation to body size. *Am. J. Phys. Anthropol.* 90:229–236.

Dainton, M. 2001. Did our ancestors knuckle-walk? *Nature* 410:324–325.

Dainton, M., and G.A. Macho. 1999. Did knuckle-walking evolve twice? *J. Hum. Evol.* 36:171–194.

Damasio, A.R., and D. Tranel. 1993. Nouns and verbs are retrieved with differently distributed neural systems. *Proc. Natl. Acad. Sci. U.S.A.* 90:4957–4960.

Daniels, D. 1983. The evolution of concealed ovulation and self-deception. *Ethol. Sociobiol.* 4:69–87.

Daniels, F., P.W. Post, and B.E. Johnson. 1972. Theories of the role of pigment in the evolution of human races. In V. Riley, ed., *Pigmentation: Its Genesis and Control*. New York: Appleton-Century-Crofts. Pp. 13–22.

Dart, R.A. 1949. The predatory implemental technique of *Australopithecus*. *Am. J. Phys. Anthropol.* 7(1):1–38.

Dart, R.A. 1957. The osteodontoceratic culture of *Australopithecus prometheus*. *Transvaal Mus. Mem.* 10.

Darwin, C. 1859. *The Origin of Species by Means of Natural Selection*. Reprinted in New York by The Modern Library.

Darwin, C. 1871. *The Descent of Man and Selection in Relation to Sex*. Reprinted in New York by the Modern Library.

Daum, I., H. Ackermann, M.M. Schugens, C. Reimhold, J. Dichgans, and N. Birbaumer. 1993. The cerebellum and cognitive functions in humans. *Behav. Neurosci.* 107(3):411–419.

Day, M.H., and J. Napier. 1963. The functional significance of the deep head of flexor pollicis brevis in primates. *Folia Primatol.* 1:122–134.

Deacon, T.W. 1997a. *The Symbolic Species.* New York: W.W. Norton.

Deacon, T.W. 1997b. What makes the human brain different? *Annu. Rev. Anthropol.* 26:337–357.

Dean, M.C. 1987. Growth layers and incremental markings in hard tissues: A review of the literature and some preliminary observations about enamel structure in *Paranthropus boisei. J. Hum. Evol.* 16:157–172.

Dean, M.C. 1988. Another look at the nose and the functional significance of the face and nasal mucous membrane for cooling the brain in fossil hominids. *J. Hum. Evol.* 17:715–718.

Dean, M.C. 2000. Incremental markings in enamel and dentine: What they can tell us about the way teeth grow. In M.F. Teaford, M.M. Smith, and M.W.J. Ferguson, eds., *Development, Function, and Evolution of Teeth.* Cambridge: Cambridge University Press. Pp. 119–130.

Dean, M.C., and A.D. Beynon. 1991. Histological reconstruction of crown formation times and initial root formation times in a modern human child. *Am. J. Phys. Anthropol.* 86:215–228.

Dean, M.C., and D.J. Reid. 2001. Perikymata spacing and distribution on hominid anterior teeth. *Am. J. Phys. Anthropol* 116:209–215.

Dean, M.C., A.D. Beynon, J.F. Thackery, and G.A. Macho. 1993. Histological reconstruction of dental development and age at death of a juvenile *Paranthropus robustus* specimen, SK 63, from Swartkrans, South Africa. *Am. J. Phys. Anthropol.* 91:401–419.

Dean, C., M.G. Leakey, D. Reid, F. Schrenk, G.T. Schwartz, C. Stringer, and A. Walker. 2001. Growth processes in teeth distinguish modern humans from *Homo erectus* and earlier hominids. *Nature* 414:628–631.

Deban, S.M., and D.R. Carrier. 2002. Hypaxial muscle activity during running and breathing in dogs. *J. Exp. Biol.* 205:1953–1967.

de Catanzaro, D., and E. MacNiven. 1992. Psychogenic pregnancy disruptions in mammals. *Neurosci. Biobehav. Rev.* 16:43–53.

DeGusta, D., W.H. Gilbert, and S.F. Turner. 1999. Hypoglossal canal size and hominid speech. *Proc. Natl. Acad. Sci. U.S.A.* 96:1800–1804.

Dehaene-Lambertz, G., S. Dehaene, and L. Hertz-Pannier. 2002. Functional neuroimaging of speech perception in infants. *Science* 298:2013–2015.

de Heinzelin, J., J.D. Clark, T. White, W. Hart, P. Renne, G. Woldegabriel, Y. Beyene, and E. Vrba. 1999. Environment and behavior of 2.5-million-year-old Bouri hominids. *Science* 284:625–629.

Delgado, R.A., and C.P. Van Schaik. 2000. The behavioural ecology and conservation of the orangutan (*Pongo pygmaeus*): A tale of two islands. *Evol. Anthropol.* 9(5):201–218.

Demes, B., and N. Creel. 1988. Bite force, diet, and cranial morphology of fossil hominids. *J. Hum. Evol.* 17:657–670.

Denton, D. 1982. *The Hunger for Salt.* Berlin: Springer-Verlag.

de Troyer, A., and S. Kelly. 1982. Chest wall mechanics in dogs with acute diaphragm paralysis. *J. Appl. Physiol.* 53:373–379.

Dettwyler, K.A. 1995. A time to wean: The hominid blueprint for the natural age of weaning in human populations. In P. Stuart-Macadam and K.A. Dettwyler, eds., *Breastfeeding: Biocultural Perspectives.* Hawthorne, NY: Aldine de Gruyter. Pp. 39–73.

de Visser, J.A.G.M., C.W. Zeyl, P.J. Gerrish, J.L. Blanchard, and R.E. Lenski. 1999. Diminishing returns from mutation supply rate in asexual populations. *Science* 283:404–406;

de Waal, F.B.M. 1995. Bonobo sex and society. *Sci. Am.* 272(Mar):82–88.

de Waal, F. 1997a. Bonobo dialogues. *Nat. Hist.* 106(May):22–25.

de Waal, F.B.M. 1997b. *Bonobo: The Forgotten Ape.* Berkeley, CA: University of California Press.

de Waal, F. 2001. *The Ape and the Sushi Master.* New York: Basic Books.

Dewey, K. 1997. Energy and protein requirements during lactation. *Annu. Rev. Nutr.* 17:19–36.

de Winter, W., and C.E. Oxnard. 2001. Evolutionary relationships and convergences in the structural organization of mammalian brains. *Nature* 409:710–714.

Diamond, J. 1992a. Sweet death. *Nat. Hist.* 103(Feb):2–6.

Diamond, J. 1992b. *The Third Chimpanzee: The Evolution and Future of the Human Animal.* New York: Harper Collins.

Diamond, J. 1996. Why women change. *Discover* 17(July):130–137.

Diamond, J. 2003. The double puzzle of diabetes. Nature 423:599–602.

Dietz, V. 2002. Do human bipeds use quadrupedal coordination? *Trends Neurosci.* 25(9):462–467.

Dixson, A.F. 1987. Observations on the evolution of the genitalia and copulatory behaviour in male primates. *J. Zool.* 213:423–443.

Dixson, A.F. 1998. Sexual selection and evolution of the seminal vesicles in primates. *Folia Primatol.* 69:300–306.

Doblhammer, G., and J. Oeppen. 2003. Reproduction and longevity among the British peerage: The effect of frailty and health selection. *Proc. R. Soc. Lond. B Biol. Sci.* 270:1541–1547.

Dobson, S.D., and E. Trinkaus. 2002. Cross-sectional geometry and morphology of the mandibular symphysis in Middle and Late Pleistocene *Homo. J. Hum. Evol.* 43:67–87.

Dolan, R.J. 2002. Emotion, cognition, and behavior. *Science* 298:1191–1194.

Dominy, N.J., and P.W. Lucas. 2001. Ecological importance of trichromatic vision to primates. *Nature* 410:363–366.

Dominy, N.J., J.-C. Svenning, and W.-H. Li. 2003. Historical contingency in the evoution of primate color vision. *J. Hum. Evol.* 44:25–45.

Doran, D.M. 1993a. Comparative locomotor behavior of chimpanzees and bonobos: The influence of morphology on locomotion. *Am. J. Phys. Anthropol.* 91:83–98.

Doran, D.M. 1993b. Sex differences in adult chimpanzee positional behavior: The influence of body size on locomotion and posture. *Am. J. Phys. Anthropol.* 91:99–115.

Dorozynski, A., and A. Anderson. 1991. Collagen: A new probe into prehistoric diet. *Science* 254:520–521.

Døving, K.B., and D. Trotier. 1998. Structure and function of the vomeronasal organ. *J. Exp. Biol.* 201:2913–2925.

DuBrul, E.L. 1977. Early hominid feeding mechanisms. *Am. J. Phys. Anthropol.* 47:305–320.

DuBrul, E.L., and H. Sicher. 1954. *The Adaptive Chin.* Springfield, IL: Charles C Thomas.

Duchin, L.E. 1990. The evolution of articulate speech: Comparative anatomy of the oral cavity in *Pan* and *Homo. J. Hum. Evol.* 19:687–697.

Ducrocq, S. 1998. Eocene primates from Thailand: Are Asian anthropoideans related to African ones? *Evol. Anthropol.* 7(3):97–104.

Dunbar, R.I.M. 1992. Neocortex size as a constraint on group size in primates. *J. Hum. Evol.* 20:469–493.

Dunbar, R.I.M. 1995. Neocortex size and group size in primates: A test of the hypothesis. *J. Hum. Evol.* 28:287–296.

Dunbar, R.I.M. 1998. The social brain hypothesis. *Evol. Anthropol.* 6(5):178–190.

Duncan, A.S., J. Kappleman, and L.J. Shapiro. 1994. Metatarso-phalangeal joint function and positional behavior in *Australopithecus afarensis. Am. J. Phys. Anthropol.* 93:67–81.

Eaton, S.B. 1992. Humans, lipids, and evolution. *Lipids* 27(10):814–820.

Eaton, S.B., and M. Konner. 1985. Paleolithic nutrition: A consideration of its nature and current implications. *N. Engl. J. Med.* 312:283–289.

Eaton, S.B., M. Shostak, and M. Konner. 1988. *The Paleolithic Prescription: A Program of Diet and Exercise and a Design for Living.* New York: Harper and Row.

Eaton, S.B., S.B. Eaton, III, and L. Cordain. 2002. Evolution, diet, and health. In P.S. Ungar and M.F. Teaford, eds., *Human Diet: Its Origin and Evolution.* Westport, CT: Bergin and Garvey. Pp. 7–17.

Eaton, T.H. 1960. The aquatic origin of tetrapods. *Trans. Kansas Acad. Sci.* 63(3):115–120.

Ebert, D., and W.D. Hamilton. 1996. Sex against virulence: The coevolution of parasitic diseases. *Trends Ecol. Evol.* 11(2):79–81.

Ebling, J. 1985. The mythological origin of nudity. *J. Hum. Evol.* 14(1):33–41.

Eckenhoff, J.E. 1970. The physiologic significance of the vertebral venous plexus. *Surg. Gynecol. Obstet.* 131(1):72–78.

Edwards, J.L. 1989. Two perspectives on the evolution of the tetrapod limb. *Am. Zool.* 29:235–254.

Eibl-Eibesfeldt, I. 1989. *Human Ethology*. New York: Aldine de Gruyter.

Eickhoff, R. 1988. Origin of bipedalism—When, why, how and where? *S. Afr. J. Sci.* 84:486–488.

Eidelberg, E., J.G. Walden, and L.H. Nguyen. 1981. Locomotor control in macaque monkeys. *Brain* 104:647–663.

Eisthen, H.L. 1997. Evolution of vertebrate olfactory systems. *Brain Behav. Evol.* 50:222–233.

Elbert, T., C. Panter, C. Weinbruch, B. Rockstroh, and E. Taub. 1995. Increased cortical representation of the fingers of the left hand in string players. *Science* 270:305–307.

Elia, I. 1986. *The Female Animal*. New York: Henry Holt and Co.

Ellis, K.J., R.J. Shypailo, and R.H. Schanler. 1994. Body composition of the preterm infant. *Ann. Hum. Biol.* 21(6):533–545.

Ellison, P.T. 1990. Human ovarian function and reproductive ecology: New hypotheses. *Am. Anthropol.* 92(4):933–952.

Ellison, P.T. 2001. *On Fertile Ground*. Cambridge, MA: Harvard University Press.

Enard, W., P. Khaitovich, J. Klose, S. Zöllner, F. Heissig, P. Gialvalisco, K. Nieselt-Struwe, E. Muchmore, A. Varki, R. Ravid, G.M. Doxiadis, R.E. Bontrop, and S. Pääbo. 2002. Intra- and interspecific variation in primate gene expression patterns. *Science* 296:340–343.

Endler, J.A. 1986. *Natural Selection in the Wild*. Princeton, NJ: Princeton University Press.

Enquist, M., and A. Arak. 1993. Selection of exaggerated male traits by female aesthetic senses. *Nature* 361:446–448.

Essock-Vitale, S., and R.M. Seyfarth. 1987. Intelligence and social cognition. In B.B. Smuts, D.L. Cheney, R.M. Seyfarth, R.W. Wrangham, and T.T. Struhsaker, eds., *Primate Societies*. Chicago: University of Chicago Press. Pp. 452–461.

Falk, D. 1980. Hominid brain evolution: The approach from paleoneurology. *Yrbk. Phys. Anthropol.* 23:93–107.

Falk, D. 1987. Brain lateralization and its evolution in hominids. *Yrbk. Phys. Anthropol.* 30:107–125.

Falk, D. 1990. Brain evolution in *Homo*: The "radiator" theory. *Behav. Brain Sci.* 13(2):333–381.

Fairgrieve, C. 1997. Meat eating by blue monkeys (*Cercopithecus mitis stuhlmanni*): Predation of a flying squirrel (*Anomalurus derbianus jacksonii*). *Folia Primatol.* 68:354–356.

Fedak, M.A., and H.J. Seeherman. 1979. Reappraisal of energetics of locomotion shows identical cost in bipeds and quadrupeds, including ostrich and horse. *Nature* 282:713–716.

Fedigan, L.M. 1982. *Primate Paradigms: Sex Roles and Social Bonds*. Montreal: Eden Press.

Fedigan, L.M. 1990. Vertebrate predation in *Cebus capucinus*: Meat eating in a Neotropical monkey. *Folia Primatol.* 54:196–205.

Finlay, B.L., and R.B. Darlington. 1995. Linked regularities in the development and evolution of mammalian brains. *Science* 268:1578–1584.

Finlay, B.L., R.B. Darlington, and N. Nicastro. 2001. Developmental structure in brain evolution. *Behav. Brain Sciences* 24:263–308.

Finn, C.A. 1986. Implantation, menstruation, and inflammation. *Biol Rev.* 61:313–328.

Finn, C.A. 1987. Why do women and some other primates menstruate? *Persp. Biol. Med.* 30(4):566–574.

Finn, C.A. 1998. Menstruation: A nonadaptive consequence of uterine evolution. *Q. Rev. Biol.* 73(2):163–173.

Firestein, S. 1991. A noseful of odor receptors. *Trends Neurosci.* 14(7):270–272.

Firestein, S. 2001. How the olfactory system makes sense of scents. *Nature* 413:211–218.

Fish, J.L., and C.A. Lockwood. 2003. Dietary constraints on encephalization in primates. *Am. J. Phys. Anthropol.* 120(2):171–181.

Fischer, S., M. Hallschmid, A.L. Elsner, and J. Born. 2002. Sleep forms memory for finger skills. *Proc. Natl. Acad. Sci. U.S.A.* 99(18):11987–11991.

Fisher, H.E. 1982. *The Sex Contract*. New York: William Morrow and Co.

Fisher, H.E. 1992. *Anatomy of Love: The Natural History of Monogamy, Adultery, and Divorce*. New York: W.W. Norton.

Fisher, R.E. 2000. The primate appendix: A reassessment. *Anat. Rec. (New Anat.)* 261:228–236.

Fitch, W.T., and D. Reby. 2001. The descended larynx is not uniquely human. *Proc. R. Soc. Lond. B Biol. Sci.* 268:1669–1675.

FitzGerald, C.M. 1998. Do enamel microstructures have regular time dependency? Conclusions from the literature and a large-scale study. *J. Hum. Evol.* 35:371–386.

Fleagle, J.G. 1976. Locomotion and posture of the Malayan siamang and implications for hominoid evolution. *Folia Primatol.* 26:245–269.

Fleagle, J.G., J.T. Stern, W.L. Jungers, R.L. Susman, A.K. Vangor, and J.P. Wells. 1981. Climbing: A biomechanical link with brachiation and bipedalism. *Symp. Zool. Soc. Lond.* 48:359–375.

Florence, S.L., N. Jain, and J.H. Kaas. 1997. Plasticity of somatosensory cortex in primates. *Semin. Neurosci.* 9:3–12.

Foley, R.A., and S. Elton. 1998. Time and energy: the ecological context for the evolution of bipedalism. In E. Strasser, J. Fleagle, A. Rosenberger, and H. McHenry, eds., *Primate Locomotion: Recent Advances*. New York: Plenum Pub. Pp. 419–433.

Foley, R.A., and P.C. Lee. 1991. Ecology and energetics of encephalization in hominid evolution. *Phil. Trans. R. Soc. Lond. B Biol. Sci.* 334:223–232.

Fomon, S.J., F. Haschke, E.E. Ziegler, and S.E. Nelson. 1982. Body composition of reference children from birth to age 10 years. *Am. J. Clin. Nutr.* 35:1169–1175.

Ford, C.S., and F.A. Beach. 1951. *Patterns of Sexual Behavior*. New York: Harper and Brothers.

Formicola, V., and M. Giannecchini. 1999. Evolutionary trends of stature in Upper Paleolithic and Mesolithic Europe. *J. Hum. Evol.* 36:319–333.

Fox, P.T. 1989. Functional brain mapping with positron emission tomography. *Semin. Neurol.* 9(4):323–329.

Fragaszy, D. 2003. Making space for traditions. *Evol. Anthropol.* 12(2):61–70.

Franciscus, R.G., and E. Trinkaus. 1988. Nasal morphology and the emergence of *Homo erectus*. *Am. J. Phys. Anthropol.* 75:517–527.

Fray, R.R., and A.N. Popper. 1985. The octavolateralis system. In M. Hildebrand, D.M. Bramble, K.F. Liem, and D.B. Wake, eds., *Functional Vertebrate Morphology*. Cambridge, MA: Harvard University Press. Pp. 291–316.

Frayer, D.W. 1981. Body size, weapon use, and natural selection in the European Upper Paleolithic and Mesolithic. *Am. Anthropol.* 83:57–73.

Frayer, D.W., and C. Nicolay. 2000. Fossil evidence for the origin of speech sounds. In N.L. Wallin, B. Merker, and S. Brown, eds., *The Origins of Music*. Cambridge, MA: MIT Press. Pp. 217–234.

Fredrickson, B.L. 2003. The value of positive emotions. *Am. Sci.* 91(July–Aug):330–335.

Freedman, D.H. 1992. The aggressive egg. *Discover* 13(June):61–65.

Freeman, A.L.J., and H.Y. Wong. 1995. The evolution of self-concealed ovulation in humans. *Ethol. Sociobiol.* 16:531–533.

Freeman, D.H. 1993. In the realm of the chemical. *Discover* 14(June):69–76.

Freeman, W.J. 1991. The physiology of perception. *Sci. Am.* 264(Feb):78–85.

Frisancho, A.R. 1981. *Human Adaptation: A Functional Interpretation*. Ann Arbor, MI: University of Michigan Press.

Frisch, R.E. 2002. *Female Fertility and the Body Fat Connection*. Chicago: University of Chicago Press.

Fuentes, A. 2000. Hylobatid communities: Changing views on pair bonding and social organization in hominoids. *Yrbk. Phys. Anthropol.* 43:33–60.

Fuss, F.K. 1989. Anatomy of the cruciate ligaments and their function in extension and flexion of the human knee joint. *Am. J. Anat.* 184:165–176.

Fuss, F.K. 1991. The restraining function of the cruciate ligaments on hyperextension and hyperflexion of the human knee joint. *Anat. Rec.* 230:283–289.

Fuss, F.K. 1992. Principles and mechanisms of automatic rotation during terminal extension in the human knee joint. *J. Anat.* 180:297–304.

Galis, F., J.J.M. van Alphen, and J.A.J. Metz. 2001. Why five fingers? Evolutionary constraints on digit numbers. *Trends Ecol. Evol.* 16(11):637–646.

Gallup, G.G. 1982. Permanent breast enlargement in human females: A sociobiological analysis. *J. Hum. Evol.* 11:597–601.

Gallup, G.G., and S.D. Suarez. 1983. Optimal reproductive strategies for bipedalism. *J. Hum. Evol.* 12:193–196.

Gangestad, S.W., R. Thornhill, and C.E. Garver. 2002. Changes in women's sexual interests and their partners' mate-retention tactics across the menstrual cycle: Evidence for shifting conflicts of interest. *Proc. R. Soc. Lond. B. Biol. Sci.* 269:975–982.

Gannon, P.J., R.L. Holloway, D.C. Broadfield, and A.R. Braun. 1998. Asymmetry of chimpanzee planum temporale: Humanlike pattern of Wernicke's brain language area homolog. *Science* 279:220–222.

Gannon, P.J., N.M. Kheck, and P.R. Hof. 2001. Language areas of the hominoid brain: a dynamic communicative shift on the upper east side planum. In D. Falk and K.R. Gibson, eds. *Evolutionary Anatomy of the Primate Cerebral Cortex*. Cambridge: Cambridge University Press. Pp. 216–240.

Gans, C. 1970. Respiration in early tetrapods—The frog is a red herring? *Evolution* 24:723–734.

Gans, C. 1989. Stages in the origin of vertebrates: Analysis by means of scenarios. *Biol. Rev.* 64:221–268.

Gantt, D.G. 1982. Neogene hominoid evolution: A tooth's inside view. In B. Kurten, ed., *Teeth: Form, Function, and Evolution*. New York: Columbia University Press. Pp. 93–108.

Gantt, D.G. 1983. The enamel of Neogene hominoids. In R.L. Ciochon and R.S. Corruccini, eds., *New Interpretations of Ape and Human Ancestry*. New York: Plenum Pub. Pp. 249–298.

Gao, J.-H., L.M. Parsons, J.M. Bower, J. Xiong, J. Li, and P.T. Fox. 1996. Cerebellum implicated in sensory acquisition and discrimination rather than motor control. *Science* 272:545–547.

Gebo, D.L., L. MacLatchy, R. Kityo, A. Deino, J. Kingston, and D. Pilbeam. 1997. A hominoid genus from the early Miocene of Uganda. *Science* 276:401–404.

Gebo, D.L., M. Dagosto, K.C. Beard, T. Qi, and J. Wang. 2000. The oldest known anthropoid postcranial fossils and the early evolution of higher primates. *Nature* 404:276–278.

Gibbons, A. 1998. "Monogamous" gibbons really swing. *Science* 280:677–678.

Gibbons, A. 2002. In search of the first hominids. *Science* 295:1214–1219.

Gibson, K.R., D. Rumbaugh, and M. Beran. 2001. Bigger is better: primate brain size in relationship to cognition. In D. Falk and K.R. Gibson, eds. *Evolutionary Anatomy of the Primate Cerebral Cortex*. Cambridge: Cambridge University Press. Pp. 79–97.

Gilad, Y., C.D. Bustamonte, D. Lancet, and S. Pääbo. 2003a. Natural selection on the olfactory receptor gene family in humans and chimpanzees. *Am. J. Hum. Genet.* 73:489–501.

Gilad, Y., O. Man, S. Pääbo, and D. Lancet. 2003b. Human specific loss of olfactory genes. *Proc. Natl. Acad. Sci. U.S.A.* 100(6):3324–3327.

Gilissen, E. 2001. Structural symmetries and asymmetries in human and chimpanzee brains. In D. Falk and K.R. Gibson, eds. *Evolutionary Anatomy of the Primate Cerebral Cortex*. Cambridge: Cambridge University Press. Pp. 187–215.

Gillis, A.M. 1996. Why sleep? *BioScience* 46(6):391–393.

Gimelbrant, A.A., H. Skaletsky, and A. Chess. 2004. Selective pressures on the olfactory receptor repertoire since the human-chimpanzee divergence *Proc. Natl., Acad. Sci. U.S.A.* 101:9019–9022.

Ginsberg, J.R., and U.W. Huck. 1989. Sperm competition in mammals. *Trends Ecol. Evol.* 4(3):74–79.

Gittleman, J.L., and S.D. Thompson. 1988. Energy allocation in mammalian reproduction. *Am. Zool.* 28(3):863–875.

Glusman, G., I. Yanai, I. Rubin, and D. Lancet. 2001. The complete human olfactory subgenome. *Genome Res.* 11:685–702.

Goodall, J. 1986. *The Chimpanzees of Gombe: Patterns of Behavior*. Cambridge, MA: Belknap Press.

Goren-Inbar, N., G. Sharon, Y. Melamed, and M. Kislev. 2002. Nuts, nut cracking, and pitted stones at Gesher Benot Ya-aqov, Israel. *Proc. Natl. Acad. Sci. U.S.A.* 99(4):2455–2460.

Gottschalk, F., S. Kourosh, and B. Leveau. 1989. The functional anatomy of tensor fasciae latae and gluteus medius and minimus. *J. Anat.* 166:179–189.

Gould, K.G., L.G. Young, E.B. Smithwick, and S.R. Phythyon. 1993. Semen characteristics of the adult male chimpanzee (*Pan troglodytes*). *Am. J. Primatol.* 29:221–232.

Gould, S.J. 1977. *Ontogeny and Phylogeny*. Cambridge, MA: Harvard University Press.

Gould, S.J. 1990. An earful of jaw. *Nat. Hist.* 99(Mar):12–23.

Gould, S.J. 1991. Seven (or more) little piggies. *Nat. Hist.* 100(Jan):22–29.

Gould, S.J., and E.S. Vrba. 1982. Exaptation—A missing term in the science of form. *Paleobiology* 8(1):4–15.

Graf, W., and P.-P. Vidal. 1996. Semicircular canals and upright stance are not interrelated. *J. Hum. Evol.* 30:175–181.

Graham, C.A. 1991. Menstrual synchrony: An update and review. *Hum. Nat.* 2(4):293–311.

Graham, C.A., and W.C. McGrew. 1980. Menstrual synchrony in female undergraduates living on a coeducational campus. *Psychoneuroendocrinology* 5:245–252.

Grammar, K., B. Fink, A.P. Møller, and R. Thornhill. 2003. Darwinian aesthetics: Sexual selection and the biology of beauty. *Biol. Rev.* 78:385–407.

Gray, J.P., and L.D. Wolfe. 1983. Human female sexual cycles and the concealment of ovulation problem. *J. Soc. Biol.* 6:345–352.

Greaves, W.S. 1985. The mammalian postorbital bar as a torsion-resisting helical strut. *J. Zool.* 207:125–136.

Greene, D.L. 1970. Environmental influences on Pleistocene hominid dental evolution. *BioScience* 20(5):276–279.

Greenlaw, R.K., and J.V. Basmajian. 1975. Function of the gluteals in man. In R.H. Tuttle, ed., *Primate Functional Morphology and Evolution*. The Hague: Mouton. Pp. 271–280.

Gregory, W.K. 1928. The upright posture of man: A review of its origin and evolution. *Proc. Am. Phil. Soc.* 67:129–150.

Grine, F.E. 1981. Trophic differences between "gracile" and "robust" australopithecines: A scanning electron microscope analysis of occlusal events. *S. Afr. J. Sci.* 77(5):203–230.

Grine, F.E., and R.F. Kay. 1988. Early hominid diets from quantitative image analysis of dental microwear. *Nature* 333:765–768.

Grine, F.E., R.G. Klein, and T.P. Volman. 1991. Dating, archaeology and human fossils from the Middle Stone Age levels of Die Kelders, South Africa. *J. Hum. Evol.* 21:363–395.

Gunnell, G.F., and E.R. Miller. 2001. Origin of anthropoidea: Dental evidence and recognition of early anthropoids in the fossil record, with comments on the Asian anthropoid radiation. *Am. J. Phys. Anthropol.* 114:177–191.

Haenel, H. 1989. Phylogenesis and nutrition. *Die Nahrung* 33:867–887.

Haeusler, M., and H.M. McHenry. 2004. Body proportions in *Homo habilis* revisited. *J. Hum. Evol.* 46:433–465.

Haeusler, M., S.A. Martelli, and T. Boeni. 2002. Vertebrae numbers of the early hominid lumbar spine. *J. Evol.* 43: 621–643.

Haile-Selassie, Y. 2001. Late Miocene hominids from the Middle Awash, Ethiopia. *Nature* 412:178–181.

Hamilton, W.J. 1973. *Life's Color Code*. New York: McGraw-Hill Book Company.

Hamrick, M.W. 1998. Functional and adaptive significance of primate pads and claws: evidence from New World anthropoids. *Am. J. Phys. Anthropol.* 106:113–127.

Hamrick, M.W., and S.E. Inouye. 1995. Thumbs, tools, and early humans. *Science* 268:586–587.

Hamrick, M.W., S.E. Churchill, D. Schmitt, and W.L. Hylander. 1998. EMG of the human flexor pollicis longus muscle: Implications for the evolution of hominid tool use. *J. Hum. Evol.* 34:123–136.

Harcourt, A.H., P.H. Harvey, S.G. Larson, and R.V. Short. 1981. Testis weight, body weight, and breeding system in primates. *Nature* 293:55–57.

Harding, R.S.O., and S.C. Strum. 1976. The predatory baboons of Kekopey. *Nat. Hist.* 85(Mar):46–53.

Hardy, A. 1960. Was man more aquatic in the past? *New Sci.* 7(March 17):642–645.

Harris, M. 1986. *The Sacred Cow and the Abominable Pig*. New York: Simon and Schuster.

Harrison, G.A., J.M. Tanner, D.R. Pilbeam, and P.T. Baker. 1996. *Human Biology: An Introduction to Human Evolution, Variation, Growth, and Adaptability*, 3rd ed. Oxford: Oxford University Press.

Hartmann, P. 1988. The function of the human mammary gland. *Proc. Austral. Soc. Hum. Biol.* 1:251–258.

Hartwig-Scherer, S., and R.D. Martin. 1991. Was "Lucy" more human than her "child?" Observations on early hominid postcranial skeletons. *J. Hum. Evol.* 21:439–449.

Harvey, P.H., and T.H. Clutton-Brock. 1985. Life history variation in primates. *Evolution* 39(3):559–581.

Harvey, P.H., and J.R. Krebs. 1990. Comparing brains. *Science* 249:140–146.

Harvey, P.H., and R.M. May. 1989. Out for the sperm count. *Nature* 337:508–509.

Harvey, P.H., A.F. Read, and D.E.L. Promislow. 1989. Life history variation in placental mammals: Unifying the data with theory. In P.H. Harvey and L. Partridge, eds., *Oxford Surveys in Evolutionary Biology* 6:13–31. Oxford: Oxford University Press.

Hatley, T., and J. Kappleman. 1980. Bears, pigs, and Plio-Pleistocene hominids: A case for the exploitation of belowground food resources. *Hum. Ecol.* 8:371–387.

Hawkes, K. 2003. Grandmothers and the evolution of human longevity. *Am. J. Hum. Biol.* 15:380–400.

Hawkes, K., J.F. O'Connell, and N.G. Blurton-Jones. 1997. Hazda women's time allocation, offspring provisioning, and the evolution of long postmenopausal life spans. *Curr. Anthropol.* 38:551–577.

Hawkes, K., J.F. O'Connell, N.G. Blurton-Jones, H. Alvarez, and E.L. Charnov. 1998. Grandmothering, menopause, and the evolution of human life histories. *Proc. Natl. Acad. Sci. U.S.A.* 95:1336–1339.

Hayden, B. 1981. Subsistence and ecological adaptations of modern hunter/gatherers. In R.S.O. Harding and G. Teleki, eds., *Omnivorous Primates: Gathering and Hunting in Human Evolution*. New York: Columbia University Press. Pp. 344–421.

Hedricks, C.A., W. Schramm, and J.R. Udry. 1994a. Effects of creatinine correction to urinary LH levels on the timing of the LH peak and the distribution of coitus within the human menstrual cycle. In K.L. Campbell and J.W. Wood, eds. *Human Reproductive Ecology*. New York Academy of Sciences. Pp. 204–206.

Hedricks, C.A., M. Ghiglieri, R.B. Church, J. LeFevre, and M.K. McClintock. 1994b. Hormonal and ecological contributions toward interpersonal intimacy in couples. In K.L. Campbell and J.W. Wood, eds., *Human Reproductive Ecology*. New York: New York Academy of Sciences. Pp. 207–209.

Heesy, C.P., and C.F. Ross. 2001. Evolution of activity patterns and chromatic vision in primates: Morphometrics, genetics and cladistics. *J. Hum. Evol.* 40:111–149.

Heffner, R.S., and R.B. Masterson. 1983. The role of the corticospinal tract in the evolution of human digital dexterity. *Brain Behav. Evol.* 23:165–183.

Heglund, N.C. 1985. Comparative energetics and mechanics of locomotion: How do primates fit in? In W. Jungers, ed., *Size and Scaling in Primate Biology*. New York: Plenum Pub. Pp. 319–335.

Heiple, K.G., and C.O. Lovejoy. 1971. The distal femoral anatomy of *Australopithecus*. *Am. J. Phys. Anthropol.* 35:75–81.

Heistermann, M., T. Ziegler, C.P. van Schaik, K. Launhardt, P. Winkler, and J.K. Hodges. 2001. Loss of oestrus, concealed ovulation and paternity confusion in free-ranging Hanuman langurs. *Proc. R. Soc. Lond. B Biol. Sci.* 268:2445–2451.

Helle, S., P. Käär, and J. Jokela. 2002a. Human longevity and early reproduction in pre-industrial Sami populations. *J. Evol. Biol.* 15:803–807.

Helle, S., V. Lummaa, and J. Jokela. 2002b. Sons reduced maternal longevity in preindustrial humans. *Science* 296:1085.

Helmuth, H. 1985. Biomechanics, evolution and upright stature. *Anthropol. Anz.* 43(1):1–9.

Helmuth, H. 1999. The maximum lifespan potential of Hominidae—A re-evaluation. *Homo* 50(3):283–296.

Henshilwood, C.S., J.C. Sealy, R. Yates, K. Cruz-Uribe, P. Goldberg, F.E. Grine, R.G. Klein, C. Poggepoel, K. van Niekerk, and I. Watts. 2001. Blombos Cave, Southern Cape, South Africa: Preliminary report on the 1992–1999 excavations of the Middle Stone Age Levels. *J. Archaeol. Sci.* 28:421–448.

Hepp-Reymond, M.-C. 1988. Functional organization of motor cortex and its participation in voluntary movements. In H.D. Steklis and J. Erwin, eds., *Comparative Primate Biology*, Volume 4: Neurosciences. New York: Alan R. Liss. Pp. 501–624.

Hewes, G.W. 1961. Food transport and the origin of hominid bipedalism. *Am. Anthropol.* 63:687–710.

Hewes, G.W. 1964. Hominid bipedalism: Independent evidence for the food-carrying theory. *Science* 146:416–418.

Hill, K. 1993. Life history theory and evolutionary anthropology. *Evol. Anthropol.* 2(3):78–88.

Hill, K., and A.M. Hurtado. 1991. The evolution of premature reproductive senescence and menopause in human females. *Hum. Nat.* 2(4):313–350.

Hill, K., and A.M. Hurtado. 1999. Packer and colleagues' model of menopause for humans. *Hum. Nat.* 10:199–204.

Hill, W.C.O. 1948. The caecum of primates: Its appendages, mesenteries, and blood supply. *Trans. Zool. Soc. Lond.* 26:199–257.

Hillson, S. 1986. *Teeth.* Cambridge: Cambridge University Press.

Hillson, S. 1996. *Dental Anthropology.* Cambridge: Cambridge University Press.

Hinton, G.E., D.C. Plaut, and T. Shallice. 1993. Simulating brain damage. *Sci. Am.* 269(Oct):76–82.

Hinton, R.J. 1981. Form and patterning of anterior tooth wear among aboriginal human groups. *Am. J. Phys. Anthropol.* 54:555–564.

Hirasaki, E., H. Kumakura, and S. Matano. 2000. Biomechanical analysis of vertical climbing in the spider monkey and the Japanese macaque. *Am. J. Phys. Anthropol.* 113:455–472.

Hively, W. 1999. Bruckner's anatomy. *Discover* 20(Nov):110–116.

Hladik, C.-M., and B. Simmen. 1996. Taste perception and feeding behavior in nonhuman primates and human populations. *Evol. Anthropol.* 5(2):58–71.

Hladik, C.-M., P. Pasquet, and B. Simmen. 2002. New perspectives on taste and primate evolution: The dichotomy of gustatory coding for perception of beneficent versus noxious substances as supported by correlations among human thresholds. *Am. J. Phys. Anthropol.* 117:342–348.

Hoberg, E.P., N.L. Alkire, A. de Queiroz, and A. Jones. 2000. Out of Africa: Origins of the *Taenia* tapeworms in humans. *Proc. R. Soc. Lond. B Biol. Sci.* 268:781–787.

Hoffman, J.M. 1975. Retinal pigmentation, visual acuity, and brightness levels. *Am. J. Phys. Anthropol.* 43:417–424.

Hofman, M.A. 1983a. Energy metabolism, brain size and longevity in mammals. *Q. Rev. Biol* 58(4):495–512.

Hofman, M.A. 1983b. Encephalization in hominids: Evidence for the model of punctuationalism. *Brain Behav. Evol.* 22:102–117.

Hofman, M.A. 1984. On the presumed coevolution of brain size and longevity in hominids. *J. Hum. Evol.* 13:371–376.

Hohmann, G., and B. Fruth. 1993. Field observations on meat sharing among bonobos (*Pan paniscus*). *Folia Primatol.* 60:225–229.

Holliday, R. 1996. The evolution of human longevity. *Persp. Biol. Med.* 40(1):100–107.

Holliday, T.W. 1997. Postcranial evidence of cold adaptation in European Neanderthals. *Am. J. Phys. Anthropol.* 104:245–258.

Holliday, T.W. 1999. Brachial and crural indices of European Late Upper Paleolithic and Mesolithic humans. *J. Hum. Evol.* 36:549–566.

Holliday, T.W. 2002. Body size and postcranial robusticity of European Upper Paleolithic hominins. *J. Hum. Evol.* 43:513–528.

Holliday, T.W., and A.B. Falsetti. 1995. Lower limb length in European early and modern humans in relation to mobility and climate. *J. Hum. Evol.* 29:141–153.

Holloway, R.L. 1967. Tools and teeth: Some speculations regarding canine reduction. *Am. Anthropol.* 69:63–67.

Holloway, R.L. 1968. More on tools and teeth. *Am. Anthropol.* 70:101–105.

Holloway, R.L., D.C. Broadfield, and M.S. Yuan. 2001. Revisiting australopithecine visual striate cortex: Newer data from chimpanzee and human brains suggest it could have been reduced during australopithecine times. In D. Falk and K.R. Gibson, eds. *Evolutionary Anatomy of the Primate Cerebral Cortex.* Cambridge: Cambridge University Press. Pp. 177–186.

Holy, T.E., C. Dulac, and M. Meister. 2000. Responses of vomeronasal neurons to natural stimuli. *Science* 289:1569–1572.

Hopkins, W.D., and R.D. Morris. 1993. Handedness in great apes: A review of findings. *Int. J. Primatol.* 14(1):1–25.

Hovers, E., S. Ilani, O. Bar-Yosef, and B. Vandermeersch. 2003. An early case of color symbolism. *Curr. Anthropol.* 44:491–522.

Hrdy, S.B. 1981. *The Woman that Never Evolved.* Cambridge, MA: Harvard University Press.

Hrdy, S.B. 1997. Raising Darwin's consciousness: Female sexuality and the prehominid origins of patriarchy. *Hum. Nat.* 8(1):1–49.

Hrdy, S.B. 1999. *Mother Nature: Maternal Instincts and How They Shape the Human Species.* New York: Ballantine Books.

Hrdy, S.B., and W. Bennett. 1981. Lucy's husband: What did he stand for? *Harvard Mag.* 83(July–Aug):7–9, 46.

Hrdy, S.B., and Patricia L.Whitten. 1986. Patterning of sexual behavior. In B.B. Smuts, D.L. Cheney, R.M. Seyfarth, R.W. Wrangham, and T.T. Struhsaker, eds., *Primate Societies.* Chicago: University of Chicago Press. Pp. 370–384.

Hubel, D.H., and T.N. Wiesel. 1979. Brain mechanisms of vision. *Sci. Amer.* 241(Sept):150–162.

Huber, R., M.F. Ghilardi, M. Massimini, and G. Tononi. 2004. Local sleep and learning. *Nature* 430:78–81.

Humle, T., and T. Matsuzawa. 2001. Behavioural diversity among the wild chimpanzee populations of Bossou and neighboring areas, Guinea and Côte d'Ivoire, West Africa. *Folia Primatol.* 72:57–68.

Humphrey, N.K. 1976. The social function of intellect. In P.P.G. Bateson and R.A. Hind, eds., *Growing Points in Ethology.* Cambridge: Cambridge University Press.

Hunt, K.D. 1992. Positional behavior of *Pan troglodytes* in the Mahale Mountains and Gombe Stream National Parks, Tanzania. *Am. J. Phys. Anthropol.* 87:83–105.

Hunt, K.D. 1994. The evolution of human bipedality: Ecology and functional morphology. *J. Hum. Evol.* 26:183–202.

Hunter, J.P., and J. Jernvall. 1995. The hypocone as a key innovation in mammalian evolution. *Proc. Natl. Acad. Sci. U.S.A.* 92:10718–10722.

Hurst, L.D., and J.R. Peck. 1996. Recent advances in understanding of the evolution and maintenance of sex. *Trends Ecol. Evol.* 11(2):46–52.

Huss-Ashmore, R. 1980. Fat and fertility: Demographic implications of differential fat storage. *Yrbk. Phys. Anthropol.* 23:65–91.

Hylander, W.L. 1977. The adaptive significance of the Eskimo craniofacial morphology. In A.A. Dahlberg and T.M. Graber, eds., *Orofacial Growth and Development.* The Hague: Mouton. Pp. 129–169.

Hylander, W.L. 1979. The functional significance of primate mandibular form. *J. Morphol.* 160:223–240.

Hylander, W.L., P.G. Picq, and K.R. Johnson. 1991. Masticatory stress hypothesis and the supraorbital region of primates. *Am. J. Phys. Anthropol.* 86:1–36.

Hylander, W.L., M.J. Ravosa, C.F. Ross, and K.R. Johnson. 1998. Mandibular corpus strain in primates: Further evidence for a functional link between symphyseal fusion and jaw-adductor muscle force. *Am. J. Phys. Anthropol.* 107:257–271.

Hylander, W.L., M.J. Ravosa, Callum F. Ross, C.E. Wall, and K.R. Johnson. 2000. Symphyseal fusion and jaw-adductor muscle force: An EMG study. *Am. J. Phys. Anthropol.* 112:469–492.

Isaac, G.L. 1982. Models of human evolution. *Science* 217:295–296.

Ishida, H., T. Kimura, and M. Okada. 1975. Patterns of bipedal walking in anthropoid primates. *Symp. 5th Congr. Int. Primatol. Soc.* Pp. 287–301.

Jablonski, N.G., and G. Chaplin. 1993. Origin of habitual terrestrial bipedalism in the ancestor of the Hominidae. *J. Hum. Evol.* 24:259–280.

Jablonski, N.G., and G. Chaplin. 2000. The evolution of human skin coloration. *J. Hum. Evol.* 39:57–106.

Jablonski, N.G., and G. Chaplin. 2002. Skin deep. *Sci. Am.* 287(Oct):74–81.

Jablonski, N.G., M.J. Whitfort, N. Roberts-Smith, and X. Qinqi. 2000. The influence of life history and diet on the distribution of catarrhine primates during the Pleistocene in eastern Asia. *J. Hum. Evol.* 39:131–157.

Jackendorff, R. 1994. *Patterns in the Mind: Language and Human Nature.* New York: Basic Books.

Jacob, S., B. Zelano, A. Gungor, D. Abbott, R. Naclerio, and M.K. McClintock. 2000. Location and gross morphology of the nasopalatine duct in human adults. *Arch. Otolaryngol. Head Neck Surg.* 126:741–748.

Jacobs, G.H. 1993. The distribution and nature of colour vision among the mammals. *Biol. Rev.* 68:413–471.

Jacobs, K.H. 1985. Evolution in the postcranial skeleton of late glacial and early postglacial European hominids. *Z. Morphol. Anthropol.* 75(3):307–326.

Jaeger, J.-J., U A.N. Soe., U A.K. Aung, M. Benammi, Y. Chaimanee, R.-M. Ducroq, T. Tun, U T. Thein, and S. Ducrocq. 1998. New Myanmar middle Eocene anthropoids. An Asian origin for catarrhines? *C. R. Acad. Sci. III* 321:953–959.

Janis, C.M., and M. Fortelius. 1988. On the means whereby mammals achieve increased functional durability of their dentitions with special reference to limiting factors. *Biol. Rev.* 63:197–230.

Jeffrey, N., and F. Spoor. 2002. Brain size and the human cranial base: A prenatal perspective. *Am. J. Phys. Anthropol.* 118:324–340.

Jéquier, E., P.-H. Gygax, P. Pittet, and A. Vannotti. 1974. Increased thermal body insulation: Relationship to the development of obesity. *J. Appl. Physiol.* 36(6):674–678.

Jerison, H.J. 1973. *Evolution of the Brain and Intelligence.* New York: Academic Press.

Jiang, H., M. Moreau, V.J. Raso, G. Russell, and K. Bagnall. 1995. A comparison of spinal ligaments—Differences between bipeds and quadrupeds. *J. Anat.* 187:85–91.

Johanson, D., and B. Edgar. 1996. *From Lucy to Language.* New York: Simon and Schuster.

Jolly, A. 1966. Lemur social behavior and primate intelligence. *Science* 153:501–506.

Jolly, C.J. 1970. The seed-eaters: A new model of hominid differentiation based on a baboon analogy. *Man* 5:5–26.

Jones, D. 1995. Sexual selection, physical attractiveness, and facial neoteny. *Curr. Anthropol* 36(1):723–748.

Jones, D., and K. Hill. 1993. Criteria of facial attractiveness in five populations. *Hum. Nat.* 4(3):271–296.

Jones-Engel, L.E., and K.A. Bard. 1996. Precision grips in young chimpanzees. *Am. J. Primatol.* 39:1–15.

Jungers, W.L. 1988. New estimates of body size in australopithecines. In F.E. Grine, ed., *Evolutionary History of the "Robust" Australopithecines.* New York: Aldine de Gruyter. Pp. 115–125.

Kaifu, Y., K. Kasai, G.C. Townsend, and L.C. Richards. 2003. Tooth wear and the "design" of the human dentition: A perspective from evolutionary medicine. *Yrbk. Phys. Anthropol.* 46:47–61.

Kalil, R.E. 1989. Synapse formation in the developing brain. *Sci. Am.* 261(Dec):76–85.

Kaplan, H.S., and A.J. Robson. 2002. The emergence of humans: The coevolution of intelligence and longevity with intergenerational transfers. *Proc. Natl. Acad. Sci. U.S.A.* 99:10221–10226.

Kaplan, H., K. Hill, J. Lancaster, and A.M. Hurtado. 2000. A theory of human life history evolution: Diet, intelligence, and longevity. *Evol. Anthropol.* 9(4):156–185.

Kaskan, P.M., and B.L. Finlay. 2001. Encephalization and its developmental structure: How many ways can a brain get big? In D. Falk and K.R. Gibson, eds. *Evolutionary Anatomy of the Primate Cerebral Cortex.* Cambridge: Cambridge University Press. Pp. 14–29.

Katz, L.C., and C.J. Shatz. 1996. Synaptic activity and the construction of cortical circuits. *Science* 274:1133–1138.

Katzenberg, M.A. 1992. Advances in stable isotope analysis of prehistoric bones. In S.R. Saunders and M.A. Katzenberg, eds., *Skeletal Biology of Past Peoples*: Research Methods. New York: Wiley-Liss. Pp. 105–119.

Kavanau, J.L. 2002. REM and NREM sleep as natural accompaniments of the evolution of warm-bloodedness. *Neurosci. Biobehav. Rev.* 26:889–906.

Kay, R.F. 1981. The nut-crackers—A new theory of the adaptations of the Ramapithecinae. *Am. J. Phys. Anthropol.* 55:141–151.

Kay, R.F., C. Ross, and B.A. Williams. 1997. Anthropoid origins. *Science* 275:797–804.

Kay, R.F., M. Cartmill, and M. Balow. 1998. The hypoglossal canal and the origin of human vocal behavior. *Proc. Natl. Acad. Sci. U.S.A.* 95:5417–5419.

Kebler, A., M. Vorobyev, and D. Osorio. 2003. Animal colour vision—Behavioural tests and physiological concepts. *Biol. Rev.* 78:81–118.

Keith, A. 1923. Man's posture: Its evolution and disorders. *Br. Med. J.* (1923)1:451–454, 499–502, 545–548, 587–590, 624–626, 669–672.

Kennedy, G.E. 2003. Palaeolithic grandmothers? Life history theory and early *Homo*. *J. R. Anthropol. Inst.* 9:549–572.

Ker, R.F., M.B. Bennett, S.R. Bibby, R.C. Kester, and R.M. Alexander. 1987. The spring in the arch of the human foot. *Nature* 325:147–149.

Kermack, K. 1989. Hearing in early mammals. *Nature* 341:568–569.

Keverne, E.B. 1999. The vomeronasal organ. *Science* 286:716–720.

Key, C.A. 2000. The evolution of human life history. *World Archaeol.* 31(3):329–350.

King, B.F. 1993. Development and structure of the placenta and fetal membranes of nonhuman primates. *J. Exp. Zool.* 266:528–540.

Kingdon, J. 2003. *Lowly Origin: Where, When, and Why Our Ancestors First Stood Up*. Princeton, NJ: Princeton University Press.

Kirchengast, S., and J. Huber. 2001. Fat distribution patterns in young amenorrheic females. *Hum. Nat.* 12(2):123–140.

Kirkpatrick, M., and C.D. Jenkins. 1999. Genetic segregation and the maintenance of sexual reproduction. *Nature* 339:300–301.

Kirkwood, T.B.L., and S.N. Austad. 2000. Why do we age? *Nature* 408:233–238.

Kirkwood, T.B.L., P. Kapahi, and D.P. Shanley. 2000. Evolution, stress, and longevity. *J. Anat.* 197:587–590.

Klein, R.G. 1983. The stone age prehistory of southern Africa. *Annu. Rev. Anthropol.* 12:25–48.

Klein, R.G. 1999. *The Human Career*, 2nd ed. Chicago: University of Chicago Press.

Knoll, A.H., and S.B. Carroll. 1999. Early animal evolution: Emerging views from comparative biology and geology. *Science* 284:2129–2137.

Kohl, J.V., and R.T. Francoeur. 1995. *The Scent of Eros*. New York: Continuum.

Köhler, M., and S. Moyà-Solà. 1997. Ape-like or hominid-like? The positional behavior of *Oreopithecus bambolii* reconsidered. *Proc. Natl. Acad. Sci. U.S.A.* 94:11747–11750.

Kolb, H. 2003. How the retina works. *Am. Sci.* 91(Jan–Feb):28–35.

Kondrashov, A.S. 1988. Deleterious mutations and the evolution of sexual reproduction. *Nature* 336:435–440.

Konner, M. 2002. *The Tangled Wing: Biological Constraints on the Human Spirit*, 2nd ed. New York: W.H. Freeman.

Kortlandt, A. 1979. How might the early hominids have defended themselves against large predators and food competitors? *J. Hum. Evol.* 9:79–112.

Kouros-Mehr, H., S. Pintchovski, J. Melnyk, Y.-J. Chen, C. Friedman, B. Trask, and H. Shizuya. 2001. Identification of non-functional human VNO receptor genes provides evidence for vestigiality of the human VNO. *Chem. Senses* 26:1167–1174.

Kram, R., and C.R. Taylor. 1990. Energetics of running: A new perspective. *Nature* 346:265–267.

Kramer, P.A. 1999. Modelling the locomotor energetics of extinct hominids. *J. Exp. Biol.* 202:2807–2818.

Kramer, P.A., and G.G. Eck. 2000. Locomotor energetics and leg length in hominid bipedality. *J. Hum. Evol.* 38:651–666.

Krantz, G.S. 1963. The functional significance of the mastoid process in man. *Am. J. Phys. Anthropol.* 21:591–593.

Krantz, G.S. 1980a. Sapienization and speech. *Curr. Anthropol.* 21(6):773–792.

Krantz, G.S. 1980b. *Climatic Races and Descent Groups*. North Quincy, MA: Christopher Pub. House.

Krubitzer, L. 1995. The organization of neocortex in mammals: Are species differences really so different? *Trends Neurosci.* 18(9):408–417.

Kuhl, P.K., K.A. Williams, F. Lacerda, K.N. Stevens, and B. Lindblom. 1992. Linguistic experience alters phonetic perception in infants by six months of age. *Science* 255:606–608.

Kuhl, P.K., F.-M. Tsao, and H.-M. Liu. 2003. Foreign-language experience in infancy: Effects of short-term exposure and social interaction on phonetic learning. *Proc. Natl. Acad. Sci. U.S.A.* 100:9096–9101.

Kummer, B.K.F. 1975. Functional adaptation to posture in the pelvis of man and other primates. In R.H. Tuttle, ed., *Primate Functional Morphology and Evolution*. The Hague: Mouton. Pp. 281–290.

Kushlan, J.A. 1985. The vestiary hypothesis of human hair reduction. *J. Hum. Evol.* 14(1):29–32.

Kuzawa, C.W. 1998. Adipose tissue in human infancy and childhood: An evolutionary perspective. *Yrbk Phys. Anthropol.* 41:177–209.

Lagerkrantz, H. 1996. Stress, arousal, and gene activation at birth. *News Physiol. Sci.* 11:214–219.

Lagerkrantz, H., and T.A. Slotkin. 1986. The stress of being born. *Sci. Am.* 254(May):100–107.

Laitman, J.T. 1983. The evolution of the hominid upper respiratory system and implications for the origins of speech. In E. de Grolier, ed. *Glossogenetics: The Origin and Evolution of Language*. Chur, Switzerland: Harwood Academic Pub.

Laitman, J.T., E.S. Crelin, and G.J. Conlogue. 1977. The function of the epiglottis in monkey and man. *Yale J. Biol. Med.* 50:43–48.

Lambert, J.E. 1998. Primate digestion: Interactions among anatomy, physiology, and feeding ecology. *Evol. Anthropol.* 7(1):8–20.

Lambert, K.L. 1971. The weight-bearing function of the fibula. *J. Bone Joint Surg. Am.* 53A(3):507–513.

Lammi-Keefe, C.J., and R.G. Jensen. 1984. Lipids in human milk: A review. 2. Composition and fat-soluble vitamins. *J. Pediatr. Gastroenterol. Nutr.* 3:172–198.

Lancaster, J.B. 1978. Carrying and sharing in human evolution. *Hum. Nat.* 1(2):82–89.

Langdon, J.H. 1985. Fossils and the origin of bipedalism. *J. Hum. Evol.* 14:615–635.

Langdon, J.H. 1990. Variations in cruro-pedal musculature. *Int. J. Primatol.* 11(6):575–606.

Langdon, J.H. 1992. Lessons from Piltdown. *Creation/Evolution* 31:11–27.

Langdon, J.H. 1997. Umbrella hypotheses and parsimony in human evolution: A critique of the aquatic ape hypothesis. *J. Hum. Evol.* 33:479–494.

Langdon, J.H. 2002. Darwin meets DHA: Natural selection, diet, and brain evolution. *Am. J. Phys. Anthropol.* Suppl. 34:99.

Langlois, J.H., and L.A. Roggman. 1990. Attractive faces are only average. *Psychol. Sci.* 1(2):115–121.

Larson, E.J. 2001. *Evolution's Workshop: God and Science on the Galapagos Islands*. New York: Basic Books.

Larson, S.G. 1993. Functional morphology of the shoulder in primates. In D.L. Gebo, ed., *Postcranial Adaptation in Non-human Primates*. DeKalb, IL: Northern Illinois University Press. Pp. 45–69.

Larson, S.G. 1998. Parallel evolution in the hominoid trunk and forelimb. *Evol. Anthropol.* 6(3):87–99.

Latimer, B., and C.O. Lovejoy. 1989. The calcaneus of *Australopithecus afarensis* and its implications for the evolution of bipedality. *Am. J. Phys. Anthropol.* 78:369–386.

Latimer, B., and C.O. Lovejoy. 1990a. Hallucal tarsometatarsal joint in *Australopithecus afarensis*. *Am. J. Phys. Anthropol.* 82:135–133.

Latimer, B., and C.O. Lovejoy. 1990b. Metatarsophalangeal joints in *Australopithecus afarensis. Am. J. Phys. Anthropol.* 83:13–23.

Latimer, B., J.C. Owen, and C.O. Lovejoy. 1987. Talocrural joint in African hominoids: Implications for *Australopithecus afarensis. Am. J. Phys. Anthropol.* 74:155–175.

Laughlin, W.S. 1968. Hunting: an integrating biobehavior system and its evolutionary importance. In R.B. Lee and I. DeVore, eds., *Man the Hunter.* Chicago: Aldine. Pp. 304–320.

Laurent, G. 1999. A systems perspective on early olfactory coding. *Science* 286:723–728.

Laurin, M., M. Girondot, and A. de Ricqles. 2000. Early tetrapod evolution. *Trends Ecol. Evol.* 15(3):118–123.

Lavelle, C.L.B., R.P. Shellis, and D.F.G. Poole. 1977. *Evolutionary Changes to the Primate Skull and Dentition.* Springfield, IL: Charles C Thomas.

Lazar, P. 1986. La naissance prématurée, un lien entre la station debout et le volume crânien? *L'Anthropologie* (Paris) 90(3):439–445.

Leach, H. 2003. Human domestication reconsidered. *Curr. Anthropol.* 44(3):349–368.

Leaf, A., and P.C. Weber. 1987. A new era for science in nutrition. *Am. J. Clin. Nutr.* 45:1048–1053.

Lebedev, O.A. 1997. Fins made for walking. *Nature* 390:21–22.

LeDoux, J.E. 1994. Emotion, memory, and the brain. *Sci. Am.* 270(June):50–57.

Lee, H., and R.B. Banzett. 1997. Mechanical links between locomotion and breathing: Can you breathe with your legs? *News Physiol. Sci.* 12:273–278.

Lee, P. 1989. Comparative ethological approaches in modelling hominid behavior. *Ossa* 14:113–126.

Lee, S.-H., and M. Wolpoff. 2003. The pattern of evolution in Pleistocene human brain size. *Paleobiology* 29(2):186–196.

Lee-Thorp, J.E., J.F. Thackeray, and N.J. van der Merwe. 2000. The hunter and the hunters revisited. *J. Hum. Evol.* 39:565–576.

Lee-Thorp, J.E., N.J. van der Merwe, and C.K. Brain. 1994. Diet of *Australopithecus robustus* at Swartkrans from stable carbon isotope analysis. *J. Hum. Evol.* 27:361–372.

Le Gros Clark, W.E. 1970. *History of the Primates.* London: British Museum (Natural History).

Leidy, L.E. 1994. Biological aspects of menopause across the lifespan. *Annu. Rev. Anthropol.* 23:231–253.

Leinders-Zufall, T., A.P. Lane, A.C. Puche, W. Ma, M.V. Novotny, M.T. Shipley, and F. Zufall. 2000. Untrasensitive pheromone detection by mammalian vomeronasal neurons. *Nature* 405:792–796.

Leiner, H.C., A.L. Leiner, and R.S. Dow. 1993. Cognitive and language functions of the human cerebellum. *Trends Neurosci.* 16(11):444–454.

Leonard, W.R. 2000. Human nutritional evolution. In S. Stinson, B. Bogin, R. Huss-Ashmore, and D. O'Rourke, eds., *Human Biology: An Evolutionary and Biocultural Perspective.* New York: Wiley-Liss. Pp. 295–343.

Leonard, W.R. 2002. Food for thought. *Sci. Am.* 287(Dec):106–115.

Leonard, W.R., and M.L. Robertson. 1995. Energetic efficiency of human bipedality. *Am. J. Phys. Anthropol.* 97:335–338.

Leonard, W.R., and M.L. Robertson. 1997a. Comparative primate energetics and hominid evolution. *Am. J. Phys. Anthropol.* 102:265–281.

Leonard, W.R., and M.L. Robertson. 1997b. Rethinking the energetics of bipedalism. *Curr. Anthropol.* 38(2):304–309.

Leutenegger, W. 1972. Newborn size and pelvic dimensions of *Australopithecus. Nature* 240:568–569.

Leutenegger, W. 1974. Functional aspects of pelvic morphology in simian primates. *J. Hum. Evol.* 3:207–222.

Leutnegger, W. 1977. A functional interpretation of the sacrum of *Australopithecus africanus. S. Afr. J. Sci.* 73:308–310.

Leutenegger, W. 1982. Encephalization and obstetrics in primates with particular reference to human evolution. In E. Armstrong and D. Falk, eds., *Primate Brain Evolution.* New York: Plenum Pub. Pp. 85–95.

Leutenegger, W. 1987. Neonatal brain size and neurocranial dimensions in Pliocene hominids: implications for obstetrics. *J. Hum. Evol.* 16:291–296.

Liberman, A.M., and I.G. Mattingly. 1989. A specialization for speech perception. *Science* 243:489–493.

Lieberman, D.E. 2000. Ontogeny, homology, and phylogeny in the hominid craniofacial skeleton: The problem of the browridge. In P. O'Higgins and M.J. Cohn, eds., *Development, Growth, and Evolution: Implications for the Study of the Hominid Skeleton*. San Diego: Academic Press. Pp. 85–122.

Lieberman, D.E., and A.W. Crompton. 2000. Why fuse the mandibular symphysis? A comparative analysis. *Am. J. Phys. Anthropol.* 112:517–540.

Lieberman, D.E., and R.C. McCarthy. 1999. The ontogeny of cranial base angulation in humans and chimpanzees and its implications for reconstructing pharyngeal dimensions. *J. Hum. Evol.* 36:487–517.

Lieberman, D.E., C.F. Ross, and M.J. Ravosa. 2000. The primate cranial base: Ontogeny, function, and integration. *Yrbk. Phys. Anthropol.* 43:117–169.

Lieberman, D.E., B.M. McBratney, and G. Krovitz. 2002. The evolution and development of cranial form in *Homo sapiens*. *Proc. Natl. Acad. Sci. U.S.A.* 99(3):1134–1139.

Lieberman, P. 1993. On the Kebara KMH2 hyoid and Neanderthal speech. *Curr. Anthropol.* 34(2):172–175.

Lieberman, P. 1994. Hyoid bone position and speech: Reply to Dr. Arensburg et al. (1990). *Am. J. Phys. Anthropol.* 94:275–278.

Lieberman, P. 1998. *Eve Spoke: Human Language and Human Evolution*. New York: W.W. Norton & Co.

Lieberman, P. 2002. On the nature and evolution of the neural bases of language. *Yrbk. Phys. Anthropol.* 45:36–62.

Lieberman, P., and E.S. Crelin. 1971. On the speech of Neanderthal man. *Linguistic Inquiry* 2(2):203–222.

Lieberman, P., E.S. Crelin, and D.H. Klatt. 1972. Phonetic ability and related anatomy of the newborn and adult human, Neanderthal Man, and the chimpanzee. *Am. Anthropol.* 74:287–307.

Lieberman, P., J.T. Laitman, J.S. Reidenberg, and P.J. Gannon. 1992. The anatomy, physiology, acoustics and perception of speech: Essential elements in analysis of the evolution of human speech. *J. Hum. Evol.* 23:447–467.

Liman, E.R., and H. Innan. 2003. Relaxed selective pressure on an essential component of pheromone transduction in primate evolution. *Proc. Natl. Acad. Sci. U.S.A.* 100(6):3328–3332.

Lindburg, D.G. 1982. Primate obstetrics: The biology of birth. *Am. J. Primatol.* Suppl. 1:193–199.

Lindemann, B. 2000. A taste for umami. *Nat. Neurosci.* 3(2):99–100.

Lindemann, B. 2001. Receptors and transduction in taste. *Nature* 413:219–225.

Linden, E. 1974. *Apes, Men, and Language*. New York: Saturday Review Press.

Lindley, J.M., and G.A. Clark. 1990. Symbolism and modern human origins. *Curr. Anthropol.* 31(3):233–261.

Lockwood, C.A., W.H. Kimbel, and D.C. Johanson. 2000. Temporal trends and metric variation in the mandibles and dentition of *Australopithecus afarensis*. *J. Hum. Evol.* 39:23–55.

Loomis, W.F. 1967. Skin-pigment regulation of vitamin-D biosynthesis in man. *Science* 157:501–506.

Lovejoy, C.O. 1981. The origin of man. *Science* 211:341–350.

Lovejoy, C.O. 1988. Evolution of human walking. *Sci. Am.* 259(Nov):118–125.

Lovejoy, C.O., K.G. Heiple, and A.H. Burstein. 1973. The gait of *Australopithecus*. *Am. J. Phys. Anthropol.* 38:757–780.

Lovejoy, C.O., M.J. Cohn, and T.D. White. 1999. Morphological analysis of the mammalian postcranium: A developmental perspective. *Proc. Natl. Acad. Sci. U.S.A.* 96(23):13247–13252.

Lovejoy, C.O., M.J. Cohn, and T.D. White. 2000. The evolution of mammalian morphology: A developmental perspective. In P. O'Higgins and M.J. Cohn, eds., *Development, Growth, and Evolution: Implications for the Study of the Hominid Skeleton*. San Diego, CA: Academic Press. Pp. 41–55.

Lovejoy, C.O., R.S. Meindl, J.C. Ohman, K.G. Heiple, and T.D. White. 2002. The Maka femur and its bearing on the antiquity of human walking: Applying contemporary concepts of morphogenesis to the human fossil record. *Am. J. Phys. Anthropol.* 119:97–133.

Low, B.S., R.D. Alexander, and K.M. Noonan. 1987. Human hips, breasts and buttocks: Is fat deceptive? *Ethol. Sociobiol.* 8:249–257.

Lucas, P.W., B.W. Darvell, P.K.D. Lee, T.D.B., Yuen, and M.F. Choong. 1998. Colour cues for leaf food selection by long-tailed macaques (*Macaca fascicularis*) with a new suggestion for the evolution of trichromatic colour vision. *Folia Primatol.* 69:139–152.

Lucas, P.W., N.J. Dominy, P. Riba-Hernandez, K.E. Stoner, N. Yamasita, E. Loría-Calderón, W. Petersen-Pereira, Y. Rojas-Durán, R. Salas-Pena, S. Solis-Madrigal, D. Osorio, and B.W. Darvell. 2003. Evolution and function of routine trichromatic vision in primates. Evolution 57(11):2636–2643.

Luckett, W.P. 1975. Ontogeny of the fetal membranes and placenta: Their bearing on primate phylogeny. In W.P. Luckett and F.S. Szalay, eds., *Phylogeny of the Primates: A Multidisciplinary Approach.* New York: Plenum Pub. Pp. 157–182.

Lummaa, V., and T. Clutton-Brock. 2002. Early development, survival and reproduction in humans. *Trends Ecol. Evol.* 17(3):141–147.

Luo, M., M.S. Fee, and L.C. Katz. 2003. Encoding pheromone signals in the accessory olfactory bulb of behaving mice. *Science* 299:1196–1201.

Macciarelli, R., and L. Bondioli. 1986. Post-Pleistocene reduction in human dental structure: A reappraisal in terms of increasing population density. *Hum. Evol.* 1(5):405–418.

Macciarelli, R., V. Galichon, L. Bondioli, and P.V. Tobias. 1996. Hip bone trabecular architecture and locomotor behavior in South African australopithecines. *International Union of Prehistoric and Protohistoric Sciences. Proc. XIII Congress.* Vol. 2:35–41.

Macciarelli, L. Bondioli, R., V. Galichon, and P.V. Tobias. 1999. Hip bone trabecular architecture shows uniquely distinctive locomotor behaviour in South African australopithecines. *J. Hum. Evol.* 36:211–236.

MacLarnon, A.M., and G.P. Hewitt. 1999. The evolution of human speech: The role of enhanced breathing control. *Am. J. Phys. Anthropol.* 109:341–363.

MacLarnon, A.M., R.D. Martin, D.J. Chivers, and C.M. Hladik. 1986. Some aspects of gastro-intestinal allometry in primates and other mammals. In Michel Sakka, ed., *Definition et Origines de l'Homme.* Paris: Editions du Centre National de la Recherche Scientifique. Pp. 293–302.

MacLatchy, L.M. 1996. Another look at the australopithecine hip. *J. Hum. Evol.* 31:455–476.

MacLatchy, L.M. 2004. The oldest ape. *Evol. Anthropol.* 13:90–103.

MacLeod, C.E., K. Zilles, A. Schleicher, J.K. Rilling, and K.R. Gibson. 2003. Expansion of the neocerebellum in Hominoidea. *J. Hum. Evol.* 44:401–429.

Maess, B., S. Koelsch, T.C. Gunter, and A.D. Friederici. 2001. Musical syntax is processed in Broca's area: An MEG study. *Nat. Neurosci.* 4:540–543.

Malainey, M.E., R. Przybylski, and B.L. Sherriff. 2001. One person's food: How and why fish avoidance may affect the settlement and subsistence patterns of hunter-gatherers. *Am. Antiquity* 66(1):141–161.

Malcom, G.T., A.K. Bhattacharyya, M. Velez-Duran, M.A. Guzman, M.C. Oalmann, and J.P. Strong. 1989. Fatty acid composition of adipose tissue in humans: Differences between subcutaneous sites. *Am. J. Clin. Nutr.* 50:288–291.

Mann, A. 1975. *Some Paleodemographic Aspects of the South African Australopithecines.* Philadelphia: University of Pennsylvania Press.

Mann, R., and V.T. Inman. 1964. Phasic activity of intrinsic muscles of the foot. *J. Bone Joint Surg. Am.* 46A(3):469–481.

Mannino, M.A., and K.D. Thomas. 2002. Depletion of a resource? The impact of prehistoric human foraging on intertidal mollusc communities and its significance for human settlement, mobility, and dispersal. *World Archaeol.* 33(3):452–474.

Maquet, P. 2001. The role of sleep in learning and memory. *Science* 294:1048–1052.

Marchant, L.F., and W.C. McGrew. 1991. Laterality of function in apes: A meta-analysis of methods. *J. Hum. Evol.* 21:425–438.

Margulis, L., and D. Sagan. 1986. *Origins of Sex.* New Haven, CT: Yale University Press.

Marlowe, F. 1997. The nubility hypothesis: The human breast as an honest signal of residual reproductive value. *Hum. Nat.* 9(3):263–271.

Marshack, A. 1989. Evolution of the human capacity: The symbolic evidence. *Yrbk. Phys. Anthropol.* 32:1–34.

Martin, E. 1991. The egg and the sperm: How science has constructed a romance based on stereotypical male-female roles. Signs: *J. Women Cult. Soc.* 16(3):485–501.

Martin, R.D. 1983a. Relative brain size and metabolic rate in terrestrial vertebrates. *Nature* 293:57–60.

Martin, R.D. 1983b. *Human Brain Evolution in an Ecological Context*. Fifty-Second James Arthur Lecture on the Evolution of the Human Brain. New York: American Museum of Natural History.

Martin, R.D. 1988. Evolution of the brain in early hominids. *Ossa* 14:49–62.

Martin, R.D. 1990. *Primate Origins and Evolution: A Phylogenetic Reconstruction*. Princeton, NJ: Princeton University Press.

Martin, R.D., and A.M. MacLarnon. 1990. Reproductive patterns in primates and other mammals: The dichotomy between altricial and precocial offspring. In C.J. DeRousseau, ed., *Primate Life History and Evolution*. New York: Wiley-Liss. Pp. 47–79.

Martin, R.D., and R.M. May. 1981. Outward signs of breeding. *Nature* 293:7–9.

Martin, R.D., D.J. Chivers, A.M. MacLarnon, and C.M. Hladik. 1985. Gastrointestinal allometry in primates and other mammals. In William L. Jungers, ed., *Size and Scaling in Primate Biology*. New York: Plenum Pub. Pp. 61–89.

Marzke, M.W. 1986. Tool use and the evolution of hominid hands and bipedality. In J.G. Else and P.C. Lee, eds., *Primate Evolution*. Cambridge: Cambridge University Press. Pp. 203–209.

Marzke, M.W. 1987. The third metacarpal styloid process in humans: Origin and functions. *Am. J. Phys. Anthropol.* 73:415–431.

Marzke, M.W. 1992. Evolution of the power ("squeeze") grip and its morphological correlates in hominids. *Am. J. Phys. Anthropol.* 89:283–298.

Marzke, M.W. 1997. Precision grips, hand morphology, and tools. *Am. J. Phys. Anthropol.* 102:91–110.

Marzke, M.W., and R.F. Marzke. 1987. The third metacarpal styloid process in humans: Origin and functions. *Am. J. Phys. Anthropol.* 73:415–431.

Marzke, M.W., and K.L. Wullstein. 1996. Chimpanzee and human grips: A new classification with a focus on evolutionary morphology. *Int. J. Primatol.* 17(1):117–139.

Marzke, M.W., N. Toth, K. Schick, S. Reece, B. Setinberg, K. Hunt, R.I. Linsheid, and K.-N. An. 1998. EMG study of hand muscle recruitment during hard hammer percussion manufacture of Oldowan tools. *Am. J. Phys. Anthropol.* 105:315–332.

Masland, R.H. 2001. The fundamental plan of the retina. *Nat. Neurosci.* 4(9):877–886.

Matsuoka, L.Y., J. Wortsman, J.G. Haddad, P. Kolm, and B.W. Hollis. 1991. Racial pigmentation and the cutaneous synthesis of vitamin D. *Arch. Dermatol.* 127:536–538.

May, R.M. 1978. Human reproduction reconsidered. *Nature* 272:491–495.

McCarthy, R.A., and E.K. Warrington. 1988. Evidence for modality-specific meaning systems in the brain. *Nature* 334:428–430.

McClintock, M.K. 1971. Menstrual synchrony and suppression. *Nature* 229:244–245.

McCrossin, M. 2002. Functional morphology of the metacarpophalangeal joints of *Kenyapithecus* and *Australopithecus*: Implications for the adaptive history of locomotion among African apes and humans. (Abstract.) *Am. J. Phys. Anthropol.* Suppl. 34:109–110.

McGrew, W.C. 1995. Thumbs, tools, and early humans. *Science* 268:586.

McGrew, W.C. 2001. The other faunivory: Primate insectivory and early human diet. In C.B. Stanford and H.T. Bunn, eds., *Meat-Eating and Human Evolution*. Oxford: Oxford University Press. Pp. 160–178.

McGrew, W.C., and L.F. Marchant. 1996. On which side of the apes? Ethological study of laterality of hand use. In W.C. McGrew, L.F. Marchant, and T. Nishida, eds., *Great Ape Societies*. Cambridge: Cambridge University Press. Pp. 255–272.

McHenry, H.M. 1975. The ischium and hip extensor mechanism in human evolution. *Am. J. Phys. Anthropol.* 43:39–46.

McHenry, H.M. 1984. Relative cheek-tooth size in *Australopithecus. Am. J. Phys. Anthropol.* 64(3):297–304.

McHenry, H.M. 1986. The first bipeds: A comparison of the *A. afarensis* and *A. africanus* postcranium and implications for the evolution of bipedalism. *J. Hum. Evol.* 15:177–191.

McHenry, H.M. 1988. New estimates of body weight in early hominids and their significance to encephalization and megadontia in "robust" australopithecines. In F.E. Grine, ed., *Evolutionary History of the "Robust" Australopithecines.* New York: Aldine de Gruyter. Pp. 133–148.

McHenry, H.M. 1991. Femoral lengths and stature in Plio-Pleistocene hominids. *Am. J. Phys. Anthropol.* 85:149–158.

McHenry, H.M. 1992. Body size and proportions in early hominids. *Am. J. Phys. Anthropol.* 87:407–431.

McHenry, H.M., and L.R. Berger. 1998. Body proportions in *Australopithecus afarensis* and *A. africanus* and the origin of the genus *Homo. J. Hum. Evol.* 35:1–22.

McHenry, H.M., and K. Coffing. 2000. *Australopithecus* to *Homo*: Transformations in body and mind. *Annu. Rev. Anthropol.* 29:125–146.

McLaughlin, S., and R.F. Margolskee. 1994. The sense of taste. *Am. Sci.* 82(Nov–Dec):538–545.

McNab, B.K., and J.F. Eisenberg. 1989. Brain size and its relation to the rate of metabolism in mammals. *Am. Nat.* 133(2):157–167.

Mednick, S.C., K. Nakayama, J.L. Cantero, M. Atienza, A.A. Levin, N. Pathak, and R. Stickgold. 2004. The restorative effect of naps on perceptual deterioration. *Nat. Neurosci.* 5:677–681.

Menashe, I., O. Man, D. Lancet, and Y. Gilad. 2003. Different noses for different people. *Nat. Gen.* 34(2):143–144.

Merbs, C.F. 1983. Patterns of activity-induced pathology in a Canadian Inuit population. *Archaeological Survey of Canada* Paper 119. Ottawa: National Museum of Man.

Merbs, S.L., and J. Nathans. 1992a. Absorption spectra of human cone pigments. *Nature* 356:433–435.

Merbs, S.L., and J. Nathans. 1992b. Absorption spectra of the hybrid pigments responsible for anomalous color vision. *Science* 258:464–466.

Merchant, K., and R. Martorell. 1988. Frequent reproductive cycling: Does it lead to nutritional depletion of mothers? *Prog. Food Nutr. Sci.* 12:339–369.

Meredith, M. 2001. Human vomeronasal organ function: A critical review of best and worst cases. *Chem. Senses* 26:433–445.

Merker, B. 1984. A note on hunting and hominid origins. *Am. Anthropol.* 86:112–114.

Middleton, F.A., and P.L. Strick. 1994. Anatomical evidence for cerebellar and basal ganglial involvement in higher cognitive function. *Science* 266:458–461.

Milton, K. 1981. Distribution patterns of tropical plant foods as an evolutionary stimulus to primate mental development. *Am. Anthropol.* 83:534–548.

Milton, K. 1986. Digestive physiology in primates. *News Physiol. Sci.* 1:76–79.

Milton, K. 1987. Primate diets and gut morphology: Implications for hominoid evolution. In M. Harris and E. Ross, eds., *Food and Evolution: Toward a Theory of Human Food Habits.* Philadelphia: Temple University Press. Pp. 93–115.

Milton, K. 1991. Comparative aspects of diet in Amazonian forest-dwellers. *Phil. Trans. R. Soc. Lond. B Biol. Sci.* 334:253–263.

Milton, K. 1993. Diet and primate evolution. *Sci. Am.* 269(Aug):86–93.

Milton, K. 1999. A hypothesis to explain the role of meat-eating in human evolution. *Evol. Anthropol.* 8(1):11–11–21.

Minugh-Purvis, N., and K.J. McNamara, eds. 2002. *Human Evolution through Developmental Change.* Baltimore, MD: Johns Hopkins University Press.

Mishkin, M., and T. Appenzeller. 1987. The anatomy of memory. *Sci. Am.* 256(June):80–89.

Mitani, J.C., T. Hasegawa, J. Gros-Louis, P. Marler, and R. Byrne. 1992. Dialects in wild chimpanzees? *Am. J. Primatol.* 27:233–243.

Mittra, E.S., A. Fuentes, and W.C. McGrew. 1997. Lack of hand preference in wild Hanuman langurs (*Presbytis entellus*). *Am. J. Phys. Anthropol.* 103:455–461.

Møller, A.P. 1988. Ejaculate quality, testes size, and sperm competition in primates. *J. Hum. Evol.* 17:479–488.

Møller, A.P., and T.R. Birkhead. 1989. Copulation behavior in mammals: Evidence that sperm competition is widespread. *Biol. J. Linnean Soc.* 38:119–131.

Mollon, J.D. 1989. " 'Tho' she kneel'd in that place where they grew . . ." *J. Exp. Biol.* 146:21–38.

Mombaerts, P. 1999. Seven-transmembrane proteins as odorant and chemosensory receptors. *Science* 286:707–711.

Mombaerts, P. 2001. The human repertoire of odorant receptor genes and pseudogenes. *Annu. Rev. Genomics Hum. Gen.* 2:493–510.

Montagna, W. 1985. The evolution of human skin(?) *J. Hum. Evol.* 14:3–22.

Montagna, W., and P.F. Parakkal. 1974. *The Structure and Function of Skin*, 3rd ed. New York: Academic Press.

Montagu, A. 1961. Neonatal and infant immaturity in man. *JAMA* 178(1):56–57.

Montagu, A. 1962. The functions of man's dictribution of hair. *JAMA* 182:161–162.

Montagu, A. 1964. Natural selection and man's relative hairlessness. *JAMA* 187:356–357.

Montagu, A. 1971. *Touching: The Human Significance of the Skin*. New York: Columbia University Press.

Montgomery, G. 1989. The mind in motion. *Discover* (Mar):58–68.

Monti-Bloch, L., C. Jennings-White, and D.L. Berliner. 1998. The human vomeronasal system. *Ann. N. Y. Acad. Sci.* 855:373–389.

Moore, H.D.M., M. Martin, and T.R. Birkhead. 1999. No evidence for killer sperm or other selective interactions between human spermatozoa in ejaculates of different males *in vitro*. *Proc. R. Soc. Lond. B Biol. Sci.* 266:2343–2350.

Morgan, E. 1972. *The Descent of Woman*. New York: Bantam.

Morgan, E. 1982. *The Aquatic Ape*. New York: Stein and Day.

Morgan, E. 1986. Lucy's child. *New Sci.* 112(25 Dec):13–15.

Morgan, E. 1990. *The Scars of Evolution*. New York: Oxford University Press.

Morgan, E. 1997. *The Aquatic Ape Hypothesis*. London: Souvenir Press.

Mori, K., H. Nagao, and Y. Yoshihara. 1999. The olfactory bulb: Coding and processing of odor molecule information. *Science* 286:711–715.

Mori, K., H. von Campenhausen, and Y. Yoshihara. 2000. Zonal organization of the mammalian main and accessory olfactory systems. *Proc. R. Soc. Lond. B Biol. Sci.* 355:1801–1812.

Morris, D. 1967. *The Naked Ape*. New York: Dell Pub.

Morris, J.M., D.B. Lucas, and B. Bresler. 1961. Role of the trunk in stability of the spine. *J. Bone Joint Surg. Am.* 43-A(3):327–351.

Morton, D.J. 1924. Evolution of the longitudinal arch of the human foot. *J. Bone Joint Surg.* 6:56–90.

Morton, D.J. 1926. Evolution of man's erect posture. *J. Morphol. Physiol.* 43:147–179.

Moskowitz, B.A. 1988. The acquisition of language. *Sci. Am.* 259(Nov):92–108.

Murray, F.G. 1934. Pigmentation, sunlight, and nutritional disease. *Am. Anthropol.* 36:438–445.

Nakatsukasa, M., C.V. Ward, A. Walker, M.F. Teaford, Y. Kunimatsu, and N. Ogihara. 2004. Tail loss in *Proconsul helsoni*. *J. Hum.Evol.* 46:777–784.

Napier, J.R. 1961. Prehensility and opposability in the hands of primates. *Symp. Zool. Soc. Lond.* 5:115–132.

Napier, J.R. 1962. The evolution of the hand. *Sci. Am.* 207(Dec.):56–62.

Napier, J.R. 1993. *Hands*. Revised by R.H. Tuttle. Princeton, NJ: Princeton University Press.

Napier, J.R., and P.H. Napier. 1967 *A Handbook of Living Primates*. New York: Academic Press.

Nathanielsz, P.W. 1996. The timing of birth. *Am. Sci.* 84(Nov–Dec):562–569.

Nathans, J. 1989. The genes for color vision. *Sci. Am.* 260(Feb.):42–49.

Nathans, J. 1994. In the eye of the beholder: Visual pigments and inherited variation in human vision. *Cell* 78:357–360.

Neel, J.V., A.B. Weder, and S. Julius. 1998. Type II diabetes, essential hypertension, and obesity as "syndromes of impaired genetic homeostasis": The "thrifty genotype" hypothesis enters the 21st century. *Perspect. Biol. Med.* 42:44–74.

Neer, R.M. 1975. The evolutionary significance of vitamin D, skin pigment, and ultraviolet light. *Am. J. Phys. Anthropol.* 43:409–416.

Nef, P. 1998. How we smell: The molecular and cellular bases of olfaction. *News Physiol. Sci.* 13:1–5.

Neitz, M., and J. Neitz. 1995. Numbers and ratios of visual pigment genes for normal red-green color vision. *Science* 267:1013–1016.

Nesse, R.M. 1990. Evolutionary explanations of emotions. *Hum. Nat.* 1(3):261–289.

Nesse, R.M. 1991. What good is feeling bad? The evolutionary benefits of psychic pain. *Sciences* 31(Nov/Dec):30–37.

Nesse, R.M. 1999. The evolution of hope and despair. *Soc. Res.* 66(2):429–469.

Nettle, D. 2002. Women's height, reproductive success and the evolution of sexual dimorphism in modern humans. *Proc. R. Soc. Lond. B Biol. Sci.* 269:1919–1923.

Neville, M.C., and M.F. Picciano. 1997. Regulation of milk lipid secretion and composition. *Annu. Rev. Nutr.* 17:159–184.

Newman, R.W. 1970. Why man is such a sweaty and thirsty naked animal: a speculative review. *Hum. Biol.* 42(1):12–27.

Nishimura, T., A. Mikami, J. Suzuki, and T. Matsuzawa. 2003. Descent of the larynx in chimpanzee infants. *Proc. Natl. Acad. Sci. U.S.A.* 100(12):6930–6933.

Noback, C.R. 1975. The visual system of primates in phylogenetic studies. In W.P. Luckett and F.S. Szalay, eds., *Phylogeny of the Primates.* New York: Plenum Pub. Pp. 199–218.

Northcutt, R.G. 2002. Understanding vertebrate brain evolution. *Integr. Comp. Biol.* 42:743–756.

Northcutt, R.G., and J.H. Klass. 1995. The emergence and evolution of mammalian neocortex. *Trends Neurosci.* 18(9):373–379.

Nowak, R. 1992. The pronghorn's prowess. *Discover* 19(Dec):16.

Nudo, R.J., E.J. Plautz, and G.W. Milliken. 1997. Adaptive plasticity in primate motor cortex as a consequence of behavioral experience and neuronal injury. *Semin. Neurosci.* 9:13–23.

O'Connell, J.F., K. Hawkes, and N.G. Blurton-Jones. 1999. Grandmothering and the evolution of *Homo erectus. J. Hum. Evol.* 36:461–485.

O'Connor, B.L. 1979. Normal amplitudes of radioulnar pronation and supination in several genera of anthropoid primates. *Am. J. Phys. Anthropol.* 51:39–44.

O'Connor, K., D.J. Holman, and J.W. Wood. 2001. Menstrual cycle variability and the perimenopause. *Am. J. Hum. Biol.* 13:465–478.

O'Dea, K. 1991. Traditional diet and food preferences of Australian Aboriginal hunter-gatherers. *Phil. Trans. R. Soc. Lond. B Biol. Sci.* 334:233–241.

Oeppen, J., and J.W. Vaupel. 2002. Broken limits to life expectancy. *Science* 296:1029–1031.

Ohman, J.C., M. Slanina, G. Baker, and R.P. Mensforth. 1995. Thumbs, tools, and early humans. *Science* 268:587–589.

Ohman, J.C., T.J. Krochta, C.O. Lovejoy, R.P. Mensforth, and B. Latimer. 1997. Cortical bone distribution in the femoral neck of hominoids: Implications for the locomotion of *Australopithecus afarensis. Am. J. Phys. Anthropol.* 104:117–131.

Okada, M., H. Ishida, and T. Kimura. 1976. Biomechanical features of bipedal gait in human and nonhuman primates. In P.V. Komi, ed., *Biomechanics* V. (Proc. 5th Int. Cong. Biomech. Vol. IA). Baltimore, MD: University Park Press. Pp. 303–310.

Owerkowicz, T., C.G. Farmer, J.W. Hicks, and E.L. Brainerd. 1999. Contribution of gular pumping to lung ventilation in monitor lizards. *Science* 284:1661–1663.

Packer, C., M. Tatar, and A. Collins. 1998. Reproductive cessation in female mammals. *Nature* 392:807–811.

Pagel, M., and W. Bodmer. 2003. A naked ape would have fewer parasites. *Proc. R. Soc. Lond. B Biol. Sci.* 270(Suppl. 1):S117–S119.

Palmer, A.R. 2002. Chimpanzee right-handedness reconsidered: Evaluating the evidence with funnel plots. *Am. J. Phys. Anthropol.* 118:191–199.

Palumbi, S.R. 2001. *The Evolution Explosion: How Humans Cause Rapid Evolutionary Change.* New York: W.W. Norton.

Panchen, A.L. 1989. Ears and vertebrate evolution. *Nature* 342:342–343.

Panger, M.A. 1998. Hand preference in free-ranging white-throated capuchins (*Cebus capucinus*) in Costa Rica. *Int. J. Primatol.* 19(1):133–163.

Panger, M.A., A.S. Brooks, B.G. Richmond, and B. Wood. 2002. Older than the Oldowan? Rethinking the emergence of hominin tool use. *Evol. Anthropol.* 11(6):235–245.

Parish, A.R., and F.B.M. de Waal. 2000. The other "closest living relative:" How bonobos (*Pan paniscus*) challenge traditional assumptions about females, dominance, intra- and intersexual interactions, and hominid evolution. *Ann. N.Y. Acad. Sci.* 907:97–113.

Parker, S.T., and K.R. Gibson, eds. 1990. *"Language" and Intelligence in Monkeys and Apes: Comparative Developmental Perspectives.* Cambridge: Cambridge University Press.

Parker, S.T., and M.L. McKinney. 1999. *Origins of Intelligence: The Evolution of Cognitive Development in Monkeys, Apes, and Humans.* Baltimore, MD: Johns Hopkins University Press.

Parkington, J. 2001. Milestones: The impact of the systematic exploitation of marine foods on human evolution. In P.V. Tobias, M.A. Rath, J. Moggi-Cecchi, and G.A. Doyle, eds., *Humanity from African Naissance to Coming Millennia.* Florence: Firenze University Press. Pp. 327–336.

Passingham, R. 1982. *The Human Primate.* Oxford: W.H. Freeman.

Patterson, F., and E. Linden. 1981. *The Education of Koko.* New York: Holt, Rinehart and Winston.

Paul, A., J. Kuester, and D. Podzuweit. 1993. Reproductive senescence and terminal investment in female Barbary macaques (*Macaca sylvanus*) at Salem. *Int. J. Primatol.* 14(1):105–124.

Pavelka, M.S.M., and L.M. Fedigan. 1991. Menopause: A comparative life history perspective. *Yrbk. Phys. Anthropol.* 34:13–38.

Pawlowski, B. 1998. Why are human newborns so big and fat? *Hum. Evol.* 13:65–72.

Pawlowski, B. 1999. Loss of oestrus and concealed ovulation in human evolution. *Curr. Anthropol.* 40:257–275.

Pawlowski, B. 2003. Variable preferences for sexual dimorphism in height as a strategy for increasing the pool of potential partners in humans. *Proc. R. Soc. Lond. B Biol. Sci.* 270:709–712.

Pawlowski, B., R.I.M. Dunbar, and A. Lipowicz. 2000. Tall men have more reproductive success. *Nature* 403:156.

Peacock, N.R. 1994. Comparative and cross-cultural approaches to the study of human female reproductive failure. In C.J. DeRousseau, ed., *Primate Life History and Evolution.* New York: Wiley-Liss. Pp. 195–220.

Pearson, O.M. 2000a. Postcranial remains and the origin of modern humans. *Evol. Anthropol.* 9(6):229–247.

Pearson, O.M. 2000b. Activity, climate, and postcranial robusticity. *Curr. Anthropol.* 41(4):569–607.

Peccei, J.S. 1995. The origin and evolution of menopause: The altriciality-lifespan hypothesis. *Ethol. Sociobiol.* 16:425–449.

Peccei, J.S. 2001a. A critique of the grandmother hypotheses: Old and new. *Am. J. Hum. Biol.* 13:434–452.

Peccei, J.S. 2001b. Menopause: Adaptation or epiphenomenon? *Evol. Anthropol.* 10(2):43–57.

Pecorari, D. 2002. Motherhood, metabolic changes and evolution. *Minerva Ginecol.* 54:239–244.

Pepperberg, I.M. 1999. *The Alex Studies: Cognitive and Communicative Abilities of Grey Parrots.* Cambridge, MA: Harvard University Press.

Père, M.-C. 2003. Materno-fetal exchanges and utilisation of nutrients by the foetus: Comparison between species. *Reprod. Nutr. Dev.* 43:1–15.

Perls, T.T., and R.C. Fretts. 2001. The evolution of menopause and human life span. *Ann. Hum. Biol.* 28(3):237–245.

Perls, T.T., J. Wilmoth, R. Levenson, M. Drinkwater, M. Cohen, H. Bogan, E. Joyce, S. Brewster, L. Kunkel, and A. Puca. 2002. Life-long sustained mortality advantage in siblings of centenarians. *Proc. Natl. Acad. Sci. U.S.A.* 99:8442–8447.

Perrett, D.I., K.A. May, and S. Yoshikawa. 1994. Facial shape and judgements of female attractiveness. *Nature* 368:239–242.

Perry, S., and J.H. Manson. 2003. Traditions in monkeys. *Evol. Anthropol.* 12(2):71–81.

Pert, C.B. 1986. The wisdom of the receptors: Neuropeptides, the emotions, and bodymind. *Advances* 3(3):8–16.

Pert, C.B., M.R. Ruff, R.J. Weber, and M. Kerkenham. 1985. Neuropeptides and their receptors: A psychosomatic network. *J. Immunol.* 135(2):820S–826S.

Pérusse, D. 1993. Cultural and reproductive success in industrial societies: Testing the relationship at the proximate and ultimate levels. *Behav. Brain Sci.* 16:267–322.

Pérusse, D. 1994. Mate choice in modern societies: Testing hypotheses with behavioral data. *Hum. Nat.* 5(3):255–278.

Peters, A.D., and S.P. Otto. 2003. Liberating genetic variance through sex. *BioEssays* 25:533–537.

Petersen, S., A. Gotfredsen, F.U. Knudsen. 1988. Lean body mass in small for gestational age and appropriate for gestational age infants. *J. Pediatr.* 113(5):886–889.

Petersen, S.E., P.T. Fox, A.Z. Snyder, and M.E. Raichle. 1990. Activation of extrastriate and frontal cortical areas by visual words and word-like stimuli. *Science* 249:1041–1044.

Petito, L.A., S. Holowka, L.E. Sergio, and D. Ostry. 2001. Language rhythms in baby hand movements. *Nature* 411:35.

Pianka, E.R. 1970. On r- and K-selection. *Am. Nat.* 104:592–597.

Pichaud, F., A. Briscoe, and C. Desplan. 1999. Evolution of color vision. *Curr. Opin. Neurobiol.* 9:622–627.

Pickering, T.R. 2001. Taphonomy of the Swartkrans hominid postcrania and its bearing on issues of meat-eating and fire management. In C.B. Stanford and H.T. Bunn, eds., *Meat-Eating and Human Evolution.* Oxford: Oxford University Press. Pp. 33–51.

Picq, P. 1990. The diet of *Australopithecus afarensis*: An attempted reconstruction. *Anthropol. Annu. 1990.* Pp. 99–101.

Picq, P. 1994. Craniofacial size and proportions and the functional significance of the supraorbital region in primates. *Z. Morphol. Anthropol.* 80(1):51–63.

Picq, P.G., and W.L. Hylander. 1989. Endo's stress analysis of the primate skull and the functional significance of the supraorbital region. *Am. J. Phys. Anthropol.* 79:393–398.

Pinker, S. 1998. The evolution of the human language faculty. In N.G. Jablonski and L.C. Aiello, eds., The Origin and Diversification of Language. *Mem. Calif. Acad. Sci.* 24. Pp. 117–126.

Pitts, M., and M. Roberts. 2000. *Fairweather Eden.* New York: Fromm International.

Plata-Salaman, C.R. 1991. Immunoregulators in the nervous system. *Neurosci. Biobehav. Rev.* 15:185–215.

Plavcan, J.M. 2000. Inferring social behavior from sexual dimorphism in the fossil record. *J. Hum. Evol.* 39:327–344.

Ploog, D. 2002. Is the neural basis of vocalisation different in non-human primates and *Homo sapiens*? In T.J. Crow, ed., *The Speciation of Modern* Homo Sapiens. *Proc. Br. Acad.* 106:120–135.

Plummer, T., J. Ferraro, P. Ditchfield, and L. Bishop. 2001. Current research on Oldowan hominid activities at Kanjera South, Kenya. (Abstract.) *Am. J. Phys. Anthropol.* Suppl. 32:120.

Plutchik, R. 2001. The nature of emotions. *Am. Sci.* 89(July/Aug):344–350.

Pond, C.M. 1987. Fat and figures. *New Sci.* 114(4 June):62–66.

Pond, C.M. 1991. Adipose tissue in human evolution. In M. Roede, J. Wind, J.M. Patrick, and V. Reynolds, eds., *The Aquatic Ape: Fact or Fiction?* London: Souvenir Press. Pp. 193–220.

Pond, C. 1998. *The Fats of Life.* Cambridge: Cambridge University Press.

Pond, C.M. 1999. Physiological specialization of adipose tissue. *Prog. Lipid Res.* 38:225–248.

Pond, CM., and C.A. Mattacks. 1987. The anatomy of adipose tissue in captive Macaca monkeys and its implications for human biology. *Folia Primatol.* 48:164–185.

Pond, CM., and C.A. Mattacks. 1989. Biochemical correlates of the structural allometry and site-specific properties of mammalian adipose tissue. *Comp. Biochem. Physiol. A Mol. Integr. Physiol.* 92A(3):455–463.

Pond, C.M., and M.A. Ramsey. 1990. Allometry of the distribution of adipose tissue in Carnivora. *Can. J. Zool.* 70:342–347.

Poppele, R., and G. Bosco. 2003. Sophisticated spinal contributions to motor control. *Trends Neurosci.* 26(5):269–276.

Porter, A.M.W. 1993. Sweat and thermoregulation in hominids. Comments prompted by the publications of P.E. Wheeler 1984–1993. *J. Hum. Evol.* 25:417–423.

Post, P.W., F. Daniels, and R.T. Binford, Jr. 1975. Cold injury and the evolution of "white" skin. *Hum. Biol.* 47(1):65–80.

Potts, R. 1984. Home bases and early hominids. *Am. Sci.* 72(July):338–347.

Potts, R. 1996. *Humanity's Descent.* New York: Avon Books.

Potts, R. 1998a. Variability selection in hominid evolution. *Evol. Anthropol.* 7(3):81–96.

Potts, R. 1998b. Environmental hypotheses of hominin evolution. *Yrbk. Phys. Anthropol.* 41:93–136.

Premack, D. 2004. Is language the key to human intelligence? *Science* 303:318–320.

Prentice, A.M., and A. Prentice. 1988. Energy costs of lactation. *Annu. Rev. Nutr.* 8:63–79.

Preuss, T.M. 2001. The discovery of cerebral diversity: An unwelcome scientific revolution. In D. Falk and K.R. Gibson, eds., *Evolutionary Anatomy of the Primate Cerebral Cortex.* Cambridge: Cambridge University Press. Pp. 138–164.

Profet, M. 1993. Menstruation as a defense against pathogens transported by sperm. *Q. Rev. Biol.* 68(3):335–386.

Promislow, D.E.L., and P.H. Harvey. 1990. Living fast and dying young: A comparative analysis of life-history variation among mammals. *J. Zool. Soc. Lond. B Biol. Sci.* 220:417–437.

Prost, J.H. 1980. Origin of bipedalism. *Am. J. Phys. Anthropol.* 52:175–189.

Puech, P.-F., and H. Albertini. 1984. Dental microwear and mechanisms in early hominids from Laetoli and Hadar. *Am. J. Phys. Anthropol.* 65:87–91.

Putz, R.L.V., and M. Müller-Gerbl. 1996. The vertebral column—A phylogenetic failure? *Clin. Anat.* 9:205–212.

Quadagno, D.M., H.E. Shubeita, J. Deck, and D. Francoeur. 1981. Influence of male social contacts, exercise, and all-female living conditions on the menstrual cycle. *Psychoneuroendocrinology* 6(3):239–244.

Radetsky, P. 1995. Gut thinking. *Discover* 16(May):76–81.

Rak, Y. 1983. *The Australopithecine Face.* New York: Academic Press.

Rak, Y. 1991. Lucy's pelvic anatomy: Its role in bipedal gait. *J. Hum. Evol.* 20:283–290.

Rakic, P. 2001. Neurocreationism—Making new cortical maps. *Science* 294:1011–1012.

Rakic, P., and D.R. Kornack. 2001. Neocortical expansion and elaboration during primate evolution: A view from neuroembryology. In D. Falk and K.R. Gibson, eds., *Evolutionary Anatomy of the Primate Cerebral Cortex.* Cambridge: Cambridge University Press. Pp. 30–56.

Rasmussen, D.T. 1990. Primate origins: Lessons from a Neotropical marsupial. *Am. J. Primatol.* 22:263–277.

Ravey, M. 1978. Bipedalism: An early warning system for Miocene hominoids. *Science* 199:372.

Ravosa, M.J. 1988. Browridge development in Cercopithecidae: A test of two models. *Am. J. Phys. Anthropol.* 76:535–555.

Ravosa, M.J. 1991a. Interspecific perspective on mechanical and nonmechanical models of primate circumorbital morphology. *Am. J. Phys. Anthropol.* 86:369–396.

Ravosa, M.J. 1991b. Structural allometry of the prosimian mandibular corpus and symphysis. *J. Hum. Evol.* 20:3–20.

Ravosa, M.J. 1997. Anthropoid origins and the modern symphysis. *Folia Primatol.* 70:65–78.

Ravosa, M.J., V.E. Noble, W.L. Hylander, K.R. Johnson, and E.M. Kowalski. 2000. Masticatory stress, orbital orientation, and the evolution of the primate postorbital bar. *J. Hum. Evol.* 38:667–693.

Reader, S.M., and K.N. Laland. 2002. Social intelligence, innovation, and enhanced brain size in primates. *Proc. Natl. Acad. Sci. U.S.A.* 99(7):4436–4441.

Regan, B.C., C. Julliot, B. Simmen, F. Vienot, P. Charles-Dominique, and J.D. Mollon. 2001. Fruits, foliage and the evolution of primate colour vision. *Phil. Trans. R. Soc. Lond. B Biol. Sci.* 356:229–283.

Reid, D.J., G.T. Schwartz, C. Dean, and M.S. Chandrasekera. 1998. A histological reconstruction of dental development in the common chimpanzee, *Pan troglodytes. J. Hum. Evol.* 35:427–448.

Reitenbaugh, C., and C.-S. Goodby. 1989. Beyond the thrifty gene: Metabolic implications of prehistoric migration into the New World. *Med. Anthropol.* 11:227–236.

Relethford, J.H. 1997. Hemispheric difference in human skin color. *Am. J. Phys. Anthropol.* 104:449–457.

Rendell, L., and H. Whitehead. 2001. Culture in whales and dolphins. *Behav. Brain Sci.* 24:309–382.

Reynolds, P.C. 1981. *On the Evolution of Human Behavior: The Argument from Animals to Man.* Berkeley, CA: University of California Press.

Rhine, R.J., G.W. Norton, G.M. Wynn, R.D. Wynn, and H.B. Rhine. 1986. Insect and meat eating among infant and adult baboons (*Papio cynocephalus*) of Mikumi National Park, Tanzania. *Am. J. Phys. Anthropol.* 70:105–118.

Rhodes, G., L.A. Zebrowitz, A. Clark, S.M. Kalick, A. Hightower, and R. McKay. 2001. Do facial averageness and symmetry signal health? *Evol. Hum. Behav.* 22:31–46.

Rice, S.H. 2002. The role of heterochrony in primate brain evolution. In N. Minugh-Purvis and K.J. McNamara, eds., *Human Evolution through Developmental Change.* Baltimore, MD: Johns Hopkins University Press. Pp. 154–170.

Rice, W.R. 2002. Experimental tests of the adaptive significance of sexual recombination. *Nat. Gen.* 3:241–251.

Rice, W.R., and A.K. Chippindale. 2001. Sexual recombination and the power of natural selection. *Science* 294:555–559.

Richards, M.P., P.B. Pettitt, M.C. Stiner, and E. Trinkaus. 2001. Stable isotope evidence for increasing dietary breadth in the European mid-Upper Paleolithic. *Proc. Natl. Acad. Sci. U.S.A.* 98(11):6528–6532.

Richmond, B.G., and D.S. Strait. 2000. Evidence that humans evolved from a knuckle-walking ancestor. *Nature* 410:382–385.

Richmond, B.G., D.R. Begun, and D.S. Strait. 2001. Origin of human bipedalism: The knuckle-walking hypothesis revisited. *Yrbk. Phys. Anthropol.* 44:79–105.

Richmond, B.G., L.C. Aiello, and B.A. Wood. 2002. Early hominin limb proportions. *J. Hum. Evol.* 43:529–548.

Ricklan, D.E. 1987. Functional anatomy of the hand of *Australopithecus africanus. J. Hum. Evol.* 16:643–664.

Ridley, M. 1994. *The Red Queen.* New York: Macmillan Pub. Co.

Risnes, S. 1998. Growth tracks in dental enamel. *J. Hum. Evol.* 35:331–350.

Roberts, D.F. 1978. *Climate and Human Variability.* Menlo Park, CA: Cummings Pub.

Roberts, T.J., R. Kram, P.G. Weyand, and C.R. Taylor. 1998a. Energetics of bipedal running. I. Metabolic cost of generating force. *J. Exp. Biol.* 201:2745–2751.

Roberts, T.J., M.S. Chen, and C.R. Taylor. 1998a. Energetics of bipedal running. II. Limb design and running mechanics. *J. Exp. Biol.* 201:2753–2762.

Robins, A.H. 1991. *Biological Perspectives on Human Pigmentation.* Cambridge: Cambridge University Press.

Robinson, J.T. 1956. The dentition of the Australopithecinae. *Transvaal Mus. Mem.* 9:1–179.

Robinson, J.T. 1962. The origin and adaptive radiation of the australopithecines. In G. Kurth, ed., *Evolution und Hominisation.* Stuttgart: Gustav Fischer Verlag. Pp. 120–140.

Robson, S.L. 2004. Breast milk, diet, and large human brains. *Curr. Anthropol.* 45:419–425.

Rodman, P., and H.M. McHenry. 1980. Bioenergetics and the origin of hominid bipedalism. *Am. J. Phys. Anthropol.* 52:103–106.

Rodriguez, I., and P. Mombaerts. 2002. Novel human vomeronasal receptor-like genes reveal species-specific families. *Curr. Biol.* 12(12):R409–R411.

Rogers, A.P., D. Iltis, and S. Wooding. 2004. Genetic variation at the MC1R locus and the time since loss of human body hair. *Curr. Anthropol.* 45(1):105–107.

Rose, M.D. 1975. Functional proportions of primate lumbar vertebral bodies. *J. Hum. Evol.* 4:21–38.

Rose, M.D. 1976. Bipedal behavior of olive baboons (*Papio anubis*) and its relevance to an understanding of the evolution of human bipedalism. *Am. J. Phys. Anthropol.* 44:247–262.

Rose, M.D. 1984. Food acquisition and the evolution of positional behavior: The case of bipedalism. In D.J. Chivers, B.A. Wood, and A. Bilsborough, eds., *Food Acquisition and Processing in Primates.* New York: Plenum Pub. Pp. 509–524.

Rose, M.D. 1993. Functional anatomy of the elbow and forearm in primates. In D.L. Gebo, ed., *Postcranial Adaptation in Non-human Primates.* DeKalb, IL: Northern Illinois University Press. Pp. 70–95.

Rosenberg, K.R. 1992. The evolution of modern human childbirth. *Yrbk. Phys. Anthropol.* 35:89–124.

Rosenberg, K.R., and W. Trevathan. 1996. Bipedalism and human birth: The obstetrical dilemma revisited. *Evol. Anthropol.* 4(5):161–168.

Rosenberg, K.R., and W.R. Trevathan. 2001. The evolution of human birth. *Sci. Am.* 285(Nov):72–77.

Ross, C. 1995. Muscular and osseous anatomy of the primate anterior temporal fossa and the function of the postorbital septum. *Am. J. Phys. Anthropol.* 98:275–306.

Ross, C.F. 2000. Into the light: The origin of Anthropoidea. *Annu. Rev. Anthropol.* 29:147–194.

Ross, C.F., and M. Henneberg. 1995. Basicranial flexion, relative brain size, and facial kyphosis in *Homo sapiens* and some fossil hominids. *Am. J. Phys. Anthropol.* 98:575–593.

Ross, C.F., and M.J. Ravosa. 1993. Basicranial flexion, relative brain size, and facial kyphosis in non-human primates. *Am. J. Phys. Anthropol.* 91:305–324.

Roth, G.S., M.A. Lane, D.K. Ingram, J.A. Mattison, D. Elahi, J.D. Tobin, D. Muller, and E.J. Metter. 2002. Biomarkers of caloric restriction may predict longevity in humans. *Science* 297:811.

Rouquier, S., A. Blancher, and D. Giorgi. 2000. The olfactory receptor gene repertoire in primates and mouse: Evidence for reduction of the funcitonal fraction in primates. *Proc. Natl. Acad. Sci. U.S.A.* 97(6):2870–2874.

Routtenberg, A. 1978. The reward system of the brain. *Sci. Am.* 239(Nov):154–164.

Rowe, M.H. 2002. Trichromatic color vision in primates. *News Physiol. Sci.* 17:93–98.

Rowe, T. 1996. Coevolution of the mammalian middle ear and the neocortex. *Science* 273:651–654.

Ruben, J.A. 1989. Activity physiology and evolution of the vertebrate skeleton. *Am. Zool.* 29:195–203.

Ruben, J.A., W.J. Hillenius, N.R. Geist, A. Leitch, T.D. Jones, P.J. Currie, J.R. Horner, and G. Espe. 1996. The metabolic status of some Late Cretaceous dinosaurs. *Science* 273:1204–1207.

Ruff, C.B. 1993. Climatic adaptation and hominid evolution: The thermoregulatory imperative. *Evol. Anthropol.* 2(2):53–60.

Ruff, C.B. 1994. Morphological adaptation to climate in modern and fossil hominids. *Yrbk. Phys. Anthropol.* 37:65–107.

Ruff, C.B. 1995. Biomechanics of the hip and birth in early *Homo. Am. J. Phys. Anthropol.* 98:527–574.

Ruff, C.B., E. Trinkaus, A. Walker, and C.S. Larsen. 1993. Postcranial robusticity in *Homo.* I: Temporal trends and mechanical interpretation. *Am. J. Phys. Anthropol.* 91:21–53.

Russell, M.J., G.M. Switz, and K. Thompson. 1980. Olfactory influences on the human menstrual cycle. *Pharmacol. Biochem. Behav.* 13:737–738.

Ryan, A.S., and D.C. Johanson. 1989. Anterior dental microwear in *Australopithecus afarensis*: Comparisons with human and nonhuman primates. *J. Hum. Evol.* 18:235–268.

Sacher, G.A. 1975. Maturation and longevity in relation to cranial capacity in hominid evolution. In R.H. Tuttle, ed., *Primate Morphology and Evolution.* The Hague: Mouton. Pp. 417–441.

Saffran, J., A. Senghas, and J.C. Trueswell. 2001. The acquisition of language by children. *Proc. Natl. Acad. Sci. U.S.A.* 98(23):12874–12875.

Salamonsen, L.A. 1998. Current concepts of the mechanisms of menstruation: A normal process of tissue destruction. *Trends Endocrinol. Metab.* 9(8):305–309.

Sam, M., S. Vora, B. Malnic, W. Ma, M.V. Novotny, and L.B. Buck. 2001. Odorants may arouse instinctive behaviours. *Nature* 412:142.

Sanders, W.J. 1998. Comparative morphometric study of the australopithecine vertebral series Stw-H8/H41. *J. Hum. Evol.* 34:249–302.

Sansom, I.J., M.P. Smith, H.A. Armstrong, and M.M. Smith. 1992. Presence of the earliest vertebrate hard tissues in conodonts. *Science* 256:1308–1311.

Sapolsky, R.M. 1990. Stress in the wild. *Sci. Am.* 262(Jan):116–123.

Savage-Rumbaugh, S. 1994. Ape at the brink. *Discover* 15(Sept):91–96.

Savage-Rumbaugh, S., and R. Lewin. 1994. *Kanzi: The Ape at the Brink of the Human Mind.* New York: John Wiley and Sons.

Savage-Rumbaugh, E.S., and B.J. Wilkerson. 1978. Socio-sexual behavior in *Pan paniscus* and *Pan troglodytes*: A comparative study. *J. Hum. Evol.* 7:327–344.

Savage-Rumbaugh, S., S.G. Shanker, and T.J. Taylor. 1998. *Apes, Language, and the Human Mind.* New York: Oxford University Press.

Sawaguchi, T. 1988. Correlations of cerebral indices for "extra" cortical parts and ecological variables in primates. *Brain Behav. Evol.* 32:129–140.

Sawaguchi, T. 1989. Relationships between cerebral indices for "extra" cortical parts and ecological categories in anthropoids. *Brain Behav. Evol.* 34:281–293.

Schilling, T.F., and P.V. Thorogood. 2000. Development and evolution of the vertebrate skull. In P. O'Higgins and M.J. Cohn, eds., *Development, Growth, and Evolution: Implications for the Study of the Hominid Skeleton.* San Diego: Academic Press. Pp. 57–83.

Schnapf, J.L., and D.A. Baylor. 1987. How photoreceptors respond to light. *Sci. Am.* 256(Apr):40–47.

Schröder, I. 1992. Human sexual behavior, social organization, and fossil evidence: A reconsideration of human evolution. *Homo* 43(3):263–277.

Schulkin, J. 1991. *Sodium Hunger: The Search for a Salty Taste.* Cambridge: Cambridge University Press.

Schultz, A.H. 1931. The density of hair in primates. *Hum. Biol.* 3(3):303–321.

Schultz, A.H. 1955. The position of the occipital condyles and of the face relative to the skull base in primates. *Am. J. Phys. Anthropol.* 13:97–120.

Schultz, W. 2000. Multiple reward signals in the brain. *Nat. Rev. Neurosci.* 1:199–207.

Schwartz, G.G., and L.A. Rosenblum. 1981. Allometry of primate hair density and the evolution of human hairlessness. *Am. J. Phys. Anthropol.* 55:9–12.

Schwarcz, H.P., and M.J. Schoeninger. 1991. Stable isotope analysis in human nutritional ecology. *Yrbk. Phys. Anthropol.* 34:283–321.

Schwartz, J.H., and I. Tattersall. 2000. The human chin revisited: what is it and who has it? *J. Hum. Evol.* 38:367–409.

Scott, S.K., and I.S. Johnsruhe. 2003. The neuroanatomical and functional organization of speech perception. *Trends Neurosci.* 26(2):100–107.

Sealy, J.C., and N.J. van der Merwe. 1985. Isotope assessment of Holocene human diets in the south-western Cape: South Africa. *Nature* 315:138–140.

Semendeferi, K., A. Lu, N. Schenker, and H. Damasio. 2002. Humans and great apes share a large frontal cortex. *Nat. Neurosci.* 5:272–276.

Senut, B., and C. Tardieu. 1985. Functional aspects of Plio-Pleistocene hominid limb bones: Implications for taxonomy and phylogeny. In E. Delson, ed., *Ancestors: The Hard Evidence.* New York: Alan R. Liss, Inc. Pp. 193–201.

Senut, B., M. Pickford, D. Gommery, P. Mein, K. Cheboi, and Y. Coppens. 2001. First hominid from the Miocene (Lukeino Formation, Kenya). *C. R. Acad. Sci. IIa* 332:137–144.

Seyfarth, R.M. 1986. Vocal communication and its relation to language. In B.B. Smuts, D.L. Cheney, R.M. Seyfarth, R.W. Wrangham, and T.T. Struhsaker, eds., *Primate Societies*. Chicago: University of Chicago Press. Pp. 440–451.

Shanley, D.P., and T.B.L. Kirkwood. 2001. Evolution of the human menopause. *BioEssays* 23:282–287.

Shapiro, L. 1993. Evaluation of "unique" aspects of human vertebral bodies and pedicles with a consideration of *Australopithecus africanus*. *J. Hum. Evol.* 25:433–470.

Shapiro, L.J., and W.L. Jungers. 1988. Back muscle function during bipedal walking in chimpanzee and gibbon: Implications for the evolution of human locomotion. *Am. J. Phys. Anthropol.* 77:201–212.

Sharma, K., A.E. Leonard, K. Lettieri, and S.L. Pfaff. 2000. Genetic and epigenetic mechanisms contribute to motor neuron pathfinding. *Science* 406:515–519.

Sherwood, C.C., D.H. Broadfield, R.L. Holloway, P.J. Gannon, and P.R. Hof. 2003. Variability of Broca's area homologue in African great apes: Implications for language evolution. *Anat. Rec.* 271A:276–285.

Shipman, P. 1984. Scavenger hunt. *Nat. Hist.* 93(Apr):20–27.

Shipman, P. 1986. Scavenging or hunting in early hominids: Theoretical framework and tests. *Am. Anthropol.* 88:27–43.

Short, G.B. 1975. Iris pigmentation and photopic visual activity: A preliminary study. *Am. J. Phys. Anthropol.* 43:425–434.

Short, R.V. 1976. Definition of the problem: The evolution of human reproduction. *Proc. R. Soc. Lond. B Biol. Sci.* 195:3–24.

Short, R.V. 1979. Sexual selection and its component parts, somatic and genital selection, as illustrated by man and the great apes. *Adv. Stud. Behav.* 9:131–158.

Shreeve, J. 1996. Sunset on the Savanna. *Discover* 17(July):116–125.

Shrewsbury, M.M., and R.K. Johnson. 1983. Form, function, and evolution of the distal phalanx. *J. Hand Surg.* 8(4):475–479.

Shrewsbury, M.M., and A. Sonek. 1986. Precision holding in humans, non-human primates, and Plio-Pleistocene hominids. *Hum. Evol.* 1(3):233–242.

Shrewsbury, M.M., M.W. Marzke, R.L. Linsheid, and S.P. Reece. 2003. Comparative morphology of the pollical distal phalanx. *Am. J. Phys. Anthropol.* 121:30–47.

Shyue, S.-K., D. Hewett-Emmett, H.G. Sperling, D.M. Hunt, J.K. Bowmaker, J.D. Mollon, and W.-H. Li. 1995. Adaptive evolution of color vision genes in higher primates. *Science* 269:1265–1267.

Siegel, J.M. 2001. The REM sleep-memory consolidation hypothesis. *Science* 294:1058–1063.

Siegel, J.M., P.R. Manger, R. Nienhuis, H.M. Fahringer, and J.D. Perrigrew. 1998. Monotremes and the evolution of rapid eye movement sleep. *Phil. Trans. R. Soc. Lond. B Biol. Sci.* 353(1372):1147–1157.

Siegel, J.M., P.R. Manger, R. Nienhuis, H.M. Fahringer, T. Shalita, and J.D. Perrigrew. 1999. Sleep in the platypus. *Neuroscience* 91(1):391–400.

Sillen, A. 1992. Strontium-calcium ratios (Sr/Ca) of *Australopithecus robustus* and associated fauna from Swartkrans. *J. Hum. Evol.* 23:495–516.

Sillen, A., and M. Kavanaugh. 1982. Strontium and paleodietary research: A review. *Yrbk. Phys. Anthropol.* 25:67–90.

Sillen, A., G. Hall, and R. Armstrong. 1995. Strontium calcium ratios (Sr/Ca) and strontium isotopic ratios ($^{87}Sr/^{86}Sr$) of *Australopithecus robustus* and *Homo* sp. from Swartkrans. *J. Hum. Evol.* 28:277–285.

Sillén-Tullberg, B., and A.P. Møller. 1993. The relationship between concealed ovulation and mating systems in anthropoid primates: A phylogenetic analysis. *Am. Nat.* 141(1):1–25.

Siller, S. 2001. Sexual selection and the maintenance of sex. *Science* 411:689–692.

Simons, E.L. 1994. A whole new world of ancestors: Eocene anthropoideans from Africa. *Evol. Anthropol.* 3(4):128–139.

Simopoulos, A.P. 1990. Genetics and evolution: Or what your genes can tell you about nutrition. *World Rev. Nutr. Diet.* 63:25–34.

Simopoulos, A.P. 1991. Omega-3 fatty acids in health and disease and in growth and development. *Am. J. Clin. Nutr.* 54 (3):438–463.

Simpson, S.W., C.O. Lovejoy, and R.S. Meindl. 1990. Hominoid dental maturation. *J. Hum. Evol.* 19:285–297.

Simpson, S.W., C.O. Lovejoy, and R.S. Meindl. 1991. Relative dental development in hominoids and its failure to predict somatic growth velocity. *Am. J. Phys. Anthropol.* 86:113–120.

Simpson, S.W., C.O. Lovejoy, and R.S. Meindl. 1992. Further evidence on relative dental maturation and somatic developmental rate in hominoids. *Am. J. Phys. Anthropol.* 87:29–38.

Sinclair, A.R.E., M.D. Leakey, and M. Norton-Griffiths. 1986. Migration and hominid bipedalism. *Nature* 324:307–308.

Singh, D. 1993a. Adaptive significance of female physical attractiveness: Role of waist-to-hip ratio. *J. Pers. Soc. Psychol.* 65(2):293–307.

Singh, D. 1993b. Body shape and women's attractiveness. *Hum. Nat.* 4(3):297–321.

Singh, D., and P.M. Bronstad. 2001. Female body odour is a potential cue to ovulation. *Proc. R. Soc. Lond. B Biol. Sci.* 268:797–801.

Singh, D., and R.K. Young. 1995. Body weight, waist-to-hip ratio, breasts, and hips: Role in judgments of female attractiveness and desirability of relationships. *Ethol. Sociobiol.* 16:483–507.

Skerry, T. 2000. Biomechanical influences on skeletal growth and development. In P. O'Higgins and M.J. Cohn, eds., *Development, Growth, and Evolution: Implications for the Study of the Hominid Skeleton.* San Diego: Academic Press. Pp. 29–39.

Skinner, M. 1991. Bee brood consumption: An alternative explanation for hypervitaminosis A in KNM-ER 1808 (*Homo erectus*) from Koobi Fora, Kenya. *J. Hum. Evol.* 20:493–503.

Slijper, E.J. 1946. Comparative biologic-anatomical investigations on the vertebral column and spinal musculature of mammals. *Verh. Kon. Ned. Akad. Wetensch.*, Afd. Natuurkunde. Tweedie Sectie 42(5):1–128.

Small, M.F. 1989. Aberrant sperm and the evolution of human mating patterns. *Animal Behav.* 38(3):544–546.

Small, M.F. 1991. Sperm wars. *Discover* 12(July):48–53.

Small, M.F. 1996. "Revealed" ovulation in humans? *J. Hum. Evol.* 30:483–488.

Smith, B.H. 1986. Dental development in *Australopithecus* and early *Homo*. *Nature* 323:327–330.

Smith, B.H. 1989. Dental development as a measure of life history in primates. *Evolution* 43(3):683–688.

Smith, B.H. 1991. Dental development and the evolution of life history in Hominidae. *Am. J. Phys. Anthropol.* 86:157–174.

Smith, B.H. 1992. Life history and the evolution of human maturation. *Evol. Anthropol.* 1(4):134–142.

Smith, B.H. 1994. Patterns of dental development in *Homo*, *Australopithecus*, *Pan*, and *Gorilla*. *Am. J. Phys. Anthropol.* 94:307–325.

Smith, B.H., and R.L. Tompkins. 1995. Toward a life history of the Hominidae. *Annu. Rev. Anthropol.* 24:257–279.

Smith, D.V., and R.F. Margolskee. 2001. Making sense of taste. *Sci. Am.* 264(Mar):32–39.

Smith, E.E., and J. Jonides. 1999. Storage and executive processes in the frontal lobes. *Science* 283:1657–1661.

Smith, M.M., and M.I. Coates. 2000. Evolutionary origins of teeth and jaws: Developmental models and phylogenetics patterns. In M.F. Teaford, M.M. Smith, and M.W.J. Ferguson, eds. *Development, Function, and Evolution of Teeth.* Cambridge: Cambridge University Press. Pp. 133–151.

Smith, P. 1982. Dental reduction: Selection or drift? In B. Kurten, ed., *Teeth: Form, Function, and Evolution.* New York: Columbia University Press. Pp. 366–379.

Smith, R. 1999. The timing of birth. *Sci. Am.* 280(Mar):68–75.

Smith, R.L. 1984. Human sperm competition. In R.L. Smith, ed., *Sperm Competition and the Evolution of Animal Mating Systems.* New York: Academic Press. Pp. 601–659.

Smith, R.J., P.J. Gannon, and B.H. Smith. 1995. Ontogeny of australopithecines and early *Homo*: Evidence from cranial capacity and dental evolution. *J. Hum. Evol.* 29:155–168.

Smith, T.D., M.I. Siegel, A.M. Burrows, M.P. Mooney, A.R. Burdi, P.A. Fabrizio, and F.R. Clemente. 1998. Searching for the vomeronasal organ of adult humans: Preliminary findings on location, structure, and size. *Microsc. Res. Tech.* 41:483–491.

Smith, T.D., M.I. Siegel, C.J. Bonar, K.P. Bhatnagar, M.P. Mooney, A.M. Burrows, M.K. Smith, and L.M. Maico. 2001a. The existence of the vomeronasal organ in postnatal chimpanzees and evidence for its homology with that of humans. *J. Anat.* 198:77–82.

Smith, T.D., M.I. Siegel, and K.P. Bhatnagar. 2001b. Reappraisal of the vomeronasal system of catarrhine primates: Ontogeny, morphology, functionality, and persisting questions. *Anat. Rec. (New Anat.)* 265:176–192.

Smuts, B.B. 1985. *Sex and Friendship in Baboons*. New York: Aldine Publishing Co.

Snowdon, C.T. 1990. Language capacities of nonhuman animals. *Yrbk. Phys. Anthropol.* 33:215–243.

Soligo, C., and A.E. Muller. 1999. Nails and claws in primate evolution. *J. Hum. Evol.* 36:97–114.

Soltis, J. 2002. Do primate females gain nonprocreative benefits by mating with multiple males? Theoretical and empirical considerations. *Evol. Anthropol.* 11(5):187–197.

Southgate, D.A.T. 1991. Nature and variability of human food consumption. *Phil. Trans. R. Soc. Lond. B Biol. Sci.* 334:281–288.

Spanagel, R., and F. Weiss. 1999. The dopamine hypothesis of reward: Past and current status. *Trends Neurosci.* 22(11):521–527.

Spehr, M., G. Gisselmann, A. Poplawski, J.A. Riffell, C.H. Wetzel, R.K. Zimmer, and H. Hatt. 2003. Identification of a testicular odorant receptor mediating human sperm chemotaxis. *Science* 299:2054–2058.

Spencer, F. 1990. *Piltdown: A Scientific Forgery*. New York: Oxford University Press.

Spencer, M.A., and B. Demes. 1993. Biomechanical analysis of masticatory system configuration in Neanderthals and Inuits. *Am. J. Phys. Anthropol.* 91:1–20.

Speth, J.D. 1991. Protein selection and avoidance strategies of contemporary and ancestral foragers: Unresolved issues. *Phil. Trans. R. Soc. Lond. B Biol. Sci.* 334:265–270.

Sponheimer, M., and J.A. Lee-Thorp. 1999. Isotopic evidence for the diet of an early hominid *Australopithecus africanus*. *Science* 283:368–370.

Spoor, F., and F. Zonneveld. 1998. Comparative review of the human bony labyrinth. *Yrbk. Phys. Anthropol.* 41:211–251.

Spoor, F., B. Wood, and F. Zonneveld. 1994. Implications of early hominid labyrinthine morphology for evolution of human bipedal locomotion. *Nature* 369:645–648.

Squire, L.R., and S. Zola-Morgan. 1991. The medial temporal lobe memory system. *Science* 253:1380–1386.

Stanford, C.B. 1995. To catch a colobus. *Nat. Hist.* 104(Jan):48–55.

Stanford, C.B. 1996. The hunting ecology of wild chimpanzees: Implications for the evolutionary ecology of Pliocene hominids. *Am. Anthropol.* 98(1):96–113.

Stanford, C.B. 1999. *The Hunting Apes: Meat Eating and the Origins of Human Behavior*. Princeton, NJ: Princeton University Press.

Stanford, C.B., J. Wallis, H. Matama, and J. Goodall. 1994. Patterns of predation by chimpanzees on red colobus monkeys in Gombe National Park 1982–1991. *Am. J. Phys. Anthropol.* 94:213–228.

Stanley, S.M. 1998. *Children of the Ice Age: How a Global Catastrophe Allowed Humans to Evolve*. New York: W.H. Freeman.

Starks, P.T., and C.A. Blackie. 2000. The relationship between serial monogamy and rape in the United States (1960–1995). *Proc. R. Soc. Lond. B Biol. Sci.* 267:1259–1263.

Stearns, S.C. 1976. Life-history tactics: A review of the ideas. *Q. Rev. Biol.* 1(1):3–47.

Steegmann, A.T., F.J. Cerny, and T.W. Holliday. 2002. Neandertal cold adaptation: Physiological and energetic factors. *Am. J. Hum. Biol.* 14:566–583.

Steele, J. 1989. Hominid evolution and primate social cognition. *J. Hum. Evol.* 18:421–432.

Stern, J.T. 1972. Anatomical and functional specializations of the human gluteus maximus. *Am. J. Phys. Anthropol.* 36:315–340.

Stern, J.T. 1975. Before bipedality. *Yrbk. Phys. Anthropol.* 19:59–68.

Stern, J.T. 2000. Climbing to the top: A personal memoir of *Australopithecus afarensis*. *Evol. Anthropol.* 9(3):113–133.

Stern J.T., and S.G. Larson. 2001. Telemetered electromyography of the supinators and pronators of the forearm in gibbons and chimpanzees: Implications for the fundamental positional behavior adaptation of hominoids. *Am. J. Phys. Anthropol.* 115:253–268.

Stern, J.T., and R.L. Susman. 1981. Electromyography of the gluteal muscles in *Hylobates*, *Pongo*, and *Pan*: Implications for the evolution of hominid bipedality. *Am. J. Phys. Anthropol.* 55:153–166.

Stern, J.T., and R.L. Susman. 1983. The locomotor anatomy of *Australopithecus afarensis*. *Am. J. Phys. Anthropol.* 60:279–317.

Stern, K., and M.K. McClintock. 1998. Regulation of ovulation by human pheromones. *Nature* 392:177–179.

Sternberg, E.M. 2000. *The Balance Within: The Science Connecting Health and Emotions*. New York: W.H. Freeman.

Steudel, K. 1996. Limb morphology, bipedal gait, and the energetics of hominid locomotion. *Am. J. Phys. Anthropol.* 99:345–355.

Steudel-Numbers. K.L. 2003. The energetic cost of locomotion: Humans and primates compared to generalized endotherms. *J. Hum. Evol.* 44:255–262.

Stickgold, R., J.A. Hobson, R. Fosse, and M. Fosse. 2001. Sleep, learning and dreams: Off-line memory reprocessing. *Science* 294:1052–1057.

Stiner M.C., N.D. Munro, T.A. Surovell, E. Tchernov, and O. Ben-Yosef. 1999. Paleolithic population growth pulses evidenced by small animal exploitation. *Science* 283:190–194.

Stiner M.C., N.D. Munro, and T.A. Surovell. 2000. The tortoise and the hare: Small-game use, the broad-spectrum revolution, and Paleolithic demography. *Curr. Anthropol.* 41:39–73.

Stoddart, D.M. 1990. *The Scented Ape: The Biology and Culture of Human Odour*. Cambridge: Cambridge University Press.

Stowers, L., T.E. Holy, M. Meister, C. Dulac, and G. Koentges. 2002. Loss of sex discrimination and male-male aggression in mice deficient for TRP2. *Science* 295:1493–1500.

Strait, D.S. 1999. The scaling of basicranial flexion and length. *J. Hum. Evol.* 37:701–719.

Strait, D.S. 2001. Integration, phylogeny, and the hominid cranial base. *Am. J. Phys. Anthropol.* 114:273–297.

Strassman, B.I. 1981. Sexual selection, paternal care, and concealed ovulation in humans. *Ethol. Sociobiol.* 2:31–40.

Strassman, B.I. 1996a. The evolution of endometrial cycles and menstruation. *Q. Rev. Biol.* 71(2):181–220.

Strassman, B.I. 1996b. Energy economy in the evolution of menstruation. *Evol. Anthropol.* 5(5):157–164.

Strassman, B.I. 1997. The biology of menstruation in *Homo sapiens*: Total lifetime menses, fecundity, and nonsynchrony in a natural fertility population. *Curr. Anthropol.* 38(1):123–129.

Streeter, S.A., and D.H. McBurney. 2003. Waist-hip ratio and attractiveness: New evidence and a critique of "a critical test." *Evol. Hum. Behav.* 24:88–98.

Stryer, L. 1987. The molecules of visual excitation. *Sci. Am.* 257(July):42–50.

Sullivan, S.L. 2002. Mammalian chemosensory receptors. *NeuroReport* 13(1):A9–A17.

Surridge, A.K., D. Osorio, and N.I. Mundy. 2003. Evolution and selection of trichromatic vision in primates. *Trends Ecol. Evol.* 18(4):198–205.

Susman, R.L. 1984. The locomotor behavior of *Pan paniscus* in the Lomako Forest. In R.L. Susman, ed., *The Pygmy Chimpanzee: Evolutionary Biology and Behavior*. New York: Press. Pp. 369–393.

Susman, R.L. 1988. Hand of *Paranthropus robustus* from Member 1, Swartkrans: Fossil evidence for tool behavior. *Science* 240:781–784.

Susman, R.L. 1994. Fossil evidence for early hominid tool use. *Science* 265:1570–1573.

Susman, R.L. 1998. Hand function and tool behavior in early hominids. *J. Hum. Evolution* 35:23–46.

Susman, R.L., and T.M. Brain. 1988. New first metatarsal (SKX 5017) from Swartkrans and the gait of *Paranthropus robustus*. *Am. J. Phys. Anthropol.* 77:7–15.

Susman, R.L., J.T. Stern, and W.L. Jungers. 1984. Arboreality and bipedality in the Hadar hominids. *Folia Primatolog.* 43:113–156.

Sussman, R.W. 1991. Primate origins and the evolution of angiosperms. *Am. J. Primatol.* 23:209–223.

Sutherland, S. 1990. A nommocnu impairment. *Nature* 346:217–218.

Suwa, G., B.A. Wood, and T.D. White. 1994. Further analysis of mandibular molar crown and cusp areas in Pleistocene and early Pliocene hominids. *Am. J. Phys. Anthropol.* 93:407–426.

Swaddle, J.P., and I.C. Cuthill. 1995. Asymmetry and human facial attractiveness: Symmetry may not always be beautiful. *Proc. R. Soc. Lond. B Biol. Sci.* 261:111–116.

Swan, S.H., and E.P. Elkin. 1999. Declining semen quality: Can the past inform the present? *BioEssays* 21:614–621.

Symons, D. 1979. *The Evolution of Human Sexuality.* Oxford: Oxford University Press.

Szalay, F.S., and R.K. Costello. 1991. Evolution of permanent estrus displays in hominids. *J. Hum. Evol.* 20:439–464.

Tague, R.G., and C.O. Lovejoy. 1986. The obstetric pelvis of A.L. 288-1 (Lucy). *J. Hum. Evol.* 15:237–255.

Takahata, Y., H. Ihobe, and G. Idani. 1996. Comparing copulations of chimpanzees and bonobos: do females exhibit proceptivity or receptivity? In W.C. McGrew, L.F. Marchant, and T. Nishida, eds., *Great Ape Societies.* Cambridge: Cambridge University Press. Pp. 146–155.

Takami, S. 2002. Recent progress in the neurobiology of the vomeronasal organ. *Microsc. Res. Tech.* 58:228–250.

Tardieu, C. 1981. Morpho-functional analysis of the articular surface of the knee-joint in primates. In A.B. Chiarelli and R.S. Corruccini, eds., *Primate Evolutionary Biology.* Berlin: Springer-Verlag. Pp. 68–80.

Taylor, C.R., and V.J. Rowntree. 1973. Running on two or on four legs: Which consumes more energy? *Science* 179:186–187.

Taylor, K.I., and M. Regard. 2003. Language in the right hemisphere: Contributions from reading studies. *News. Physiol. Sci.* 18:257–261.

Teaford M.F., P.S. Ungar. 2000. Diet and the evolution of the earliest human ancestors. *Proc. Natl. Acad. Sci. U.S.A.* 97(25):13506–13511.

Temerin, L.A., and J.G.H. Cant. 1983. The evolutionary divergence of Old World monkeys and apes. *Am. Nat.* 122:335–351.

ten Donkelaar, H.J. 2001. Evolution of vertebrate motor systems. In G. Roth and M.F. Wullimann, eds., *Brain Evolution and Cognition.* New York: John Wiley and Sons. Pp. 77–112.

Terrace, H.S., L.A. Petitto, R.J. Sanders, and T.G. Beaver. 1979. Can an ape create a sentence? *Science* 206:891–902.

Tessier-Lavigne, M., and C.S. Goodman. 1996. The molecular biology of axon guidance. *Science* 274:1123–1133.

Thackeray, J.F. 1995. Do strontium/calcium ratios from early Pleistocene hominids from Swartkrans reflect physiological differences in males and females? *J. Hum. Evol.* 29:401–404.

Thieme, H. 1997. Lower Paleolithic hunting spears from Germany. *Nature* 385:807–810.

Thompson, K.S. 1991. Where did tetrapods come from? *Am. Sci.* 79(Nov–Dec):488–490.

Thompson-Handler, N., R.K. Malenky, and N. Badrian. 1984. Sexual behavior of *Pan paniscus* under natural conditions in the Lomako Forest, Equateur, Zaire. In R.L. Susman, ed., *The Pygmy Chimpanzee: Evolutionary Biology and Behavior.* New York: Plenum Pub. Pp. 347–368.

Thong, Y.H. 1982. Obesity in evolutionary perspective. *Med. Hypotheses* 8:431–435.

Thorne, N., and H. Amrein. 2003. Vomeronasal organ: Pheromone recognition with a twist. *Curr. Biol.* 13:R220–R222.

Thornhill, R., and S.W. Gangestad. 1993. Human facial beauty. *Hum. Nat.* 4(3):237–269.

Thornhill, R., and N.W. Thornhill. 1983. Human rape: An evolutionary analysis. *Ethol. Sociobiol.* 4:37–73.

Tobias, P.V., 1987. The brain of *Homo habilis*: A new level of organization in cerebral evolution. *J. Hum. Evol.* 16:741–761.

Tobias, P.V., 1995. The brain of the first hominids. In J.-P. Changeux and Jean Chavaillon, eds., *Origins of the Human Brain.* Oxford: Oxford University Press. Pp. 61–81.

Tootell, R.B.H., A.M. Dale, M.I. Sereno, and R. Malach. 1996. New images from human visual cortex. *Trends Neurosci.* 19(11):481–489.

Toth, N. 1985. Archaeological evidence for preferential right-handedness in the lower and middle Pleistocene and its possible implications. *J. Hum. Evol.* 14:607–614.

Travis, J. 1999. Making sense of scents. *Sci. News* 155:236–238.

Trevathan, W.R. 1987. *Human Birth: An Evolutionary Perspective.* New York: Aldine de Gruyter.

Trevathan, W., and K. Rosenberg. 2000. The shoulders follow the head: Postcranial constraints on human childbirth. *J. Hum. Evol.* 39:583–586.

Trinkaus, E. 1986. Bodies, brawn, brains, and noses: Human ancestors and human predation. In M.H. Nitecki and D.V. Nitecki, eds., *The Evolution of Human Hunting.* New York: Plenum Pub.

Trinkaus, E. 1997. Appendicular robusticity and the paleobiology of modern human emergence. *Proc. Natl. Acad. Sci. U.S.A.* 94:13367–13373.

Trinkaus, E. 2000. The "robusticity transition" revisited. In C.B. Stringer, R.N.E. Barton, and J.C. Finley, eds., *Neanderthals on the Edge.* Oxford: Oxbow Books. Pp. 227–236.

Trinkaus, E., and M.H. Wolpoff. 1992. Brain size in post-habiline archaic *Homo.* (Abstract.) *Am. J. Phys. Anthropol.* (Suppl. 14):163.

Trinkaus, E., S.E. Churchill, and C.B. Ruff. 1994. Postcranial robusticity in *Homo.* II. Humeral bilateral asymmetry and bone plasticity. *Am. J. Phys. Anthropol.* 93:1–34.

Trivers, R.L. 1974. Parent-offspring conflict. *Am. Zool.* 14:249–264.

Trotier, D., C. Eloit, M. Wasser, G. Talmain, J.L. Bension, K.B. Døving, and J. Ferrand. 2000. The vomeronasal cavity in adult humans. *Chem. Senses* 25:369–380.

Tuljapurkar, S., N. Li, and C. Boe. 2000. A universal pattern of mortality decline in the G7 countries. *Nature* 405:789–792.

Turke, P.W. 1984. Effects of ovulatory concealment and synchrony on protohominid mating systems and parental roles. *Ethol. Sociobiol.* 5:33–44.

Turke, P.W. 1988. Concealed ovulation, menstrual synchrony, and paternal investment. In E.E. Filsinger, ed., *Biosocial Perspectives on the Family.* Newbury Park, CA: Sage Pub. Pp. 119–136.

Turner, A., and B. Wood. 1993. Comparative palaeontological context for the evolution of the early hominid masticatory system. *J. Hum. Evol.* 24:301–318.

Turner, C.G. 1989. Teeth and prehistory in Asia. *Sci. Am.* 260(Feb):88–96.

Tuttle, R.H. 1967. Knuckle-walking and the evolution of hominoid hands. *Am. J. Phys. Anthropol.* 26:171–206.

Tuttle, R.H. 1974. Darwin's apes, dental apes, and the descent of man: Normal science in evolutionary anthropology. *Curr. Anthropol.* 15:389–398.

Tuttle, R.H. 1975. Knuckle-walking and knuckle-walkers: A commentary on some recent perspectives on hominoid evolution. In R.H. Tuttle, ed., *Primate Functional Morphology and Evolution.* The Hague: Mouton. Pp. 203–209.

Tuttle, R.H. 1981. Evolution of hominid bipedalism and prehensile capabilities. *Phil. Trans. R. Soc. Lond. B Biol. Sci.* 292:89–94.

Udry, J.R., and N.M. Morris. 1968. Distribution of coitus in the menstrual cycle. *Nature* 220:593–596.

van der Merwe, N.J., J.F. Thackery, J.A. Lee-Thorp, and J. Luyt. 2003. The carbon isotope ecology and diet of *Australopithecus africanus* at Sterkfontein, South Africa. *J. Hum. Evol.* 44:581–597.

Van Essen, D.C., C.H. Anderson, and D.J. Felleman. 1992. Information processing in the primate visual system: An integrated systems perspective. *Science* 255:419–423.

Van Reenen, J.F. 1966. Dental features of a low-caries primitive population. *J. Dent. Res.* 45(Suppl. to 3):703–713.

van Shaik, C.P., M. Ancrenaz, G. Borgen, B. Galdikas, C.D. Knott, I. Singleton, A. Suzuki, S.S. Utami, and M. Merrill. 2003. Orangutan culture and the evolution of material culture. *Science* 299:102–105.

Vancata, V. 1978. Reconstruction of the locomotor behaviour of the earliest hominids. *Anthropologie* 16(3):271–276.

Verhaegen, M.J.B. 1985. The aquatic ape theory: Evidence and a possible scenario. *Med. Hypotheses* 16:17–32.

Vigliocco, G. 2000. Language processing: The anatomy of meaning and syntax. *Curr. Biol.* 10(2):R78–R80.

Vilensky, J.A. 1989. Primate quadrupedalism: How and why does it differ from that of typical quadrupeds? *Brain Behav. Evol.* 34:357–364.

Vining, D.R. 1986. Social versus reproductive success: The central theoretical problem of human sociobiology. *Behav. Brain Sci.* 9:167–216.

Vrba, E.S. 1998. Multiphasic growth models and the evolution of prolonged growth exemplified by human brain evolution. *J. Theor. Biol.* 190:227–239.

Wade, G.N., and J.E. Schneider. 1992. Metabolic fuels and reproduction in female mammals. *Neurosci. Biobehav. Rev.* 16:235–272.

Wagner, P.D. 1987. The lungs during exercise. *News Physiol. Sci.* 2:6–10.

Wagner, U., S. Gais, H. Haider, R. Verlager, and J. Born. 2004. Sleep inspires insight. *Nature* 427:352–355.

Walker, A. 1973. New *Australopithecus* femora from East Rudolf, Kenya. *J. Hum. Evol.* 2:545–555.

Walker, A. 1981. Dietary hypotheses and human evolution. *Phil. Trans. R. Soc. Lond B Biol. Sci.* 292:57–64.

Walker, A., and P. Shipman. 1996. *The Wisdom of Bones*. London: Weidenfeld and Nicolson.

Walker, A., and M. Teaford. 1989. The hunt for *Proconsul. Sci. Am.* 260(Jan.):76–82.

Walker, A., M.R. Zimmerman, and R.E.F. Leakey. 1982. A possible case of hypervitaminosis A in *Homo erectus. Nature* 296:248–250.

Wallman, J. 1992. *Aping Language*. Cambridge: Cambridge University Press.

Wang, W.J., R.H. Crompton, Y. Li, and M.M. Gunther. 2003. Energy transformation during erect and "bent-hip, bent-knee" walking by humans with implications for the evolution of bipedalism. *J. Hum. Evol.* 44:563–579.

Wang, Y., U. Hu., J. Meng, and C. Li. 2001. An ossified Meckel's cartilage in two Cretaceous mammals and the origin of the mammalian middle ear. *Science* 294:357–361.

Ward, C.V. 1993. Torso morphology and locomotion in *Proconsul nyanzae. Am. J. Phys. Anthropol.* 92:291–328.

Ward, C.V. 2002. Interpreting the posture and locomotion of *Australopithecus afarensis*: Where do we stand? *Yrbk. Phys. Anthropol.* 45:185–215.

Ward, C.V., A. Walker, and M.F. Teaford. 1991. Proconsul did not have a tail. *J. Hum. Evol.* 21:215–220.

Washburn, S.L. 1960. Tools and human evolution. *Sci. Am.* 203(Sept.):62–75.

Washburn, S.L., and C.S. Lancaster. 1968. The evolution of hunting. In R.B. Lee and I. DeVore, eds., *Man the Hunter*. Chicago: Aldine. Pp. 293–303.

Wasser, S.K. 1990. Infertility, abortion, and biotechnology: When it's not nice to fool Mother Nature. *Hum. Nat.* 1(1):3–24.

Wasser, S.K., and D.P. Barash. 1983. Reproductive suppression among female mammals: Implications for biomedicine and sexual selection theory. *Q. Rev. Biol.* 58:513–538.

Wassermann, H.P. 1965. Human pigmentation and environmental adaptation. *Arch. Environ. Health* 11:691–694.

Wassermann, H.P. 1969. Melanin pigment and the environment. In R.D.G.P. Simons and J. Marshall, eds., 1969, *Essays on Tropical Dermatology*. Amsterdam: Excerpta Medica Foundation. Vol. 1:7–16.

Watson, L. 2000. *Jacobson's Organ and the Remarkable Nature of Smell*. New York: W.W. Norton.

Watson, P.J., and R. Thornhill. 1994. Fluctuating asymmetry and sexual selection. *Trends Ecol. Evol.* 9(1):21–25.

Watts, D.P., and J.C. Mitani. 2002. Hunting behavior of chimpanzees at Ngogo, Kibale National Park, Uganda. *Int. J. Primatol.* 23(1):1–28.

Weaver, T.D. 2003. The shape of the Neandertal femur is primarily the consequence of a hyperpolar body form. *Proc. Natl. Acad. Sci. U.S.A.* 100(12):6926–6929.

Webb, D., and S. Fabiny. 2001. Human hands are adapted for bipedalism, not tool use. (Abstract.) *Am. J. Phys. Anthropol.* Suppl. 32:161–162.

Weidenreich, F. 1941. The brain and its role in the phylogenetic transformation of the human skull. *Trans. Am. Phil. Soc.* 31(5):321–442.

Weiner, J. 1995. *The Beak of the Finch: A Story of Evolution in Our Time*. New York: Alfred A. Knopf.

Weiss, K. 2002. How the eye got its brain. *Evol. Anthropol.* 11(6):215–219.

Weller, L., and A. Weller. 1993. Human menstrual synchrony: A critical assessment. *Neurosci. Biobehav. Rev.* 17:427–439.

Weller, L., A. Weller, and S. Roizman. 1999. Human menstrual synchrony in families and among close friends: Examining the importance of mutual exposure. *J. Comp. Psych.* 113(3):261–268.

Wescott, R.W. 1969. Human uprightness and primate display. *Am. Anthropol.* 69:738.

Westendorp, R.G.L., and T.B.L. Kirkwood. 1998. Human longevity at the cost of reproductive success. *Nature* 396:743–746.

Wetsman, A., and F. Marlowe. 1999. How universal are preferences for female waist-to-hip ratio? Evidence from the Hadza of Tanzania. *Evol. Hum. Behav.* 20:219–228.

Whalen, D.H., and A.M. Liberman. 1987. Speech perception takes precedence over nonspeech perception. *Science* 237:169–171.

Wheeler, P.E. 1984. The evolution of bipedality and loss of functional body hair in hominids. *J. Hum. Evol.* 13:91–98.

Wheeler, P.E. 1985. The loss of functional body hair in man: The influence of thermal environment, body form, and bipedality. *J. Hum. Evol.* 14(1):23–28.

Wheeler, P.E. 1991a. The thermoregulatory advantages of hominid bipedalism in open equatorial environments: The contribution of increased convective heat loss and cutaneous evaporative cooling. *J. Hum. Evol.* 21:107–115.

Wheeler, P.E. 1991b. The influence of bipedalism on the energy and water budgets of early hominids. *J. Hum. Evol.* 21:117–136.

Wheeler, P.E. 1992a. The thermoregulatory advantages of large body size for hominids foraging in savannah environments. *J. Hum. Evol.* 23:351–362.

Wheeler, P.E. 1992b. The influence of the loss of functional body hair on the water budgets of early hominids. *J. Hum. Evol.* 23:379–388.

Wheeler, P.E. 1993. The influence of stature and body form on hominid energy and water budgets: A comparison of *Australopithecus* and early *Homo* physiques. *J. Hum. Evol.* 24:13–28.

Wheeler, P.E. 1994. The thermoregulatory advantages of heat storage and shade-seeking behavior to hominids foraging in equatorial savannah environments. *J. Hum. Evol.* 26:339–350.

Whinnett, A., and N.I. Mundy. 2003. Isolation of novel olfactory receptor genes in marmosets (*Callithrix*): Insights into pseudogene formation and evidence for functional degeneracy in non-human primates. *Gene* 304:87–96.

White, M.D., and M. Cabanac. 1995a. Nasal mucosal vasodilation in response to passive hyperthermia in humans. *Eur. J. Appl. Physiol.* 70:207–212.

White, M.D., and M. Cabanac. 1995b. Core temperature thresholds for hyperpnea during passive hyperthermia in humans. *Eur. J. Appl. Physiol.* 71:71–76.

White, M.D., and M. Cabanac. 1995c. Respiratory heat loss and core temperatures during submaximal exercise. *J. Therm. Biol.* 20(6):489–496.

White, M.D., and M. Cabanac. 1996. Exercise hyperpnea and hyperthermia in humans. *J. Appl. Physiol.* 81(3):1249–1254.

Whiten, A., and C. Boesch. 2001. The cultures of chimpanzees. *Sci. Am.* 284(Jan):60–67.

Whiten, A., J. Goodall, W.C. McGrew, T. Nichida, V. Reynolds, Y. Sugiyama, C.E.G. Tutin, R.W. Wrangham, and C. Boesch. 1999. Culture in chimpanzees. *Nature* 399:682–688.

Whiten, A., V. Horner, and S. Marshall-Pescini. 2003. Cultural panthropology. *Evol. Anthropol.* 12(2):92–105.

Whiting, B.A., and R.A. Barton. 2003. The evolution of the cortico-cerebellar complex in primates: Anatomical connections predict patterns of correlated evolution. *J. Hum. Evol.* 44:3–10.

Wildman, D.E., M. Uddin, G. Liu, L.I. Grossman, and M. Goodman. 2003. Implications of natural selection in shaping 99.4% nonsynonymous DNA identity between humans and chimpanzees: Enlarging genus *Homo*. *Proc. Natl. Acad. Sci. U.S.A.* 100:7181–7188.

Wilson, E.O. 1978. *On Human Nature*. New York: Bantam.

Wilson, H.C. 1992. A critical review of menstrual synchrony research. *Psychoneuroendocrinology* 17(6):565–591.

Wise, P.M., K.M. Krajnak, and M.L. Kashon. 1996. Menopause: The aging of multiple pacemakers. *Science* 273:67–70.

Wolf, K. 2002. Visual ecology: Coloured fruit is what the eye sees best. *Curr. Biol.* 12:R253–R255.

Wolfe, L.D., J.P. Gray, J.G. Robinson, L.S. Lieberman, and E.H. Peters. 1982. Models of human evolution. *Science* 217:302.

Wolpoff, M.H. 1969. The effect of mutations under conditions of reduced selection. *Soc. Biol.* 16:11–23.

Wolpoff, M.H. 1975. Some aspects of human mandibular evolution. In J.A. McNamara, ed., *Determinants of Mandibular Form and Growth*. Craniofacial Growth Series, Monograph No. 4. Ann Arbor, MI: Center for Human Growth and Development, University of Michigan. Pp. 1–64.

Wolpoff, M.H. 1978. Some implications of relative biomechanical neck length in hominid femora. *Am. J. Phys. Anthropol.* 48:143–148.

Wood, B. 1992. Origin and evolution of the genus *Homo*. *Nature* 355:783–790.

Wood, B., and M. Collard. 1999. The human genus. *Science* 284:65–71.

Wood, B., and B.G. Richmond. 2000. Human evolution: Taxonomy and paleobiology. *J. Anat.* 196:19–60.

Wood, J.D. 1991. Communication between minibrain in gut and enteric immune system. *News Physiol. Sci.* 6(2):64–69.

Wood, J.W. 1982. Models of human evolution. *Science* 217:296–302.

Wrangham, R.W. 1993. The evolution of sexuality in chimpanzees and bonobos. *Hum. Nat.* 4(1):47–79.

Wrangham R.W., J.H. Jones, G. Laden, D. Pilbeam, and N. Conklin-Brittain. 1999. The raw and the stolen. *Curr. Anthropol.* 40(5):567–577.

Wu, C. 1999. Unraveling the mystery of melanin. *Sci. News* 156:190–191.

Young, I.S., R.M. Alexander, A.J. Woakes, P.J. Butler, and L.l. Anderson. 1992. The synchronization of ventilation and locomotion in horses (*Equus caballus*). *J. Exp. Biol.* 166:19–31.

Young, R.A. 1976. Fat, energy and mammalian survival. *Am. Zool.* 16:609–710.

Young, R.W. 2003. Evolution of the human hand: The role of throwing and clubbing. *J. Anat.* 202:165–174.

Zatorre, R.J., A.C. Evans, E. Meyer, and A. Gjedde. 1992. Lateralization of phonetic and pitch discrimination in speech processing. *Science* 256:846–849.

Zhang, B., Y. Zhang, and H.F. Rosenberg. 2002. Adaptive evolution of a duplicated pancreatic ribonuclease gene in a leaf-eating monkey. *Nat. Gen.* 30:411–415.

Zhang, J. 2003. Evolution by gene duplication: An update. *Trends Ecol. Evol.* 18(6):292–298.

Zhang, J., and D.M. Webb. 2003. Evolutionary deterioration of the vomeronasal pheromone transduction pathway in catarrhine primates. *Proc. Natl. Acad. Sci. U.S.A.* 100:8337–8341.

Zihlman, A.L., and B.A. Cohn. 1988. The adaptive response of human skin to the savanna. *Hum. Evol.* 3(5):397–409.

Zimmer, C. 1994. The importance of noses. *Discover* 15(Aug):24–25.

Zimmer, C. 1998. *At the Water's Edge: Macroevolution and the Transformation of Life*. New York: Free Press.

Zimmer, C. 2002. The rise and fall of the nasal empire. *Nat. Hist.* 111(June):32–35.

Zola-Morgan, S., and L.R. Squire. 1993. Neuroanatomy of memory. *Annu. Rev. Neurosci.* 16:547–563.

Zou, Z., L.F. Horowitz, J.-P. Montmayeur, S. Snapper, and L.B. Buck. 2001. Genetic racing reveals a stereotyped sensory map in the olfactory cortex. *Nature* 414:173–179.

INDEX

Note: *Italicized* page numbers indicate terms found in illustrations.